Greenhill Books

DICTIONARY OF MILITARY TERMS

DICTIONARY OF MILITARY TERMS

US Department of Defense

Introduction by Charles Messenger

Greenhill Books, London
Stackpole Books, Pennsylvania

This edition of the *Dictionary of Military Terms*
first published 1995 by Greenhill Books
Lionel Leventhal Limited, Park House, 1 Russell Gardens
London NW11 9NN
and
Stackpole Books, 5067 Ritter Road
Mechanicsburg, PA 17055, USA

British Library Cataloguing in Publication Data
United States Department of Defense
Dictionary of Military Terms. – 3Rev.ed
I. Title
355.003

ISBN 1–85367–217–3

Library of Congress Cataloging-in-Publication Data available

Publishing History
The US *Department of Defense Dictionary of Military and
Associated Terms* was first published by the Joint Chiefs of
Staff, Washington, DC, 20318-0001, as Joint Pub 1-02. This
edition is revised and updated, and includes a new Introduction
by Charles Messenger, plus Appendixes with US acronyms and
abbreviations, and NATO terms in English.

Printed and bound in Great Britain by
Bookcraft (Bath) Ltd., Midsomer Norton, Somerset

INTRODUCTION

Every walk of life has its own language or jargon, and the armed forces are no different from any other profession or speciality in this respect. This jargon is often a form of shorthand for describing what are sometimes highly complex concepts. But within these fields of endeavor there are sub-branches that develop their own language, and this language is often not understood by others working in the same overall area.

This is especially true of the language used by armies, navies, and air forces. Yet all three branches of the armed forces often have to work very closely with one another in joint operations ('in which elements of more than one Service of the same nation participate'). This has been a state of affairs that has existed since 1914, when the air arm first began to play a serious role in war, and reached a peak in 1942–45 with the numerous landings on hostile coasts in both the European and Pacific theaters. Post-1945, and in spite of the Inchon landings in Korea in 1950 and Suez in 1956, the concept of 'jointery' lost much of its priority, at least where all three services were concerned. Indeed, on both sides of the Atlantic the main emphasis was on ground–air co-operation. In the past decade, however, the Falklands, Grenada, and more recently the Gulf War, Somalia, and former Yugoslavia have brought a renewal of emphasis on joint operations and hence the necessity for this publication.

When compiling a dictionary of specialist terms it is very easy to fall into the trap of producing jargon-riddled definitions which make the reader no wiser. This is a pitfall that the authors of this work have successfully avoided. Indeed, the vast majority of definitions are models of clarity which could be readily understood by someone without any military connections. The terms are also very wide ranging. Thus, we have the mine warfare term 'otter', with which fishermen will be familiar as a means of freeing a trapped line, 'moderate risk (nuclear)', the air force's 'stern attack', and land warfare's 'contact point'. Indeed, close study of the dictionary gives a fascinating insight into how modern war is fought.

Introduction

There is, however, another important aspect to this work. This concerns combined operations, which in British 1939–45 parlance meant joint operations, but now refers to the involvement of multi-national forces. A number of terms are asterisked, indicating that they are standardized NATO terms. Throughout the Alliance's 45 years of existence it has been beset by problems of uniformity or standardization. Different weapons and ammunition, force structures, and, indeed, national interests have often caused difficulties, to a much greater extent than in its long-time rival the Warsaw Pact, which was much more rigidly controlled from Moscow than NATO was from Washington DC. In order to overcome these problems NATO instituted among its members a system of Standardization Agreements (STANAGs), which ranged from the design of aircraft fuel filler pipes, enabling a NATO aircraft to be refuelled on any member nation's military airfield, to nuclear, chemical or biological attack reporting proformae.

While it was agreed from the outset that English should be NATO's official language, it soon became apparent that different military expressions had varying shades of meaning among Alliance members. This was proved time and again on maneuvers, when misunderstandings resulted in confusion. Hence NATO embarked on a program of standard definitions of military terms under the umbrella of STANAG 3680. This has proved to be a long and sometimes uphill exercise. Indeed, Annex A of this publication lists NATO-only terms whose definitions are not used by US forces operating outside the Alliance. An example of this is 'concept of operations'. The NATO definition of this is 'a clear and concise statement of the line of action chosen by a commander in order to accomplish his mission'. The domestic US interpretation is, however, much broader, encompassing not just his 'intent', but his 'assumptions in regard to an operation or series of operations' as well. Another example is 'counter-air'. While both the USAF and NATO agree that its object is to gain air superiority, the NATO definition states that this is achieved through operations 'against the enemy's air offensive and defensive capability'. The USAF version merely speaks of destroying or neutralizing enemy air forces, but the US Navy and Marines have another term, 'antiair warfare', which is more defensive and merely designed to reduce the air and missile threat to an 'acceptable level'. Indeed, this is an instance where confusion between the two terms could result in either too little airpower being dedicated to creating the desired air environment or too much at the expense of other perceived vital air roles.

Technical language does not, however, stand still. In the military field new concepts and newly emerging technology propagate new terms. Furthermore existing definitions often need to be refined to remove previously unidentified points of ambiguity. Thus a work of this nature needs to be constantly updated. There are, however, two caveats that must be mentioned. Firstly, there is a tendency to use jargon for jargon's sake, either to impress others with the user's knowledge of the subject or out of sheer laziness of thought. There is also the 'change for the sake of change' syndrome often brought about by officers and officials who wish to leave a permanent mark. In both cases the motives are entirely selfish and often have a degrading effect on overall efficiency. Any supervisory panel for a work like this dictionary must constantly ask itself whether proposed new terms are essential to the furtherance of clarity in communication. Military history is littered with examples of needless casualties being incurred, and worse, because of failures to communicate clearly. The object of this dictionary and others like it is to help overcome this problem and they are therefore important not just to soldiers, sailors, and airmen, but to anyone concerned with the business of defense. This DOD Dictionary is a model of its type and I warmly recommend it.

Charles Messenger
1995

PUBLISHER'S NOTE

This revised and updated edition of the *Dictionary of Military Terms* has been prepared under the direction of the US Joint Chiefs of Staff, and co-ordinated with the US Office of the Secretary of Defense, the Military Services, and Defense agencies. It sets forth standard military terminology to govern the joint activities and performance of the US Armed Forces in joint and multi-national operations, and its use is mandatory in the US military establishment.

The terms included are those of general military or associated significance not adequately defined in standard dictionaries. Technical or specialized terms are given if they can be defined in easily understood language and if their inclusion is of general military or associated significance. Weapon terms are limited to modern weapons.

This new edition is set out in three parts:

1. The main dictionary contains all terms and definitions approved for use within the US Department of Defense. This includes terms and definitions that are approved for both DOD and NATO use, indicated by an asterisk in parentheses: (*).

2. Appendix A contains the English version of all currently approved NATO-only terms.

3. Appendix B contains a listing of current abbreviations and acronyms approved for use within the US Department of Defense.

Greenhill Books
1995

A

A-4--See **Skyhawk.**

A-6--See **Intruder.**

A-7--See **Corsair II.**

A-10--See **Thunderbolt II.**

AADC--See **area air defense commander.**

ABCCC--See **airborne battlefield command and control center.**

abeam(*)--Bearing approximately 090 degrees or 270 degrees relative; at right angles to the longitudinal axis of a vehicle.

abeam replenishment(*)--The transfer at sea of personnel and/or supplies by rigs between two or more ships proceeding side by side.

abort(*)--1. To terminate a mission for any reason other than enemy action. It may occur at any point after the beginning of the mission and prior to its completion. 2. To discontinue aircraft takeoff or missile launch.

abrasion(*)--In photography, a scratch or mark produced mechanically on an emulsion surface or film base.

absolute altimeter(*)--Radio or similar apparatus that is designed to indicate the true vertical height of an aircraft above the terrain.

absolute altitude(*)--The height of an aircraft directly above the surface or terrain over which it is flying. See also **altitude.**

absolute dud--A nuclear weapon which, when launched at or emplaced on a target, fails to explode.

absolute filter(*)--A filter capable of cutting off 100% by weight of solid particles greater than a stated micron size.

absorbed dose(*)--The amount of energy imparted by nuclear (or ionizing) radiation to unit mass of absorbing material. The unit is the rad.

AC-130--See **Hercules.**

acceptability--Operation plan review criterion. The determination whether the contemplated course of action is worth the cost in manpower, material, and time involved; is consistent with the law of war; and militarily and politically supportable. See also **adequacy; completeness; feasibility; suitability.**

access to classified information--The ability and opportunity to obtain knowledge of classified information. Persons have access to classified information if they are permitted to gain knowledge of the information or if they are in a place where they would be expected to gain such knowledge. Persons do not have access to classified information by being in a place where classified information is kept if security measures prevent them from gaining knowledge of the information.

accidental attack--An unintended attack which occurs without deliberate national design as a direct result of a random event, such as a mechanical failure, a simple human error, or an unauthorized action by a subordinate.

accidental war--Not to be used. See **accidental attack.**

accompanying supplies--Unit supplies that deploy with forces.

accountability--The obligation imposed by law or lawful order or regulation on an officer or other person for keeping accurate record of property, documents, or funds. The person having this obligation may or may not have actual possession of the property, documents, or funds. Accountability is concerned primarily with records, while responsibility is concerned primarily with custody, care, and safekeeping. See also **responsibility.**

accounting line designator--A five-character code consisting of the target desired ground zero designator and the striking command suffix to indicate a specific nuclear strike by a specified weapon delivery system on a target objective to the operation plan. Also called **ALD.**

accreditation--In computer modeling and simulation, an official determination that a model or simulation is acceptable for a specific purpose. See also **configuration management; independent review; validation; verification.**

accuracy of fire(*)--The precision of fire expressed by the closeness of a grouping of shots at and around the center of the target.

accuracy of information--See **evaluation.**

acknowledgment(*)--A message from the addressee informing the originator that his or her communication has been received and is understood.

aclinic line--See **magnetic equator.**

ACO--See **airspace control order.**

acoustical surveillance--Employment of electronic devices, including sound-recording, -receiving, or -transmitting equipment, for the collection of information.

acoustic circuit(*)--A mine circuit which responds to the acoustic field of a target. See also **mine.**

acoustic intelligence(*)--Intelligence derived from the collection and processing of acoustic phenomena.

acoustic jamming--The deliberate radiation or reradiation of mechanical or electroacoustic signals with the objectives of obliterating or obscuring signals which the enemy is attempting to receive and of deterring enemy weapon systems. See also **barrage jamming; electronic warfare; jamming; spot jamming.**

acoustic mine(*)--A mine with an acoustic circuit which responds to the acoustic field of a ship or sweep. See also **mine.**

acoustic minehunting(*)--The use of a sonar to detect mines or mine-like objects which may be on or protruding from the seabed, or buried.

acoustic warfare(*)--Action involving the use of underwater acoustic energy to determine, exploit, reduce or prevent hostile use of the underwater acoustic spectrum and actions which retain friendly use of the underwater acoustic spectrum. There are three divisions within acoustic warfare: 1. acoustic warfare support measures. That aspect of acoustic warfare involving actions to search for, intercept, locate, record and analyze radiated acoustic energy in water for purpose of exploiting such radiations. The use of acoustic warfare support measures involves no intentional underwater acoustic emission and is generally not detectable by the enemy. 2. acoustic warfare countermeasures. That aspect of acoustic warfare involving actions taken to prevent or reduce an enemy's effective use of the underwater acoustic spectrum. Acoustic warfare countermeasures involve intentional underwater acoustic emissions for deception and jamming. 3. acoustic warfare counter-countermeasures. That aspect of acoustic warfare involving actions taken to ensure friendly effective use of the underwater acoustic spectrum despite the enemy's use of underwater acoustic warfare.

Acoustic warfare counter-countermeasures involve anti-acoustic warfare support measures and anti-acoustic warfare countermeasures, and may not involve underwater acoustic emissions.

acoustic warfare counter-countermeasures--See **acoustic warfare Part 3.**

acoustic warfare countermeasures--See **acoustic warfare Part 2.**

acoustic warfare support measures--See **acoustic warfare Part 1.**

ACP--See **airspace control plan.**

acquire--1. When applied to acquisition radars, the process of detecting the presence and location of a target in sufficient detail to permit identification. 2. When applied to tracking radars, the process of positioning a radar beam so that a target is in that beam to permit the effective employment of weapons. See also **target acquisition.**

acquire (radar)--See **acquire.**

acquisition--See **collection (acquisition).**

action agent--In intelligence usage, one who has access to, and performs actions against, the target.

action deferred--Tactical action on a specific track is being withheld for better tactical advantage. Weapons are available and commitment is pending.

action information center--See **air defense control center; combat information center.**

activate(*)--1. To put into existence by official order a unit, post, camp, station, base or shore activity which has previously been constituted and designated by name or number, or both, so that it can be organized to function in its as-

signed capacity. (DOD) 2. To prepare for active service a naval ship or craft which has been in an inactive or reserve status. See also **commission; constitute.**

activation detector(*)--A device used to determine neutron flux or density by virtue of the radioactivity induced in it as a result of neutron capture.

active aircraft--Aircraft currently and actively engaged in supporting the flying missions either through direct assignment to operational units or in the preparation for such assignment or reassignment through any of the logistic processes of supply, maintenance, and modification. See also **aircraft.**

active air defense(*)--Direct defensive action taken to nullify or reduce the effectiveness of hostile air action. It includes such measures as the use of aircraft, air defense weapons, weapons not used primarily in an air defense role, and electronic warfare. See also **air defense.**

active communications satellite--See **communications satellite.**

active defense--The employment of limited offensive action and counterattacks to deny a contested area or position to the enemy. See also **passive defense.**

active duty--Full-time duty in the active military service of the United States. This includes members of the Reserve Components serving on active duty or full-time training duty, but does not include full-time National Guard duty. Also called **AD.** See also **active duty for training; inactive duty training.**

active duty for special work--A tour of active duty for reserve personnel authorized from military and reserve personnel appropriations for work on active or reserve component programs. This includes annual screening, training camp operations, training ship operations, and unit conversion to new weapon systems when such duties are essential. Active duty for special work may also be authorized to support study groups, training sites and exercises, short-term projects, and doing administrative or support functions. By policy, active duty for special work tours are normally limited to 139 days or less in one fiscal year. Tours exceeding 180 days are accountable against active duty end strength.

active duty for training--A tour of active duty which is used for training members of the Reserve Components to provide trained units and qualified persons to fill the needs of the Armed Forces in time of war or national emergency and such other times as the national security requires. The member is under orders which provide for return to non-active status when the period of active duty for training is completed. It includes annual training, special tours of active duty for training, school tours, and the initial duty for training performed by nonprior service enlistees.

active homing guidance(*)--A system of homing guidance wherein both the source for illuminating the target and the receiver for detecting the energy reflected from the target as the result of the illumination are carried within the missile. See also **guidance.**

active material(*)--Material, such as plutonium and certain isotopes of uranium, which is capable of supporting a fission chain reaction.

active mine(*)--A mine to be actuated by the reflection from a target of a signal emitted by the mine.

active status--Status of all Reserves except those on an inactive status list or in the Retired Reserve. Reservists in an active status may train for points and/or pay and may be considered for promotion.

activity--1. A unit, organization, or installation performing a function or mission, e.g., reception center, redistribution center, naval station, naval shipyard. 2. A function or mission, e.g., recruiting, schooling. See also **establishment.**

actual ground zero(*)--The point on the surface of the Earth at, or vertically below or above, the center of an actual nuclear detonation. See also **desired ground zero; ground zero.**

actuate(*)--To operate a mine-firing mechanism by an influence or a series of influences in such a way that all the requirements of the mechanism for firing, or for registering a target count, are met.

actuator(*)--A mechanism that furnishes the force required to displace a control surface or other control element.

acute radiation dose(*)--Total ionizing radiation dose received at one time and over a period so short that biological recovery cannot occur.

ACV--See **air cushion vehicle.**

AD--See **active duty.**

additional training assemblies--Inactive duty training periods authorized for selected individuals to participate in specialized training or in support of training. These are in addition to the training periods an individual attends as a part of unit training.

adequacy--Operation plan review criterion. The determination whether the scope and concept of a planned operation are sufficient to accomplish the task assigned. See also **acceptability; completeness; feasibility; suitability.**

adjust--An order to the observer or spotter to initiate an adjustment on a designated target.

adjust fire(*)--In artillery and naval gunfire support: 1. An order or request to initiate an adjustment of fire. 2. A method of control transmitted in the call for fire by the observer or spotter to indicate that he will control the adjustment.

adjustment--See **adjustment of fire.**

adjustment of fire(*)--Process used in artillery and naval gunfire to obtain correct bearing, range, and height of burst (if time fuzes are used) when engaging a target by observed fire. See also **spot.**

administration--1. The management and execution of all military matters not included in strategy and tactics. 2. Internal management of units.

administrative airlift service--The airlift service normally provided by specifically identifiable aircraft assigned to organizations or commands for internal administration.

administrative chain of command(*)--The normal chain of command for administration. See also **chain of command; operational chain of command.**

administrative control(*)--Direction or exercise of authority over subordinate or other organizations in respect to administrative matters such as personnel management, supply, services, and other matters not included in the operational missions of the subordinate or other organizations. See also **combatant command (command authority); command; control; full command; operational command (no longer used by DOD); operational control; tactical control.**

administrative escort(*)--A warship or merchant ship under naval control, carrying a convoy commodore and his staff, serving as a platform for simultaneous communication with an operational control authority and a coastal convoy.

administrative landing--An unopposed landing involving debarkation from vehicles which have been administratively loaded. See also **administrative loading; administrative movement; logistics over-the-shore operations.**

administrative lead time--The interval between initiation of procurement action and letting of contract or placing of order. See also **procurement lead time.**

administrative loading(*)--A loading system which gives primary consideration to achieving maximum utilization of troop and cargo space without regard to tactical considerations. Equipment and supplies must be unloaded and sorted before they can be used. See also **loading.** Also called **commercial loading.**"

administrative map--A map on which is graphically recorded information pertaining to administrative matters, such as supply and evacuation installations, personnel installations, medical facilities, collecting points for stragglers and enemy prisoners of war, train bivouacs, service and maintenance areas, main supply roads, traffic circulation, boundaries, and other details necessary to show the administrative situation. See also **map.**

administrative march--See **administrative movement.**

administrative movement(*)--A movement in which troops and vehicles are arranged to expedite their movement and conserve time and energy when no enemy interference, except by air, is anticipated. (DOD) Also called **administrative march.**

administrative order(*)--An order covering traffic, supplies, maintenance, evacuation, personnel, and other administrative details.

administrative shipping--Support shipping that is capable of transporting troops and cargo from origin to destination, but which cannot be loaded or unloaded without non-organic personnel and/or equipment; e.g., stevedores, piers, barges, boats. See also **administrative loading; administrative movement.**

advance--A request from a spotter to indicate the desire that the illuminating projectile burst earlier in relation to the subsequent bursts of high explosive projectiles.

advanced base(*)--A base located in or near a theater of operations whose primary mission is to support military operations.

advanced fleet anchorage--A secure anchorage for a large number of naval ships, mobile support units and auxiliaries located in or near a theater of operations. See emergency anchorage.

advanced landing field(*)--An airfield, usually having minimum facilities, in or near an objective area. See also **airfield.**

advanced operations base--In special operations, a small temporary base established near or within a joint special operations area to command, control, and/or support training or tactical operations. Facilities are normally austere. The base may be ashore or afloat. If ashore, it may include an airfield or unimproved airstrip, a pier, or an anchorage. An advanced operations base is normally controlled and/or supported by a main operations base or a forward operations base. Also called **AOB.** See also **forward operations base; main operations base.**

advance force(*)--A temporary organization within the amphibious task force which precedes the main body to the objective area. Its function is to participate in preparing the objective for the main assault by conducting

such operations as reconnaissance, seizure of supporting positions, minesweeping, preliminary bombardment, underwater demolitions, and air support.

advance guard--Detachment sent ahead of the main force to ensure its uninterrupted advance; to protect the main body against surprise; to facilitate the advance by removing obstacles, and repairing roads and bridges; and to cover the deployment of the main body if it is committed to action.

advance guard reserve--Second of the two main parts of an advance guard, the other being the advance guard support. It protects the main force and is itself protected by the advance guard support. Small advance guards do not have reserves.

advance guard support--First of the two main parts of an advance guard, the other being the advance guard reserve. It is made up of three smaller elements, in order from front to rear, the advance guard point, the advance party, and the support proper. The advance guard support protects the advance guard reserve.

advance to contact(*)--An offensive operation designed to gain or reestablish contact with the enemy. See also **approach march.**

adverse weather--Weather in which military operations are generally restricted or impeded. See also **marginal weather.**

advisory area(*)--A designated area within a flight information region where air traffic advisory service is available.

Aegis--A totally integrated shipboard weapon system that combines computers, radars, and missiles to provide a defense umbrella for surface shipping. The system is capable of automatically detecting, tracking, and destroying airborne, seaborne, and land-launched weapons.

aerial picket--See **air picket.**

aerial port--An airfield that has been designated for the sustained air movement of personnel and materiel, and to serve as an authorized port for entrance into or departure from the country in which located.

aerial port squadron--An Air Force organization which operates and provides the functions assigned to aerial ports, including processing personnel and cargo, rigging for airdrop, packing parachutes, loading equipment, preparing air cargo and load plans, loading and securing aircraft, ejecting cargo for inflight delivery, and supervising units engaged in aircraft loading and unloading operations.

aerodynamic missile(*)--A missile which uses aerodynamic forces to maintain its flight path, generally employing propulsion guidance. See also **ballistic missile; guided missile.**

aeromedical evacuation--The movement of patients under medical supervision to and between medical treatment facilities by air transportation.

aeromedical evacuation control center(*)--The control facility established by the commander of an air transport division, air force, or air command. It operates in conjunction with the command movement control center and coordinates overall medical requirements with airlift capability. It also assigns medical missions to the appropriate aeromedical evacuation elements in the system and monitors patient movement activities.

aeromedical evacuation control officer--An officer of the air transport force or air command controlling the flow of patients by air.

aeromedical evacuation system--A system which provides: a. control of patient movement by air transport; b. specialized medical attendants and equipment for inflight medical care; c. facilities on or in the vicinity of air strips and air bases, for the limited medical care of in-transit patients entering, en route via, or leaving the system; and d. communication with originating, destination, and en route medical facilities concerning patient transportation.

aeromedical evacuation unit--An operational medical organization concerned primarily with the management and control of patients being transported via an aeromedical evacuation system or system echelon. See also **forward aeromedical evacuation.**

aeronautical chart--A specialized representation of mapped features of the Earth, or some part of it, produced to show selected terrain, cultural and hydrographic features, and supplemental information required for air navigation, pilotage, or for planning air operations.

aeronautical information overprint(*)--Additional information which is printed or stamped on a map or chart for the specific purpose of air navigation.

aeronautical plotting chart(*)--A chart designed for the graphical processes of navigation.

aerospace--Of, or pertaining to, Earth's envelope of atmosphere and the space above it; two separate entities considered as a single realm for activity in launching, guidance, and control of vehicles that will travel in both entities.

aerospace control operations--The employment of air forces, supported by ground and naval forces, as appropriate, to achieve military objectives in vital aerospace areas. Such operations include destruction of enemy aerospace and surface-to-air forces, interdiction of enemy aerospace operations, protection of vital air lines of communication, and the establishment of local military superiority in areas of air operations.

aerospace defense--1. All defensive measures designed to destroy attacking enemy aircraft, missiles, and space vehicles after they leave the Earth's surface, or to nullify or reduce the effectiveness of such attacks. 2. An inclusive term encompassing air defense and space defense. See also **air defense; space defense.**

aerospace projection operations--See **land, sea, or aerospace projection operations.**

affiliation with the Department of Defense--Persons, groups of persons, or organizations are considered to be affiliated with the Department of Defense if they are: a. employed by, or contracting with, the Department of Defense or any activity under the jurisdiction of the Department of Defense, whether on a full-time, part-time, or consultative basis; b. members of the Armed Forces on active duty, National Guard members, or those in a reserve or retired status; c. residing on, authorized access to, or conducting or operating any business or other function at any DOD installation or facility; d. authorized access to defense information; e. participating in other authorized DOD programs; or f. applying or being considered for any status described above.

afloat support(*)--A form of logistic support outside the confines of a harbor in which fuel, ammunition, and supplies are provided for operating forces either underway or at anchor. See also **floating base support.**

AFSOB--See **Air Force special operations base.**

AFSOC--See **Air Force special operations component.**

AFSOD--See **Air Force special operations detachment.**

AFSOE--See **Air Force special operations element.**

AFSOF--See **Air Force special operations forces.**

afterburning(*)--1. The characteristic of some rocket motors to burn irregularly for some time after the main burning and thrust has ceased. 2. The process of fuel injection and combustion in the exhaust jet of a turbojet engine (aft or to the rear of the turbine).

afterwinds--Wind currents set up in the vicinity of a nuclear explosion directed toward the burst center, resulting from the updraft accompanying the rise of the fireball.

agency(*)--In intelligence usage, an organization or individual engaged in collecting and/or processing information. See also **agent; intelligence cycle; source.** Also called **collection agency.**

agent--In intelligence usage, one who is authorized or instructed to obtain or to assist in obtaining information for intelligence or counterintelligence purposes.

agent authentication--The technical support task of providing an agent with personal documents, accoutrements, and equipment which have the appearance of authenticity as to claimed origin and which support and are consistent with the agent's cover story.

agent net--An organization for clandestine purposes which operates under the direction of a principal agent.

age of moon(*)--The elapsed time, usually expressed in days, since the last new moon.

aggressor forces--1. Forces engaged in aggressive military action. 2. In the context of training exercises, the "enemy" created to add realism in training maneuvers and exercises.

This method replaces the less realistic system of fictional "red" and "blue" armies.

AGM-28A--See **Hound Dog.**

AGM-45--See **Shrike.**

AGM-53--See **Condor.**

AGM-65--See **Maverick.**

AGM-69--See **short range attack missile.**

AGM-78--See **Standard Arm.**

AGM-84A--See **Harpoon.**

agonic line(*)--A line drawn on a map or chart joining points of zero magnetic declination for a specified year date. See also **magnetic declination.**

agreed point(*)--A predetermined point on the ground, identifiable from the air, and used when aircraft assist in fire adjustment.

AH-1J--See **Sea Cobra.**

AIM-7--See **Sparrow.**

AIM-9--See **Sidewinder.**

AIM-54A--See **Phoenix.**

air(*)--In artillery and naval gunfire support, a spotting, or an observation, by a spotter or an observer to indicate that a burst or group of bursts occurred before impact.

air alert--See **airborne alert; air defense warning conditions; alert; ground alert.**

air attack--1. coordinated--A combination of two or more types of air attack (dive, glide, low-level) in one strike, using one or more types of aircraft. 2. deferred--A procedure in which

attack groups rendezvous as a single unit. It is used when attack groups are launched from more than one station with their departure on the mission being delayed pending further orders. 3. divided--A method of delivering a coordinated air attack which consists of holding the units in close tactical concentration up to a point, then splitting them to attack an objective from different directions.

airborne(*)--1. In relation to personnel, troops especially trained to effect, following transport by air, an assault debarkation, either by parachuting or touchdown. 2. In relation to equipment, pieces of equipment that have been especially designed for use by airborne troops during or after an assault debarkation. It also designates some aeronautical equipment used to accomplish a particular mission. 3. When applied to materiel, items that form an integral part of the aircraft. 4. The state of an aircraft, from the instant it becomes entirely sustained by air until it ceases to be so sustained. A lighter-than-air aircraft is not considered to be airborne when it is attached to the ground, except that moored balloons are airborne whenever sent aloft. See also **air transportable unit.**

airborne alert(*)--A state of aircraft readiness wherein combat-equipped aircraft are airborne and ready for immediate action. See also **fighter cover.** (DOD) It is designed to reduce reaction time and to increase the survivability factor. See also **combat air patrol; ground alert; fighter cover.**

airborne assault--See **assault phase.**

airborne assault weapon--An unarmored, mobile, full-tracked gun providing a mobile antitank capability for airborne troops. Can be air-dropped.

airborne battlefield command and control center--A United States Air Force aircraft equipped with communications, data link, and display equipment; it may be employed as an airborne command post or a communications and intelligence relay facility. Also called **ABCCC.**

airborne command post(*)--A suitably equipped aircraft used by the commander for the control of his or her forces.

airborne early warning--The detection of enemy air or surface units by radar or other equipment carried in an airborne vehicle, and the transmitting of a warning to friendly units.

airborne early warning and control(*)--Air surveillance and control provided by airborne early warning aircraft which are equipped with search and height-finding radar and communications equipment for controlling weapon systems. See also **air picket.**

airborne force(*)--A force composed primarily of ground and air units organized, equipped, and trained for airborne operations. See also **force(s).**

airborne interception equipment(*)--A fire control system, including radar equipment, installed in interceptor aircraft used to effect air interception.

airborne lift--The total capacities expressed in terms of personnel and cargo that are, or can be, carried by available aircraft in one trip.

airborne operation--An operation involving the air movement into an objective area of combat forces and their logistic support for execution of a tactical or a strategic mission. The means employed may be any combination of airborne units, air transportable units, and types of transport aircraft, depending on the mission and the overall situation. See also **assault; assault phase.**

airborne order--A command and authorization for flight when a predetermined time greater than five minutes is established for aircraft to become airborne.

airborne radio relay--Airborne equipment used to relay radio transmission from selected originating transmitters.

airborne sensor operator--An individual trained to operate sensor equipment aboard aircraft and to perform limited interpretations of collected information produced in flight.

airborne tactical data system--An airborne early warning system capable of integration into the tactical data system environment. It provides an automated, operator-controlled capability for collecting, displaying, evaluating, and disseminating tactical information via tactical digital information links. It is part of the Naval Tactical Data System (NTDS). Also called **ATDS.** See also **tactical digital information link.**

airborne troops--Those ground units whose primary mission is to make assault landings from the air. See also **troops.**

air-breathing missile--A missile with an engine requiring the intake of air for combustion of its fuel, as in a ramjet or turbojet. To be contrasted with the rocket missile, which carries its own oxidizer and can operate beyond the atmosphere.

airburst(*)--An explosion of a bomb or projectile above the surface as distinguished from an explosion on contact with the surface or after penetration. See also **air; types of burst.**

air cargo(*)--Stores, equipment or vehicles, which do not form part of the aircraft, and are either part or all of its payload. See also **cargo.**

air cartographic camera(*)--A camera having the accuracy and other characteristics essential for air survey or cartographic photography. Also called **mapping camera.**

air cartographic photography(*)--The taking and processing of air photographs for mapping and charting purposes.

air command--A major subdivision of the Air Force; for operational purposes, it normally consists of two or more air forces. See also **command.**

air control--See **air controller; air traffic control center; airway; area control center; combat zone; control and reporting center; control area; controlled airspace; control zone; interceptor controller; tactical air control center; tactical air controller; terminal control area.**

air controller(*)--An individual especially trained for and assigned the duty of the control (by use of radio, radar, or other means) of such aircraft as may be allotted to him for operation within his area. See also **air traffic controller; air weapons controller; tactical air controller.**

air corridor(*)--A restricted air route of travel specified for use by friendly aircraft and established for the purpose of preventing friendly aircraft from being fired on by friendly forces.

aircraft--See **active aircraft; inactive aircraft inventory; nonprogram aircraft; program aircraft; reserve aircraft; supporting aircraft; unit aircraft.**

aircraft arresting barrier(*)--A device, not dependent on an aircraft hook, used to engage and absorb the forward momentum of an emergency landing or an aborted takeoff. See also **aircraft arresting system.**

aircraft arresting cable(*)--That part of an aircraft arresting gear which spans the runway surface or flight deck landing area and is engaged by the aircraft arresting system. Also called **aircraft arresting wire.**

aircraft arresting gear(*)--A device used to engage hook-equipped aircraft to absorb the forward momentum of a routine or emergency landing or aborted takeoff. See also **aircraft arresting system.**

aircraft arresting hook(*)--A device fitted to an aircraft to engage arresting gear. See also **aircraft arresting system.**

aircraft arresting system(*)--A series of components used to engage an aircraft and absorb the forward momentum of a routine or emergency landing or an aborted takeoff. See also **aircraft arresting barrier; aircraft arresting gear; aircraft arresting hook.**

aircraft arresting wire--See **aircraft arresting cable.** See also **aircraft arresting system.**

aircraft arrestment(*)--Controlled stopping of an aircraft by external means.

aircraft block speed--True airspeed in knots under zero wind conditions adjusted in relation to length of sortie to compensate for takeoff, climbout, letdown, instrument approach, and landing.

aircraft captain--See **aircraft commander.**

aircraft commander(*)--The aircrew member designated by competent authority as being in command of an aircraft and responsible for its safe operation and accomplishment of the assigned mission.

aircraft control and warning system--A system established to control and report the movement of aircraft. It consists of observation facilities (radar, passive electronic, visual, or other means), control center, and necessary communications.

aircraft cross-servicing(*)--Services performed on an aircraft by an organization other than that to which the aircraft is assigned, according to an established operational aircraft cross-servicing requirement, and for which there may be a charge. Aircraft cross-servicing has been divided into two categories: **a. Stage A cross-servicing**: The servicing of aircraft on airfields/ships which enables flights to be made to another airfield/ship. The servicing includes refueling, replenishment of fluids and gases, drag chutes (if applicable), starting facilities, and ground handling. **b. Stage B cross-servicing**: The servicing of aircraft on airfields/ships which enables the aircraft to be flown on an operational mission. The servicing includes all Stage A services plus the loading of weapons and/or film, including the processing and interpretation of any exposed film from the previous mission. See also **aircraft transient servicing.**

aircraft loading table(*)--A data sheet used by the force unit commander containing information as to the load that actually goes into each aircraft.

aircraft mission equipment(*)--Equipment that must be fitted to an aircraft to enable it to fulfill a particular mission or task. Also called **aircraft role equipment.**

aircraft modification(*)--A change in the physical characteristics of aircraft, accomplished either by a change in production specifications or by alteration of items already produced.

aircraft monitoring and control--That equipment installed in aircraft to permit monitoring and control of safing, arming, and fuzing functions of nuclear weapons or nuclear weapon systems.

aircraft piracy--Any seizure or exercise of control, by force or violence or threat of force or violence or by any other form of intimidation and with wrongful intent, of an aircraft within the special aircraft jurisdiction of the United States.

aircraft repair--The process of restoring aircraft or aircraft material to a serviceable condition.

aircraft role equipment--See **aircraft mission equipment.**

aircraft scrambling(*)--Directing the immediate takeoff of aircraft from a ground alert condition of readiness.

aircraft store(*)--Any device intended for internal or external carriage and mounted on aircraft suspension and release equipment, whether or not the item is intended to be separated in flight from the aircraft. Aircraft stores are classified in two categories as follows: **a. Expendable store**--An aircraft store normally separated from the aircraft in flight such as a missile, rocket, bomb, nuclear weapon, mine, torpedo, pyrotechnic device, sonobuoy, signal underwater sound device, or other similar items. **b. Nonexpendable store**--An aircraft store which is not normally separated from the aircraft in flight such as a tank (fuel and spray), line--source disseminator, pod (refueling, thrust augmentation, gun, electronic attack, data link, etc.), multiple rack, target, cargo drop container, drone or other similar items. See also **payload.**

aircraft tiedown--Securing aircraft when parked in the open to restrain movement due to the weather or condition of the parking area. See also **aircraft picketing.**

aircraft transient servicing(*)--Services performed on an aircraft by an organization other than that to which the aircraft is assigned, and for which there may be a charge. See also **aircraft cross-servicing.**

aircraft utilization--Average numbers of hours during each 24-hour period that an aircraft is actually in flight.

aircraft vectoring(*)--The directional control of in-flight aircraft through transmission of azimuth headings.

air cushion vehicle--A vehicle capable of being operated so that its weight, including its payload, is wholly or significantly supported on a continuously generated cushion or "bubble" of air at higher than ambient pressure. Also called **ACV.** (Note: NATO uses the term "ground effect machine".)

air data computer--See **central air data computer.**

air defense--All defensive measures designed to destroy attacking enemy aircraft or missiles in the Earth's envelope of atmosphere, or to nullify or reduce the effectiveness of such attack. See also **active air defense; aerospace defense; passive air defense.**

air defense action area(*)--An area and the airspace above it within which friendly aircraft or surface-to-air weapons are normally given precedence in operations except under specified conditions. See also **air defense operations area.**

air defense area--1. overseas--A specifically defined airspace for which air defense must be planned and provided. 2. United States--Airspace of defined dimensions designated by the appropriate agency within which the ready control of airborne vehicles is required in the interest of national security during an air defense emergency.

air defense artillery--Weapons and equipment for actively combatting air targets from the ground.

air defense battle zone--A volume of airspace surrounding an air defense fire unit or defended area, extending to a specified altitude and range, in which the fire unit commander will engage and destroy targets not identified as friendly under criteria established by higher headquarters.

air defense control center(*)--The principal information, communications, and operations center from which all aircraft, antiaircraft operations, air defense artillery, guided missiles, and air warning functions of a specific area of air defense responsibility are supervised and coordinated. Also called **air defense operations center.** See also **combat information center.**

air defense direction center--An installation having the capability of performing air surveillance, interception, control, and direction of allocated air defense weapons within an assigned sector of responsibility. It may also have an identification capability.

air defense division--A geographic subdivision of an air defense region. See also **air defense sector.**

air defense early warning--See **early warning.**

air defense emergency--An emergency condition, declared by the Commander in Chief, North American Air Defense Command, that exists when attack upon the continental United States, Alaska, Canada, or United States installations; in Greenland by hostile aircraft or missiles is considered probable, is imminent, or is taking place.

air defense ground environment(*)--The network of ground radar sites and command and control centers within a specific theater of operations which are used for the tactical control of air defense operations.

air defense identification zone--Airspace of defined dimensions within which the ready identification, location, and control of airborne vehicles are required. Also called **ADIZ.** See also **air defense operations area.**

air defense operations area--An area and the airspace above it within which procedures are established to minimize mutual interference between air defense and other operations; it may include designation of one or more of the following: air defense action area, air defense area; air defense identification zone, and/or firepower umbrella. See also **air defense action area; air defense identification zone; firepower umbrella; positive identification and radar advisory zone.**

air defense operations center--See **air defense control center.**

air defense operations team--A team of United States Air Force ground environment personnel assigned to certain allied air defense control and warning units/elements.

air defense readiness--An operational status requiring air defense forces to maintain higher than ordinary preparedness for a short period of time.

air defense region(*)--A geographical subdivision of an air defense area.

air defense sector(*)--A geographical subdivision of an air defense region. See also **air defense division.**

air defense suppression--In air operations, actions taken to degrade fixed and mobile surface--based components of enemy air defense systems so that offensive air forces may effectively attack a target.

air defense warning conditions--A degree of air raid probability according to the following code. The term air defense division/sector referred to herein may include forces and units afloat and/or deployed to forward areas, as applicable. **Air defense warning yellow**--attack by hostile aircraft and/or missiles is probable. This means that hostile aircraft and/or missiles are en route toward an air defense division/sector, or unknown aircraft and/or missiles suspected to be hostile are en route toward or are within an air defense division/sector. **Air defense warning red**--attack by hostile aircraft and/or missiles is imminent or is in progress. This means that hostile aircraft and/or missiles are within an air defense division/sector or are in the immediate vicinity of an air defense division/sector with high probability of entering the division/sector. **Air defense warning white**--attack by hostile aircraft and/or missiles is improbable. May be called either before or after air defense warning yellow or red. The initial declaration of air defense emergency will automatically establish a condition of air defense warning other than white for purposes of security control of air traffic.

air delivery--See **airdrop; air landed; air movement; air supply.**

air delivery container--A sling, bag, or roll, usually of canvas or webbing, designed to hold supplies and equipment for air delivery.

air delivery equipment--Special items of equipment, such as parachutes, air delivery containers, platforms, tie downs, and related items, used in air delivery of personnel, supplies, and equipment.

air direct delivery--The strategic air movement of cargo or personnel from an airlift point of embarkation to a point as close as practicable to the user's specified final destination, thereby minimizing transshipment requirements. Air direct delivery eliminates the traditional Air Force two step strategic and theater airlift transshipment mission mix.

air division--A unit or its headquarters, on a level of command above wing level, composed of two or more combat wings, but sometimes adapted to other organizational structures.

airdrop--The unloading of personnel or materiel from aircraft in flight. See also **airdrop platform; air movement; free drop; free fall; high- velocity drop; low-velocity drop.**

airdrop platform(*)--A base on which vehicles, cargo, or equipment are loaded for airdrop or low-altitude extraction. See also **airdrop.**

air employment/allocation plan--The means by which subordinate commanders advise the joint force commander of planned employment/allocation of organic or assigned assets, of any expected excess sorties, or of any additional air support requirements.

air facility--An installation from which air operations may be or are being conducted. See also **facility.**

airfield--An area prepared for the accommodation (including any buildings, installations, and equipment), landing and takeoff of aircraft. See also **advanced landing field; alternative airfield; departure airfield; landing area; landing point; landing site; landing zone; main airfield; redeployment airfield; regroup airfield.** DOD Note: In all entries involving **airfield** or **aerodrome**, the US uses **airfield**, and NATO uses **aerodrome**. The terms are synonymous.

airfield traffic(*)--All traffic on the maneuvering area of an airfield and all aircraft flying in the vicinity of an airfield.

air fire plan--A plan for integrating and coordinating tactical air support of ground forces with other fire support.

Air Force Component Headquarters--The field headquarters facility of the Air Force commander charged with the overall conduct of Air Force operations. It is composed of the command section and appropriate staff elements.

Air Force special operations base--That base, airstrip, or other appropriate facility that provides physical support to Air Force special operations forces. The facility may be used solely to support Air Force special operations forces or may be a portion of a larger base supporting other operations. As a supporting facility, it is distinct from the forces operating from or being supported by it. Also called **AFSOB.**

Air Force special operations component--The Air Force component of a joint force special operations component. Also called **AFSOC.** See also **Army special operations component; Navy special operations component.**

Air Force special operations detachment--A squadron-size headquarters, which could be a composite organization composed of different Air Force special operations assets. The detachment is normally subordinate to an Air Force special operations component, joint special operations task force, or joint task force, depending upon size and duration of the operation. Also called **AFSOD.**

Air Force special operations element--An element-size Air Force special operations headquarters. It is normally subordinate to an Air Force special operations component or detachment, depending upon size and duration of the operation. Also called **AFSOE.**

Air Force special operations forces--Those active and reserve component Air Force forces designated by the Secretary of Defense that are specifically organized, trained, and equipped to conduct and support special operations. Also called **AFSOF.**

airframe--1. The structural components of an airplane, including the framework and skin of such parts as the fuselage, empennage, wings, landing gear (minus tires), and engine mounts. 2. The framework, envelope, and cabin of an airship. 3. The assembled principal structural components, less propulsion system, control, electronic equipment, and payload, of a missile.

air ground operations system(*)--An Army/Air Force system providing the ground commander with the means for receiving, processing and forwarding the requests of subordinate ground commanders for air support missions and for the rapid dissemination of information and intelligence.

airhead(*)--1. A designated area in a hostile or threatened territory which, when seized and held, ensures the continuous air landing of troops and materiel and provides the maneuver space necessary for projected operations. Normally it is the area seized in the assault phase of an airborne operation. 2. A designated location in an area of operations used as a base for supply and evacuation by air. See also **beachhead; bridgehead.**

air intercept control common--A tactical air-to--ground radio frequency, monitored by all air intercept control facilities within an area, which is used as a backup for other discrete tactical control frequencies.

air interception--To effect visual or electronic contact by a friendly aircraft with another aircraft. Normally, the air intercept is conducted in the following five phases: **a. climb phase**--Airborne to cruising altitude. **b. maneuver phase**--Receipt of initial vector to target

until beginning transition to attack speed and altitude. **c. transition phase**--Increase or decrease of speed and altitude required for the attack. **d. attack phase**--Turn to attack heading, acquire target, complete attack, and turn to breakaway heading. **e. recovery phase**--Breakaway to landing. See also **broadcast controlled air interception; close controlled air interception.**

air intercept zone--A subdivided part of the destruction area in which it is planned to destroy or defeat the enemy airborne threat with interceptor aircraft. See also **destruction area.**

air interdiction(*)--Air operations conducted to destroy, neutralize, or delay the enemy's military potential before it can be brought to bear effectively against friendly forces at such distance from friendly forces that detailed integration of each air mission with the fire and movement of friendly forces is not required.

air landed(*)--Moved by air and disembarked, or unloaded, after the aircraft has landed or while a helicopter is hovering. See also **air movement.**

air-launched ballistic missile--A ballistic missile launched from an airborne vehicle.

air liaison officer--An officer (aviator/pilot) attached to a ground unit who functions as the primary advisor to the ground commander on air operation matters.

airlift capability--The total capacity expressed in terms of number of passengers and/or weight/cubic displacement of cargo that can be carried at any one time to a given destination by available airlift. See also **airlift requirement; allowable load; payload.**

airlift control center--An operations center where the detailed planning, coordinating, and tasking for tactical airlift operations are accomplished.

This is the focal point for communications and the source of control and direction for the tactical airlift forces. Also called **ALCC.**

airlift requirement(*)--The total number of passengers and/or weight/cubic displacement of cargo required to be carried by air for a specific task. See also **airlift capability.**

airlift service--The performance or procurement of air transportation and services incident thereto required for the movement of persons, cargo, mail, or other goods.

air logistic support--Support by air landing or airdrop, including air supply, movement of personnel, evacuation of casualties and enemy prisoners of war, and recovery of equipment and vehicles.

air logistic support operation(*)--An air operation, excluding an airborne operation, conducted within a theater to distribute and recover personnel, equipment and supplies.

airmiss--See **near miss.**

air mission--See **mission.**

air mission intelligence report--A detailed report of the results of an air mission, including a complete intelligence account of the mission.

airmobile forces(*)--The ground combat, supporting and air vehicle units required to conduct an airmobile operation.

airmobile operation(*)--An operation in which combat forces and their equipment move about the battlefield by aircraft to engage in ground combat.

airmobility(*)--A capability of airmobile forces which permits them to move by air while retaining the ability to engage in ground combat.

Air Mobility Command--The Air Force Component Command of the US Transportation Command. Also called **AMC.**

air movement(*)--Air transport of units, personnel, supplies and equipment including airdrops and air landings. See also **airdrop; air landed; free drop; high velocity drop; low velocity drop.**

air movement column--In airborne operations, the lead formation and the serials following, proceeding over the same flight path at the same altitude.

air movement table(*)--A table prepared by a ground force commander in coordination with an air force commander. This form, issued as an annex to the operation order: a. Indicates the allocation of aircraft space to elements of the ground units to be airlifted; b. Designates the number and type of aircraft in each serial; c. Specifies the departure area, time of loading, and takeoff.

air/naval gunfire liaison company--An organization composed of Marine and Navy personnel specially qualified for shore control of naval gunfire and close air support. Also called **ANGLICO.**

air observation--See **air observer.**

air observation post--See **observation post.**

air observer(*)--An individual whose primary mission is to observe or take photographs from an aircraft in order to adjust artillery fire or obtain military information.

air observer adjustment--The correcting of gunfire from an aircraft. See also **spot.**

air offensive--Sustained operations by strategic and/or tactical air weapon systems against hostile air forces or surface targets.

air operations center--See **tactical air control center.**

air photographic reconnaissance(*)--The obtaining of information by air photography, divided into three types: a. Strategic photographic reconnaissance; b. Tactical photographic reconnaissance; and c. Survey/cartographic photography-air photography taken for survey/cartographical purposes and to survey/cartographic standards of accuracy. It may be strategic or tactical.

air picket(*)--An airborne early warning aircraft positioned primarily to detect, report, and track approaching enemy aircraft or missiles and to control intercepts. See also **airborne early warning and control.** Also called **aerial picket.**

air plot(*)--1. A continuous plot used in air navigation of a graphic representation of true headings steered and air distances flown. 2. A continuous plot of the position of an airborne object represented graphically to show true headings steered and air distances flown. 3. Within ships, a display that shows the positions and movements of an airborne object relative to the plotting ship.

airport--See **airfield.**

air portable(*)--Denotes materiel which is suitable for transport by an aircraft loaded internally or externally, with no more than minor dismantling and reassembling within the capabilities of user units. This term must be qualified to show the extent of air portability. See also **load.**

airport surface detection equipment--Short-range radar displaying the airport surface. Aircraft and vehicular traffic operating on runways, taxiways, and ramps, moving or stationary, may be observed with a high degree of resolution.

airport surveillance radar--Radar displaying range and azimuth that is normally employed in a terminal area as an aid to approach- and departure-control.

airport traffic area--Unless otherwise specifically designated, that airspace within a horizontal radius of five statute miles from the geographic center of any airport at which a control tower is operating, extending from the surface up to, but not including, an altitude of 3,000 feet above the elevation of the airport.

air position(*)--The calculated position of an aircraft assuming no wind effect.

air priorities committee(*)--A committee set up to determine the priorities of passengers and cargo. See also **air transport allocations board.**

air raid reporting control ship(*)--A ship to which the air defense ship has delegated the duties of controlling air warning radar and air raid reporting.

air reconnaissance--The acquisition of intelligence information by employing visual observation and/or sensors in air vehicles.

air reconnaissance liaison officer--An Army officer especially trained in air reconnaissance and imagery interpretation matters who is attached to a tactical air reconnaissance unit. This officer assists and advises the air commander and staff on matters concerning ground operations and informs the supported ground commander on the status of air reconnaissance requests.

air request net--A high frequency, single sideband, nonsecure net monitored by all tactical air control parties (TACPs) and the air support operations center (ASOC) that allows immediate requests to be transmitted from a TACP at any Army echelon directly to the ASOC for rapid response.

air route(*)--The navigable airspace between two points, identified to the extent necessary for the application of flight rules.

air route traffic control center--The principal facility exercising en route control of aircraft operating under instrument flight rules within its area of jurisdiction. Approximately 26 such centers cover the United States and its possessions. Each has a communication capability to adjacent centers.

air sovereignty--A nation's inherent right to exercise absolute control and authority over the airspace above its territory. See also **air sovereignty mission.**

air sovereignty mission--The integrated tasks of surveillance and control, the execution of which enforces a nation's authority over its territorial airspace. See also **air sovereignty.**

airspace control--See **airspace control in the combat zone.**

airspace control area(*)--Airspace which is laterally defined by the boundaries of the area of operations. The airspace control area may be subdivided into airspace control sub-areas.

airspace control authority(*)--The commander designated to assume overall responsibility for the operation of the airspace control system in the airspace control area.

airspace control boundary(*)--The lateral limits of an airspace control area, airspace control sub-area, high density airspace control zone, or airspace restricted area.

airspace control center--The airspace control authority's primary airspace control facility, including assigned Service component, host nation, and/or allied personnel and equipment.

airspace control facility--Any of the several Service component, host nation, or allied facilities that provide airspace control in the combat zone.

airspace control in the combat zone--A process used to increase combat effectiveness by promoting the safe, efficient, and flexible use of airspace. Airspace control is provided in order to prevent fratricide, enhance air defense operations, and permit greater flexibility of operations. Airspace control does not infringe on the authority vested in commanders to approve, disapprove, or deny combat operations. Also called **combat airspace control; airspace control.**

airspace control order--An order implementing the airspace control plan that provides the details of the approved requests for airspace control measures. It is published either as part of the air tasking order or as a separate document. Also called **ACO.**

airspace control plan--The document approved by the joint force commander that provides specific planning guidance and procedures for the airspace control system for the joint force area of responsibility. Also called **ACP.**

airspace control sector--A subelement of the airspace control area, established to facilitate the control of the overall area. Airspace control sector boundaries normally coincide with air defense organization subdivision boundaries. Airspace control sectors are designated in accordance with procedures and guidance contained in the airspace control plan in consideration of Service component, host nation, and allied airspace control capabilities and requirements. See also **airspace control area.**

airspace control system(*)--An arrangement of those organizations, personnel, policies, procedures, and facilities required to perform airspace control functions.

airspace management--The coordination, integration, and regulation of the use of airspace of defined dimensions.

airspace reservation--The airspace located above an area on the surface of the land or water, designated and set apart by Executive Order of the President or by a state, commonwealth, or territory, over which the flight of aircraft is prohibited or restricted for the purpose of national defense or for other governmental purposes.

airspace restrictions(*)--Special restrictive measures applied to segments of airspace of defined dimensions.

air space warning area--See **danger area.**

airspeed--The speed of an aircraft relative to its surrounding air mass. The unqualified term "airspeed" can mean any one of the following: **a. calibrated airspeed**--Indicated airspeed corrected for instrument installation error. **b. equivalent airspeed**--Calibrated airspeed corrected for compressibility error. **c. indicated airspeed**--The airspeed shown by an airspeed indicator. **d. true airspeed**--Equivalent airspeed corrected for error due to air density (altitude and temperature).

airspeed indicator(*)--An instrument which displays the indicated airspeed of the aircraft derived from inputs of pitot and static pressures.

air spot--The correcting adjustment of gunfire based on air observation.

air staging unit(*)--A unit situated at an airfield and concerned with reception, handling, servicing, and preparation for departure of aircraft and control of personnel and cargo.

air station(*)--In photogrammetry, the point in space occupied by the camera lens at the moment of exposure.

air strike--An attack on specific objectives by fighter, bomber, or attack aircraft on an offensive mission. May consist of several air organizations under a single command in the air.

air strike coordinator--The air representative of the force commander in a target area, who is responsible for directing all aircraft in the target area and coordinating their efforts to achieve the most effective use of air striking power.

air strip(*)--An unimproved surface which has been adapted for takeoff or landing of aircraft, usually having minimum facilities. See also **airfield.**

air superiority(*)--That degree of dominance in the air battle of one force over another which permits the conduct of operations by the former and its related land, sea and air forces at a given time and place without prohibitive interference by the opposing force.

air supply(*)--The delivery of cargo by airdrop or air landing.

air support(*)--All forms of support given by air forces on land or sea. See also **call mission; close air support; immediate air support; indirect air support; preplanned air support; tactical air support.**

air support operations center(*)--An agency of a tactical air control system collocated with a corps headquarters or an appropriate land force headquarters, which coordinates and directs close air support and other tactical air support. See also **tactical air control center.**

air support radar team--A subordinate operational component of a tactical air control system which provides ground-controlled precision flight path guidance and weapons release. See also **armstrong.**

air supremacy(*)--That degree of air superiority wherein the opposing air force is incapable of effective interference.

air surface zone(*)--A restricted area established for the purpose of preventing friendly surface vessels and aircraft from being fired upon by friendly forces and for permitting antisubmarine operations, unrestricted by the operation of friendly submarines. See also **restricted area.**

air surveillance(*)--The systematic observation of air space by electronic, visual or other means, primarily for the purpose of identifying and determining the movements of aircraft and missiles, friendly and enemy, in the air space under observation. See also **satellite and missile surveillance; surveillance.**

air surveillance officer(*)--An individual responsible for coordinating and maintaining an accurate, current picture of the air situation within an assigned airspace area.

air survey camera--See **air cartographic camera.**

air survey photography--See **air cartographic photography.**

air target chart--A display of pertinent air target intelligence on a specialized graphic base. It is designed primarily to support operations against designated air targets by various weapon systems.

Air Target Materials Program--A DOD program under the management control of the Defense Mapping Agency established for and limited to the production of medium- and large-scale map, chart, and geodetic products which support worldwide targeting requirements of the unified and specified commands, the Military Departments, and allied participants. It encompasses the determination of production and coverage requirements, standardization of products, establishment of production priorities and schedules, and the production, distribution, storage, and release/exchange of products included under it.

air target mosaic--A large-scale mosaic providing photographic coverage of an area and permitting comprehensive portrayal of pertinent target detail. These mosaics are used for intelligence study and in planning and briefing for air operations.

air terminal--A facility on an airfield that functions as an air transportation hub and accommodates the loading and unloading of airlift aircraft and the intransit processing of traffic. The airfield may or may not be designated an aerial port.

air-to-air guided missile(*)--An air-launched guided missile for use against air targets. See also **guided missile.**

air-to-surface guided missile(*)--An air-launched guided missile for use against surface targets. See also **guided missile.**

air traffic control and landing systems--Department of Defense facilities, personnel, and equipment (fixed, mobile, and seaborne) with associated avionics to provide safe, orderly, and expeditious aerospace vehicle movements worldwide. Also called **ATCALS.**

air traffic control center(*)--A unit combining the functions of an area control center and a flight information center. See also **area control center; flight information region.**

air traffic control clearance(*)--Authorization by an air traffic control authority for an aircraft to proceed under specified conditions.

air traffic control facility--Any of the component airspace control facilities primarily responsible for providing air traffic control services and, as required, limited tactical control services.

air traffic controller--An air controller especially trained for and assigned to the duty of airspace management and traffic control of airborne objects. See also **air controller.**

air traffic control service(*)--A service provided for the purpose of: a. preventing collisions: (1) between aircraft; and (2) on the maneuvering area between aircraft and obstructions; and b. expediting and maintaining an orderly flow of air traffic.

air traffic identification--The use of electronic devices, operational procedures, visual observation, and/or flight plan correlation for the purpose of identifying and locating aircraft flying within the airspace control area.

air traffic section--The link between the staging post and the local air priority committee. It is the key to the efficient handling of passengers and cargo at a staging post. It must include load control (including Customs and Immigrations facilities), freight, and mail sections.

air transportable unit(*)--A unit other than airborne, whose equipment is adapted for air movement. See also **airborne; airborne operation.**

air transport allocations board(*)--The joint agency responsible within the theater for the establishment of airlift priorities and for space allocation of available aircraft capabilities allotted to the theater. See also **air priorities committee.**

air transported operations--The movement by aircraft of troops and their equipment for an operation.

air transport group--A task organization of transport aircraft units that provides air transport for landing force elements or provides logistic support.

air transport liaison officer(*)--An officer attached for air transport liaison duties to a headquarters or unit. See also **ground liaison officer.**

air transport liaison section(*)--A subunit of the movement control organization deployed to airfields and responsible for the control of service movement at the airfield in connection with air movement operations and exercises.

air transport operations--See **strategic air transport operations; tactical air transport operations.**

airway(*)--A control area or portion thereof established in the form of a corridor marked with radio navigational aids. See also **air control.**

airways station--A ground communication installation established, manned, and equipped to communicate with aircraft in flight, as well as with other designated airways installations, for the purpose of expeditious and safe movements of aircraft. These stations may or may not be located on designated airways.

air weapons controller--An individual especially trained for and assigned to the duty of employing and controlling air weapon systems against airborne and surface objects.

ALD--See **accounting line designator; available-to-load date.**

alert(*)--1. Readiness for action, defense or protection. 2. A warning signal of a real or threatened danger, such as an air attack. 3. The period of time during which troops stand by in response to an alarm. 4. To forewarn; to prepare for action. See also **airborne alert.** (DOD) 5. A warning received by a unit or a headquarters which forewarns of an impending operational mission. See also **air defense warning conditions; ground alert; warning order.**

alert force--Specified forces maintained in a special degree of readiness.

alerting service(*)--A service provided to notify appropriate organizations regarding aircraft in need of search and rescue aid, and assist such organizations as required.

all available--A command or request to obtain the fire of all artillery able to deliver effective fire on a given target.

allocation--In a general sense, distribution of limited resources among competing requirements for employment. Specific allocations (e.g., air sorties, nuclear weapons, forces, and transportation) are described as allocation of air sorties, nuclear weapons, etc. See also **allocation (air); allocation (nuclear); allocation (transportation); apportionment.**

allocation (air)--The translation of the apportionment into total numbers of sorties by aircraft type available for each operation/task. See also **allocation.**

allocation (nuclear)--The apportionment of specific numbers and types of nuclear weapons to a commander for a stated time period as a planning factor for use in the development of war plans. (Additional authority is required for the actual deployment of allocated weapons to locations desired by the commander to support the war plans. Expenditures of these weapons are not authorized until released by proper authority.)

allocation (transportation)--Apportionment by designated authority of available transport capability to users.

allotment--The temporary change of assignment of tactical air forces between subordinate commands. The authority to allot is vested in the commander having combatant command (command authority). See also **combatant command (command authority).**

all out war--Not to be used. See **general war.**

allowable load(*)--The total load that an aircraft can transport over a given distance, taking into account weight and volume. See also **airlift capability; airlift requirement; combat load; load; payload; standard load.**

all-purpose hand-held weapon--A lightweight, hand-held, small arms weapon capable of projecting munitions required to engage both area- and point-type targets.

all-source intelligence--1. Intelligence products and/or organizations and activities that incorporate all sources of information, including, most frequently, human resources intelligence, imagery intelligence, measurement and signature intelligence, signals intelligence, and open source data, in the production of finished intelligence. 2. In intelligence collection, a phrase that indicates that in the satisfaction of intelligence requirements, all collection, processing, exploitation, and reporting systems and resources are identified for possible use and those most capable are tasked. See also **intelligence.**

all-weather air defense fighter(*)--A fighter aircraft with equipment and weapons which enable it to engage airborne targets in all weather conditions, day and night.

alphabet code--See phonetic alphabet.

ALSS--See naval advanced logistic support site.

alternate airfield(*)--An airfield specified in the flight plan to which a flight may proceed when it becomes inadvisable to land at the airfield of intended landing. An alternate airfield may be the airfield of departure.

alternate command authority--One or more predesignated officers empowered by the commander through predelegation of authority to act under stipulated emergency conditions in the accomplishment of previously defined functions.

alternate command post--Any location designated by a commander to assume command post functions in the event the command post becomes inoperative. It may be partially or fully equipped and manned or it may be the command post of a subordinate unit.

alternate headquarters--An existing headquarters of a component or subordinate command which is predesignated to assume the responsibilities and functions of another headquarters under prescribed emergency conditions.

alternative airfield(*)--An airfield with minimal essential facilities for use as an emergency landing ground, or when main or redeployment airfields are out of action, or as required for tactical flexibility. See also airfield.

altimeter(*)--An instrument which measures vertical distance with respect to a reference level.

altitude(*)--The vertical distance of a level, a point or an object considered as a point, measured from mean sea level. See also absolute altitude; critical altitude; density altitude; drop altitude; elevation; height; high alti-
tude; minimum safe altitude; pressure altitude; transition altitude; true altitude.

altitude acclimatization(*)--A slow physiological adaptation resulting from prolonged exposure to significantly reduced atmospheric pressure.

altitude chamber--See hypobaric chamber.

altitude datum(*)--The arbitrary level from which vertical displacement is measured. The datum for height measurement is the terrain directly below the aircraft or some specified datum; for pressure altitude, the level at which the atmospheric pressure is 29.92 inches of mercury (1013.2 m.bs); and for true altitude, mean sea level. See also altitude.

altitude delay(*)--Synchronization delay introduced between the time of transmission of the radar pulse and the start of the trace on the indicator, for the purpose of eliminating the altitude hole on the plan position indicator-type display.

altitude height--See altitude datum.

altitude hold(*)--In a flight control system, a control mode in which the barometric altitude existing at time of engagement is maintained automatically.

altitude hole(*)--The blank area at the origin of a radial display, on a radar tube presentation, the center of the periphery of which represents the point on the ground immediately below the aircraft. In side-looking airborne radar, this is known as the altitude slot.

altitude separation--See vertical separation.

altitude sickness--The syndrome of depression, anorexia, nausea, vomiting, and collapse, due to decreased atmospheric pressure, occurring in an individual exposed to an altitude beyond that to which acclimatization has occurred.

altitude slot--See **altitude hole.**

altitude tint--See **hypsometric tinting.**

ambulatory patient--See **walking patient.**

AMC--See **Air Mobility Command.**

ammo (plus, minus, zero)--In air intercept, a code meaning I have amount of ammunition indicated left (type may be specified. For example: **ammo plus**--I have more than half my ammunition left. **ammo minus**--I have less than half my ammunition left. **ammo zero**--I have no ammunition left.)

ammunition--See **munition.**

ammunition and toxic material open space(*)--An area especially prepared for storage of explosive ammunition and toxic material. For reporting purposes, it does not include the surrounding area restricted for storage because of safety distance factors. It includes barricades and improvised coverings. See also **storage.**

ammunition controlled supply rate--In Army usage, the amount of ammunition estimated to be available to sustain operations of a designated force for a specified time if expenditures are controlled at that rate. It is expressed in terms of rounds per weapon per day for ammunition items fired by weapons, and in terms of units of measure per organization per day for bulk allotment ammunition items. Tactical commanders use this rate to control expenditures of ammunition during tactical operations at planned intervals. It is issued through command channels at each level. It is determined based on consideration of the required supply rates submitted by subordinate commanders and ammunition assets available.

ammunition lot(*)--A quantity of homogeneous ammunition, identified by a unique lot number, which is manufactured, assembled, or renovated by one producer under uniform conditions and which is expected to function in a uniform manner.

ammunition supply point--See **distribution point.**

amphibious assault--The principal type of amphibious operation that involves establishing a force on a hostile or potentially hostile shore. See also **assault; assault phase.**

amphibious assault area--See **landing area.**

amphibious assault landing--See **amphibious operation, Part e.**

amphibious assault ship (general purpose)--A naval ship designed to embark, deploy, and land elements of a landing force in an assault by helicopters, landing craft, amphibious vehicles, and by combinations of these methods. Designated LHA.

amphibious chart(*)--A special naval chart designed to meet special requirements for landing operations and passive coastal defense, at a scale of 1:25,000 or larger, and showing foreshore and coastal information in greater detail than a combat chart.

amphibious command ship(*)--A naval ship from which a commander exercises control in amphibious operations. (DOD) Also called **LCC.**

amphibious construction battalion--A permanently commissioned naval unit, subordinate to the commander, naval beach group, designed to provide an administrative unit from which personnel and equipment are formed in tactical elements and made available to appropriate commanders to operate pontoon causeways, transfer barges, warping tugs, and assault bulk fuel systems, and to meet salvage requirements of the naval beach party.

amphibious control group(*)--Personnel, ships, and craft designated to control the waterborne ship-to-shore movement in an amphibious operation.

amphibious demonstration(*)--A type of amphibious operation conducted for the purpose of deceiving the enemy by a show of force with the expectation of deluding the enemy into a course of action unfavorable to him.

amphibious force(*)--1. A naval force and landing force, together with supporting forces that are trained, organized and equipped for amphibious operations. 2. In naval usage, the administrative title of the amphibious type command of a fleet.

amphibious group--A command within the amphibious force, consisting of the commander and staff, designed to exercise operational control of assigned units in executing all phases of a division-size amphibious operation.

amphibious lift(*)--The total capacity of assault shipping utilized in an amphibious operation, expressed in terms of personnel, vehicles, and measurement or weight tons of supplies.

amphibious objective area(*)--A geographical area, delineated in the initiating directive, for purposes of command and control within which is located the objective(s) to be secured by the amphibious task force. This area must be of sufficient size to ensure accomplishment of the amphibious task force's mission and must provide sufficient area for conducting necessary sea, air, and land operations.

amphibious objective study--A study designed to provide basic intelligence data of a permanent or semipermanent nature required for planning amphibious operations. Each study deals with a specific area, the selection of which is based on strategic location, susceptibility to seizure by amphibious means, and other considerations.

amphibious operation--An attack launched from the sea by naval and landing forces, embarked in ships or craft involving a landing on a hostile or potentially hostile shore. As an entity, the amphibious operation includes the following phases: a. planning--The period extending from issuance of the initiating directive to embarkation. b. embarkation--The period during which the forces, with their equipment and supplies, are embarked in the assigned shipping. c. rehearsal--The period during which the prospective operation is rehearsed for the purpose of: (1) testing adequacy of plans, the timing of detailed operations, and the combat readiness of participating forces; (2) ensuring that all echelons are familiar with plans; and (3) testing communications. d. movement--The period during which various components of the amphibious task force move from points of embarkation to the objective area. e. assault--The period between the arrival of the major assault forces of the amphibious task force in the objective area and the accomplishment of the amphibious task force mission.

amphibious planning--The process of planning for an amphibious operation, distinguished by the necessity for concurrent, parallel, and detailed planning by all participating forces; and wherein the planning pattern is cyclical in nature, composed of a series of analyses and judgments of operational situations, each stemming from those that have preceded.

amphibious raid(*)--A type of amphibious operation involving swift incursion into or temporary occupation of an objective followed by a planned withdrawal. See also **amphibious operation.**

amphibious reconnaissance(*)--An amphibious landing conducted by minor elements, normally involving stealth rather than force of arms, for the purpose of securing information, and usually followed by a planned withdrawal.

amphibious reconnaissance unit--A unit organized, equipped, and trained to conduct and support amphibious reconnaissance missions. An amphibious reconnaissance unit is made up of a number of amphibious reconnaissance teams.

amphibious shipping--Organic Navy ships specifically designed to transport, land, and support landing forces in amphibious assault operations and capable of being loaded or unloaded by naval personnel without external assistance in the amphibious objective area.

amphibious squadron(*)--A tactical and administrative organization composed of amphibious assault shipping to transport troops and their equipment for an amphibious assault operation.

amphibious striking forces--Forces capable of projecting military power from the sea upon adjacent land areas for initiating and/or conducting operations there in the face of enemy opposition.

amphibious task force--The task organization formed for the purpose of conducting an amphibious operation. The amphibious task force always includes Navy forces and a landing force, with their organic aviation, and may include Military Sealift Command-provided ships and Air Force forces when appropriate. Also called **ATF.**

amphibious tractor--See **amphibious vehicle.**

amphibious transport dock--A ship designed to transport and land troops, equipment, and supplies by means of embarked landing craft, amphibious vehicles, and helicopters. Designated as **LPD.**

amphibious transport group--A subdivision of an amphibious task force, composed primarily of transport ships. The size of the transport group will depend upon the scope of the operation. Ships of the transport group will be combat--loaded to support the landing force scheme of maneuver ashore. A transport unit will usually be formed to embark troops and equipment to be landed over a designated beach or to embark all helicopter-borne troops and equipment.

amphibious vehicle(*)--A wheeled or tracked vehicle capable of operating on both land and water. See also **landing craft; vehicle.**

amphibious vehicle availability table--A tabulation of the type and number of amphibious vehicles available primarily for assault landings and for support of other elements of the operation.

amphibious vehicle employment plan--A plan showing in tabular form the planned employment of amphibious vehicles in landing operations, including their employment after the initial movement to the beach.

amphibious vehicle launching area(*)--An area, in the vicinity of and to seaward of the line of departure, to which landing ships proceed and launch amphibious vehicles.

amphibious withdrawal(*)--A type of amphibious operation involving the extraction of forces by sea in naval ships or craft from a hostile or potentially hostile shore. See also **amphibious operation.**

amplifying report--See **contact report.**

analysis(*)--In intelligence usage, a step in the processing phase of the intelligence cycle in which information is subjected to review in order to identify significant facts for subsequent interpretation. See also **intelligence cycle.**

anchor--See **sinker.**

anchor cable(*)--In air transport, a cable in an aircraft to which the parachute static lines or strops are attached.

anchored--In air intercept, a code meaning, "Am orbiting a visible orbit point."

anchor line extension kit(*)--A device fitted to an aircraft equipped with removable clamshell doors to enable paratroopers to exit from the rear.

angels--In air intercept and close air support, a code meaning aircraft altitude (in thousands of feet).

angle of convergence(*)--The angle subtended by the eyebase of an observer at the point of focus.

angle of depression(*)--1. The angle in a vertical plane between the horizontal and a descending line. 2. In air photography, the angle between the optical axis of an obliquely mounted air camera and the horizontal. Also called **depression angle.** See also **tilt angle.**

angle of safety(*)--The minimal permissible angular clearance, at the gun, of the path of a projectile above the friendly troops. It is the angle of clearance corrected to ensure the safety of the troops. See also **elevation of security.**

angle of view(*)--1. The angle between two rays passing through the perspective center (rear nodal point) of a camera lens to two opposite corners of the format. 2. In photogrammetry, twice the angle whose tangent is one-half the length of the diagonal of the format divided by the calibrated focal length.

angle T(*)--In artillery and naval gunfire support, the angle formed by the intersection of the gun-target line and the observer-target line.

ANGLICO--See **air/naval gunfire liaison company.**

annex--A document appended to an operation order or other document to make it clearer or to give further details.

annotated print(*)--A photograph on which interpretation details are indicated by words or symbols.

annotation(*)--A marking placed on imagery or drawings for explanatory purposes or to indicate items or areas of special importance.

annual screening--One day of active duty for training required each year for Individual Ready Reserve members so the Services can keep current on each member's physical condition, dependency status, military qualifications, civilian occupational skills, availability for service, and other information.

annual training--The minimal period of training reserve members must perform each year to satisfy the training requirements associated with their reserve component assignment.

antenna mine(*)--In naval mine warfare, a contact mine fitted with antennae which, when touched by a steel ship, sets up galvanic action to fire the mine. See also **mine.**

antiair warfare--A US Navy/US Marine Corps term used to indicate that action required to destroy or reduce to an acceptable level the enemy air and missile threat. It includes such measures as the use of interceptors, bombers, antiaircraft guns, surface-to-air and air-to-air missiles, electronic attack, and destruction of the air or missile threat both before and after it is launched. Other measures which are taken to minimize the effects of hostile air action are cover, concealment, dispersion, deception (including electronic), and mobility. See also **counter air.**

antiarmor helicopter(*)--A helicopter armed primarily for use in the destruction of armored targets. Also called **antitank helicopter.**

anticountermining device(*)--A device fitted in an influence mine designed to prevent its actuation by shock.

anticrop agent(*)--A living organism or chemical used to cause disease or damage to selected food or industrial crops.

anticrop operation(*)--The employment of anticrop agents in military operations to destroy the enemy's source of selected food or industrial crops. See also **antiplant agent; herbicide.**

anti-G suit--A device worn by aircrew to counteract the effects on the human body of positive acceleration.

antilift device(*)--A device arranged to detonate the mine to which it is attached, or to detonate another mine or charge nearby, if the mine is disturbed.

antimateriel agent(*)--A living organism or chemical used to cause deterioration of, or damage to, selected materiel.

antimateriel operation(*)--The employment of antimateriel weapons or agents in military operations.

antipersonnel mine (land mine warfare)--A mine designed to cause casualties to personnel. See also **mine.**

antiplant agent--A microorganism or chemical that will kill, disease, or damage plants. See also **anticrop agent; herbicide.**

antiradiation missile(*)--A missile which homes passively on a radiation source. See also **guided missile.**

antirecovery device(*)--In naval mine warfare, any device in a mine designed to prevent an enemy discovering details of the working of the mine mechanism.

antisubmarine action(*)--An operation by one or more antisubmarine ships or aircraft, or a combination of both, against a particular enemy submarine.

antisubmarine air close support--Air operations for the antisubmarine warfare protection of a supported force. These operations are normally carried out within 80 nautical miles of the force, but this limit may be varied at the discretion of the controlling officer in tactical command.

antisubmarine air distant support--Antisubmarine air support at a distance from, but directly related to, specific convoys or forces.

antisubmarine air search attack unit--The designation given to one or more aircraft separately organized as a tactical unit to search for and destroy submarines.

antisubmarine barrier(*)--The line formed by a series of static devices or mobile units arranged for the purpose of detecting, denying passage to, or destroying hostile submarines. See also **antisubmarine patrol.**

antisubmarine operation--Operation contributing to the conduct of antisubmarine warfare.

antisubmarine patrol(*)--The systematic and continuing investigation of an area or along a line to detect or hamper submarines, used when the direction of submarine movement can be established. See also **antisubmarine barrier.**

antisubmarine rocket--A surface ship-launched, rocket-propelled, nuclear depth charge or homing torpedo. Designated as **RUR-5A.** Also called **ASROC.**

antisubmarine screen(*)--An arrangement of ships and/or aircraft for the protection of a screened unit against attack by a submarine.

antisubmarine search(*)--Systematic investigation of a particular area for the purpose of locating a submarine known or suspected to be somewhere in the area. Some types of search are also used in locating the position of a distress incident.

antisubmarine support operation(*)--An operation conducted by an antisubmarine force in the area around a force or convoy, in areas through which the force or convoy is passing, or in defense of geographic areas. Support operations may be completely coordinated with those of the force or convoy, or they may be independent operations coordinated only to the extent of providing operational intelligence and information.

antisubmarine torpedo--A submarine-launched, long-range, high-speed, wire-guided, deep-diving, wakeless torpedo capable of carrying a nuclear warhead for use in antisubmarine and antisurface ship operations. Also called **AS-TOR.**

antisubmarine warfare(*)--Operations conducted with the intention of denying the enemy the effective use of submarines.

antisubmarine warfare forces--Forces organized primarily for antisubmarine action. May be composed of surface ships, aircraft, submarines, or any combination of these, and their supporting systems.

antisurface air operation(*)--An air operation conducted in an air/sea environment against enemy surface forces.

antisweep device(*)--Any device incorporated in the mooring of a mine or obstructor, or in the mine circuits to make the sweeping of the mine more difficult.

antisweeper mine(*)--A mine which is laid or whose mechanism is designed or adjusted with the specific object of damaging mine countermeasures vessels. See also **mine.**

antitank helicopter--See **antiarmor helicopter.**

antitank mine(*)--A mine designed to immobilize or destroy a tank. See also **mine.**

antiterrorism--Defensive measures used to reduce the vulnerability of individuals and property to terrorist acts, to include limited response and containment by local military forces. Also called **AT.** See also **antiterrorism awareness; counterterrorism; proactive measures; terrorism.**

antiterrorism awareness--Fundamental knowledge of the terrorist threat and measures to reduce personal vulnerability to terrorism. See also **antiterrorism.**

antivignetting filter(*)--A filter bearing a deposit which is graduated in density to correct for the uneven illumination given by certain lenses, particularly wide angle types.

antiwatching device(*)--A device fitted in a moored mine which causes it to sink should it watch, so as to prevent the position of the mine or minefield being disclosed. See also **watching mine.**

AO--See **area of operations.**

AOB--See **advanced operations base.**

AOR--See **area of responsibility.**

apogee--The point at which a missile trajectory or a satellite orbit is farthest from the center of the gravitational field of the controlling body or bodies.

apparent horizon(*)--The visible line of demarcation between land/sea and sky.

apparent precession(*)--The apparent deflection of the gyro axis, relative to the Earth, due to the rotating effect of the Earth and not due to any applied force. Also called **apparent wander.**

apparent wander--See **apparent precession.**

appendix--A subsidiary addition to a main paper. Details essential to the main paper but too bulky or numerous to include therein are usually embodied in appendices.

applicable materiel assets--That portion of the total acceptable materiel assets that meets the military or other characteristics as defined by the responsible Military Service and that is in the right condition and location to satisfy a specific military requirement.

application--1. The system or problem to which a computer is applied. Reference is often made to an application as being either of the computational type, wherein arithmetic computations predominate, or of the data processing type, wherein data handling operations predominate. 2. In the intelligence context, the direct extraction and tailoring of information from an existing foundation of intelligence and near real time reporting. It is focused on and meets specific, narrow requirements, normally on demand.

applied research--Research concerned with the practical application of knowledge, material, and/or techniques directed toward a solution to an existent or anticipated military requirement. See also **basic research; research.**

apportionment--In the general sense, distribution for planning of limited resources among competing requirements. Specific apportionments (e.g., air sorties and forces for planning) are described as apportionment of air sorties and forces for planning, etc. See also **allocation; apportionment (air).**

apportionment (air)--The determination and assignment of the total expected effort by percentage and/or by priority that should be devoted to the various air operations and/or geographic areas for a given period of time.

appreciations--Personal conclusions, official estimates, and assumptions about another party's intentions, military capabilities, and activities used in planning and decisionmaking. **a. desired appreciations**--Adversary personal conclusions and official estimates, valid or invalid, that result in adversary behaviors and official actions advantageous to friendly interests and objectives. **b. harmful appreciations**--Adversary personal conclusions, official estimates, or assumptions, valid or invalid, that result in adversary behaviors and official actions harmful to friendly interests and objectives.

approach clearance--Authorization for a pilot conducting flight in accordance with instrument flight rules to commence an approach to an airport.

approach end of runway(*)--That end of the runway nearest to the direction from which the final approach is made.

approach lane(*)--An extension of a boat lane from the line of departure toward the transport area. It may be terminated by marker ships, boats, or buoys.

approach march(*)--Advance of a combat unit when direct contact with the enemy is imminent. Troops are fully or partially deployed. The approach march ends when ground contact with the enemy is made or when the attack position is occupied. See also **advance to contact.**

approach schedule(*)--The schedule which indicates, for each scheduled wave, the time of departure from the rendezvous area, from the line of departure, and from other control points and the time of arrival at the beach.

approach sequence(*)--The order in which two or more aircraft are cleared for an approach.

approach time--The time at which an aircraft is expected to commence approach procedure.

apron--A defined area, on an airfield, intended to accommodate aircraft for purposes of loading or unloading passengers or cargo, refueling, parking, or maintenance.

architecture--A framework or structure that portrays relationships among all the elements of the subject force, system, or activity.

archive--When used in the context of deliberate planning, the directed command will remove the referenced operation plan, operation plan in concept format, or concept summary and any associated Joint Operation Planning and Execution System automated data processing files from its library of active plans. All material will be prepared for shipment to appropriate archive facilities in accordance with Joint Pub 5-03.1 and appropriate command directives. See also **maintain; retain.**

area--See advisory area; aircraft dispersal area; aircraft marshalling area; air defense action area; alighting area; area control center; assembly area; caution area; closed area; concentration area; control area; danger area; defensive coastal area; embarkation area; fire support area; homogeneous area; impact area; initial approach area; landing area; maneuvering area; maritime area; naval support area; objective area; prohibited area; run-up area; signal area; staging area; submarine patrol area; terminal control area; transit area. See also **zone.**

area air defense commander--Within a unified command, subordinate unified command, or joint task force, the commander will assign overall responsibility for air defense to a single commander. Normally, this will be the component commander with the preponderance of air defense capability and the command, control, and communications capability to plan and execute integrated air defense operations. Representation from the other components involved will be provided, as appropriate, to the area air defense commander's headquarters. Also called **AADC.**

area assessment--The commander's prescribed collection of specific information that commences upon employment and is a continuous operation. It confirms, corrects, refutes, or adds to previous intelligence acquired from area studies and other sources prior to employment.

area bombing(*)--Bombing of a target which is in effect a general area rather than a small or pinpoint target.

area command(*)--A command which is composed of those organized elements of one or more of the armed services, designated to operate in a specific geographical area, which are placed under a single commander. See also **command; functional command.**

area control center(*)--A unit established to provide air traffic control service to controlled flights in control areas under its jurisdiction.

See also **air traffic control center; flight information region.**

area coordination group--A composite organization, including representatives of local military, paramilitary, and other governmental agencies and their US counterparts, responsible for planning and coordinating internal defense and development operations.

area damage control(*)--Measures taken before, during, or after hostile action or natural or manmade disasters, to reduce the probability of damage and minimize its effects. See also **damage control; disaster control; rear area security.**

area of influence(*)--A geographical area wherein a commander is directly capable of influencing operations by maneuver or fire support systems normally under the commander's command or control.

area of intelligence responsibility(*)--An area allocated to a commander in which the commander is responsible for the provision of intelligence within the means at the commander's disposal. See also **area of interest; area of responsibility.**

area of interest(*)--That area of concern to the commander, including the area of influence, areas adjacent thereto, and extending into enemy territory to the objectives of current or planned operations. This area also includes areas occupied by enemy forces who could jeopardize the accomplishment of the mission.

area of militarily significant fallout(*)--Area in which radioactive fallout affects the ability of military units to carry out their normal mission.

area of northern operations--A region of variable width in the Northern Hemisphere that lies north of the 50 degrees isotherm--a line along which the average temperature of the warmest 4-month period of the year does not exceed 50 degrees Fahrenheit. Mountain regions located outside of this area are included in this category of operations provided these same temperature conditions exist.

area of operations(*)--That portion of an area of war necessary for military operations and for the administration of such operations. Also called **AO.**

area of responsibility(*)--1. A defined area of land in which responsibility is specifically assigned to the commander of the area for the development and maintenance of installations, control of movement, and the conduct of tactical operations involving troops under the commander's control, along with parallel authority to exercise these functions. 2. In naval usage, a predefined area of enemy terrain for which supporting ships are responsible for covering by fire on known targets or targets of opportunity and by observation. Also called **AOR.**

area of war--That area of land, sea, and air which is, or may become, directly involved in the operations of war.

area operations(*)--In maritime usage, operations conducted in a geographical area and not related to the protection of a specific force.

area oriented--Personnel or units whose organizations, mission, training, and equipping are based on projected operational deployment to a specific geographic or demographic area.

area radar prediction analysis--Radar target intelligence study designed to provide radar significant data for use in the preparation of radar target predictions.

area search--Visual reconnaissance of limited or defined areas.

area target(*)--A target consisting of an area rather than a single point.

areodesy--The branch of mathematics which determines, by observation and measurement, the exact positions of points and the figures and areas of large portions of the surface of the planet Mars, or the shape and size of the planet Mars.

areodetic--Of, pertaining to, or determined by areodesy.

armament delivery recording--Motion picture, still photography, and video recordings showing the delivery and impact of ordnance. This differs from reconnaissance imagery in that it records the act of delivery and impact and normally is done by the weapon system delivering the ordnance. Armament delivery recording is used primarily for evaluating strike effectiveness and for combat crew training. It is also one of the principal sources of over-the--target documentation in force employments, and may be used for public affairs purposes.

armed forces--The military forces of a nation or a group of nations. See also **force(s)**.

armed forces censorship--The examination and control of personal communications to or from persons in the Armed Forces of the United States and persons accompanying or serving with the Armed Forces of the United States. See also **censorship**.

armed forces courier--An officer or enlisted member in the grade of E-7 or above, of the US Armed Forces, assigned to perform Armed Forces Courier Service duties and identified by possession of an Armed Forces Courier Service Identification Card (ARF-COS Form 9). See also **courier**.

Armed Forces Courier Service--A joint service of the Departments of the Army, the Navy, and

the Air Force, with the Chief of Staff, US Army, as Executive Agent. The courier service provides one of the available methods for the secure and expeditious transmission of material requiring protected handling by military courier.

armed forces courier station--An Army, Navy, or Air Force activity, approved by the respective military department and officially designated by Headquarters, Armed Forces Courier Service, for the acceptance, processing, and dispatching of Armed Forces Courier Service material.

Armed Forces of the United States--A term used to denote collectively all components of the Army, Navy, Air Force, Marine Corps, and Coast Guard. See also **United States Armed Forces.**

armed helicopter(*)--A helicopter fitted with weapons or weapon systems.

armed mine(*)--A mine from which all safety devices have been withdrawn and, after laying, all automatic safety features and/or arming delays have operated. Such a mine is ready to function after receipt of a target signal, influence, or contact. See also **mine.**

armed reconnaissance--A mission with the primary purpose of locating and attacking targets of opportunity, i.e., enemy materiel, personnel, and facilities, in assigned general areas or along assigned ground communications routes, and not for the purpose of attacking specific briefed targets.

Armed Services Medical Regulating Office--A joint activity reporting directly to the Commander in Chief, US Transportation Command, the Department of Defense single manager for the regulation of movement of Uniformed Services patients. The Armed Services Medical Regulating Office authorizes transfers to medical treatment facilities of the Military Departments or the Department of Veterans Affairs and coordinates inter-theater and inside continental United States patient movement requirements with the appropriate transportation component commands of US Transportation Command. Also called **ASMRO.**

armed sweep(*)--A sweep fitted with cutters or other devices to increase its ability to cut mine moorings.

arming--As applied to explosives, weapons, and ammunition, the changing from a safe condition to a state of readiness for initiation.

arming delay device(*)--A device fitted in a mine to prevent it being actuated for a preset time after laying.

arming lanyard--See **arming wire.**

arming pin(*)--A safety device which is inserted into a fuze to prevent the arming cycle from starting until its removal.

arming system--That portion of a weapon which serves to ready (arm), safe, or re-safe (disarm) the firing system and fuzing system and which may actuate devices in the nuclear system.

arming wire(*)--A cable, wire or lanyard routed from the aircraft to an expendable aircraft store in order to initiate the arming sequence for the store upon release from the aircraft, when the armed release condition has been selected; it also prevents arming initiation prior to store release and during safe jettison. Also called **arming lanyard.** See also **safety wire.**

armored personnel carrier--A lightly armored, highly mobile, full-tracked vehicle, amphibious and air-droppable, used primarily for transporting personnel and their individual equipment during tactical operations. Production modifications or application of special kits permit use as a mortar carrier, command post, flame thrower, antiaircraft artillery chassis, or limited recovery vehicle.

armored reconnaissance airborne assault vehicle--A lightly armored, mobile, full-tracked vehicle serving as the main reconnaissance vehicle in infantry and airborne operations, and as the principal assault weapon of airborne troops.

arms control--A concept that connotes: a. any plan, arrangement, or process, resting upon explicit or implicit international agreement, governing any aspect of the following: the numbers, types, and performance characteristics of weapon systems (including the command and control, logistics support arrangements, and any related intelligence-gathering mechanism); and the numerical strength, organization, equipment, deployment, or employment of the Armed Forces retained by the parties (it encompasses disarmament); and b. on some occasions, those measures taken for the purpose of reducing instability in the military environment.

arms control agreement--The written or unwritten embodiment of the acceptance of one or more arms control measures by two or more nations.

arms control agreement verification--A concept that entails the collection, processing, and reporting of data indicating testing or employment of proscribed weapon systems, including country of origin and location, weapon and payload identification, and event type.

arms control measure--Any specific arms control course of action.

armstrong--The term, peculiar to the Air Support Radar Team, indicating both the command and response for arming and fuzing circuit activation.

Army Air Defense Command Post--The tactical headquarters of an Army air defense commander.

Army air-ground system--The Army system which provides for interface between Army and tactical air support agencies of other Services in the planning, evaluating, processing, and coordinating of air support requirements and operations. It is composed of appropriate staff members, including G-2 air and G-3 air personnel, and necessary communication equipment.

Army base--A base or group of installations for which a local commander is responsible, consisting of facilities necessary for support of Army activities including security, internal lines of communication, utilities, plants and systems, and real property for which the Army has operating responsibility. See also **base complex**.

Army corps--A tactical unit larger than a division and smaller than a field army. A corps usually consists of two or more divisions together with auxiliary arms and services. See also **field army**.

Army group--Several field armies under a designated commander.

Army service area--The territory between the corps rear boundary and the combat zone rear boundary. Most of the Army administrative establishment and service troops are usually located in this area. See also **rear area**.

Army special operations component--The Army component of a joint force special operations component. Also called **ARSOC**. See also **Air Force special operations component; Navy special operations component**.

Army special operations forces--Those active and reserve component Army forces designated by

the Secretary of Defense that are specifically organized, trained, and equipped to conduct and support special operations. Also called **ARSOF**.

Army tactical data link 1--See **tactical digital information link**.

arresting barrier--See **aircraft arresting barrier**.

arresting gear--See **aircraft arresting gear**.

ARSOC--See **Army special operations component**.

ARSOF--See **Army special operations forces**.

artificial horizon--See **attitude indicator**.

artillery fire plan table(*)--A presentation of planned targets giving data for engagement. Scheduled targets are fired in a definite time sequence. The starting time may be on call, at a prearranged time or at the occurrence of a specific event.

artillery survey control point(*)--A point at which the coordinates and the altitude are known and from which the bearings/azimuths to a number of reference objects are also known.

ASMRO--See **Armed Forces Medical Regulating Office**.

aspect angle--The angle between the longitudinal axis of the target (projected rearward) and the line of sight to the interceptor measured from the tail of the target.

ASROC--See **antisubmarine rocket**.

assault--1. The climax of an attack, closing with the enemy in hand-to-hand fighting. 2. In an amphibious operation, the period of time between the arrival of the major assault forces of the amphibious task force in the objective area and the accomplishment of the amphibious task force mission. 3. To make a short, violent, but well-ordered attack against a local objective, such as a gun emplacement, a fort, or a machine gun nest. 4. A phase of an airborne operation beginning with delivery by air of the assault echelon of the force into the objective area and extending through attack of assault objectives and consolidation of the initial airhead. See also **assault phase; landing attack.**

assault aircraft(*)--A powered aircraft that moves assault troops and/or cargo into an objective area.

assault area--In amphibious operations, that area that includes the beach area, the boat lanes, the lines of departure, the landing ship areas, the transport areas, and the fire support areas in the immediate vicinity of the boat lanes.

assault area diagram--A graphic means of showing, for amphibious operations, the beach designations, boat lanes, organization of the line of departure, scheduled waves, landing ship area, transport areas, and the fire support areas in the immediate vicinity of the boat lanes.

assault craft(*)--A landing craft or amphibious vehicle primarily employed for landing troops and equipment in the assault waves of an amphibious operation.

assault craft unit--A permanently commissioned naval organization, subordinate to the commander, naval beach group, that contains landing craft and crews necessary to provide lighterage required in an amphibious operation.

assault echelon(*)--The element of a force that is scheduled for initial assault on the objective area.

assault fire--1. That fire delivered by attacking troops as they close with the enemy. 2. In artillery, extremely accurate, short-range destruction fire at point targets.

assault follow-on echelon--In amphibious operations, that echelon of the assault troops, vehicles, aircraft equipment, and supplies which, though not needed to initiate the assault, is required to support and sustain the assault. In order to accomplish its purpose, it is normally required in the objective area no later than five days after commencement of the assault landing. See also **assault; follow-up.**

assault phase(*)--1. In an amphibious operation, the period of time between the arrival of the major assault forces of the amphibious task force in the objective area and the accomplishment of their mission. 2. In an airborne operation, a phase beginning with delivery by air of the assault echelon of the force into the objective area and extending through attack of assault objectives and consolidation of the initial airhead. See also **assault.**

assault schedule--See **landing schedule.**

assault shipping(*)--Shipping assigned to the amphibious task force and utilized for transporting assault troops, vehicles, equipment, and supplies to the objective area.

assault wave--See **wave.**

assembly(*)--In logistics, an item forming a portion of an equipment, that can be provisioned and replaced as an entity and which normally incorporates replaceable parts or groups of parts. See also **component; part; subassembly.**

assembly anchorage(*)--An anchorage intended for the assembly and onward routing of ships.

assembly area(*)--1. An area in which a command is assembled preparatory to further action. 2. In a supply installation, the gross area used for collecting and combining components into complete units, kits, or assemblies.

assessment--1. Analysis of the security, effectiveness, and potential of an existing or planned intelligence activity. 2. Judgment of the motives, qualifications, and characteristics of present or prospective employees or "agents."

asset (intelligence)--Any resource--person, group, relationship, instrument, installation, or supply---at the disposition of an intelligence organization for use in an operational or support role. Often used with a qualifying term such as agent asset or propaganda asset.

assign(*)--1. To place units or personnel in an organization where such placement is relatively permanent, and/or where such organization controls and administers the units or personnel for the primary function, or greater portion of the functions, of the unit or personnel. 2. To detail individuals to specific duties or functions where such duties or functions are primary and/or relatively permanent. See also **attach.**

associated support(*)--In naval air operations, assistance provided by a force or unit to another force or unit that is under independent tactical control, neither being subordinate to the other. See also **direct support; support.**

assumed azimuth--The assumption of azimuth origins as a field expedient until the required data are available.

assumed grid--A grid constructed using an arbitrary scale superimposed on a map, chart, or photograph for use in point designation without regard to actual geographic location. See also **grid.**

assumption--A supposition on the current situation or a presupposition on the future course of events, either or both assumed to be true in the absence of positive proof, necessary to enable the commander in the process of planning to complete an estimate of the situation and make a decision on the course of action.

astern fueling(*)--The transfer of fuel at sea during which the receiving ship(s) keep(s) station astern of the delivering ship.

ASTOR--See **antisubmarine torpedo.**

asymmetrical sweep(*)--A sweep whose swept path under conditions of no wind or cross-tide is not equally spaced either side of the sweeper's track.

AT--See **antiterrorism.**

ATCALS--See **air traffic control and landing systems.**

ATF--See **amphibious task force.**

atmosphere--The air surrounding the Earth. See also **ionosphere; stratosphere; tropopause; troposphere.**

atmospheric environment--The envelope of air surrounding the Earth, including its interfaces and interactions with the Earth's solid or liquid surface.

at my command(*)--In artillery and naval gunfire support, the command used when it is desired to control the exact time of delivery of fire.

atomic air burst--See **airburst.**

atomic defense--See **nuclear defense.**

atomic demolition munition--A nuclear device designed to be detonated on or below the ground surface, or under water as a demolition munition against material-type targets to block, deny, and/or canalize the enemy.

atomic underground burst--See **nuclear underground burst.**

atomic underwater burst--See **nuclear underwater burst.**

atomic warfare--See **nuclear warfare.**

atomic weapon--See **nuclear weapon.**

at priority call(*)--A precedence applied to the task of an artillery unit to provide fire to a formation/unit on a guaranteed basis. Normally observer, communications, and liaison are not provided. An artillery unit in "direct support" or "in support" may simultaneously be placed "at priority call" to another unit or agency for a particular task and/or for a specific period of time.

attach--1. The placement of units or personnel in an organization where such placement is relatively temporary. 2. The detailing of individuals to specific functions where such functions are secondary or relatively temporary, e.g., attached for quarters and rations; attached for flying duty. See also **assign.**

attached airlift service--The airlift service provided to an organization or command by an airlift unit attached to that organization.

attachment--See **attach.**

attack aircraft carrier--A warship designed to support and operate aircraft, engage in attacks on targets afloat or ashore, and engage in sustained operations in support of other forces. Designated as **CV or CVN.** CVN is nuclear powered.

attack altitude--The altitude at which the interceptor will maneuver during the attack phase of an air intercept.

attack assessment--An evaluation of information to determine the potential or actual nature and objectives of an attack for the purpose of providing information for timely decisions. See also **damage estimation.**

attack cargo ship--A naval ship designed or converted to transport combat-loaded cargo in an assault landing. Capabilities as to carrying landing craft, speed of ship, armament, and size of hatches and booms are greater than those of comparable cargo ship types. Designated as **LKA.**

attack carrier striking forces--Naval forces, the primary offensive weapon of which is carrier-based aircraft. Ships, other than carriers, act primarily to support and screen against submarine and air threat and secondarily against surface threat.

attack condition alpha--Considers there is inadequate warning of attack, and the command post or headquarters of a decision authority becomes ineffective prior to the performance of essential functions.

attack condition bravo--Considers there is sufficient effective warning of impending attack to relocate personnel required to perform essential functions to alternate command facilities.

attack group(*)--A subordinate task organization of the navy forces of an amphibious task force. It is composed of assault shipping and supporting naval units designated to transport, protect, land, and initially support a landing group.

attack heading--1. The interceptor heading during the attack phase that will achieve the desired track-crossing angle. 2. The assigned magnetic compass heading to be flown by aircraft during the delivery phase of an air strike.

attack helicopter(*)--A helicopter specifically designed to employ various weapons to attack and destroy enemy targets.

attacking--In air intercept, a term meaning, "Am commencing attacking run with weapon indicated" (size may be given).

attack origin--1. The location or source from which an attack was initiated. 2. The nation initiating an attack. See also **attack assessment.**

attack pattern--The type and distribution of targets under attack. See also **attack assessment; target pattern.**

attack position--The last position occupied by the assault echelon before crossing the line of departure. See also **forming up place.**

attack size--The number of weapons involved in an attack. See also **attack assessment.**

attack speed--The speed at which the interceptor will maneuver during the attack phase of an air intercept.

attack timing--The predicted or actual time of bursts, impacts, or arrival of weapons at their intended targets.

attack warning/attack assessment--Not to be used. See **separate definitions for tactical warning and for attack assessment.**

attenuation(*)--1. Decrease in intensity of a signal, beam, or wave as a result of absorption of energy and of scattering out of the path of a detector, but not including the reduction due to geometric spreading, i.e., the inverse square of distance effect. 2. In mine warfare, the reduction in intensity of an influence as distance

from the source increases. 3. In camouflage and concealment, the process of making an object or surface less conspicuous by reducing its contrast to the surroundings and/or background. Also called **tone down.**

attenuation factor(*)--The ratio of the incident radiation dose or dose rate to the radiation dose or dose rate transmitted through a shielding material. This is the reciprocal of the transmission factor.

attitude(*)--1. The position of a body as determined by the inclination of the axes to some frame of reference. If not otherwise specified, this frame of reference is fixed to the Earth. **(NATO)** 2. The grid bearing of the long axis of a target area. See intelligence cycle.

attitude director indicator(*)--An attitude indicator which displays command signals from the flight director computer.

attitude indicator(*)--An instrument which displays the attitude of the aircraft by reference to sources of information which may be contained within the instrument or be external to it. When the sources of information are self-contained, the instrument may be referred to as an artificial horizon.

attrition(*)--The reduction of the effectiveness of a force caused by loss of personnel and materiel.

attrition minefield(*)--In naval mine warfare, a field intended primarily to cause damage to enemy ships. See also **minefield.**

attrition rate(*)--A factor, normally expressed as a percentage, reflecting the degree of losses of personnel or materiel due to various causes within a specified period of time.

attrition reserve aircraft--Aircraft procured for the specific purpose of replacing the anticipated losses of aircraft because of peacetime and/or wartime attrition.

attrition sweeping(*)--The continuous sweeping of minefields to keep the risk of mines to all ships as low as possible.

augmentation forces--Forces to be transferred to the operational control of a supported commander during the execution of an operation.

authenticate--A challenge given by voice or electrical means to attest to the authenticity of a message or transmission.

authentication--1. A security measure designed to protect a communications system against acceptance of a fraudulent transmission or simulation by establishing the validity of a transmission, message, or originator. 2. A means of identifying individuals and verifying their eligibility to receive specific categories of information. 3. Evidence by proper signature or seal that a document is genuine and official.

authenticator--A symbol or group of symbols, or a series of bits, selected or derived in a prearranged manner and usually inserted at a predetermined point within a message or transmission for the purpose of attesting to the validity of the message or transmission.

autocode format--An abbreviated and formatted message header used in conjunction with the Mobile Cryptologic Support Facility (MCSF) to energize the automatic communications relay functions of the MCSF providing rapid exchange of data through the system.

automated data handling--See **automatic data handling.**

automatic approach and landing(*)--A control mode in which the aircraft's speed and flight path are automatically controlled for approach,

flare-out, and landing. See also **ground controlled approach procedure.**

automatic data handling(*)--A generalization of automatic data processing to include the aspect of data transfer.

automatic data processing(*)--1. Data processing largely performed by automatic means. 2. That branch of science and technology concerned with methods and techniques relating to data processing largely performed by automatic means.

automatic flight control system(*)--A system which includes all equipment to control automatically the flight of an aircraft or missile to a path or attitude described by references internal or external to the aircraft or missile.

automatic levelling--A flight control system feature which returns an aircraft to level flight attitude in roll and pitch.

automatic message processing system--Any organized assembly of resources and methods used to collect, process, and distribute messages largely by automatic means.

automatic pilot--That part of an automatic flight control system which provides attitude stabilization with respect to internal references.

automatic resupply--A resupply mission fully planned before insertion of a special operations team into the operations area that occurs at a prearranged time and location, unless changed by the operating team after insertion. See also **emergency resupply; on-call resupply.**

automatic search jammer(*)--An intercept receiver and jamming transmitter system which searches for and jams signals automatically which have specific radiation characteristics.

Automatic Secure Voice Communications Network--A worldwide, switched, secure voice network developed to fulfill DOD long-haul, secure voice requirements. Also called **AUTOSEVOCOM**.

automatic supply--A system by which certain supply requirements are automatically shipped or issued for a predetermined period of time without requisition by the using unit. It is based upon estimated or experience-usage factors.

automatic throttle--A flight control system feature which actuates an aircraft throttle system based on its own computation and feedback from appropriate data sources.

automatic toss(*)--In a flight control system, a control mode in which the toss bombing maneuver of an aircraft is controlled automatically.

automatic trim--A flight control system feature which adjusts the trim of an aircraft in flight.

Automatic Voice Network--A major subsystem of the Defense Switched Network, which replaced the Automatic Voice Network as the principal long-haul, nonsecure voice communications network within the Defense Communications System. Also called **AUTOVON**. See also **Defense Switched Network**.

automation--1. The implementation of processes by automatic means. 2. The conversion of a procedure, a process, or equipment to automatic operation.

autonomous operation--In air defense, the mode of operation assumed by a unit after it has lost all communications with higher echelons. The unit commander assumes full responsibility for control of weapons and engagement of hostile targets.

AUTOVON--See **Automatic Voice Network.**

AV-8--See **Harrier.**

available payload--The passenger and/or cargo capacity expressed in weight and/or space available to the user.

available-to-load date--A day, relative to C-day in a time-phased force and deployment data, that unit and nonunit equipment and forces can begin loading on an aircraft or ship at the port of embarkation. Also called **ALD.**

avenue of approach--An air or ground route of an attacking force of a given size leading to its objective or to key terrain in its path.

average speed(*)--The average distance traveled per hour, calculated over the whole journey, excluding specifically ordered halts.

aviation combat element--See **Marine air-ground task force.**

aviation life support equipment--See **life support equipment.**

aviation medicine(*)--The special field of medicine which is related to the biological and psychological problems of flight.

axial route--A route running through the rear area and into the forward area. See also **route.**

axis of advance--A line of advance assigned for purposes of control; often a road or a group of roads, or a designated series of locations, extending in the direction of the enemy.

azimuth--Quantities may be expressed in positive quantities increasing in a clockwise direction, or in X, Y coordinates where south and west are negative. They may be referenced to true north or magnetic north depending on the particular weapon system used.

azimuth angle(*)--An angle measured clockwise in the horizontal plane between a reference direction and any other line.

azimuth guidance(*)--Information which will enable the pilot or autopilot of an aircraft to follow the required track.

azimuth resolution(*)--The ability of radar equipment to separate two reflectors at similar ranges but different bearings from a reference point. Normally the minimum separation distance between the reflectors is quoted and expressed as the angle subtended by the reflectors at the reference point.

B

B-52--See **Stratofortress.**

background count(*)--The evidence or effect on a detector of radiation, other than that which it is desired to detect, caused by any agent. In connection with health protection, the background count usually includes radiations produced by naturally occurring radioactivity and cosmic rays.

background radiation(*)--Nuclear (or ionizing) radiations arising from within the body and from the surroundings to which individuals are always exposed.

back order--The quantity of an item requisitioned by ordering activities that is not immediately available for issue but is recorded as a stock commitment for future issue.

back-scattering(*)--Radio wave propagation in which the direction of the incident and scattered waves, resolved along a reference direction (usually horizontal), are oppositely directed. A signal received by back-scattering is often referred to as "back-scatter."

back tell(*)--The transfer of information from a higher to a lower echelon of command. See also **track telling.**

back-up(*)--In cartography, an image printed on the reverse side of a map sheet already printed on one side. Also the printing of such images.

backup aircraft authorization--Aircraft over and above the primary aircraft authorized to permit scheduled and unscheduled maintenance, modifications, and inspections and repair without reduction of aircraft available for the operational mission. No operating resources are allocated for these aircraft in the Defense budget. See also **primary aircraft authorization.**

backup aircraft inventory--The aircraft designated to meet the backup authorization. See also **primary aircraft inventory.**

balance--A concept as applied to an arms control measure that connotes: a. adjustments of armed forces and armaments in such a manner that one state does not obtain military advantage over other states agreeing to the measure; and b. internal adjustments by one state of its forces in such manner as to enable it to cope with all aspects of remaining threats to its security in a post arms control agreement era.

balanced stock(s)--1. That condition of supply when availability and requirements are in equilibrium for specific items. 2. An accumulation of supplies in quantities determined necessary to meet requirements for a fixed period.

balance station zero--See **reference datum.**

bale cubic capacity(*)--The space available for cargo measured in cubic feet to the inside of the cargo battens, on the frames, and to the underside of the beams. In a general cargo of mixed commodities, the bale cubic applies. The stowage of the mixed cargo comes in contact with the cargo battens and as a general rule does not extend to the skin of the ship.

balisage(*)--The marking of a route by a system of dim beacon lights enabling vehicles to be driven at near day-time speed, under blackout conditions.

ballistic missile(*)--Any missile which does not rely upon aerodynamic surfaces to produce lift and consequently follows a ballistic trajectory when thrust is terminated. See also **aerodynamic missile; guided missile.**

ballistic missile early warning system--An electronic system for providing detection and early warning of attack by enemy inter-continental ballistic missiles.

ballistics(*)--The science or art that deals with the motion, behavior, appearance, or modification of missiles or other vehicles acted upon by propellants, wind, gravity, temperature, or any other modifying substance, condition, or force.

ballistic trajectory(*)--The trajectory traced after the propulsive force is terminated and the body is acted upon only by gravity and aerodynamic drag.

ballistic wind--That constant wind that would have the same effect upon the trajectory of a bomb or projectile as the wind encountered in flight.

balloon barrage--See **barrage, Part 2.**

balloon reflector(*)--In electronic warfare, a balloon-supported confusion reflector to produce fraudulent echoes.

band pass--The number of cycles per second expressing the difference between the limiting frequencies at which the desired fraction (usually half power) of the maximal output is obtained. Term applies to all types of amplifiers.

bank angle(*)--The angle between the aircraft's normal axis and the Earth's vertical plane containing the aircraft's longitudinal axis.

bare base--A base having minimum essential facilities to house, sustain, and support operations to include, if required, a stabilized runway, taxiways, and aircraft parking areas. A bare base must have a source of water that can be made potable. Other requirements to operate under bare base conditions form a necessary part of the force package deployed to the bare base. See also **base.**

barometric altitude(*)--The altitude determined by a barometric altimeter by reference to a pressure level and calculated according to the standard atmosphere laws. It corresponds to the difference in altitude between the altimeter pressure level and the mean sea level pressure.

barrage--1. A prearranged barrier of fire, except that delivered by small arms, designed to protect friendly troops and installations by impeding enemy movements across defensive lines or areas. 2. A protective screen of balloons that are moored to the ground and kept at given heights to prevent or hinder operations by enemy aircraft. This meaning also called **balloon barrage.** 3. A type of electronic attack intended for simultaneous jamming over a wide area of frequency spectrum. See also **barrage jamming; electronic warfare; fire.**

barrage fire(*)--Fire which is designed to fill a volume of space or area rather than aimed specifically at a given target. See also **fire.**

barrage jamming--Simultaneous electromagnetic jamming over a broad band of frequencies. See also **jamming.**

barrier--A coordinated series of obstacles designed or employed to channel, direct, restrict, delay, or stop the movement of an opposing force and to impose additional losses in personnel, time, and equipment on the opposing force. Barriers can exist naturally, be manmade, or a combination of both.

barrier combat air patrol--One or more divisions or elements of fighter aircraft employed between a force and an objective area as a barrier across the probable direction of enemy attack. It is used as far from the force as control conditions permit, giving added protection against raids that use the most direct routes of approach. See also **combat air patrol.**

barrier forces--Air, surface, and submarine units and their supporting systems positioned across the likely courses of expected enemy transit for early detection and providing rapid warning, blocking, and destruction of the enemy.

barrier, obstacle, and mine warfare plan--A comprehensive, coordinated plan which includes responsibilities, general location of unspecified and specific barriers, obstacles, and minefields, special instructions, limitations, coordination, and completion times. The plan may designate locations of obstacle zones or belts. It is normally prepared as an annex to a campaign plan, operation plan, or operation order.

bar scale--See **graphic scale; scale.**

base(*)--1. A locality from which operations are projected or supported. 2. An area or locality containing installations which provide logistic or other support. See also **emergency fleet operating base; establishment.** (DOD) 3. Home airfield or home carrier. See also **base of operations; facility.**

base cluster--In base defense operations, a collection of bases, geographically grouped for mutual protection and ease of command and control.

base cluster commander--In base defense operations, the senior officer in the base cluster (excluding medical officers, chaplains, and commanders of transient units), with responsibility for coordinating the defense of bases within the base cluster and for integrating defense plans of bases into a base cluster defense plan.

base cluster operations center--A command and control facility that serves as the base cluster commander's focal point for defense and security of the base cluster.

base command(*)--An area containing a military base or group of such bases organized under one commander. See also **command.**

base commander--In base defense operations, the officer assigned to command a base.

base complex--See **Army base; installation complex; Marine base; naval base; naval or marine (air) base.** See also **noncontiguous facility.**

base defense--The local military measures, both normal and emergency, required to nullify or reduce the effectiveness of enemy attacks on, or sabotage of, a base, to ensure that the maximum capacity of its facilities is available to US forces.

base defense forces--Troops assigned or attached to a base for the primary purpose of base defense and security, and augmentees and selectively armed personnel available to the base commander for base defense from units performing primary missions other than base defense.

base defense operations center--A command and control facility established by the base commander to serve as the focal point for base security and defense. It plans, directs, integrates, coordinates, and controls all base defense efforts, and coordinates and integrates into area security operations with the rear area operations center/rear tactical operations center.

base defense zone--An air defense zone established around an air base and limited to the engagement envelope of short-range air defense weapons systems defending that base. Base defense zones have specific entry, exit, and identification, friend or foe procedures established. Also called **BDZ.**

base development(*)--The improvement or expansion of the resources and facilities of an area or a location to support military operations.

base development plan--A plan for the facilities, installations, and bases required to support military operations.

base element--See **base unit**.

base line(*)--1. (surveying) A surveyed line established with more than usual care, to which surveys are referred for coordination and correlation. 2. (photogrammetry) The line between the principal points of two consecutive vertical air photographs. It is usually measured on one photograph after the principal point of the other has been transferred. 3. (radio navigation systems) The shorter arc of the great circle joining two radio transmitting stations of a navigation system. 4. (triangulation) The side of one of a series of coordinated triangles the length of which is measured with prescribed accuracy and precision and from which lengths of the other triangle sides are obtained by computation.

base map(*)--A map or chart showing certain fundamental information, used as a base upon which additional data of specialized nature are compiled or overprinted. Also, a map containing all the information from which maps showing specialized information can be prepared. See also **chart base; map**.

base of operations--An area or facility from which a military force begins its offensive operations, to which it falls back in case of reverse, and in which supply facilities are organized.

base period--That period of time for which factors were determined for use in current planning and programming.

base section--An area within the communications zone in an area of operations organized to provide logistic support to forward areas.

base surge(*)--A cloud which rolls out from the bottom of the column produced by a subsurface burst of a nuclear weapon. For underwater bursts, the surge is, in effect, a cloud of liquid droplets which has the property of flowing almost as if it were a homogeneous fluid. For subsurface land bursts the surge is made up of small solid particles but still behaves like a fluid.

base unit--1. Unit of organization in a tactical operation around which a movement or maneuver is planned and performed. 2. Base element.

basic cover--Coverage of any installation or area of a permanent nature with which later coverage can be compared to discover any changes that have taken place.

basic encyclopedia--A compilation of identified installations and physical areas of potential significance as objectives for attack.

basic intelligence--Fundamental intelligence concerning the general situation, resources, capabilities, and vulnerabilities of foreign countries or areas which may be used as reference material in the planning of operations at any level and in evaluating subsequent information relating to the same subject.

basic load(*)--The quantity of supplies required to be on hand within, and which can be moved by, a unit or formation. It is expressed according to the wartime organization of the unit or formation and maintained at the prescribed levels.

basic military route network(*)--Axial, lateral, and connecting routes designated in peacetime by the host nation to meet the anticipated military movements and transport requirements, both allied and national. See also **transport network.**

basic psychological operations study--A document which describes succinctly the characteristics of a country, geographical area, or region which are most pertinent to psychological operations, and which can serve as an immediate reference for the planning and conduct of psychological operations.

basic research--Research directed toward the increase of knowledge, the primary aim being a greater knowledge or understanding of the subject under study. See also **applied research; research.**

basic stocks(*)--Stocks to support the execution of approved operational plans for an initial predetermined period. See also **stocks; sustaining stocks.**

basic stopping power(*)--The probability, expressed as a percentage, of a single vehicle being stopped by mines while attempting to cross a minefield.

basic tactical organization--The conventional organization of landing force units for combat, involving combinations of infantry, supporting ground arms, and aviation for accomplishment of missions ashore. This organizational form is employed as soon as possible following the landing of the various assault components of the landing force.

basic undertakings--The essential things, expressed in broad terms, that must be done in order to implement the commander's concept successfully. These may include military, diplomatic, economic, psychological, and other measures. See also **strategic concept.**

basis of issue--Authority which prescribes the number of items to be issued to an individual, a unit, a military organization, or for a unit piece of equipment.

bathymetric contour--See **depth contour.**

battalion landing team--In an amphibious operation, an infantry battalion normally reinforced by necessary combat and service elements; the basic unit for planning an assault landing. Also called **BLT.**

battery(*)--1. Tactical and administrative artillery unit or subunit corresponding to a company or similar unit in other branches of the Army. 2. All guns, torpedo tubes, searchlights, or missile launchers of the same size or caliber or used for the same purpose, either installed in one ship or otherwise operating as an entity.

battery center(*)--A point on the ground, the coordinates of which are used as a reference indicating the location of the battery in the production of firing data. Also called **chart location of the battery.**

battery (troop) left (right)--A method of fire in which weapons are discharged from the left (right), one after the other, at five second intervals.

battle damage assessment--The timely and accurate estimate of damage resulting from the application of military force, either lethal or non-lethal, against a predetermined objective. Battle damage assessment can be applied to the employment of all types of weapon systems (air, ground, naval, and special forces weapon systems) throughout the range of military operations. Battle damage assessment is primarily an intelligence responsibility with required inputs and coordination from the operators. Battle damage assessment is composed of physical damage assessment, functional damage assessment, and target system assessment. Also called **BDA.** See also **bomb damage assessment; combat assessment.**

battle damage repair(*)--Essential repair, which may be improvised, carried out rapidly in a battle environment in order to return damaged or disabled equipment to temporary service.

battlefield coordination element--An Army liaison provided by the Army component commander to the Air Operations Center (AOC) and/or to the component designated by the joint force commander to plan, coordinate, and deconflict air operations. The battlefield coordination element processes Army requests for tactical air support, monitors and interprets the land battle situation for the AOC, and provides the necessary interface for exchange of current intelligence and operational data. Also called **BCE**.

battlefield illumination(*)--The lighting of the battle area by artificial light, either visible or invisible to the naked eye. See also **artificial daylight; artificial moonlight; indirect illumination**.

battlefield psychological activities(*)--Planned psychological activities conducted as an integral part of combat operations and designed to bring psychological pressure to bear on enemy forces and civilians under enemy control in the battle area, to assist in the achievement of the tactical objectives.

battlefield surveillance(*)--Systematic observation of the battle area for the purpose of providing timely information and combat intelligence. See also **surveillance**.

battle force--A standing operational naval task force organization of carriers, surface combatants, and submarines assigned to numbered fleets. A battle force is subdivided into battle groups.

battle group--A standing naval task group consisting of a carrier or battleship, surface combatants, and submarines as assigned in direct support, operating in mutual support with the task of destroying hostile submarine, surface, and air forces within the group's assigned area of responsibility and striking at targets along hostile shore lines or projecting fire power inland.

battle map--A map showing ground features in sufficient detail for tactical use by all forces, usually at a scale of 1:25,000. See also **map**.

battle reserves--Reserve supplies accumulated by an army, detached corps, or detached division in the vicinity of the battlefield, in addition to unit and individual reserves. See also **reserve supplies**.

BCE--See **battlefield coordination element**.

BDA--See **battle damage assessment; bomb damage assessment**.

BDZ--See **base defense zone**.

beach--1. The area extending from the shoreline inland to a marked change in physiographic form or material, or to the line of permanent vegetation (coastline). 2. In amphibious operations, that portion of the shoreline designated for landing of a tactical organization.

beach capacity(*)--An estimate, expressed in terms of measurement tons, or weight tons, of cargo that may be unloaded over a designated strip of shore per day. See also **clearance capacity; port capacity**.

beach group--See **naval beach group; shore party**.

beachhead--A designated area on a hostile or potentially hostile shore that, when seized and held, ensures the continuous landing of troops and materiel, and provides maneuver space requisite for subsequent projected operations ashore.

beach landing site--A geographic location selected for across-the-beach infiltration, exfiltration, or resupply operations. Also called **BLS.**

beach marker--A sign or device used to identify a beach or certain activities thereon for incoming waterborne traffic. Markers may be panels, lights, buoys, or electronic devices.

beachmaster--The naval officer in command of the beachmaster unit of the naval beach group.

beachmaster unit--A commissioned naval unit of the naval beach group designed to provide to the shore party a naval component known as a beach party which is capable of supporting the amphibious landing of one division (reinforced). See also **beach party; shore party.**

beach minefield(*)--A minefield in the shallow water approaches to a possible amphibious landing beach. See also **minefield.**

beach organization--In an amphibious operation, the planned arrangement of personnel and facilities to effect movement, supply, and evacuation across beaches and in the beach area for support of a landing force.

beach party--The naval component of the shore party. See also **beachmaster unit; shore party.**

beach party commander--The naval officer in command of the naval component of the shore party.

beach photography--Vertical, oblique, ground, and periscope coverage at varying scales to provide information of offshore, shore, and inland areas. It covers terrain which provides observation of the beaches and is primarily concerned with the geological and tactical aspects of the beach.

beach reserves(*)--In an amphibious operation, an accumulation of supplies of all classes established in dumps in beachhead areas. See also **reserve supplies.**

beach support area--In amphibious operations, the area to the rear of a landing force or elements thereof, established and operated by shore party units, which contains the facilities for the unloading of troops and materiel and the support of the forces ashore; it includes facilities for the evacuation of wounded, enemy prisoners of war, and captured materiel.

beach survey--The collection of data describing the physical characteristics of a beach; that is, an area whose boundaries are a shoreline, a coastline, and two natural or arbitrary assigned flanks.

beach width--The horizontal dimensions of the beach measured at right angles to the shoreline from the line of extreme low water inland to the landward limit of the beach (the coastline).

beacon--A light or electronic source which emits a distinctive or characteristic signal used for the determination of bearings, courses, or location. (DOD, NATO) See crash locator beacon; fan marker beacon; localizer; meaconing; personal locator beacon; radio beacon; submarine locator acoustic beacon; Z marker beacon.

beacon double--In air intercept, a code meaning, "Pilot select double pulse mode on your tracking beacon."

beacon off--In air intercept, a code meaning, "Turn off your tracking beacon."

beacon on--In air intercept, a code meaning, "Turn on your tracking beacon."

beam attack--In air intercept, an attack by an interceptor aircraft attack which terminates with a heading crossing angle greater than 45 degrees but less than 135 degrees. See also **heading crossing angle.**

beam rider--A missile guided by an electronic beam.

beam width--The angle between the directions, on either side of the axis, at which the intensity of the radio frequency field drops to one-half the value it has on the axis.

bearing(*)--The horizontal angle at a given point measured clockwise from a specific datum point to a second point. See also **grid bearing; relative bearing; true bearing.**

beaten zone--The area on the ground upon which the cone of fire falls.

beleaguered--See **missing.**

bent--In air intercept and close air support, a code meaning, "Equipment indicated is inoperative (temporarily or indefinitely)." Cancelled by "Okay."

besieged--See **missing.**

bilateral infrastructure(*)--Infrastructure which concerns only two NATO members and is financed by mutual agreement between them (e.g., facilities required for the use of forces of one NATO member in the territory of another). See also **infrastructure.**

billet--1. Shelter for troops. 2. To quarter troops. 3. A personnel position or assignment which may be filled by one person.

binary chemical munition(*)--A munition in which chemical substances, held in separate containers, react when mixed or combined as a result of being fired, launched or otherwise initiated to produce a chemical agent. See also **munition; chemical munition; multi-agent munition.**

binding(*)--The fastening or securing of items to a movable platform called a pallet. See also **palletized unit load.**

bingo--1. When originated by pilot, means, "I have reached minimal fuel for safe return to base or to designated alternate." 2. When originated by controlling activity, means, "Proceed to alternate airfield or carrier as specified."

bingo field--Alternate airfield.

bin storage--Storage of items of supplies and equipment in an individual compartment or subdivision of a storage unit in less than bulk quantities. See also **bulk storage; storage.**

biographical intelligence--That component of intelligence which deals with individual foreign personalities of actual or potential importance.

biological agent--A microorganism that causes disease in personnel, plants, or animals or causes the deterioration of materiel. See also **biological operation; biological weapon; chemical agent.**

biological ammunition(*)--A type of ammunition, the filler of which is primarily a biological agent.

biological defense(*)--The methods, plans, and procedures involved in establishing and executing defensive measures against attacks using biological agents.

biological environment(*)--Conditions found in an area resulting from direct or persisting effects of biological weapons.

biological half-time--See **half-life.**

biological operation(*)--Employment of biological agents to produce casualties in personnel or animals and damage to plants or materiel; or defense against such employment.

biological warfare--See **biological operation.**

biological weapon(*)--An item of materiel which projects, disperses, or disseminates a biological agent including arthropod vectors.

black--In intelligence handling, a term used in certain phrases (e.g., living black, black border crossing) to indicate reliance on illegal concealment rather than on cover.

black list--An official counterintelligence listing of actual or potential enemy collaborators, sympathizers, intelligence suspects, and other persons whose presence menaces the security of friendly forces.

black propaganda--Propaganda which purports to emanate from a source other than the true one. See also **propaganda.**

blast(*)--The brief and rapid movement of air, vapor or fluid away from a center of outward pressure, as in an explosion or in the combustion of rocket fuel; the pressure accompanying this movement. This term is commonly used for "explosion," but the two terms may be distinguished.

blast effect--Destruction of or damage to structures and personnel by the force of an explosion on or above the surface of the ground. Blast effect may be contrasted with the cratering and ground-shock effects of a projectile or charge that goes off beneath the surface.

blast line--A horizontal radial line on the surface of the Earth originating at ground zero on which measurements of blast from an explosion are taken.

blast wave--A sharply defined wave of increased pressure rapidly propagated through a sur-

rounding medium from a center of detonation or similar disturbance.

blast wave diffraction(*)--The passage around and envelopment of a structure by the nuclear blast wave.

bleeding edge(*)--That edge of a map or chart on which cartographic detail is extended to the edge of the sheet.

blind bombing zone(*)--A restricted area (air, land, or sea) established for the purpose of permitting air operations, unrestricted by the operations or possible attack of friendly forces.

blind transmission--Any transmission of information that is made without expectation of acknowledgement.

blip(*)--The display of a received pulse on a cathode ray tube.

blister agent(*)--A chemical agent which injures the eyes and lungs, and burns or blisters the skin. Also called **vesicant agent.**

blocking and chocking(*)--The use of wedges or chocks to prevent the inadvertent shifting of cargo in transit.

blocking position(*)--A defensive position so sited as to deny the enemy access to a given area or to prevent his advance in a given direction.

block shipment--A method of shipment of supplies to overseas areas to provide balanced stocks or an arbitrary balanced force for a specific number of days, e.g., shipment of 30 days' supply for an average force of 10,000 individuals.

block stowage loading(*)--A method of loading whereby all cargo for a specific destination is stowed together. The purpose is to facilitate rapid off-loading at the destination, with the least possible disturbance of cargo intended for other points. See also **loading.**

blood agent(*)--A chemical compound, including the cyanide group, that affects bodily functions by preventing the normal utilization of oxygen by body tissues.

blood chit--A small cloth chart depicting an American Flag and a statement in several languages to the effect that anyone assisting the bearer to safety will be rewarded.

blood chit (intelligence)--See **blood chit.**

blow--To expose, often unintentionally, personnel, installations, or other elements of a clandestine organization or activity.

blowback(*)--1. Escape, to the rear and under pressure, of gases formed during the firing of the weapon. Blowback may be caused by a defective breech mechanism, a ruptured cartridge case, or a faulty primer. 2. Type of weapon operation in which the force of expanding gases acting to the rear against the face of the bolt furnishes all the energy required to initiate the complete cycle of operation. A weapon which employs this method of operation is characterized by the absence of any breech-lock or bolt-lock mechanism.

BLS--See **beach landing site.**

Blue Bark--US military personnel, US citizen civilian employees of the Department of Defense, and the dependents of both categories who travel in connection with the death of an immediate family member. It also applies to designated escorts for dependents of deceased military members. Furthermore, the term is used to designate the personal property shipment of a deceased member.

blue commander(*)--The officer designated to exercise operational control over blue forces for a specific period during an exercise.

boat diagram--In the assault phase of an amphibious operation, a diagram showing the positions of individuals and equipment in each boat.

boat group--The basic organization of landing craft. One boat group is organized for each battalion landing team (or equivalent) to be landed in the first trip of landing craft or amphibious vehicles.

boat group commander--An officer assigned to be embarked in a control boat who is responsible for discipline and organization within the boat group to complete the assigned mission.

boat lane(*)--A lane for amphibious assault landing craft, which extends seaward from the landing beaches to the line of departure. The width of a boat lane is determined by the length of the corresponding beach.

boat space--The space and weight factor used to determine the capacity of boats, landing craft, and amphibious vehicles. With respect to landing craft and amphibious vehicles, it is based on the requirements of one person with individual equipment. The person is assumed to weigh 224 pounds and to occupy 13.5 cubic feet of space. See also **man space.**

boattail(*)--The conical section of a ballistic body that progressively decreases in diameter toward the tail to reduce overall aerodynamic drag.

boat wave--See **wave.**

bogey--An air contact which is unidentified but assumed to be enemy. (Not to be confused with unknown.) See also **friendly; hostile.**

bomb damage assessment--The determination of the effect of all air attacks on targets (e.g., bombs, guided missiles, rockets, or strafing). Also called **BDA**. See also **battle damage assessment; combat assessment.**

bomb disposal unit--See **explosive ordnance disposal unit.**

bomber--See **intermediate-range bomber aircraft; long-range bomber aircraft; medium--range bomber aircraft.**

bomb impact plot--A graphic representation of the target area, usually a pre-strike air photograph, on which prominent dots are plotted to mark the impact or detonation points of bombs dropped on a specific bombing attack.

bombing angle(*)--The angle between the vertical and a line joining the aircraft to what would be the point of impact of a bomb released from it at that instant.

bombing height(*)--In air operations, the height above ground level at which the aircraft is flying at the moment of ordnance release. Bombing heights are classified as follows:
very low: below 100 feet;
low: from 100 to 2,000 feet;
medium: from 2,000 to 10,000 feet;
high: from 10,000 to 50,000 feet;
very high: 50,000 feet and above.

bombing run(*)--In air bombing, that part of the flight that begins, normally from an initial point, with the approach to the target, includes target acquisition, and ends normally at the weapon release point.

bomb line--See **fire support coordination line.**

bomb release line(*)--An imaginary line around a defended area or objective over which an aircraft should release its bomb in order to obtain a hit or hits on an area or objective.

bomb release point(*)--The point in space at which bombs must be released to reach the desired point of detonation.

bonding(*)--In electrical engineering, the process of connecting together metal parts so that they make low resistance electrical contact for direct current and lower frequency alternating currents. See also **earthing; grounding.**

booby trap(*)--An explosive or nonexplosive device or other material, deliberately placed to cause casualties when an apparently harmless object is disturbed or a normally safe act is performed.

booster(*)--1. A high-explosive element sufficiently sensitive so as to be actuated by small explosive elements in a fuze or primer and powerful enough to cause detonation of the main explosive filling. 2. An auxiliary or initial propulsion system which travels with a missile or aircraft and which may or may not separate from the parent craft when its impulse has been delivered. A booster system may contain, or consist of, one or more units.

boost phase--That portion of the flight of a ballistic missile or space vehicle during which the booster and sustainer engines operate. See also **midcourse phase; reentry phase; terminal phase.**

border(*)--In cartography, the area of a map or chart lying between the neatline and the surrounding framework.

border break(*)--A cartographic technique used when it is required to extend a portion of the cartographic detail of a map or chart beyond the sheetlines into the margin.

border crosser(*)--An individual, living close to a frontier, who normally has to cross the frontier frequently for legitimate purposes.

boresafe fuze(*)--Type of fuze having an interrupter in the explosive train that prevents a projectile from exploding until after it has cleared the muzzle of a weapon. See also **fuze.**

bottom mine(*)--A mine with negative buoyancy which remains on the seabed. Also called **ground mine.** See also **mine.**

bound(*)--1. In land warfare, a single movement, usually from cover to cover, made by troops often under enemy fire. (DOD) 2. Distance covered in one movement by a unit that is advancing by bounds.

boundary--A line which delineates surface areas for the purpose of facilitating coordination and deconfliction of operations between adjacent units, formations, or areas. See also **airspace control boundary.**

bouquet mine(*)--In naval mine warfare, a mine in which a number of buoyant mine cases are attached to the same sinker, so that when the mooring of one mine case is cut, another mine rises from the sinker to its set depth. See also **mine.**

BQM-34--See **Firebee.**

bracketing(*)--A method of adjusting fire in which a bracket is established by obtaining an over and a short along the spotting line, and then successively splitting the bracket in half until a target hit or desired bracket is obtained.

branch--1. A subdivision of any organization. 2. A geographically separate unit of an activity which performs all or part of the primary functions of the parent activity on a smaller scale. Unlike an annex, a branch is not merely an overflow addition. 3. An arm or service of the Army.

breakaway(*)--1. The onset of a condition in which the shock front moves away from the exterior of the expanding fireball produced by the explosion of a nuclear weapon. (DOD) 2. After completion of attack, turn to heading as directed.

break off--In close air support, a command utilized to immediately terminate an attack.

breakoff position(*)--The position at which a leaver or leaver section breaks off from the main convoy to proceed to a different destination.

break-up(*)--1. In detection by radar, the separation of one solid return into a number of individual returns which correspond to the various objects or structure groupings. This separation is contingent upon a number of factors including range, beam width, gain setting, object size and distance between objects. 2. In imagery interpretation, the result of magnification or enlargement which causes the imaged item to lose its identity and the resultant presentation to become a random series of tonal impressions. Also called **split-up.**

brevity code(*)--A code which provides no security but which has as its sole purpose the shortening of messages rather than the concealment of their content.

bridgehead--An area of ground held or to be gained on the enemy's side of an obstacle. See also **airhead; beachhead.**

bridgehead line(*)--The limit of the objective area in the development of the bridgehead. See also **objective area.**

briefing(*)--The act of giving in advance specific instructions or information.

brigade--A unit usually smaller than a division to which are attached groups and/or battalions and smaller units tailored to meet anticipated requirements.

broadcast-controlled air interception(*)--An interception in which the interceptor is given a continuous broadcast of information concerning an enemy raid and effects interception without further control. See also **air interception; close-controlled air interception.**

Bronco--A light, twin turboprop, twinseat observation and support aircraft. May be equipped with machine guns and light ordnance for close air support missions. Designated as **OV-10.**

buffer distance(*)--In nuclear warfare: 1. The horizontal distance which, when added to the radius of safety, will give the desired assurance that the specified degree of risk will not be exceeded. The buffer distance is normally expressed quantitatively in multiples of the delivery error. 2. The vertical distance which is added to the fallout safe-height of burst in order to determine a desired height of burst which will provide the desired assurance that militarily significant fallout will not occur. It is normally expressed quantitatively in multiples of the vertical error.

bug--1. A concealed microphone or listening device or other audiosurveillance device. 2. To install means for audiosurveillance.

bugged--Room or object which contains a concealed listening device.

buildup(*)--The process of attaining prescribed strength of units and prescribed levels of vehicles, equipment, stores, and supplies. Also may be applied to the means of accomplishing this process.

bulk cargo--That which is generally shipped in volume where the transportation conveyance is the only external container; such as liquids, ore, or grain.

bulk petroleum product(*)--A liquid petroleum product transported by various means and stored in tanks or containers having an individual fill capacity greater than 250 liters.

bulk storage--1. Storage in a warehouse of supplies and equipment in large quantities, usually in original containers, as distinguished from bin storage. 2. Storage of liquids, such as petroleum products in tanks, as distinguished from drum or packaged storage. See also **bin storage; storage.**

burial--See **emergency burial; group burial; trench burial.** See also **graves registration.**

burn--1. Deliberately expose the true status of a person under cover. 2. The legitimate destruction and burning of classified material, usually accomplished by the custodian as prescribed in regulations.

burned--Used to indicate that a clandestine operator has been exposed to the operation (especially in a surveillance) or that reliability as a source of information has been compromised.

burn notice--An official statement by one intelligence agency to other agencies, domestic or foreign, that an individual or group is unreliable for any of a variety of reasons.

burnout(*)--The point in time or in the missile trajectory when combustion of fuels in the rocket engine is terminated by other than programmed cutoff.

burnout velocity(*)--The velocity attained by a missile at the point of burnout.

burn-through range--The distance at which a specific radar can discern targets through the external interference being received.

buster--In air intercept, a code meaning, "Fly at maximal continuous speed (or power)."

C

C-5--See **Galaxy**.

C-130--See **Hercules**.

C-141--See **Starlifter**.

C2-protection--See **command and control warfare**.

C4 systems--See **command, control, communications, and computer systems**.

CA--See **combat assessment**.

calibrated focal length(*)--An adjusted value of the equivalent focal length, so computed as to equalize the positive and negative values of distortion over the entire field used in a camera. See also **focal length**.

call fire--Fire delivered on a specific target in response to a request from the supported unit. See also **fire**.

call for fire(*)--A request for fire containing data necessary for obtaining the required fire on a target.

call mission(*)--A type of air support mission which is not requested sufficiently in advance of the desired time of execution to permit detailed planning and briefing of pilots prior to takeoff. Aircraft scheduled for this type of mission are on air, ground, or carrier alert, and are armed with a prescribed load.

call sign(*)--Any combination of characters or pronounceable words, which identifies a communication facility, a command, an authority, an activity, or a unit; used primarily for establishing and maintaining communications. See also **collective call sign; indefinite call sign;** international call sign; net call sign; tactical call sign; visual call sign; voice call sign.

camera axis(*)--An imaginary line through the optical center of the lens perpendicular to the negative photo plane.

camera axis direction(*)--Direction on the horizontal plane of the optical axis of the camera at the time of exposure. This direction is defined by its azimuth expressed in degrees in relation to true/magnetic north.

camera calibration(*)--The determination of the calibrated focal length, the location of the principal point with respect to the fiducial marks and the lens distortion effective in the focal plane of the camera referred to the particular calibrated focal length.

camera cycling rate(*)--The frequency with which camera frames are exposed, expressed as cycles per second.

camera nadir--See **photo nadir**.

camera station (photogrammetry)--See **air station (photogrammetry)**.

camouflage(*)--The use of natural or artificial material on personnel, objects, or tactical positions with the aim of confusing, misleading, or evading the enemy. See also **countersurveillance**.

camouflage detection photography(*)--Photography utilizing a special type of film (usually infrared) designed for the detection of camouflage. See also **false color film**.

camouflet(*)--The resulting cavity in a deep underground burst when there is no rupture of the surface. See also **crater**.

camp--A group of tents, huts, or other shelter set up temporarily for troops, and more permanent than a bivouac. A military post, temporary or permanent, may be called a camp.

campaign--A series of related military operations aimed at accomplishing a strategic or operational objective within a given time and space. See also **campaign plan.**

campaign plan--A plan for a series of related military operations aimed to achieve strategic and operational objectives within a given time and space. See also **campaign.**

canalize--To restrict operations to a narrow zone by use of existing or reinforcing obstacles or by fire or bombing.

cancel(*)--In artillery and naval gunfire support, the term, "cancel," when coupled with a previous order, other than an order for a quantity or type of ammunition, rescinds that order.

cancel check firing--The order to rescind check firing.

cancel converge--The command used to rescind converge.

cannibalize(*)--To remove serviceable parts from one item of equipment in order to install them on another item of equipment.

cannot observe(*)--A type of fire control which indicates that the observer or spotter will be unable to adjust fire, but believes a target exists at the given location and is of sufficient importance to justify firing upon it without adjustment or observation.

CAP--See **crisis action planning.**

capability--The ability to execute a specified course of action. (A capability may or may not be accompanied by an intention.)

capacity load (Navy)--The maximum quantity of all supplies (ammunition; petroleum, oils, and lubricants; rations; general stores; maintenance stores; etc.) which each vessel can carry in proportions prescribed by proper authority. See also **combat load (air); wartime load.**

capsule(*)--1. A sealed, pressurized cabin for extremely high altitude or space flight which provides an acceptable environment for man, animal, or equipment. 2. An ejectable sealed cabin having automatic devices for safe return of the occupants to the surface.

captive firing(*)--A firing test of short duration, conducted with the missile propulsion system operating while secured to a test stand.

captured--See **missing.**

cardinal point effect(*)--The increased intensity of a line or group of returns on the radarscope occurring when the radar beam is perpendicular to the rectangular surface of a line or group of similarly aligned features in the ground pattern.

caretaker status--A nonoperating condition in which the installations, materiel, and facilities are in a care and limited preservation status. Only a minimum of personnel is required to safeguard against fire, theft, and damage from the elements.

cargo(*)--Commodities and supplies in transit. See also **air cargo; dangerous cargo; essential cargo; immediately vital cargo; unwanted cargo; valuable cargo; wanted cargo.** See also **loading; chemical ammunition cargo; flatted cargo; general cargo; heavy-lift cargo; high explosive cargo; inflammable cargo; perishable cargo; special cargo; troop space cargo; vehicle cargo.**

cargo carrier--Highly mobile, air transportable, unarmored, full-tracked cargo and logistic carrier capable of swimming inland waterways and accompanying and resupplying self-propelled artillery weapons. Designated as **M548**.

cargo classification (combat loading)--The division of military cargo into categories for combat loading aboard ships. See also **cargo.**

cargo outturn message--A brief message report transmitted within 48 hours of completion of ship discharge to advise both the Military Sealift Command and the terminal of loading of the condition of the cargo, including any discrepancies in the form of overages, shortages, or damages between cargo as manifested and cargo as checked at time of discharge.

cargo outturn report--A detailed report prepared by a discharging terminal to record discrepancies in the form of over, short, and damaged cargo as manifested, and cargo checked at a time and place of discharge from ship.

cargo sling(*)--A strap, chain, or other material used to hold cargo items securely which are to be hoisted, lowered, or suspended.

cargo tie-down point--A point on military materiel designed for attachment of various means for securing the item for transport.

cargo transporter--A reusable metal shipping container designed for worldwide surface and air movement of suitable military supplies and equipment through the cargo transporter service.

carpet bombing(*)--The progressive distribution of a mass bomb load upon an area defined by designated boundaries, in such manner as to inflict damage to all portions thereof.

carriage--See **gun carriage.**

carrier air group(*)--Two or more aircraft squadrons formed under one commander for administrative and tactical control of operations from a carrier.

carrier striking force(*)--A naval task force composed of aircraft carriers and supporting combatant ships capable of conducting strike operations.

CARVER--A special operations forces acronym used throughout the targeting and mission planning cycle to assess mission validity and requirements. The acronym stands for criticality, accessibility, recuperability, vulnerability, effect, and recognizability.

CAS--See **close air support.**

case--1. An intelligence operation in its entirety. 2. Record of the development of an intelligence operation, including personnel, modus operandi, and objectives.

casual--See **transient.**

casualty--Any person who is lost to the organization by having been declared dead, duty status - whereabouts unknown, missing, ill, or injured. See also **casualty category; casualty status; casualty type; duty status - whereabouts unknown; hostile casualty; nonhostile casualty.**

casualty category--A term used to specifically classify a casualty for reporting purposes based upon the casualty type and the casualty status. Casualty categories include killed in action, died of wounds received in action, and wounded in action. See also **casualty; casualty status; casualty type; duty status - whereabouts unknown; missing.**

casualty receiving and treatment ship--In amphibious operations, a ship designated to receive, provide treatment for, and transfer casualties.

casualty status--A term used to classify a casualty for reporting purposes. There are seven casualty statuses: (1) deceased, (2) duty status - whereabouts unknown, (3) missing, (4) very seriously ill or injured, (5) seriously ill or injured, (6) incapacitating illness or injury, and (7) not seriously injured. See also **casualty; casualty category; casualty type; deceased; duty status - whereabouts unknown; incapacitating illness or injury; missing; not seriously injured; seriously ill or injured; very seriously ill or injured.**

casualty type--A term used to identify a casualty for reporting purposes as either a hostile casualty or a nonhostile casualty. See also **casualty; casualty category; casualty status; hostile casualty; nonhostile casualty.**

catalytic attack--An attack designed to bring about a war between major powers through the disguised machinations of a third power.

catalytic war--Not to be used. See **catalytic attack.**

catapult(*)--A structure which provides an auxiliary source of thrust to a missile or aircraft; must combine the functions of directing and accelerating the missile during its travel on the catapult; serves the same functions for a missile as does a gun tube for a shell.

categories of data--In the context of perception management and its constituent approaches, data obtained by adversary individuals, groups, intelligence systems, and officials. Such data fall in two categories: **a. information**--A compilation of data provided by protected or open sources that would provide a substantially complete picture of friendly intentions, capabilities, or activities. **b. indicators**--Data derived from open sources or from detectable actions that adversaries can piece together or interpret to reach personal conclusions or official estimates concerning friendly intentions, capabilities, or activities. Note: In operations security, actions that convey indicators exploitable by adversaries, but that must be carried out regardless, to plan, prepare for, and execute activities, are called "observables." See also **operations security.**

CATF--See **commander, amphibious task force.**

causeway launching area--An area located near the line of departure but clear of the approach lanes, where ships can launch pontoon causeways.

CAVU--Ceiling and visibility unlimited.

CD--See **counterdrug.**

C-day--See **times.**

cease engagement(*)--In air defense, a fire control order used to direct units to stop the firing sequence against a designated target. Guided missiles already in flight will continue to intercept. See also **engage; hold fire.**

cease fire--A command given to air defense artillery units to refrain from firing on, but to continue to track, an airborne object. Missiles already in flight will be permitted to continue to intercept.

cease loading(*)--In artillery and naval gunfire support, the command used during firing of two or more rounds to indicate the suspension of inserting rounds into the weapon.

ceiling--The height above the earth's surface of the lowest layer of clouds or obscuration phenomena that is reported as "broken," "overcast," or "obscured" and not classified as "thin" or "partial."

celestial guidance--The guidance of a missile or other vehicle by reference to celestial bodies. See also **guidance.**

celestial sphere(*)--An imaginary sphere of infinite radius concentric with the Earth, on which all celestial bodies except the Earth are imagined to be projected.

cell--Small group of individuals who work together for clandestine or subversive purposes.

cell system--See net, chain, cell system.

censorship--See armed forces censorship; civil censorship; field press censorship; military censorship; national censorship; primary censorship; prisoner of war censorship; secondary censorship.

center of burst--See mean point of impact.

centers of gravity--Those characteristics, capabilities, or localities from which a military force derives its freedom of action, physical strength, or will to fight.

centigray(*)--A unit of absorbed dose of radiation (one centigray equals one rad).

central air data computer(*)--A device which computes altitude, vertical speed, air speed and mach number from inputs of pitot and static pressure and temperature.

central control officer--The officer designated by the amphibious task force commander for the overall coordination of the waterborne ship-to--shore movement. The central control officer is embarked in the central control ship.

centralized control(*)--In air defense, the control mode whereby a higher echelon makes direct target assignments to fire units. See also decentralized control.

centrally managed item--An item of materiel subject to inventory control point (wholesale level) management.

central procurement--The procurement of material, supplies, or services by an officially designated command or agency with funds specifically provided for such procurement for the benefit and use of the entire component, or, in the case of single managers, for the Military Departments as a whole.

central war--Not to be used. See general war.

CG--See guided missile cruiser.

CH-53A--See Sea Stallion.

chaff--Radar confusion reflectors, which consist of thin, narrow metallic strips of various lengths and frequency responses, used to reflect echoes for confusion purposes. See also rope; rope--chaff; window.

chain--See net, chain, cell system.

chain of command(*)--The succession of commanding officers from a superior to a subordinate through which command is exercised. Also called command channel. See also administrative chain of command; operational chain of command.

Chairman of the Joint Chiefs of Staff Instruction--A replacement document for all types of correspondence containing Chairman of the Joint Chiefs of Staff (CJCS) policy and guidance that does not involve the employment of forces. An instruction is of indefinite duration and is applicable to external agencies or both the Joint Staff and external agencies. It remains in effect until superseded, rescinded, or otherwise canceled. CJCS Instructions, unlike joint publications, will not contain joint doctrine and/or joint tactics, techniques, and procedures. Also called CJCSI. See also guidance; joint doctrine; joint publication; joint tactics, techniques, and procedures.

Chairman of the Joint Chiefs of Staff Memorandum of Policy--A statement of policy approved by the Chairman of the Joint Chiefs of Staff and issued for the guidance of the Services, the combatant commands, and the Joint Staff. Also called **CJCS Memorandum of Policy.**

chalk commander(*)--The commander of all troops embarked under one chalk number. See also **chalk number; chalk troops.**

chalk number(*)--The number given to a complete load and to the transporting carrier. See also **chalk commander; chalk troops.**

chalk troops(*)--A load of troops defined by a particular chalk number. See also **chalk commander; chalk number.**

challenge(*)--Any process carried out by one unit or person with the object of ascertaining the friendly or hostile character or identity of another. See also **countersign; password; reply.**

change of operational control--The date and time (Coordinated Universal Time) at which the responsibility for operational control of a force or unit passes from one operational control authority to another. Also called **CHOP.**

channel--Used in conjunction with a predetermined letter, number, or code word to reference a specific radio frequency.

channel airlift--Common-user airlift service provided on a scheduled basis between two points.

Chaparral--A short-range, low-altitude, surface-to-air, Army air defense artillery system. Designated as **MIM-72.** See also **Sidewinder.**

characteristic actuation probability(*)--The average probability of a mine of a given type being actuated by one run of the sweep within the characteristic actuation width.

characteristic actuation width(*)--The width of path over which mines can be actuated by a single run of the sweep gear.

characteristic detection probability(*)--The ratio of the number of mines detected on a single run to the number of mines which could have been detected within the characteristic detection width.

characteristic detection width(*)--The width of path over which mines can be detected on a single run.

characterization (evaluation)--A biographical sketch of an individual or a statement of the nature and intent of an organization or group.

charge(*)--1. The amount of propellant required for a fixed, semi-fixed, or separate loading projectile, round or shell. It may also refer to the quantity of explosive filling contained in a bomb, mine or the like. 2. In combat engineering, a quantity of explosive, prepared for demolition purposes.

charged demolition target(*)--A demolition target on which all charges have been placed and which is in the states of readiness, either state 1 - safe, or state 2 - armed. See also **state of readiness - state 1 safe; state of readiness - state 2 armed.**

chart base(*)--A chart used as a primary source for compilation or as a framework on which new detail is printed. Also called **topographic base.** See also **base map.**

chart index--See **map index.**

chart location of the battery--See **battery center.**

chart series--See **map; map series.**

chart sheet--See **map; map sheet.**

CHB--See **Navy Cargo Handling Battalion.**

check firing(*)--In artillery and naval gunfire support, a command to cause a temporary halt in firing.

checkout(*)--A sequence of functional, operational, and calibrational tests to determine the condition and status of a weapon system or element thereof.

checkpoint(*)--1. A predetermined point on the surface of the Earth used as a means of controlling movement, a registration target for fire adjustment, or reference for location. 2. Center of impact; a burst center. 3. Geographical location on land or water above which the position of an aircraft in flight may be determined by observation or by electrical means. 4. A place where military police check vehicular or pedestrian traffic in order to enforce circulation control measures and other laws, orders, and regulations.

check port/starboard--In air intercept, a term meaning, "Alter heading ____ degrees to port/starboard momentarily for airborne radar search and then resume heading."

check sweeping(*)--In naval mine warfare, sweeping to check that no moored mines are left after a previous clearing operation.

chemical agent(*)--A chemical substance which is intended for use in military operations to kill, seriously injure, or incapacitate personnel through its physiological effects. Excluded from consideration are riot control agents, herbicides, smoke, and flame. See also **biological agent.**

chemical agent cumulative action--The building up, within the human body, of small ineffective doses of certain chemical agents to a point where eventual effect is similar to one large dose.

chemical ammunition(*)--A type of ammunition, the filler of which is primarily a chemical agent. See also **cargo.**

chemical ammunition cargo--Cargo such as white phosphorous munitions (shell and grenades). See also **cargo.**

chemical, biological, and radiological operation(*)--A collective term used only when referring to a combined chemical, biological, and radiological operation.

chemical defense(*)--The methods, plans and procedures involved in establishing and executing defensive measures against attack utilizing chemical agents. See also **NBC defense.**

chemical dose(*)--The amount of chemical agent, expressed in milligrams, that is taken or absorbed by the body.

chemical environment(*)--Conditions found in an area resulting from direct or persisting effects of chemical weapons.

chemical horn(*)--In naval mine warfare, a mine horn containing an electric battery, the electrolyte for which is in a glass tube protected by a thin metal sheet. Also called **Hertz Horn.**

chemical monitoring(*)--The continued or periodic process of determining whether or not a chemical agent is present. See also **chemical survey.**

chemical operations(*)--Employment of chemical agents to kill, injure, or incapacitate for a significant period of time, personnel or animals, and deny or hinder the use of areas, facilities, or material; or defense against such employment.

chemical survey(*)--The directed effort to determine the nature and degree of chemical hazard in an area and to delineate the perimeter of the hazard area.

chemical warfare--All aspects of military operations involving the employment of lethal and incapacitating munitions/agents and the warning and protective measures associated with such offensive operations. Since riot control agents and herbicides are not considered to be chemical warfare agents, those two items will be referred to separately or under the broader term "chemical," which will be used to include all types of chemical munitions/agents collectively. The term "chemical warfare weapons" may be used when it is desired to reflect both lethal and incapacitating munitions/agents of either chemical or biological origin. Also called **CW**. See also **chemical operations, herbicide, riot control agent.**

chemical warfare agent--See **chemical agent.**

chicks--Friendly fighter aircraft.

Chief Army, Navy, Air Force, or Marine Corps Censor--An officer appointed by the commander of the Army, Navy, Air Force, or Marine Corps component of a unified command to supervise all censorship activities of that Service.

chief of staff--The senior or principal member or head of a staff, or the principal assistant in a staff capacity to a person in a command capacity; the head or controlling member of a staff, for purposes of the coordination of its work; a position, that in itself is without inherent power of command by reason of assignment, except that which is invested in such a position by delegation to exercise command in another's name. In the Army and Marine Corps, the title is applied only to the staff on a brigade or division level or higher. In lower units, the corresponding title is executive officer. In the

Air Force, the title is applied normally in the staff on an Air Force level and above. In the Navy, the title is applied only on the staff of a commander with rank of commodore or above. The corresponding title on the staff of a commander of rank lower than commodore is chief staff officer, and in the organization of a single ship, executive officer.

CHOP--See **change of operational control.**

chronic radiation dose(*)--A dose of ionizing radiation received either continuously or intermittently over a prolonged period of time. A chronic radiation dose may be high enough to cause radiation sickness and death but if received at a low dose rate a significant portion of the acute cellular damage will be repaired. See also **acute radiation dose; radiation dose; radiation dose rate.**

chuffing(*)--The characteristic of some rockets to burn intermittently and with an irregular noise.

CI--See **counterintelligence.**

CIC--See **combat information center.**

CINC's required date--The original date relative to C-day, specified by the combatant commander for arrival of forces or cargo at the destination; shown in the time-phased force and deployment data to assess the impact of later arrival. Also called **CRD.**

cipher--Any cryptographic system in which arbitrary symbols or groups of symbols, represent units of plain text of regular length, usually single letters, or in which units of plain text are rearranged, or both, in accordance with certain predetermined rules. See also **cryptosystem.**

circuit--1. An electronic path between two or more points, capable of providing a number of channels. 2. A number of conductors connected together for the purpose of carrying an electrical current.

circuitry--A complex of circuits describing interconnection within or between systems.

circular error probable--An indicator of the delivery accuracy of a weapon system, used as a factor in determining probable damage to a target. It is the radius of a circle within which half of a missile's projectiles are expected to fall. Also called **CEP.** See also **delivery error; deviation; dispersion error; horizontal error.**

CIRVIS--Communications instructions for reporting vital intelligence sightings.

civic action--See **military civic action.**

civil affairs--The activities of a commander that establish, maintain, influence, or exploit relations between military forces and civil authorities, both governmental and nongovernmental, and the civilian populace in a friendly, neutral, or hostile area of operations in order to facilitate military operations and consolidate operational objectives. Civil affairs may include performance by military forces of activities and functions normally the responsibility of local government. These activities may occur prior to, during, or subsequent to other military actions. They may also occur, if directed, in the absence of other military operations.

civil affairs agreement--An agreement which governs the relationship between allied armed forces located in a friendly country and the civil authorities and people of that country. See also **civil affairs.**

civil censorship--Censorship of civilian communications, such as messages, printed matter, and films, entering, leaving, or circulating within areas or territories occupied or controlled by armed forces. See also **censorship.**

civil damage assessment--An appraisal of damage to a nation's population, industry, utilities, communications, transportation, food, water, and medical resources to support planning for national recovery. See also **damage assessment.**

civil defense--All those activities and measures designed or undertaken to: a. minimize the effects upon the civilian population caused or which would be caused by an enemy attack on the United States; b. deal with the immediate emergency conditions which would be created by any such attack; and c. effectuate emergency repairs to, or the emergency restoration of, vital utilities and facilities destroyed or damaged by any such attack.

civil defense emergency--See **domestic emergencies.**

civil defense intelligence--The product resulting from the collection and evaluation of information concerning all aspects of the situation in the United States and its territories that are potential or actual targets of any enemy attack including, in the preattack phase, the emergency measures taken and estimates of the civil populations' preparedness. In the event of an actual attack, a description of conditions in the affected area with emphasis on the extent of damage, fallout levels, and casualty and resource estimates. The product is required by civil and military authorities for use in the formulation of decisions, the conduct of operations, and the continuation of the planning processes.

civil disturbance(*)--Group acts of violence and disorder prejudicial to public law and order. See also **domestic emergencies.**

civil disturbance readiness conditions--Required conditions of preparedness to be attained by military forces in preparation for deployment to an objective area in response to an actual or threatened civil disturbance.

civil disturbances--See **domestic emergencies.**

civilian internee--1. A civilian who is interned during armed conflict or occupation for security reasons or for protection or because he has committed an offense against the detaining power. 2. A term used to refer to persons interned and protected in accordance with the Geneva Convention relative to the Protection of Civilian Persons in Time of War, 12 August 1949 (Geneva Convention). See also **Prisoner of War.**

civilian internee camp--An installation established for the internment and administration of civilian internees.

civil nuclear power--A nation which has potential to employ nuclear technology for development of nuclear weapons but has deliberately decided against doing so. See also **nuclear power.**

civil requirements--The computed production and distribution of all types of services, supplies, and equipment during periods of armed conflict or occupation to ensure the productive efficiency of the civilian economy and to provide civilians the treatment and protection to which they are entitled under customary and conventional international law.

civil reserve air fleet--A program in which the Department of Defense uses aircraft owned by a US entity or citizen. The aircraft are allocated by the Department of Transportation to augment the military airlift capability of the Department of Defense (DOD). These aircraft are allocated, in accordance with DOD requirements, to segments, according to their capabilities, such as Long-Range International (cargo and passenger), Short-Range International, Domestic, Alaskan, Aeromedical, and other segments as may be mutually agreed upon by the Department of Defense and the Department of Transportation. The Civil Reserve Air Fleet (CRAF) can be incrementally activated by the Department of Defense in three stages in response to defense-oriented situations, up to and including a declared national emergency or war, to satisfy DOD airlift requirements. When activated, CRAF aircraft are under the mission control of the Department of Defense while remaining a civil resource under the operational control of the responsible US entity or citizen. Also called **CRAF. a. CRAF Stage I.** This stage involves DOD use of civil air resources that air carriers will furnish to the Department of Defense to support substantially expanded peacetime military airlift requirements. The Commander, Air Mobility Command, may authorize activation of this stage and assume mission control of those airlift assets committed to CRAF Stage I. **b. CRAF Stage II.** This stage involves DOD use of civil air resources that the air carriers will furnish to Department of Defense in a time of defense airlift emergency. The Secretary of Defense, or designee, may authorize activation of this stage permitting the Commander, Air Mobility Command, to assume mission control of those airlift assets committed to CRAF Stage II. **c. CRAF Stage III.** This stage involves DOD use of civil air resources owned by a US entity or citizen that the air carriers will furnish to the Department of Defense in a time of declared national defense-oriented emergency or war, or when otherwise necessary for the national defense. The aircraft in this stage are allocated by the Secretary of Transportation to the Secretary of Defense. The Secretary of Defense may authorize activation of this stage permitting the Commander, Air Mobility Command, to assume mission control of those airlift assets committed to CRAF Stage III.

civil transportation--The movement of persons, property, or mail by civil facilities, and the resources (including storage, except that for agricultural and petroleum products) necessary to accomplish the movement. (Excludes transportation operated or controlled by the military, and petroleum and gas pipelines.)

CJCSI--See **Chairman of the Joint Chiefs of Staff Instruction.**

CJCS Memorandum of Policy--See **Chairman of the Joint Chiefs of Staff Memorandum of Policy.**

clandestine operation--An operation sponsored or conducted by governmental departments or agencies in such a way as to assure secrecy or concealment. A clandestine operation differs from a covert operation in that emphasis is placed on concealment of the operation rather than on concealment of identity of sponsor. In special operations, an activity may be both covert and clandestine and may focus equally on operational considerations and intelligence--related activities. See also **covert operation; overt operation.**

clara--In air intercept, a code meaning, "Radar scope is clear of contacts other than those known to be friendly."

classification--The determination that official information requires, in the interests of national security, a specific degree of protection against unauthorized disclosure, coupled with a designation signifying that such a determination has been made. See also **security classification.**

classification of bridges and vehicles--See **military load classification.**

classified contract--Any contract that requires or will require access to classified information by the contractor or the employees in the performance of the contract. (A contract may be classified even though the contract document itself is not classified.)

classified information--Official information which has been determined to require, in the interests of national security, protection against unauthorized disclosure and which has been so designated.

classified matter(*)--Official information or matter in any form or of any nature which requires protection in the interests of national security. See also **unclassified matter.**

clean aircraft--1. An aircraft in flight configuration, versus landing configuration, i.e., landing gear and flaps retracted, etc. 2. An aircraft that does not have external stores.

cleansing station--See **decontamination station.**

clear--1. To approve or authorize, or to obtain approval or authorization for: a. a person or persons with regard to their actions, movements, duties, etc.; b. an object or group of objects, as equipment or supplies, with regard to quality, quantity, purpose, movement, disposition, etc.; and c. a request, with regard to correctness of form, validity, etc. 2. To give one or more aircraft a clearance. 3. To give a person a security clearance. 4. To fly over an obstacle without touching it. 5. To pass a designated point, line, or object. The end of a column must pass the designated feature before the latter is cleared. 6. a. To operate a gun so as to unload it or make certain no ammunition remains; and b. to free a gun of stoppages. 7. To clear an engine; to open the throttle of an idling engine to free it from carbon. 8. To clear the air to gain either temporary or permanent air superiority or control in a given sector.

clearance capacity--An estimate expressed in terms of measurement or weight tons per day of the cargo that may be transported inland from a beach or port over the available means of inland communication, including roads, railroads, and inland waterways. The estimate is based on an evaluation of the physical characteristics of the transportation facilities in the area. See also **beach capacity; port capacity.**

clearance diving(*)--The process involving the use of divers for locating, identifying and disposing of mines.

clearance rate(*)--The area which would be cleared per unit time with a stated minimum percentage clearance, using specific minehunting and/or minesweeping procedures.

clearing operation--An operation designed to clear or neutralize all mines and obstacles from a route or area.

clearway(*)--A defined rectangular area on the ground or water at the end of a runway in the direction of takeoff and under control of the competent authority, selected or prepared as a suitable area over which an aircraft may make a portion of its initial climb to a specified height.

clear weather air defense fighter(*)--A fighter aircraft with equipment and weapons which enable it to engage airborne targets by day and by night, but in clear weather conditions only.

CLF--See **commander, landing force.**

climb mode(*)--In a flight control system, a control mode in which aircraft climb is automatically controlled to a predetermined program.

clinic--A medical treatment facility primarily intended and appropriately staffed and equipped to provide outpatient medical service for non-hospital type patients. Examination and treatment for emergency cases are types of services rendered. A clinic is also intended to perform certain nontherapeutic activities related to the health of the personnel served, such as physical examinations, immunizations, medical adminis-

tration, and other preventive medical and sanitary measures necessary to support a primary military mission. A clinic will be equipped with the necessary supporting services to perform the assigned mission. A clinic may be equipped with beds (normally fewer than 25) for observation of patients awaiting transfer to a hospital and for care of cases which cannot be cared for on an outpatient status, but which do not require hospitalization. Patients whose expected duration of illness exceeds 72 hours will not normally occupy clinic beds for periods longer than necessary to arrange transfer to a hospital.

clock code position--The position of a target in relation to an aircraft or ship with dead-ahead position considered as 12 o'clock.

close air support--Air action by fixed- and rotary--wing aircraft against hostile targets which are in close proximity to friendly forces and which require detailed integration of each air mission with the fire and movement of those forces. Also called **CAS.** See also **air interdiction; air support; immediate mission request; preplanned mission request.**

close-controlled air interception(*)--An interception in which the interceptor is continuously controlled to a position from which the target is within visual range or radar contact. See also **air interception; broadcast-controlled air-interception.**

closed area(*)--A designated area in or over which passage of any kind is prohibited. See also **prohibited area.**

close-hold plan--Operation plan with access to operation plan information extremely limited to specifically designated Worldwide Military Command and Control System user IDs and terminal Ids during initial course of action development before the involvement of outside commands, agencies, combatant commanders,

Services, or the Joint Staff. See also **limited--access plan.**.

close support(*)--That action of the supporting force against targets or objectives which are sufficiently near the supported force as to require detailed integration or co-ordination of the supporting action with the fire, movement, or other actions of the supported force. See also **direct support; general support; mutual support; support.**

close support area--Those parts of the ocean operating areas nearest to, but not necessarily in, the objective area. They are assigned to naval support carrier battle groups, surface action groups, surface action units, and certain logistic combat service support elements.

close supporting fire(*)--Fire placed on enemy troops, weapons, or positions which, because of their proximity, present the most immediate and serious threat to the supported unit. See also **supporting fire.**

closure--In transportation, the process of a unit arriving at a specified location. It begins when the first element arrives at a designated location, e.g., port of entry/port of departure, intermediate stops, or final destination, and ends when the last element does likewise. For the purposes of studies and command post exercises, a unit is considered essentially closed after 95 percent of its movement requirements for personnel and equipment are completed.

closure minefield(*)--In naval mine warfare, a minefield which is planned to present such a threat that waterborne shipping is prevented from moving.

closure shortfall--The specified movement requirement or portion thereof that did not meet scheduling criteria and/or movement dates.

cloud amount(*)--The proportion of sky obscured by cloud, expressed as a fraction of sky covered.

cloud chamber effect--See **condensation cloud.**

cloud cover(*)--See cloud amount.

cloud top height--The maximal altitude to which a nuclear mushroom cloud rises.

cluster(*)--1. Fireworks signal in which a group of stars burns at the same time. 2. Group of bombs released together. A cluster usually consists of fragmentation or incendiary bombs. 3. Two or more parachutes for dropping light or heavy loads. 4. In land mine warfare, a component of a pattern-laid minefield. It may be antitank, antipersonnel or mixed. It consists of one to five mines and no more than one antitank mine. 5. Two or more engines coupled together so as to function as one power unit. 6. In naval mine warfare, a number of mines laid in close proximity to each other as a pattern or coherent unit. They may be of mixed types. 7. In minehunting, designates a group of mine-like contacts.

cluster bomb unit(*)--An aircraft store composed of a dispenser and submunitions.

clutter--Permanent echoes, cloud, or other atmospheric echo on radar scope; as contact has entered scope clutter. See also **radar clutter.**

COA--See **course of action.**

coalition force--A force composed of military elements of nations that have formed a temporary alliance for some specific purpose.

coarse mine(*)--In naval mine warfare, a relatively insensitive influence mine.

coassembly--With respect to exports, a cooperative arrangement (e.g., US Government or company with foreign government or company) by which finished parts, components, assemblies, or subassemblies are provided to an eligible foreign government, international organization, or commercial producer for the assembly of an end-item or system. This is normally accomplished under the provisions of a manufacturing license agreement per the US International Traffic in Arms Regulation (-ITAR) and could involve the implementation of a government-to-government memorandum of understanding.

coastal convoy(*)--A convoy whose voyage lies in general on the continental shelf and in coastal waters.

coastal frontier--A geographic division of a coastal area, established for organization and command purposes in order to ensure the effective coordination of military forces employed in military operations within the coastal frontier area.

coastal frontier defense--The organization of the forces and materiel of the armed forces assigned to provide security for the coastal frontiers of the continental United States and its overseas possessions.

coastal refraction(*)--The change of the direction of travel of a radio ground wave as it passes from land to sea or from sea to land. Also called **land effect or shoreline effect.**

coastal sea control--The employment of forces to ensure the unimpeded use of an offshore coastal area by friendly forces and, as appropriate, to deny the use of the area to enemy forces.

coast-in point--The point of coastal penetration heading inbound to a target or objective.

coastwise traffic--Sea traffic between continental United States ports on the Atlantic coast, Gulf

coast, and Great Lakes, or between continental United States ports on the Pacific coast.

cocking circuit(*)--In mine warfare, a subsidiary circuit which requires actuation before the main circuits become alive.

COCOM--See **combatant command (command authority).**

code--1. Any system of communication in which arbitrary groups of symbols represent units of plain text of varying length. Codes may be used for brevity or for security. 2. A cryptosystem in which the cryptographic equivalents (usually called "code groups") typically consisting of letters or digits (or both) in otherwise meaningless combinations are substituted for plain text elements which are primarily words, phrases, or sentences. See also **cryptosystem.**

code word(*)--1. A word that has been assigned a classification and a classified meaning to safeguard intentions and information regarding a classified plan or operation. 2. A cryptonym used to identify sensitive intelligence data.

cold war--A state of international tension wherein political, economic, technological, sociological, psychological, paramilitary, and military measures short of overt armed conflict involving regular military forces are employed to achieve national objectives.

collaborative purchase--A method of purchase whereby, in buying similar commodities, buyers for two or more departments exchange information concerning planned purchases in order to minimize competition between them for commodities in the same market. See also **purchase.**

collapse depth(*)--The design depth, referenced to the axis of the pressure hull, beyond which the hull structure or hull penetrations are presumed to suffer catastrophic failure to the point of total collapse.

collate--1. The grouping together of related items to provide a record of events and facilitate further processing. 2. To compare critically two or more items or documents concerning the same general subject; normally accomplished in the processing phase in the intelligence cycle.

collateral mission--A mission other than those for which a force is primarily organized, trained, and equipped, that the force can accomplish by virtue of the inherent capabilities of that force.

collecting point--A point designated for the assembly of personnel casualties, stragglers, disabled materiel, salvage, etc., for further movement to collecting stations or rear installations.

collection--See **intelligence cycle, Subpart b.**

collection (acquisition)--The obtaining of information in any manner, including direct observation, liaison with official agencies, or solicitation from official, unofficial, or public sources.

collection agency--Any individual, organization, or unit that has access to sources of information and the capability of collecting information from them. See also **agency.**

collection coordination facility line number--An arbitrary number assigned to contingency intelligence reconnaissance objectives by the Defense Intelligence Agency collection coordination facility to facilitate all-source collection.

collection management(*)--In intelligence usage, the process of converting intelligence requirements into collection requirements, establishing, tasking or coordinating with appropriate collection sources or agencies, monitoring results and retasking, as required. See also **intelligence; intelligence cycle.**

collection operations management--The authoritative direction, scheduling, and control of specific collection operations and associated processing, exploitation, and reporting resources. Also called **COM.** See also **collection management; collections requirements management.**

collection plan(*)--A plan for collecting information from all available sources to meet intelligence requirements and for transforming those requirements into orders and requests to appropriate agencies. See also **information; information requirements; intelligence cycle.**

collection requirement--An established intelligence need considered in the allocation of intelligence resources to fulfill the essential elements of information and other intelligence needs of a commander.

collection requirements management--The authoritative development and control of collection, processing, exploitation, and/or reporting requirements that normally result in either the direct tasking of assets over which the collection manager has authority, or the generation of single-discipline tasking requests to collection management authorities at a higher, lower, or lateral echelon to accomplish the collection mission. Also called **CRM.** See also **collection management; collection operations management.**

collective call sign(*)--Any call sign which represents two or more facilities, commands, authorities, or units. The collective call sign for any of these includes the commander thereof and all subordinate commanders therein. See also **call sign.**

collective nuclear, biological and chemical protection(*)--Protection provided to a group of individuals in a nuclear, biological and chemical environment which permits relaxation of individual nuclear, biological and chemical protection.

collision course interception--An interception which is accomplished by the constant heading of both aircraft.

collocation(*)--The physical placement of two or more detachments, units, organizations, or facilities at a specifically defined location.

colored beach--That portion of usable coastline sufficient for the assault landing of a regimental landing team or similar sized unit. In the event that the landing force consists of a single battalion landing team, a colored beach will be used and no further subdivision of the beach is required. See also **numbered beach.**

column cover(*)--Cover of a column by aircraft in radio contact therewith, providing for its protection by reconnaissance and/or attack of air or ground targets which threaten the column.

column formation(*)--A formation in which elements are placed one behind the other.

column gap(*)--The space between two consecutive elements proceeding on the same route. It can be calculated in units of length or in units of time measured from the rear of one element to the front of the following element.

column length(*)--The length of the roadway occupied by a column or a convoy in movement. See also **road space.**

COM--See **collection operations management.**

combat air patrol(*)--An aircraft patrol provided over an objective area, over the force protected, over the critical area of a combat zone, or over an air defense area, for the purpose of intercepting and destroying hostile aircraft before they reach their target. See also **airborne alert; barrier combat air patrol; DAD-**

CAP; force combat air patrol; patrol; rescue combat air patrol; target combat air patrol.

combat airspace control--See **airspace control in the combat zone.**

combatant command--One of the unified or specified combatant commands established by the President. See also **specified command; unified command.**

combatant command (command authority)--Nontransferable command authority established by title 10, United States Code, section 164, exercised only by commanders of unified or specified combatant commands unless otherwise directed by the President or the Secretary of Defense. Combatant command (command authority) is the authority of a combatant commander to perform those functions of command over assigned forces involving organizing and employing commands and forces, assigning tasks, designating objectives, and giving authoritative direction over all aspects of military operations, joint training, and logistics necessary to accomplish the missions assigned to the command. Combatant command (command authority) should be exercised through the commanders of subordinate organizations; normally this authority is exercised through the Service or functional component commander. Combatant command (command authority) provides full authority to organize and employ commands and forces as the combatant commander considers necessary to accomplish assigned missions. Also called **COCOM.** See also **combatant command; combatant commander; operational control; tactical control.**

combatant commander--A commander in chief of one of the unified or specified combatant commands established by the President. See also **combatant command; combatant command (command authority); operational control.**

combat area--A restricted area (air, land, or sea) which is established to prevent or minimize mutual interference between friendly forces engaged in combat operations. See also **combat zone.**

combat assessment--The determination of the overall effectiveness of force employment during military operations. Combat assessment is composed of three major components, (a) battle damage assessment, (b) munitions effects assessment, and (c) reattack recommendation. The objective of combat assessment is to identify recommendations for the course of military operations. The J-3 is normally the single point of contact for combat assessment at the joint force level, assisted by the joint force J-2. Also called **CA.** See also **battle damage assessment; bomb damage assessment.**

Combat Camera--Visual information documentation covering air, sea, and ground actions of armed forces in combat and combat support operations, and in related peacetime training activities such as exercises, war games, and operations. See also **visual information documentation.**

combat cargo officer--An embarkation officer assigned to major amphibious ships or naval staffs, functioning primarily as an adviser to and representative of the naval commander in matters pertaining to embarkation and debarkation of troops and their supplies and equipment. See also **embarkation officer.**

combat chart(*)--A special naval chart, at a scale of 1:50,000, designed for naval fire support and close air support during coastal or amphibious operations and showing detailed hydrography and topography in the coastal belt. See also **amphibious chart.**

combat control team--A team of Air Force personnel organized, trained, and equipped to establish and operate navigational or terminal guidance aids, communications, and aircraft control facilities within the objective area of an airborne operation.

combat engineer vehicle, full-tracked 165mm gun--An armored, tracked vehicle that provides engineer support to other combat elements. Vehicle is equipped with a heavy-duty boom and winch, dozer blade, and 165mm demolition gun. It is also armed with a 7.62mm machine gun and a 50-caliber machine gun.

combat forces--Those forces whose primary missions are to participate in combat. See also **operating forces.**

combat information--Unevaluated data, gathered by or provided directly to the tactical commander which, due to its highly perishable nature or the criticality of the situation, cannot be processed into tactical intelligence in time to satisfy the user's tactical intelligence requirements. See also **information.**

combat information center(*)--The agency in a ship or aircraft manned and equipped to collect, display, evaluate, and disseminate tactical information for the use of the embarked flag officer, commanding officer, and certain control agencies. Certain control, assistance, and coordination functions may be delegated by command to the combat information center. Also called **action information center; CIC.** See also **air defense control center.**

combat information ship--A designated ship charged with the coordination of the intership combat information center functions of the various ships in a task force so that the overall combat information available to commands will increase. This ship is normally the flagship of the task force commander. See also **fighter direction aircraft; fighter direction ship.**

combat intelligence--That knowledge of the enemy, weather, and geographical features required by a commander in the planning and conduct of combat operations.

combat loading(*)--The arrangement of personnel and the stowage of equipment and supplies in a manner designed to conform to the anticipated tactical operation of the organization embarked. Each individual item is stowed so that it can be unloaded at the required time. See also **loading.**

combat power(*)--The total means of destructive and/or disruptive force which a Military unit/formation can apply against the opponent at a given time.

combat readiness--Synonymous with operational readiness, with respect to missions or functions performed in combat.

combat ready--Synonymous with operationally ready, with respect to missions or functions performed in combat.

combat search and rescue--A specific task performed by rescue forces to effect the recovery of distressed personnel during wartime or contingency operations. Also called **CSAR.**

combat service support--The essential capabilities, functions, activities, and tasks necessary to sustain all elements of operating forces in theater at all levels of war. Within the national and theater logistic systems, it includes but is not limited to that support rendered by service forces in ensuring the aspects of supply, maintenance, transportation, health services, and other services required by aviation and ground combat troops to permit those units to accomplish their missions in combat. Combat service support encompasses those activities at all levels of war that produce sustainment to all operating forces on the battlefield.

combat service support areas--An area ashore that is organized to contain the necessary supplies, equipment, installations, and elements to provide the landing force with combat service support throughout the operation.

combat service support element--See **Marine air-ground task force.**

combat service support elements--Those elements whose primary missions are to provide service support to combat forces and which are a part, or prepared to become a part, of a theater, command, or task force formed for combat operations. See also **operating forces; service troops; troops.**

combat support--Fire support and operational assistance provided to combat elements. Combat support includes artillery, air defense artillery, engineer, military police, signal, and military intelligence support.

combat support elements--Those elements whose primary missions are to provide combat support to the combat forces and which are a part, or prepared to become a part, of a theater, command, or task force formed for combat operations. See also **operating forces.**

combat support troops--Those units or organizations whose primary mission is to furnish operational assistance for the combat elements. See also **troops.**

combat surveillance--A continuous, all-weather, day-and-night, systematic watch over the battle area to provide timely information for tactical combat operations.

combat surveillance radar--Radar with the normal function of maintaining continuous watch over a combat area.

combat survival(*)--Those measures to be taken by Service personnel when involuntarily separated from friendly forces in combat, including procedures relating to individual survival, evasion, escape, and conduct after capture.

combatting terrorism--Actions, including antiterrorism (defensive measures taken to reduce vulnerability to terrorist acts) and counterterrorism (offensive measures taken to prevent, deter, and respond to terrorism), taken to oppose terrorism throughout the entire threat spectrum.

combat trail--Interceptors in trail formation. Each interceptor behind the leader maintains position visually or with airborne radar.

combat troops--Those units or organizations whose primary mission is destruction of enemy forces and/or installations. See also **troops.**

combat vehicle--A vehicle, with or without armor, designed for a specific fighting function. Armor protection or armament mounted as supplemental equipment on noncombat vehicles will not change the classification of such vehicles to combat vehicles. See also **vehicle.**

Combat Visual Information Support Center--A visual information support facility established at a base of operations during wartime or contingency to provide limited visual information support to the base and its supported elements. Also called **CVISC.**

combat zone--1. That area required by combat forces for the conduct of operations. 2. The territory forward of the Army rear area boundary. See also **combat area; communications zone.**

combination circuit(*)--In mine warfare, a firing circuit which requires actuation by two or more influences, either simultaneously or at a pre-ordained interval, before the circuit can function. Also called **combined circuit.**

combination firing circuit(*)--An assembly comprising two independent firing systems, one non-electric and one electric, so that the firing of either system will detonate all charges. See also **dual-firing circuit.**

combination influence mine(*)--A mine designed to actuate only when two or more different influences are received either simultaneously or in a pre-determined order. Also called **combined influence mine.**

combination mission/level of effort-oriented items--Items for which requirement computations are based on the criteria used for both level of effort-oriented and mission-oriented items.

combined(*)--Between two or more forces or agencies of two or more allies. (When all allies or services are not involved, the participating nations and services shall be identified, e.g., Combined Navies.) See also **joint.**

combined airspeed indicator(*)--An instrument which displays both indicated airspeed and mach number.

combined circuit--See **combination circuit.**

combined doctrine--Fundamental principles that guide the employment of forces of two or more nations in coordinated action toward a common objective. It is ratified by participating nations. See also **doctrine; joint doctrine; multi-Service doctrine.**

combined force(*)--A military force composed of elements of two or more allied nations. See also **force(s).**

combined influence mine--See **combination influence mine.**

combined operation(*)--An operation conducted by forces of two or more allied nations acting together for the accomplishment of a single mission.

combined rescue coordination center--See **rescue coordination center.**

combined staff--A staff composed of personnel of two or more allied nations. See also **integrated staff; joint staff; parallel staff.**

combined warfare--Warfare conducted by forces of two or more allied nations in coordinated action toward common objectives.

combustor(*)--A name generally assigned to the combination of flame holder or stabilizer, igniter, combustion chamber, and injection system of a ramjet or gas turbine.

command--1. The authority that a commander in the Military Service lawfully exercises over subordinates by virtue of rank or assignment. Command includes the authority and responsibility for effectively using available resources and for planning the employment of, organizing, directing, coordinating, and controlling military forces for the accomplishment of assigned missions. It also includes responsibility for health, welfare, morale, and discipline of assigned personnel. 2. An order given by a commander; that is, the will of the commander expressed for the purpose of bringing about a particular action. 3. A unit or units, an organization, or an area under the command of one individual. 4. To dominate by a field of weapon fire or by observation from a superior position. See also **air command; area command; base command; combatant command; combatant command (command authority).**

command altitude--Altitude that must be assumed and/or maintained by the interceptor.

command and control--The exercise of authority and direction by a properly designated commander over assigned forces in the accomplish-

ment of the mission. Command and control functions are performed through an arrangement of personnel, equipment, communications, facilities, and procedures employed by a commander in planning, directing, coordinating, and controlling forces and operations in the accomplishment of the mission.

command and control system--The facilities, equipment, communications, procedures, and personnel essential to a commander for planning, directing, and controlling operations of assigned forces pursuant to the missions assigned.

command and control warfare--The integrated use of operations security (OPSEC), military deception, psychological operations (PSYOP), electronic warfare (EW), and physical destruction, mutually supported by intelligence, to deny information to, influence, degrade, or destroy adversary command and control capabilities, while protecting friendly command and control capabilities against such actions. Command and control warfare applies across the operational continuum and all levels of conflict. Also called **C2W.** C2W is both offensive and defensive: a. counter-C2--To prevent effective C2 of adversary forces by denying information to, influencing, degrading, or destroying the adversary C2 system. b. C2-protection--To maintain effective command and control of own forces by turning to friendly advantage or negating adversary efforts to deny information to, influence, degrade, or destroy the friendly C2 system. See also **command and control; electronic warfare; intelligence; military deception; operations security; psychological operations.**

command axis(*)--A line along which a headquarters will move.

command center--A facility from which a commander and his or her representatives direct operations and control forces. It is organized to gather, process, analyze, display, and disseminate planning and operational data and perform other related tasks.

command channel--See **chain of command.**

command chaplain--The senior chaplain assigned to or designated by a commander of a staff, command, or unit. See also **command chaplain of the combatant command; lay leader or lay reader; religious ministry support; religious ministry support plan; religious ministry support team; Service component command chaplain.**

command chaplain of the combatant command---The senior chaplain assigned to the staff of, or designated by, the combatant commander to provide advice on religion, ethics, and morale of assigned personnel and to coordinate religious ministries within the commander's area of responsibility. The command chaplain of the combatant command may be supported by a staff of chaplains and enlisted religious support personnel. See also **command chaplain; lay leader or lay reader; religious ministry support; religious ministry support plan; religious ministry support team; Service component command chaplain.**

command, control, communications, and computer systems--Integrated systems of doctrine, procedures, organizational structures, personnel, equipment, facilities, and communications designed to support a commander's exercise of command and control, through all phases of the operational continuum. Also called **C4 systems.**

command controlled stocks(*)--Stocks which are placed at the disposal of a designated NATO commander in order to provide him with a flexibility with which to influence the battle logistically. "Placed at the disposal of" implies responsibility for storage, maintenance, accounting, rotation or turnover, physical security, and subsequent transportation to a particular battle area.

command destruct signal(*)--A signal used to operate intentionally the destruction signal in a missile.

command detonated mine(*)--A mine detonated by remotely controlled means.

command ejection system--See **ejection systems.**

command element--See **Marine air-ground task force.**

commander, amphibious task force--The US Navy officer designated in the initiating directive as commander of the amphibious task force. Also called **CATF.**

commander, landing force--The officer designated in the initiating directive for an amphibious operation to command the landing force. Also called **CLF.**

commander(s)--See **executing commander (nuclear weapons); exercise commander; Major NATO Commanders; national commander; national force commanders; national territorial commander; releasing commander (nuclear weapons).**

commander's concept--See **concept of operations.**

commander's estimate of the situation--A logical process of reasoning by which a commander considers all the circumstances affecting the military situation and arrives at a decision as to a course of action to be taken to accomplish the mission. A commander's estimate which considers a military situation so far in the future as to require major assumptions is called a commander's long-range estimate of the situation. See also **estimate of the situation.**

command guidance(*)--A guidance system wherein intelligence transmitted to the missile from an outside source causes the missile to traverse a directed flight path.

command heading--Heading that the controlled aircraft is directed to assume by the control station.

commanding officer of troops--On a ship that has embarked units, a designated officer, usually the senior embarking unit commander, who is responsible for the administration, discipline, and training of all embarked units. Also called **COT.**

command net(*)--A communications network which connects an echelon of command with some or all of its subordinate echelons for the purpose of command and control.

command post(*)--A unit's or subunit's headquarters where the commander and the staff perform their activities. In combat, a unit's or subunit's headquarters is often divided into echelons; the echelon in which the unit or subunit commander is located or from which such commander operates is called a command post.

command post exercise(*)--An exercise in which the forces are simulated, involving the commander, his staff, and communications within and between headquarters. See also **exercise; maneuver.**

command select ejection system--See **ejection systems.**

command speed--The speed at which the controlled aircraft is directed to fly.

command-sponsored dependent--A dependent entitled to travel to overseas commands at Government expense and endorsed by the appropriate military commander to be present in a dependent's status.

commercial items--Articles of supply readily available from established commercial distribution sources which the Department of Defense or inventory managers in the Military Services have designated to be obtained directly or indirectly from such sources.

commercial loading--See **administrative loading.**

commercial vehicle--A vehicle which has evolved in the commercial market to meet civilian requirements and which is selected from existing production lines for military use.

commission--1. To put in or make ready for service or use, as to commission an aircraft or a ship. 2. A written order giving a person rank and authority as an officer in the armed forces. 3. The rank and the authority given by such an order. See also **activate; constitute.**

commit--The process of committing one or more air interceptors or surface-to-air missiles for interception against a target track.

commodity loading(*)--A method of loading in which various types of cargoes are loaded together, such as ammunition, rations, or boxed vehicles, in order that each commodity can be discharged without disturbing the others. See also **combat loading; loading.**

commodity manager--An individual within the organization of an inventory control point or other such organization assigned management responsibility for homogeneous grouping of materiel items.

commonality--A quality that applies to materiel or systems: a. possessing like and interchangeable characteristics enabling each to be utilized, or operated and maintained, by personnel trained on the others without additional specialized training. b. having interchangeable repair parts and/or components. c. applying to consumable items interchangeably equivalent without adjustment.

common business-oriented language--A specific language by which business data-processing procedures may be precisely described in a standard form. The language is intended not only as a means for directly presenting any business program to any suitable computer for which a compiler exists, but also as a means of communicating such procedures among individuals. Commonly referred to as COBOL.

common control (artillery)--Horizontal and vertical map or chart location of points in the target area and position area, tied in with the horizontal and vertical control in use by two or more units. May be established by firing, survey, or combination of both, or by assumption. See also **control point; field control; ground control.**

common infrastructure(*)--Infrastructure essential to the training of NATO forces or to the implementation of NATO operational plans which, owing to its degree of common use or interest and its compliance with criteria laid down from time to time by the North Atlantic Council, is commonly financed by NATO members. See also **infrastructure.**

common item--1. Any item of materiel which is required for use by more than one activity. 2. Sometimes loosely used to denote any consumable item except repair parts or other technical items. 3. Any item of materiel which is procured for, owned by (Service stock), or used by any Military Department of the Department of Defense and is also required to be furnished to a recipient country under the grant-aid Military Assistance Program. 4. Readily available commercial items. 5. Items used by two or more Military Services of

similar manufacture or fabrication that may vary between the Services as to color or shape (as vehicles or clothing). 6. Any part or component which is required in the assembly of two or more complete end-items.

common servicing--That function performed by one Military Service in support of another Military Service for which reimbursement is not required from the Service receiving support. See also **servicing.**

common supplies--Those supplies common to two or more Services.

common use--Services, materials, or facilities provided by a Department of Defense agency or a Military Department on a common basis for two or more Department of Defense agencies.

common use alternatives--Systems, subsystems, devices, components, and materials, already developed or under development, which could be used to reduce the cost of new systems acquisition and support by reducing duplication of research and development effort and by limiting the addition of support base.

common user airlift service--The airlift service provided on a common basis for all DOD agencies and, as authorized, for other agencies of the US Government.

common user item(*)--An item of an interchangeable nature which is in common use by two or more nations or services of a nation. See also **interchangeability.**

common-user lift--US Transportation Command--controlled lift: The pool of strategic transportation assets either government owned or chartered that are under the operational control of Air Mobility Command, Military Sealift Command, or Military Traffic Management Command for the purpose of providing common--

user transportation to the Department of Defense across the range of military operations. These assets range from common-user organic or chartered pool of common-user assets available day-to-day to a larger pool of common--user assets phased in from other sources.

common-user military land transportation---Point-to-point land transportation service operated by a single Service for common use by two or more Services.

common user network--A system of circuits or channels allocated to furnish communication paths between switching centers to provide communication service on a common basis to all connected stations or subscribers. It is sometimes described as a General Purpose Network.

common-user ocean terminals--A military installation, part of a military installation, or a commercial facility operated under contract or arrangement by the Military Traffic Management Command which regularly provides for two or more Services terminal functions of receipt, transit storage or staging, processing, and loading and unloading of passengers or cargo aboard ships.

communication deception--Use of devices, operations, and techniques with the intent of confusing or misleading the user of a communications link or a navigation system.

communication operation instructions--See **signal operation instructions.**

communications--A method or means of conveying information of any kind from one person or place to another. See also **telecommunication.**

communications center(*)--An agency charged with the responsibility for handling and controlling communications traffic. The center normally includes message center, transmitting,

and receiving facilities. See also **telecommunications center.**

communications intelligence--Technical and intelligence information derived from foreign communications by other than the intended recipients. Also called **COMINT.**

communications intelligence data base--The aggregate of technical and intelligence information derived from the interception and analysis of foreign communications (excluding press, propaganda, and public broadcast) used in the direction and redirection of communications intelligence intercept, analysis, and reporting activities.

communications mark--An electronic indicator used for directing attention to a particular object or position of mutual interest within or between command and control systems.

communications net(*)--An organization of stations capable of direct communications on a common channel or frequency.

communications network--An organization of stations capable of intercommunications, but not necessarily on the same channel.

communications satellite(*)--An orbiting vehicle, which relays signals between communications stations. There are two types: **a. Active Communications Satellite**--A satellite that receives, regenerates, and retransmits signals between stations; **b. Passive Communications Satellite**--A satellite which reflects communications signals between stations.

communications security--The protection resulting from all measures designed to deny unauthorized persons information of value which might be derived from the possession and study of telecommunications, or to mislead unauthorized persons in their interpretation of the results of such possession and study. Also called **COMSEC.** Communications security includes: a. cryptosecurity; b. transmission security; c. emission security; and d. physical security of communications security materials and information. **a. cryptosecurity**--The component of communications security that results from the provision of technically sound cryptosystems and their proper use. **b. transmission security**--The component of communications security that results from all measures designed to protect transmissions from interception and exploitation by means other than cryptanalysis. **c. emission security**--The component of communications security that results from all measures taken to deny unauthorized persons information of value that might be derived from intercept and analysis of compromising emanations from crypto-equipment and telecommunications systems. **d. physical security**--The component of communications security that results from all physical measures necessary to safeguard classified equipment, material, and documents from access thereto or observation thereof by unauthorized persons.

communications security equipment--Equipment designed to provide security to telecommunications by converting information to a form unintelligible to an unauthorized interceptor and by reconverting such information to its original form for authorized recipients, as well as equipment designed specifically to aid in, or as an essential element of, the conversion process. Communications security equipment is crypto-equipment, cryptoancillary equipment, crypto-production equipment, and authentication equipment.

communications security material--All documents, devices, equipment, or apparatus, including cryptomaterial, used in establishing or maintaining secure communications.

communications security monitoring--The act of listening to, copying, or recording transmissions of one's own circuits (or when specially agreed, e.g., in allied exercises, those of friendly forces) to provide material for communications security analysis in order to determine the degree of security being provided to those transmissions. In particular, the purposes include providing a basis for advising commanders on the security risks resulting from their transmissions, improving the security of communications, and planning and conducting manipulative communications deception operations.

communications terminal--Terminus of a communications circuit at which data can be either entered or received; located with the originator or ultimate addressee.

communications zone(*)--Rear part of theater of operations (behind but contiguous to the combat zone) which contains the lines of communications, establishments for supply and evacuation, and other agencies required for the immediate support and maintenance of the field forces. See also **combat zone; rear area.**

community relations--The relationship between military and civilian communities.

community relations program--That command function which evaluates public attitudes, identifies the mission of a military organization with the public interest, and executes a program of action to earn public understanding and acceptance. Community relations programs are conducted at all levels of command, both in the United States and overseas, by military organizations having a community relations area of responsibility. Community relations programs include, but are not limited to, such activities as liaison and cooperation with associations and organizations and their local affiliates at all levels; armed forces participation in international, national, regional, state, and local public events; installation open houses and tours; embarkations in naval ships;

orientation tours for distinguished civilians; people-to-people and humanitarian acts; cooperation with government officials and community leaders; and encouragement of armed forces personnel and their dependents to participate in activities of local schools, churches, fraternal, social, and civic organizations, sports, and recreation programs, and other aspects of community life to the extent feasible and appropriate, regardless of where they are located.

comparative cover(*)--Coverage of the same area or object taken at different times, to show any changes in details. See also **cover.**

compartmentation--1. Establishment and management of an organization so that information about the personnel, internal organization, or activities of one component is made available to any other component only to the extent required for the performance of assigned duties. 2. Effects of relief and drainage upon avenues of approach so as to produce areas bounded on at least two sides by terrain features such as woods, ridges, or ravines that limit observation or observed fire into the area from points outside the area.

Compass direction(*)--The horizontal direction expressed as an angular distance measured clockwise from compass north.

compass north(*)--The uncorrected direction indicated by the north seeking end of a compass needle. See also **magnetic north.**

compass rose(*)--A graduated circle, usually marked in degrees, indicating directions and printed or inscribed on an appropriate medium.

compatibility(*)--Capability of two or more items or components of equipment or material to exist or function in the same system or environment without mutual interference. See also **interchangeability.**

complaint-type investigation--A counterintelligence investigation in which sabotage, espionage, treason, sedition, subversive activity, or disaffection is suspected.

completeness--Operation plan review criterion. The determination that each course of action must be complete and answer the questions: who, what, when, where, and how. See also **acceptability; completeness; feasibility; suitability.**

complete round--A term applied to an assemblage of explosive and nonexplosive components designed to perform a specific function at the time and under the conditions desired. Examples of complete rounds of ammunition are: **a. separate loading**--consisting of a primer, propelling charge, and, except for blank ammunition, a projectile and a fuze. **b. fixed or semifixed**--consisting of a primer, propelling charge, cartridge case, a projectile, and, except when solid projectiles are used, a fuze. **c. bomb**--consisting of all component parts required to drop and function the bomb once. **d. missile**--consisting of a complete warhead section and a missile body with its associated components and propellants. **e. rocket**--consisting of all components necessary to function.

component(*)--In logistics, a part or combination of parts having a specific function, which can be installed or replaced only as an entity. See also **assembly; part; subassembly.**

component (materiel)--An assembly or any combination of parts, subassemblies, and assemblies mounted together in manufacture, assembly, maintenance, or rebuild.

component search and rescue controller--The designated search and rescue representative of a component commander of a unified command who is responsible in the name of the component commander for the control of component search and rescue forces committed to joint search and rescue operations. See also **search and rescue.**

composite air photography--Air photographs made with a camera having one principal lens and two or more surrounding and oblique lenses. The several resulting photographs are corrected or transformed in printing to permit assembly as verticals with the same scale.

composite warfare commander--The officer in tactical command is normally the composite warfare commander. However the composite warfare commander concept allows an officer in tactical command to delegate tactical command to the composite warfare commander. The composite warfare commander wages combat operations to counter threats to the force and to maintain tactical sea control with assets assigned; while the officer in tactical command retains close control of power projection and strategic sea control operations.

compound helicopter(*)--A helicopter with an auxiliary propulsion system which provides thrust in excess of that which the rotor(s) alone could produce, thereby permitting increased forward speeds; wings may or may not be provided to reduce the lift required from the rotor system.

compression chamber--See **hyperbaric chamber.**

compromise--The known or suspected exposure of clandestine personnel, installations, or other assets or of classified information or material, to an unauthorized person.

compromised(*)--A term applied to classified matter, knowledge of which has, in whole or in part, passed to an unauthorized person or persons, or which has been subject to risk of such passing. See also **classified matter.**

computed air release point(*)--A computed air position where the first paratroop or cargo item is released to land on a specified impact point.

computer modeling--See **accreditation; configuration management; independent review; validation; verification.**

computer simulation--See **accreditation; configuration management; independent review; validation; verification.**

concealment(*)--The protection from observation or surveillance. See also **camouflage; cover; screen.**

concentration area(*)--1. An area, usually in the theater of operations, where troops are assembled before beginning active operations. 2. A limited area on which a volume of gunfire is placed within a limited time.

concept(*)--A notion or statement of an idea, expressing how something might be done or accomplished, that may lead to an accepted procedure.

concept of intelligence operations--A verbal or graphic statement, in broad outline, of a J2's assumptions or intent in regard to intelligence support of an operation or series of operations. The concept of intelligence operations, which complements the commander's concept of operations, is contained in the intelligence annex of operation plans. The concept of intelligence operations is designed to give an overall picture of intelligence support for joint operations. It is included primarily for additional clarity of purpose. See also **concept of operations.**

concept of logistic support--A verbal or graphic statement, in a broad outline, of how a commander intends to support and integrate with a concept of operations in an operation or campaign.

concept of operations--A verbal or graphic statement, in broad outline, of a commander's assumptions or intent in regard to an operation or series of operations. The concept of operations frequently is embodied in campaign plans and operation plans; in the latter case, particularly when the plans cover a series of connected operations to be carried out simultaneously or in succession. The concept is designed to give an overall picture of the operation. It is included primarily for additional clarity of purpose. Also called **commander's concept.**

concept plan--An operation plan in concept format. Also called **CONPLAN.** See also **concept summary; operation plan.**

concept summary--A concept of operations in Joint Operation Planning and Execution System, Volume II. Used to address Joint Strategic Capabilities Plan or other Chairman of the Joint Chiefs of Staff planning tasks in a broader sense than required by a more detailed operation plan in concept format or operation plan. See also **concept plan; operation plan.**

condensation cloud--A mist or fog of minute water droplets that temporarily surrounds the fireball following a nuclear (or atomic) detonation in a comparatively humid atmosphere. The expansion of the air in the negative phase of the blast wave from the explosion results in a lowering of the temperature, so that condensation of water vapor present in the air occurs and a cloud forms. The cloud is soon dispelled when the pressure returns to normal and the air warms up again. The phenomenon is similar to that used by physicists in the Wilson cloud chamber and is sometimes called the cloud chamber effect.

condensation trail--A visible cloud streak, usually brilliantly white in color, which trails behind a missile or other vehicle in flight under certain conditions. Also called **contrail.**

Condor--An air-to-surface guided missile which provides standoff launch capability for attack aircraft. Designated as **AGM-53.**

conducting staff--See **exercise; directing staff.**

cone of silence(*)--An inverted cone-shaped space directly over the aerial towers of some forms of radio beacons in which signals are unheard or greatly reduced in volume. See also **Z marker beacon.**

configuration management--In computer modeling and simulation, a discipline applying technical and administrative oversight and control to identify and document the functional requirements and capabilities of a model or simulation and its supporting databases, control changes to those capabilities, and document and report the changes. See also **accreditation; independent review; validation; verification.**

confirmation of information (intelligence)--An information item is said to be confirmed when it is reported for the second time, preferably by another independent source whose reliability is considered when confirming information.

confused--In air intercept, a term meaning, "Individual contacts not identifiable."

confusion agent--An individual who is dispatched by the sponsor for the primary purpose of confounding the intelligence or counterintelligence apparatus of another country rather than for the purpose of collecting and transmitting information.

confusion reflector(*)--A reflector of electromagnetic radiations used to create echoes for confusion purposes. Radar confusion reflectors include such devices as chaff, rope and corner reflectors.

connecting route(*)--A route connecting axial and/or lateral routes. See also **route.**

CONPLAN--See **concept plan.**

consecutive voyage charter--A contract by which a commercial ship is chartered by the Military Sealift Command for a series of specified voyages.

consol(*)--A long-range radio aid to navigation, the emissions of which, by means of their radio frequency modulation characteristics, enable bearings to be determined.

console(*)--A grouping of controls, indicators, and similar electronic or mechanical equipment, used to monitor readiness of, and/or control specific functions of, a system, such as missile checkout, countdown, or launch operations.

consolidated vehicle table--A summary of all vehicles loaded on a ship, listed by types, and showing the units to which they belong.

consolidation--The combining or merging of elements to perform a common or related function.

consolidation of position(*)--Organizing and strengthening a newly captured position so that it can be used against the enemy.

constitute--To provide the legal authority for the existence of a new unit of the Armed Services. The new unit is designated and listed, but it has no specific existence until it is activated. See also **activate; commission.**

consumable supplies and material--See **expendable supplies and material.**

consumer--Person or agency that uses information or intelligence produced by either its own staff or other agencies.

consumer logistics--That part of logistics concerning reception of the initial product, storage, inspection, distribution, transport, maintenance (including repair and the serviceability), and disposal of materiel, and the provision of support and services. In consequence, consumer logistics includes: materiel requirements determination, follow-on support, stock control, provision or construction of facilities (excluding any materiel element and those facilities needed to support production logistics activities), movement control, codification, reliability and defect reporting, storage, transport and handling safety standards, and related training.

consumption rate(*)--The average quantity of an item consumed or expended during a given time interval, expressed in quantities by the most appropriate unit of measurement per applicable stated basis.

contact--1. In air intercept, a term meaning, "Unit has an unevaluated target." 2. In health services, an unevaluated individual who is known to have been sufficiently near an infected individual to have been exposed to the transfer of infectious material.

contact burst preclusion--A fuzing arrangement which prevents an unwanted surface burst in the event of failure of the air burst fuze.

contact lost(*)--A target tracking term used to signify that a target believed to be still within the area of visual, sonar or radar coverage is temporarily lost but the termination of track plotting is not warranted.

contact mine(*)--A mine detonated by physical contact. See also **mine.**

contact point(*)--1. In land warfare, a point on the terrain, easily identifiable, where two or more units are required to make contact. 2. In air operations, the position at which a mission leader makes radio contact with an air control agency. See also **check point; control point; coordinating point.**

contact print(*)--A print made from a negative or a diapositive in direct contact with sensitized material.

contact reconnaissance--Locating isolated units out of contact with the main force.

contact report(*)--A report indicating any detection of the enemy.

contain(*)--To stop, hold, or surround the forces of the enemy or to cause the enemy to center activity on a given front and to prevent his withdrawing any part of his forces for use elsewhere.

container anchorage terminal(*)--A sheltered anchorage (not a port) with the appropriate facilities for the transshipment of containerized cargo from containerships to other vessels.

contamination--1. The deposit, absorption, or adsorption of radioactive material, or of biological or chemical agents on or by structures, areas, personnel, or objects. See also induced radiation; residual radiation. 2. Food and-/or water made unfit for consumption by humans or animals because of the presence of environmental chemicals, radioactive elements, bacteria or organisms, the byproduct of the growth of bacteria or organisms, the decomposing material (to include the food substance itself), or waste in the food or water.

continental United States--United States territory, including the adjacent territorial waters, located within North America between Canada and Mexico. Also called CONUS.

contingency--An emergency involving military forces caused by natural disasters, terrorists, subversives, or by required military operations. Due to the uncertainty of the situation, contingencies require plans, rapid response, and special procedures to ensure the safety and

readiness of personnel, installations, and equipment. See also contingency contracting.

contingency contracting--Contracting performed in support of a peacetime contingency in an overseas location pursuant to the policies and procedures of the Federal Acquisition Regulatory System. See also contingency.

contingency plan--A plan for major contingencies that can reasonably be anticipated in the principal geographic subareas of the command.

contingency planning--The development of plans for potential crisis involving military requirements that can reasonably be expected in an area of responsibility. Contingency planning can occur anywhere within the range of military operations and may be performed deliberately or under crisis action conditions. Contingency planning for joint operations is coordinated at the national level by assigning planning tasks and relationships among the combatant commanders and apportioning or allocating them the forces and resources available to accomplish those tasks. Commanders throughout the unified chain of command may task their staffs and subordinate commands with additional contingency planning tasks beyond those specified at the national level to provide broader contingency coverage. See also joint operation planning.

contingency planning facilities list program--A joint Defense Intelligence Agency/unified and specified command program for the production and maintenance of current target documentation of all countries of contingency planning interest to US military planners.

contingency retention stock--That portion of the quantity of an item excess to the approved force retention level for which there is no predictable demand or quantifiable requirement, and which normally would be allocated as potential DOD excess stock, except for a determination that the quantity will be retained for possible contingencies for United States forces. (Category C ships, aircraft, and other items being retained as contingency reserve are included in this stratum.)

contingent effects--The effects, both desirable and undesirable, which are in addition to the primary effects associated with a nuclear detonation.

contingent zone of fire--An area within which a designated ground unit or fire support ship may be called upon to deliver fire. See also **zone of fire.**

continue port/starboard--In air intercept, a term meaning, "Continue turning port/starboard at present rate of turn to magnetic heading indicated," (3 figures) or "Continue turning port/starboard for number of degrees indicated."

continuity of command--The degree or state of being continuous in the exercise of the authority vested in an individual of the armed forces for the direction, coordination, and control of military forces.

continuity of operations--The degree or state of being continuous in the conduct of functions, tasks, or duties necessary to accomplish a military action or mission in carrying out the national military strategy. It includes the functions and duties of the commander, as well as the supporting functions and duties performed by the staff and others acting under the authority and direction of the commander.

continuous fire(*)--1. Fire conducted at a normal rate without interruption for application of adjustment corrections or for other causes. 2. In field artillery and naval gunfire support, loading and firing at a specified rate or as rapidly as possible consistent with accuracy within the prescribed rate of fire for the weapon. Firing will continue until terminated by

the command end of mission or temporarily suspended by the command cease loading or check firing.

continuous illumination fire(*)--A type of fire in which illuminating projectiles are fired at specified time intervals to provide uninterrupted lighting on the target or specified area. See also **coordinated illumination**

continuous strip camera(*)--A camera in which the film moves continuously past a slit in the focal plane, producing a photograph in one unbroken length by virtue of the continuous forward motion of the aircraft.

continuous strip imagery(*)--Imagery of a strip of terrain in which the image remains unbroken throughout its length, along the line of flight.

contour flight--See **terrain flight.**

contour interval(*)--Difference in elevation between two adjacent contour lines.

contour line(*)--A line on a map or chart connecting points of equal elevation.

contract maintenance--The maintenance of materiel performed under contract by commercial organizations (including prime contractors) on a one-time or continuing basis, without distinction as to the level of maintenance accomplished.

contract termination--As used in Defense procurement, refers to the cessation or cancellation, in whole or in part, of work under a prime contract, or a subcontract thereunder, for the convenience of, or at the option of, the government, or due to failure of the contractor to perform in accordance with the terms of the contract (default).

contrail--See **condensation trail.**

control--1. Authority which may be less than full command exercised by a commander over part of the activities of subordinate or other organizations. 2. In mapping, charting, and photogrammetry, a collective term for a system of marks or objects on the Earth or on a map or a photograph, whose positions or elevations, or both, have been or will be determined. 3. Physical or psychological pressures exerted with the intent to assure that an agent or group will respond as directed. 4. An indicator governing the distribution and use of documents, information, or material. Such indicators are the subject of intelligence community agreement and are specifically defined in appropriate regulations. See also **administrative control; operational control; tactical control.**

control and reporting center--An element of the US Air Force tactical air control system, subordinate to the tactical air control center, from which radar control and warning operations are conducted within its area of responsibility.

control and reporting post--An element of the US Air Force tactical air control system, subordinate to the control and reporting center, which provides radar control and surveillance within its area of responsibility.

control area(*)--A controlled airspace extending upwards from a specified limit above the Earth. See also **airway; controlled airspace; control zone; terminal control area.**

control group--Personnel, ships, and craft designated to control the waterborne ship-to-shore movement.

control (intelligence)--See **control (DOD) Parts 3 and 4.**

controllable mine(*)--A mine which after laying can be controlled by the user, to the extent of making the mine safe or live, or to fire the mine. See also **mine.**

controlled airspace(*)--An airspace of defined dimensions within which air traffic control service is provided to controlled flights.

controlled dangerous air cargo(*)--Cargo which is regarded as highly dangerous and which may only be carried by cargo aircraft operating within specific safety regulations.

controlled effects nuclear weapons--Nuclear weapons designed to achieve variation in the intensity of specific effects other than normal blast effect.

controlled exercise(*)--An exercise characterized by the imposition of constraints on some or all of the participating units by planning authorities with the principal intention of provoking types of interaction. See also **free play exercise.**

controlled firing area--An area in which ordnance firing is conducted under controlled conditions so as to eliminate hazard to aircraft in flight. See also **restricted area.**

controlled forces--Military or paramilitary forces under effective and sustained political and military direction.

controlled information--1. Information conveyed to an adversary in a deception operation to evoke desired appreciations. 2. Information and indicators deliberately conveyed or denied to foreign targets to evoke invalid official estimates that result in foreign official actions advantageous to US interests and objectives.

controlled item--See **regulated item.**

controlled map--A map with precise horizontal and vertical ground control as a basis. Scale, azimuth, and elevation are accurate. See also **map.**

controlled mosaic(*)--A mosaic corrected for scale, rectified and laid to ground control to provide an accurate representation of distances and direction. See also **mosaic; rectification; uncontrolled mosaic.**

controlled passing(*)--A traffic movement procedure whereby two lines of traffic travelling in opposite directions are enabled to traverse alternately a point or section of route which can take only one line of traffic at a time.

controlled port(*)--A harbor or anchorage at which entry and departure, assignment of berths, and traffic within the harbor or anchorage are controlled by military authorities.

controlled reprisal--Not to be used. See **controlled response.**

controlled response--The selection from a wide variety of feasible options of the one which will provide the specific military response most advantageous in the circumstances.

controlled route(*)--A route, the use of which is subject to traffic or movement restrictions which may be supervised. See also **route.**

controlled shipping--Shipping that is controlled by the Military Sealift Command. Included in this category are Military Sealift Command ships (United States Naval Ships), government-owned ships operated under a general agency agreement, and commercial ships under charter to the Military Sealift Command. See also **Military Sealift Command; United States Naval Ship.**

controlled war--Not to be used. See **limited war.**

control of electromagnetic radiation--A national operational plan to minimize the use of electromagnetic radiation in the United States and its possessions and the Panama Canal Zone in the event of attack or imminent threat thereof, as an aid to the navigation of hostile aircraft, guided missiles, or other devices. See also **emission control orders.**

control point(*)--1. A position along a route of march at which men are stationed to give information and instructions for the regulation of supply or traffic. 2. A position marked by a buoy, boat, aircraft, electronic device, conspicuous terrain feature, or other identifiable object which is given a name or number and used as an aid to navigation or control of ships, boats, or aircraft. 3. In making mosaics, a point located by ground survey with which a corresponding point on a photograph is matched as a check.

control zone(*)--A controlled airspace extending upwards from the surface of the Earth to a specified upper limit. See also **airway; control area; controlled airspace; terminal control area.**

CONUS--See **continental United States.**

conventional forces--Those forces capable of conducting operations using nonnuclear weapons.

conventional mines--Land mines, other than nuclear or chemical, which are not designed to self-destruct. They are designed to be emplaced by hand or mechanical means. Conventional mines can be buried or surface laid and are normally emplaced in a pattern to aid in recording. See also **mine.**

conventional planning and execution--Worldwide Military Command and Control System command and control application software and data bases that are designed to support requirements relating to joint planning mobilization and deployment, including plan development, course of action development, execution planning, execution, movement monitoring, sustainment, and redeployment from origin to destination.

conventional weapon(*)--A weapon which is neither nuclear, biological, nor chemical.

converge--A request or command used in a call for fire to indicate that the observer or spotter desires a sheaf in which the planes of fire intersect at a point.

converged sheaf--The lateral distribution of fire of two or more pieces so that the planes of fire intersect at a given point. See also **open sheaf; parallel sheaf; sheaf; special sheaf.**

convergence--See **convergence factor; grid convergence; grid convergence factor; map convergence; true convergence.**

convergence factor(*)--The ratio of the angle between any two meridians on the chart to their actual change of longitude. See also **convergence.**

convergence zone--That region in the deep ocean where sound rays, refractured from the depths, return to the surface.

conversion angle(*)--The angle between a great circle (orthodromic) bearing and a rhumb line (loxodromic) bearing of a point, measured at a common origin.

conversion scale(*)--A scale indicating the relationship between two different units of measurement. See also **scale.**

convoy(*)--1. A number of merchant ships or naval auxiliaries, or both, usually escorted by warships and/or aircraft, or a single merchant ship or naval auxiliary under surface escort, assembled and organized for the purpose of passage together. 2. A group of vehicles organized for the purpose of control and orderly movement with or without escort protection.

See also **coastal convoy; evacuation convoy; ocean convoy.**

convoy commodore(*)--A naval officer, or master of one of the ships in a convoy, designated to command the convoy, subject to the orders of the Officer in Tactical Command. If no surface escort is present, he takes entire command.

convoy dispersal point(*)--The position at sea where a convoy breaks up, each ship proceeding independently thereafter.

convoy escort(*)--1. A naval ship(s) or aircraft in company with a convoy and responsible for its protection. 2. An escort to protect a convoy of vehicles from being scattered, destroyed, or captured. See also **escort.**

convoy joiner--See **joiner.** See also **joiner convoy; joiner section.**

convoy leaver--See **leaver.** See also **leaver convoy; leaver section.**

convoy loading(*)--The loading of troop units with their equipment and supplies in vessels of the same movement group, but not necessarily in the same vessel. See also **loading.**

convoy route(*)--The specific route assigned to each convoy by the appropriate routing authority.

convoy schedule(*)--Planned convoy sailings showing the shipping lanes, assembly and terminal areas, scheduled speed, and sailing interval.

convoy speed(*)--For ships, the speed which the convoy commodore orders the guide of the convoy to make good through the water.

convoy terminal area(*)--A geographical area, designated by the name of a port or anchorage on which it is centered, at which convoys or sections of convoys arrive and from which they will be dispersed to coastal convoy systems or as independents to their final destination.

convoy through escort(*)--Those ships of the close escort which normally remain with the convoy from its port of assembly to its port of arrival.

convoy title(*)--A combination of letters and numbers that gives the port of departure and arrival, speed, and serial number of each convoy.

cooperative logistics--The logistic support provided a foreign government/agency through its participation in the US Department of Defense logistic system with reimbursement to the United States for support provided.

cooperative logistics support arrangements--The combining term for procedural arrangements (cooperative logistics arrangements) and implementing procedures (supplementary procedures) which together support, define, or implement cooperative logistic understandings between the United States and a friendly foreign government under peacetime conditions.

coordinated draft plan(*)--A plan for which a draft plan has been coordinated with the nations involved. It may be used for future planning and exercises and may be implemented during an emergency. See also **draft plan; final plan; initial draft plan; operation plan.**

coordinated exercise--See **JCS-coordinated exercise.**

coordinated procurement assignee--The agency or military Service assigned purchase responsibility for all Department of Defense requirements of a particular Federal Supply Group/class, commodity, or item.

Coordinated Universal Time--An atomic time scale that is the basis for broadcast time signals. Coordinated Universal Time (UTC) differs from International Atomic Time by an integral number of seconds; it is maintained within 0.9 seconds of UT1 (see Universal Time) by introduction of Leap Seconds. The rotational orientation of the Earth, specified by UT1, may be obtained to an accuracy of a tenth of a second by applying the UTC to the increment DUT1 (where DUT1 = UT1 - UTC) that is broadcast in code with the time signals. Also called **UTC.** See also **International Atomic Time; Leap Second; Universal Time; ZULU Time.**

coordinates(*)--Linear or angular quantities which designate the position that a point occupies in a given reference frame or system. Also used as a general term to designate the particular kind of reference frame or system such as plane rectangular coordinates or spherical coordinates. See also **cartesian coordinates; geographic coordinates; georef; grid coordinates.**

coordinating altitude--A procedural airspace control method to separate fixed- and rotary--wing aircraft by determining an altitude below which fixed-wing aircraft will normally not fly and above which rotary-wing aircraft normally will not fly. The coordinating altitude is normally specified in the airspace control plan and may include a buffer zone for small altitude deviations.

coordinating authority--A commander or individual assigned responsibility for coordinating specific functions or activities involving forces of two or more Services or two or more forces of the same Service. The commander or individual has the authority to require consultation between the agencies involved, but does not have the authority to compel agreement. In the event that essential agreement cannot be obtained, the matter shall be referred to the appointing authority.

coordinating point(*)--Designated point at which, in all types of combat, adjacent units/formations must make contact for purposes of control and coordination.

coordinating review authority--An agency appointed by a Service or combatant command to coordinate with and assist the primary review authority in doctrine development, evaluation, and maintenance efforts. Each Service or combatant command must assign a coordinating review authority. If so authorized by the appointing Service or combatant command, coordinating review authority comments provided to designated primary review authorities should represent the position of the appointing Service or combatant command with regard to the publication under development. See also **joint doctrine; joint publication; joint tactics, techniques, and procedures; lead agent; joint test publication; primary review authority.**

coproduction--1. With respect to exports, a cooperative manufacturing arrangement (e.g., US Government or company with foreign government or company) providing for the transfer of production information that enables an eligible foreign government, international organization, or commercial producer to manufacture, in whole or in part, an item of US defense equipment. Such an arrangement would include the functions of production engineering, controlling, quality assurance, and determination of resource requirements. This is normally accomplished under the provisions of a manufacturing license agreement per the US International Traffic in Arms Regulation (ITAR) and could involve the implementation of a government-to-government memorandum of understanding. 2. A cooperative manufacturing arrangement (US Government or company with foreign government or company) providing for the transfer of production infor-

mation which enables the receiving government, international organization, or commercial producer to manufacture, in whole or in part, an item of defense equipment. The receiving party could be an eligible foreign government, international organization, or foreign producer; or the US Government or a US producer, depending on which direction the information is to flow. A typical coproduction arrangement would include the functions of production engineering, controlling, quality assurance and determining of resource requirements. It may or may not include design engineering information and critical materials production and design information.

copy negative(*)--A negative produced from an original not necessarily at the same scale.

corner reflector(*)--1. A device, normally consisting of three metallic surfaces or screens perpendicular to one another, designed to act as a radar target or marker. 2. In radar interpretation, an object which, by means of multiple reflections from smooth surfaces, produces a radar return of greater magnitude than might be expected from the physical size of the object.

corps troops(*)--Troops assigned or attached to a corps, but not a part of one of the divisions that make up the corps.

correction(*)--1. In fire control, any change in firing data to bring the mean point of impact or burst closer to the target. 2. A communication proword to indicate that an error in data has been announced and that corrected data will follow.

correlation(*)--1. In air defense, the determination that an aircraft appearing on a radarscope, on a plotting board, or visually is the same as that on which information is being received from another source. 2. In intelligence usage, the process which associates and combines data on a single entity or subject from independent observations, in order to improve the reliability or credibility of the information.

correlation factor(*)--The ratio of a ground dose rate reading to a reading taken at approximately the same time at survey height over the same point on the ground.

Corsair II--A single-seat, single turbofan engine, all-weather light attack aircraft designed to operate from aircraft carriers, armed with cannon and capable of carrying a wide assortment of nuclear and/or conventional ordnance and advanced air-to-air and air-to-ground missiles. Designated as **A-7**.

COS--See **critical occupational specialty.**

cost contract--1. A contract which provides for payment to the contractor of allowable costs, to the extent prescribed in the contract, incurred in performance of the contract. 2. A cost-reimbursement type contract under which the contractor receives no fee.

cost-plus a fixed-fee contract--A cost reimbursement type contract which provides for the payment of a fixed fee to the contractor. The fixed fee, once negotiated, does not vary with actual cost but may be adjusted as a result of any subsequent changes in the scope of work or services to be performed under the contract.

cost sharing contract--A cost reimbursement type contract under which the contractor receives no fee but is reimbursed only for an agreed portion of its allowable costs.

COT--See **commanding officer of troops.**

countdown(*)--The step-by-step process leading to initiation of missile testing, launching, and firing. It is performed in accordance with a pre-designated time schedule.

counter air--A US Air Force term for air operations conducted to attain and maintain a desired degree of air superiority by the destruction or neutralization of enemy forces. Both air offensive and air defensive actions are involved. The former range throughout enemy territory and are generally conducted at the initiative of the friendly forces. The latter are conducted near or over friendly territory and are generally reactive to the initiative of the enemy air forces. See also **antiair warfare.**

counterattack(*)--Attack by part or all of a defending force against an enemy attacking force, for such specific purposes as regaining ground lost or cutting off or destroying enemy advance units, and with the general objective of denying to the enemy the attainment of his purpose in attacking. In sustained defensive operations, it is undertaken to restore the battle position and is directed at limited objectives. See also **countermove; counteroffensive.**

counterbattery fire(*)--Fire delivered for the purpose of destroying or neutralizing indirect fire weapon systems.

counter-C2--See **command and control warfare.**

counterdeception--Efforts to negate, neutralize, diminish the effects of, or gain advantage from, a foreign deception operation. Counterdeception does not include the intelligence function of identifying foreign deception operations. See also **deception.**

counterdrug--Those active measures taken to detect, monitor, and counter the production, trafficking, and use of illegal drugs. Also called **CD.**

counterespionage--That aspect of counterintelligence designed to detect, destroy, neutralize, exploit, or prevent espionage activities through identification, penetration, manipulation, deception, and repression of individuals, groups, or organizations conducting or suspected of conducting espionage activities.

counterfire(*)--Fire intended to destroy or neutralize enemy weapons. (DOD) Includes counterbattery, counterbombardment, and countermortar fire. See also **fire.**

counterforce--The employment of strategic air and missile forces in an effort to destroy, or render impotent, selected military capabilities of an enemy force under any of the circumstances by which hostilities may be initiated.

counterguerrilla warfare(*)--Operations and activities conducted by armed forces, paramilitary forces, or nonmilitary agencies against guerrillas.

counterinsurgency--Those military, paramilitary, political, economic, psychological, and civic actions taken by a government to defeat insurgency.

counterintelligence--Information gathered and activities conducted to protect against espionage, other intelligence activities, sabotage, or assassinations conducted by or on behalf of foreign governments or elements thereof, foreign organizations, or foreign persons, or international terrorist activities. Also called **CI.** See also **counterespionage; countersabotage; countersubversion; security; security intelligence.**

countermeasures--That form of military science that, by the employment of devices and/or techniques, has as its objective the impairment of the operational effectiveness of enemy activity. See also **electronic warfare.**

countermilitary--See **counterforce.**

countermine(*)--To explode the main charge in a mine by the shock of a nearby explosion of another mine or independent explosive charge.

The explosion of the main charge may be caused either by sympathetic detonation or through the explosive train and/or firing mechanism of the mine.

countermining--1. Land mine warfare--Tactics and techniques used to detect, avoid, breach, and/or neutralize enemy mines and the use of available resources to deny the enemy the opportunity to employ mines. **2. Naval mine warfare**--The detonation of mines by nearby explosions, either accidental or deliberate.

countermove(*)--An operation undertaken in reaction to or in anticipation of a move by the enemy. See also **counterattack.**

counteroffensive--A large scale offensive undertaken by a defending force to seize the initiative from the attacking force. See also **counterattack.**

counterpreparation fire(*)--Intensive prearranged fire delivered when the imminence of the enemy attack is discovered. (DOD) It is designed to: break up enemy formations; disorganize the enemy's systems of command, communications, and observation; decrease the effectiveness of artillery preparation; and impair the enemy's offensive spirit. See also **fire.**

counterreconnaissance--All measures taken to prevent hostile observation of a force, area, or place.

countersabotage--That aspect of counterintelligence designed to detect, destroy, neutralize, or prevent sabotage activities through identification, penetration, manipulation, deception, and repression of individuals, groups, or organizations conducting or suspected of conducting sabotage activities.

countersign(*)--A secret challenge and its reply. See also **challenge; password; reply.**

countersubversion--That aspect of counterintelligence designed to detect, destroy, neutralize, or prevent subversive activities through the identification, exploitation, penetration, manipulation, deception, and repression of individuals, groups, or organizations conducting or suspected of conducting subversive activities.

counterterrorism--Offensive measures taken to prevent, deter, and respond to terrorism. Also called **CT**. See also **antiterrorism; combatting terrorism; terrorism.**

country cover diagram(*)--A small scale index, by country, depicting the existence of air photography for planning purposes only.

coup de main--A offensive operation that capitalizes on surprise and simultaneous execution of supporting operations to achieve success in one swift stroke.

coupled mode--A flight control state in which an aircraft is controlled through the automatic flight control system by signals from guidance equipment.

Courier--A delayed repeater communication satellite which had the capability of storing and relaying communications using microwave frequencies. This satellite gave a limited demonstration of instantaneous microwave communications.

courier--A messenger (usually a commissioned or warrant officer) responsible for the secure physical transmission and delivery of documents and material. Generally referred to as a command or local courier. See also **armed forces courier.**

course(*)--The intended direction of movement in the horizontal plane.

course of action--1. A plan that would accomplish, or is related to, the accomplishment of a mission. 2. The scheme adopted to accomplish a task or mission. It is a product of the Joint Operation Planning and Execution System concept development phase. The supported commander will include a recommended course of action in the commander's estimate. The recommended course of action will include the concept of operations, evaluation of supportability estimates of supporting organizations, and an integrated time-phased data base of combat, combat support, and combat service support forces and sustainment. Refinement of this data base will be contingent on the time available for course of action development. When approved, the course of action becomes the basis for the development of an operation plan or operation order. Also called **COA.**

course of action development--The phase of the Joint Operation Planning and Execution System within the crisis action planning process that provides for the development of military responses and includes, within the limits of the time allowed: establishing force and sustainment requirements with actual units; evaluating force, logistic, and transportation feasibility; identifying and resolving resource shortfalls; recommending resource allocations; and producing a course of action via a commander's estimate that contains a concept of operations, employment concept, risk assessments, prioritized courses of action, and supporting data bases. See also **course of action; crisis action planning.**

cover(*)--1. The action by land, air, or sea forces to protect by offense, defense, or threat of either or both. 2. Those measures necessary to give protection to a person, plan, operation, formation, or installation from the enemy intelligence effort and leakage of information. 3. The act of maintaining a continuous receiver watch with transmitter calibrated and available, but not necessarily available for immediate use. 4. Shelter or protection, either natural or artificial. (DOD) 5. Photographs or other recorded images which show a particular area of ground. 6. A code meaning, "Keep fighters between force/base and contact designated at distance stated from force/base" (e.g., "cover bogey twenty-seven to thirty miles").

coverage(*)--1. The ground area represented on imagery, photomaps, mosaics, maps, and other geographical presentation systems. (DOD) 2. Cover or protection, as the coverage of troops by supporting fire. 3. The extent to which intelligence information is available in respect to any specified area of interest. 4. The summation of the geographical areas and volumes of aerospace under surveillance. See also **comparative cover.**

coverage index--One of a series of overlays showing all photographic reconnaissance missions covering the map sheet to which the overlays refer. See also **covertrace (reconnaissance).**

covering fire(*)--1. Fire used to protect troops when they are within range of enemy small arms. 2. In amphibious usage, fire delivered prior to the landing to cover preparatory operations such as underwater demolition or minesweeping. See also **fire.**

covering force(*)--1. A force operating apart from the main force for the purpose of intercepting, engaging, delaying, disorganizing, and deceiving the enemy before he can attack the force covered. 2. Any body or detachment of troops which provides security for a larger force by observation, reconnaissance, attack, or defense, or by any combination of these methods. See also **force(s).**

covering force area--(NATO) The area forward of the forward edge of the battle area out to the forward positions initially assigned to the covering forces. It is here that the covering forces execute assigned tasks.

covering troops--See **covering force.**

cover (intelligence)--See **cover, Part 6.**

cover (military)--Actions to conceal actual friendly intentions, capabilities, operations, and other activities by providing a plausible, yet erroneous, explanation of the observable.

cover search(*)--In air photographic reconnaissance, the process of selection of the most suitable existing cover for a specific requirement.

covert operation--An operation that is so planned and executed as to conceal the identity of or permit plausible denial by the sponsor. A covert operation differs from a clandestine operation in that emphasis is placed on concealment of identity of sponsor rather than on concealment of the operation. See also **clandestine operation; overt operation.**

covertrace(*)--One of a series of overlays showing all air reconnaissance sorties covering the map sheet to which the overlays refer.

crab angle(*)--The angle between the aircraft track or flight line and the fore and aft axis of a vertical camera, which is in line with the longitudinal axis of the aircraft.

CRAF--See **Civil Reserve Air Fleet.**

crash locator beacon(*)--An automatic emergency radio locator beacon to help searching forces locate a crashed aircraft. See also **emergency locator beacon; personal locator beacon.**

crash position indicator--See **crash locator beacon.**

crater--The pit, depression, or cavity formed in the surface of the Earth by an explosion. It may range from saucer shaped to conical, depending largely on the depth of burst. In the case of a deep underground burst, no rupture of the surface may occur. The resulting cavity is termed a camouflet.

crater depth--The maximum depth of the crater measured from the deepest point of the pit to the original ground level.

cratering charge(*)--A charge placed at an adequate depth to produce a crater.

crater radius--The average radius of the crater measured at the level corresponding to the original surface of the ground.

CRD--See **CINC's required date.**

creeping barrage(*)--A barrage in which the fire of all units participating remains in the same relative position throughout and which advances in steps of one line at a time.

creeping mine(*)--In naval mine warfare, a buoyant mine held below the surface by a weight, usually in the form of a chain, which is free to creep along the seabed under the influence of stream or current.

crest(*)--A terrain feature of such altitude that it restricts fire or observation in an area beyond, resulting in dead space, or limiting the minimum elevation, or both.

crested--A report which indicates that engagement of a target or observation of an area is not possible because of an obstacle or intervening crest.

crisis action planning--The Joint Operation Planning and Execution System process involving the time-sensitive development of joint operation plans and orders in response to an imminent crisis. Crisis action planning follows prescribed crisis action procedures to formulate and implement an effective response within the time frame permitted by the crisis. Also called

CAP. See also **deliberate planning; Joint Operation Planning and Execution System.**

critical altitude(*)--The altitude beyond which an aircraft or airbreathing guided missile ceases to perform satisfactorily. See also **altitude.**

critical information--Specific facts about friendly intentions, capabilities, and activities vitally needed by adversaries for them to plan and act effectively so as to guarantee failure or unacceptable consequences for friendly mission accomplishment.

critical intelligence--Intelligence which is crucial and requires the immediate attention of the commander. It is required to enable the commander to make decisions that will provide a timely and appropriate response to actions by the potential/actual enemy. It includes but is not limited to the following: a. strong indications of the imminent outbreak of hostilities of any type (warning of attack); b. aggression of any nature against a friendly country; c. indications or use of nuclear-biological-chemical weapons (targets); and d. significant events within potential enemy countries that may lead to modification of nuclear strike plans.

critical item--An essential item which is in short supply or expected to be in short supply for an extended period. See also **critical supplies and materials; regulated item.**

critical item list--Prioritized list, compiled from commanders' composite critical item lists, identifying items and weapon systems that assist Services and Defense Logistics Agency in selecting systems for production surge planning.

critical joint duty assignment billet--A joint duty assignment position for which, considering the duties and responsibilities of the position, it is highly important that the assigned officer be particularly trained in, and oriented toward, joint matters. Critical billets are selected by heads of joint organizations, approved by the Secretary of Defense and documented in the Joint Duty Assignment List.

critical mass--The minimum amount of fissionable material capable of supporting a chain reaction under precisely specified conditions.

critical node--An element, position, or communications entity whose disruption or destruction immediately degrades the ability of a force to command, control, or effectively conduct combat operations. See also **target critical damage point.**

critical occupational specialty--A military occupational specialty selected from among the combat arms in the Army or equivalent military specialties in the Navy, Air Force, or Marine Corps. Equivalent military specialties are those engaged in the operational art to attain strategic goals in a theater of conflict through the design, organization, and conduct of campaigns and major operations. Critical occupational specialties are designated by the Secretary of Defense. Also called **COS.**

critical point--1. A key geographical point or position important to the success of an operation. 2. In point of time, a crisis or a turning point in an operation. 3. A selected point along a line of march used for reference in giving instructions. 4. A point where there is a change of direction or change in slope in a ridge or stream. 5. Any point along a route of march where interference with a troop movement may occur.

critical safety item--A part, assembly, installation, or production system with one or more essential characteristics that, if not conforming to the design data or quality requirements, would result in an unsafe condition that could cause loss or serious damage to the end item or

major components, loss of control, or serious injury to personnel. Also called **CSI.**

critical speed(*)--A speed or range of speeds which a ship cannot sustain due to vibration or other similar phenomena.

critical supplies and materiel(*)--Those supplies vital to the support of operations, which owing to various causes are in short supply or are expected to be in short supply. See also **critical item; regulated item.**

critical sustainability items--Items described at National Stock Number level of detail, by Federal Supply Class, as part of the Logistic Factors File, that significantly affect the commander's ability to execute an operation plan.

critical zone--The area over which a bombing plane engaged in horizontal or glide bombing must maintain straight flight so that the bomb sight can be operated properly and bombs dropped accurately.

critic report--See **critical intelligence.**

CRM--See **collection requirements management.**

crossing--In air intercept, a term meaning, "Passing from _____ to _____."

crossing area(*)--A number of adjacent crossing sites under the control of one commander.

cross-loading (personnel)--A system of loading troops so that they may be disembarked or dropped at two or more landing or drop zones, thereby achieving unit integrity upon delivery. See also **loading.**

crossover point--That range in the air warfare area at which a target ceases to be an air intercept target and becomes a surface-to-air missile target.

cross-servicing--That function performed by one Military Service in support of another Military Service for which reimbursement is required from the Service receiving support. See also **servicing.**

cross tell(*)--The transfer of information between facilities at the same operational level. See also **track telling.**

cruise missile--Guided missile, the major portion of whose flight path to its target is conducted at approximately constant velocity; depends on the dynamic reaction of air for lift and upon propulsion forces to balance drag.

cruising altitude(*)--A level determined by vertical measurement from mean sea level, maintained during a flight or portion thereof.

cruising level(*)--A level maintained during a significant portion of a flight. See also **altitude.**

cryogenic liquid--Liquefied gas at very low temperature, such as liquid oxygen, nitrogen, argon.

cryptanalysis--The steps and operations performed in converting encrypted messages into plain text without initial knowledge of the key employed in the encryption.

cryptochannel--A complete system of crypto-communications between two or more holders. The basic unit for naval cryptographic communication. It includes: a. the cryptographic aids prescribed; b. the holders thereof; c. the indicators or other means of identification; d. the area or areas in which effective; e. the special purpose, if any, for which provided; and f. pertinent notes as to distribution, usage, etc. A cryptochannel is analogous to a radio circuit.

cryptographic information--All information significantly descriptive of cryptographic techniques and processes or of cryptographic systems and equipment, or their functions and capabilities, and all cryptomaterial.

cryptologic--Of or pertaining to cryptology.

cryptology--The science which deals with hidden, disguised, or encrypted communications. It includes communications security and communications intelligence.

cryptomaterial--All material, including documents, devices, equipment, and apparatus, essential to the encryption, decryption, or authentication of telecommunications. When classified, it is designated CRYPTO and subject to special safeguards.

cryptopart(*)--A division of a message as prescribed for security reasons. The operating instructions for certain cryptosystems prescribe the number of groups which may be encrypted in the systems, using a single message indicator. Cryptoparts are identified in plain language. They are not to be confused with message parts.

cryptosecurity--See **communications security.**

cryptosystem--The associated items of cryptomaterial that are used as a unit and provide a single means of encryption and decryption. See also **cipher; code; decrypt; encipher; encrypt.**

CSAR--See **combat search and rescue.**

CSI--See **critical safety item.**

CT--See **counterterrorism.**

cultivation--A deliberate and calculated association with a person for the purpose of recruitment, obtaining information, or gaining control for these or other purposes.

culture(*)--A feature of the terrain that has been constructed by man. Included are such items as roads, buildings, and canals; boundary lines, and, in a broad sense, all names and legends on a map.

curb weight--Weight of a ground vehicle including fuel, lubricants, coolant and on-vehicle materiel, excluding cargo and operating personnel.

Current Force--The force that exists today. The Current Force represents actual force structure and/or manning available to meet present contingencies. It is the basis for operations and contingency plans and orders. See also **force; Intermediate Force Planning Level; Programmed Forces.**

current intelligence--Intelligence of all types and forms of immediate interest which is usually disseminated without the delays necessary to complete evaluation or interpretation. See also **intelligence; intelligence cycle.**

curve of pursuit(*)--The curved path described by a fighter plane making an attack on a moving target while holding the proper aiming allowance.

custody--The responsibility for the control of, transfer and movement of, and access to, weapons and components. Custody also includes the maintenance of accountability for weapons and components.

customer ship(*)--The ship in a replenishment unit that receives the transferred personnel and/or supplies.

cut-off(*)--The deliberate shutting off of a reaction engine.

cutoff attack--An attack that provides a direct vector from the interceptor's position to an intercept point with the target track.

cut-off velocity(*)--The velocity attained by a missile at the point of cutoff.

cutout--An intermediary or device used to obviate direct contact between members of a clandestine organization.

cutter(*)--In naval mine warfare, a device fitted to a sweep wire to cut or part the moorings of mines or obstructors; it may also be fitted in the mooring of a mine or obstructor to part a sweep.

cutting charge(*)--A charge which produces a cutting effect in line with its plane of symmetry.

CV--See **attack aircraft carrier.**

CVN--See **attack aircraft carrier.**

D

DA--See **direct action.**

DADCAP--Dawn and dusk combat air patrol.

daily intelligence summary--A report prepared in message form at the joint force headquarters that provides higher, lateral, and subordinate headquarters with a summary of all significant intelligence produced during the previous 24-hour period. The "as of" time for information, content, and submission time for the report will be as specified by the joint force commander. Also called **DISUM.**

daily movement summary (shipping)--A tabulation of departures and arrivals of all merchant shipping (including neutrals) arriving or departing ports during a 24-hour period.

damage--See **nuclear damage (land warfare).**

damage area(*)--In naval mine warfare, the plan area around a minesweeper inside which a mine explosion is likely to interrupt operations.

damage assessment(*)--1. The determination of the effect of attacks on targets. (DOD) 2. A determination of the effect of a compromise of classified information on national security. See also **civil damage assessment; military damage assessment.**

damage control(*)--In naval usage, measures necessary aboard ship to preserve and reestablish watertight integrity, stability, maneuverability, and offensive power; to control list and trim; to effect rapid repairs of materiel; to limit the spread of, and provide adequate protection from, fire; to limit the spread of, remove the contamination by, and provide adequate protection from, toxic agents; and to provide for care of wounded personnel. See also **area damage control; disaster control.**

damage criteria--The critical levels of various effects, such as blast pressure and thermal radiation, required to achieve specified levels of damage.

damage estimation--A preliminary appraisal of the potential effects of an attack. See also **attack assessment.**

damage radius(*)--In naval mine warfare, the average distance from a ship within which a mine containing a given weight and type of explosive must detonate if it is to inflict a specified amount of damage.

damage threat(*)--The probability that a target ship passing once through a minefield will explode one or more mines and sustain a specified amount of damage.

dan--To mark a position or a sea area with dan buoys.

dan buoy--A temporary marker buoy used during minesweeping operations to indicate boundaries of swept paths, swept areas, known hazards, and other locations or reference points.

danger--Information in a call for fire to indicate that friendly forces are within 600 to 1,500 meters of the target.

danger area--1. A specified area above, below, or within which there may be potential danger. (DOD, NATO) 2. In air traffic control, an airspace of defined dimensions within which activities dangerous to the flight of aircraft may exist at specified times. See also **closed area; prohibited area; restricted area.**

danger close(*)--In artillery and naval gunfire support, information in a call for fire to indicate that friendly forces are within 600 meters of the target.

dangerous cargo(*)--Cargo which, because of its dangerous properties, is subject to special regulations for its transport. See also **cargo.**

dangerously exposed waters(*)--The sea area adjacent to a severely threatened coastline.

danger space--That space between the weapon and the target where the trajectory does not rise 1.8 meters (the average height of a standing human). This includes the area encompassed by the beaten zone. See also **beaten zone.**

dan runner(*)--A ship running a line of dan buoys.

dart--A target towed by a jet aircraft and fired at by fighter aircraft. Used for training only.

data--Representation of facts, concepts, or instructions in a formalized manner suitable for communication, interpretation, or processing by humans or by automatic means. Any representations such as characters or analog quantities to which meaning is or might be assigned.

data base--Information that is normally structured and indexed for user access and review. Data bases may exist in the form of physical files (folders, documents, etc.) or formatted automated data processing system data files.

data block(*)--Information presented on air imagery relevant to the geographical position, altitude, attitude, and heading of the aircraft and, in certain cases, administrative information and information on the sensors employed.

data code--A number, letter, character, or any combination thereof used to represent a data element or data item. For example, the data codes "E8," "03," and "06" might be used to represent the data items of sergeant, captain,

and colonel under the data element "military personnel grade."

data element--1. A basic unit of information built on standard structures having a unique meaning and distinct units or values. 2. In electronic recordkeeping, a combination of characters or bytes referring to one separate item of information, such as name, address, or age.

data item--A subunit of descriptive information or value classified under a data element. For example, the data element "military personnel grade" contains data items such as sergeant, captain, and colonel.

data link(*)--The means of connecting one location to another for the purpose of transmitting and receiving data. See also **tactical digital information link.**

data mile--A standard unit of distance

date line--See **international date line.**

date-time group--The date and time, expressed in digits and zone suffix, the message was prepared for transmission. (Expressed as six digits followed by the zone suffix; first pair of digits denotes the date, second pair the hours, third pair the minutes.)

datum(*)--Any numerical or geometrical quantity or set of such quantities which may serve as reference or base for other quantities. Where the concept is geometric, the plural form is "datums" in contrast to the normal plural "data."

datum (antisubmarine warfare)--A datum is the last known position of a submarine, or suspected submarine, after contact has been lost.

datum dan buoy(*)--In naval mine warfare, a dan buoy intended as a geographic reference or check, which needs to be more visible and more securely moored than a normal dan buoy.

datum error (antisubmarine warfare)--An estimate of the degree of accuracy in the reported position of datum.

datum (geodetic)--A reference surface consisting of five quantities: the latitude and longitude of an initial point, the azimuth of a line from that point, and the parameters of the reference ellipsoid.

datum level(*)--A surface to which elevations, heights, or depths on a map or chart are related. See also **altitude.**

datum point(*)--Any reference point of known or assumed coordinates from which calculation or measurements may be taken. See also **pinpoint.**

datum time (antisubmarine warfare)--The datum time is the time when contact with the submarine, or suspected submarine, was lost.

day air defense fighter(*)--A fighter aircraft with equipment and weapons which enable it to engage airborne targets, but in clear weather conditions and by day only. See also **fighter.**

day of supply--See **one day's supply.**

dazzle--Temporary loss of vision or a temporary reduction in visual acuity. See also **flash blindness.**

DD--See **destroyer.**

D-day--See **times.**

D-day consumption/production differential assets--As applied to the D-to-P concept, these assets are required to compensate for the inability of the production base to meet expenditure

(consumption) requirements during the D-to-P period. See also **D-to-P concept.**

D-day materiel readiness gross capability--As applied to the D-to-P concept, this capability represents the sum of all assets on hand on D-day and the gross production capability (funded and unfunded) between D-day and P-day. When this capability equals the D-to-P Materiel Readiness Gross Requirement, requirements and capabilities are in balance. See also **D-to-P concept.**

D-day pipeline assets--As applied to the D-to-P concept, these assets represent the sum of CONUS and overseas operating and safety levels and intransit levels of supply. See also **D-to-P concept.**

DDG--See **guided missile destroyer.**

DDS--See **dry deck shelter.**

DE--See **directed energy.**

deadline--To remove a vehicle or piece of equipment from operation or use for one of the following reasons: a. is inoperative due to damage, malfunctioning, or necessary repairs. The term does not include items temporarily removed from use by reason of routine maintenance, and repairs that do not affect the combat capability of the item; b. is unsafe; and c. would be damaged by further use.

dead mine(*)--A mine which has been neutralized, sterilized or rendered safe. See also **mine.**

dead space(*)--1. An area within the maximum range of a weapon, radar, or observer, which cannot be covered by fire or observation from a particular position because of intervening obstacles, the nature of the ground, or the characteristics of the trajectory, or the limitations of the pointing capabilities of the weapons. 2. An area or zone which is within range of a radio transmitter, but in which a signal is not received. 3. The volume of space above and around a gun or guided missile system into which it cannot fire because of mechanical or electronic limitations.

dead zone--See **dead space.**

debarkation--The unloading of troops, equipment, or supplies from a ship or aircraft.

debarkation net--A specially prepared type of cargo net employed for the debarkation of troops over the side of a ship.

debarkation schedule(*)--A schedule which provides for the timely and orderly debarkation of troops and equipment and emergency supplies for the waterborne ship-to-shore movement.

Decca(*)--A radio phase-comparison system which uses a master and slave stations to establish a hyperbolic lattice and provide accurate ground position-fixing facilities.

deceased--A casualty status applicable to a person who is either known to have died, determined to have died on the basis of conclusive evidence, or declared to be dead on the basis of a presumptive finding of death. The recovery of remains is not a prerequisite to determining or declaring a person deceased. See also **casualty status.**

decentralized control(*)--In air defense, the normal mode whereby a higher echelon monitors unit actions, making direct target assignments to units only when necessary to ensure proper fire distribution or to prevent engagement of friendly aircraft. See also **centralized control.**

decentralized items--Those items of supply for which appropriate authority has prescribed local management and procurement.

deception(*)--Those measures designed to mislead the enemy by manipulation, distortion, or falsification of evidence to induce him to react in a manner prejudicial to his interests. See also **counterdeception; military deception.**

deception means--Methods, resources, and techniques that can be used to convey information to a foreign power. There are three categories of deception means: **a. physical means**--Activities and resources used to convey or deny selected information to a foreign power. (Examples: military operations, including exercises, reconnaissance, training activities, and movement of forces; the use of dummy equipment and devices; tactics; bases, logistic actions, stockpiles, and repair activity; and test and evaluation activities.) **b. technical means**--Military materiel resources and their associated operating techniques used to convey or deny selected information to a foreign power through the deliberate radiation, reradiation, alteration, absorption, or reflection of energy; the emission or suppression of chemical or biological odors; and the emission or suppression of nuclear particles. **c. administrative means**--Resources, methods, and techniques designed to convey or deny oral, pictorial, documentary, or other physical evidence to a foreign power.

decision--In an estimate of the situation, a clear and concise statement of the line of action intended to be followed by the commander as the one most favorable to the successful accomplishment of the mission.

decision altitude(*)--An altitude related to the highest elevation in the touchdown zone, specified for a glide slope approach, at which a missed-approach procedure must be initiated if the required visual reference has not been established. See also **decision height.**

decision height(*)--A height above the highest elevation in the touchdown zone, specified for a glide slope approach, at which a missed-approach procedure must be initiated if the required visual reference has not been established. See also **decision altitude.**

decisive engagement--In land and naval warfare, an engagement in which a unit is considered fully committed and cannot maneuver or extricate itself. In the absence of outside assistance, the action must be fought to a conclusion and either won or lost with the forces at hand.

deck alert--See **ground alert.**

declared speed(*)--The continuous speed which a master declares his ship can maintain on a forthcoming voyage under moderate weather conditions having due regard to her present condition.

declassification--The determination that in the interests of national security, classified information no longer requires any degree of protection against unauthorized disclosure, coupled with removal or cancellation of the classification designation.

declassify(*)--To cancel the security classification of an item of classified matter. See also **downgrade.**

declination(*)--The angular distance to a body on the celestial sphere measured north or south through 90 degrees from the celestial equator along the hour circle of the body. Comparable to latitude on the terrestrial sphere. See also **magnetic declination; magnetic variation.**

decompression chamber--See **hypobaric chamber.**

decompression sickness--A syndrome, including bends, chokes, neurological disturbances, and collapse, resulting from exposure to reduced ambient pressure and caused by gas bubbles in the tissues, fluids, and blood vessels.

decontamination(*)--The process of making any person, object, or area safe by absorbing, destroying, neutralizing, making harmless, or removing, chemical or biological agents, or by removing radioactive material clinging to or around it.

decontamination station(*)--A building or location suitably equipped and organized where personnel and materiel are cleansed of chemical, biological or radiological contaminants.

decoy(*)--An imitation in any sense of a person, object, or phenomenon which is intended to deceive enemy surveillance devices or mislead enemy evaluation. Also called **dummy.**

decoy ship(*)--A ship camouflaged as a noncombatant ship with its armament and other fighting equipment hidden and with special provisions for unmasking its weapons quickly. Also called **Q-ship.**

decrypt--To convert encrypted text into its equivalent plain text by means of a cryptosystem. (This does not include solution by cryptanalysis.) Note: The term decrypt covers the meanings of decipher and decode. See also **cryptosystem.**

deep fording--The ability of a self-propelled gun or ground vehicle equipped with built-in waterproofing and/or a special waterproofing kit, to negotiate a water obstacle with its wheels or tracks in contact with the ground. See also **flotation; shallow fording.**

deep fording capability(*)--The characteristic of self-propelled gun or ground vehicle equipped with built-in waterproofing and/or a special waterproofing kit, to negotiate a water obstacle with its wheels or tracks in contact with the ground. See also **shallow fording capability.**

deep minefield(*)--An antisubmarine minefield which is safe for surface ships to cross. See also **minefield.**

deep supporting fire(*)--Fire directed on objectives not in the immediate vicinity of our forces, for neutralizing and destroying enemy reserves and weapons, and interfering with enemy command, supply, communications, and observations. See also **close supporting fire; direct supporting fire; supporting fire.**

de facto boundary(*)--An international or administrative boundary whose existence and legality is not recognized, but which is a practical division between separate national and provincial administering authorities.

DEFCON--See **defense readiness conditions.**

defector--National of a country who has escaped from the control of such country or who, being outside such jurisdiction and control, is unwilling to return thereto and is of special value to another country.

defense area(*)--For any particular command, the area extending from the forward edge of the battle area to its rear boundary. It is here that the decisive defensive battle is fought.

defense classification--See **security classification.**

defense emergency--An emergency condition that exists when: a. a major attack is made upon US forces overseas, or on allied forces in any theater and is confirmed by either the commander of a command established by the Secretary of Defense or higher authority; or b. an overt attack of any type is made upon the United States and is confirmed either by the commander of a command established by the Secretary of Defense or higher authority.

defense in depth(*)--The siting of mutually supporting defense positions designed to absorb and progressively weaken attack, prevent initial observations of the whole position by the enemy, and to allow the commander to maneuver his reserve.

defense intelligence production--The integration, evaluation, analysis, and interpretation of information from single or multiple sources into finished intelligence for known or anticipated military and related national security consumer requirements.

Defense Planning Guidance--This document, issued by the Secretary of Defense, provides firm guidance in the form of goals, priorities, and objectives, including fiscal constraints, for the development of the Program Objective Memorandums by the Military Departments and Defense agencies. Also called **DPG.**

defense readiness conditions--A uniform system of progressive alert postures for use between the Chairman of the Joint Chiefs of Staff and the commanders of unified and specified commands and for use by the Services. Defense readiness conditions are graduated to match situations of varying military severity (status of alert). Defense readiness conditions are identified by the short title DEFCON (5), (4), (3), (2), and (1), as appropriate. Also called **DEFCON.**

Defense Switched Network--The worldwide interbase telecommunications system that provides end-to-end, common-user, and dedicated voice service for the Department of Defense with the capability of incorporating data and other traffic. It is composed of several subsystems, to include: the Automatic Voice Network; Oahu Telephone System; Defense Commercial Telecommunications Network, etc. It replaced the Automatic Voice Network as the principal long-haul, nonsecure voice communications network within the Defense Communications System. Also called **DSN.** See also **Automatic Voice Network.**

defensive coastal area(*)--A part of a coastal area and of the air, land, and water area adjacent to the coastline within which defense operations may involve land, sea, and air forces.

defensive minefield(*)--1. In naval mine warfare, a minefield laid in international waters or international straits with the declared intention of controlling shipping in defense of sea communications. (DOD) 2. In land mine warfare, a minefield laid in accordance with an established plan to prevent a penetration between positions and to strengthen the defense of the positions themselves. See also **minefield.**

defensive sea area--A sea area, usually including the approaches to and the waters of important ports, harbors, bays, or sounds, for the control and protection of shipping; for the safeguarding of defense installations bordering on waters of the areas; and for provision of other security measures required within the specified areas. It does not extend seaward beyond the territorial waters. See also **maritime control area.**

defensive zone--A belt of terrain, generally parallel to the front, which includes two or more organized, or partially organized, battle positions.

defilade(*)--1. Protection from hostile observation and fire provided by an obstacle such as a hill, ridge, or bank. 2. A vertical distance by which a position is concealed from enemy observation. 3. To shield from enemy fire or observation by using natural or artificial obstacles.

defoliant operation(*)--The employment of defoliating agents on vegetated areas in support of military operations.

defoliating agent(*)--A chemical which causes trees, shrubs, and other plants to shed their leaves prematurely.

degaussing--The process whereby a ship's magnetic field is reduced by the use of electromagnetic coils, permanent magnets, or other means.

degree of risk--As specified by the commander, the risk to which friendly forces may be subjected from the effects of the detonation of a nuclear weapon used in the attack of a close-in enemy target; acceptable degrees of risk under differing tactical conditions are emergency, moderate, and negligible. See also **emergency risk (nuclear); moderate risk (nuclear); negligible risk (nuclear).**

de jure boundary(*)--An international or administrative boundary whose existence and legality is recognized.

delay--A report from the firing ship to the observer or the spotter to inform that the ship will be unable to provide the requested fire immediately. It will normally be followed by the estimated duration of the delay.

delaying action--See **delaying operation.**

delaying operation(*)--An operation in which a force under pressure trades space for time by slowing down the enemy's momentum and inflicting maximum damage on the enemy without, in principle, becoming decisively engaged.

delay release sinker(*)--A sinker which holds a moored mine on the sea-bed for a pre-determined time after laying.

delegation of authority(*)--The action by which a commander assigns part of his or her authority commensurate with the assigned task to a subordinate commander. While ultimate responsibility cannot be relinquished, delegation of authority carries with it the imposition of a measure of responsibility. The extent of the authority delegated must be clearly stated.

deliberate attack(*)--A type of offensive action characterized by preplanned coordinated employment of firepower and maneuver to close with and destroy or capture the enemy.

deliberate breaching(*)--The creation of a lane through a minefield or a clear route through a barrier or fortification, which is systematically planned and carried out.

deliberate crossing(*)--A crossing of an inland water obstacle that requires extensive planning and detailed preparations.   See also **hasty crossing.**

deliberate defense(*)--A defense normally organized when out of contact with the enemy or when contact with the enemy is not imminent and time for organization is available.   It normally includes an extensive fortified zone incorporating pillboxes, forts, and communications systems. See also **hasty defense.**

deliberate planning--The Joint Operation Planning and Execution System process involving the development of joint operation plans for contingencies identified in joint strategic planning documents.   Conducted principally in peacetime, deliberate planning is accomplished in prescribed cycles that complement other Department of Defense planning cycles and in accordance with the formally established Joint Strategic Planning System.   See also **crisis action planning; Joint Operation Planning and Execution System.**

delivering ship(*)--The ship in a replenishment unit that delivers the rig(s).

delivery error(*)--The inaccuracy associated with a given weapon system resulting in a dispersion of shots about the aiming point.   See also **circular error probable; deviation; dispersion; dispersion error; horizontal error.**

delivery forecasts--1.   Periodic estimates of contract production deliveries used as a measure of the effectiveness of production and supply availability scheduling and as a guide to corrective actions to resolve procurement or production bottlenecks. 2. Estimates of deliveries under obligation against procurement from appropriated or other funds.

delivery requirements--The stipulation which requires that an item of material must be delivered in the total quantity required by the date required and, when appropriate, overpacked as required.

demilitarized zone(*)--A defined area in which the stationing, or concentrating of military forces, or the retention or establishment of military installations of any description, is prohibited.

demolition(*)--The destruction of structures, facilities, or material by use of fire, water, explosives, mechanical, or other means.

demolition belt--A selected land area sown with explosive charges, mines, and other available obstacles to deny use of the land to enemy operations, and as a protection to friendly troops. **a. Primary.** A continuous series of obstacles across the whole front, selected by the division or higher commander. The preparation of such a belt is normally a priority engineer task. **b. Subsidiary.** A supplement to the primary belt to give depth in front or behind or to protect the flanks.

demolition chamber(*)--Space intentionally provided in a structure for the emplacement of explosive charges.

demolition firing party(*)--The party at the site which is technically responsible for the demolition. See also **demolition guard.**

demolition guard(*)--A local force positioned to ensure that a target is not captured by an enemy before orders are given for its demolition and before the demolition has been successfully fired. The commander of the demolition guard is responsible for the tactical control of all troops at the demolition site, including the demolition firing party. The commander of the demolition guard is responsible for transmitting the order to fire to the demolition firing party.

demolition kit(*)--The demolition tool kit complete with explosives. See also **demolition tool kit.**

demolition target(*)--A target of known military interest identified for possible future demolition. See also **charged demolition target; reserved demolition target; uncharged demolition target.**

demolition tool kit(*)--The tools, materials and accessories of a nonexplosive nature necessary for preparing demolition charges. See also **demolition kit.**

demonstration(*)--An attack or show of force on a front where a decision is not sought, made with the aim of deceiving the enemy. See also **amphibious demonstration; diversion; diversionary attack.**

denial measure(*)--An action to hinder or deny the enemy the use of space, personnel, or facilities. It may include destruction, removal, contamination, or erection of obstructions.

denied area--An area under enemy or unfriendly control in which friendly forces cannot expect to operate successfully within existing operational constraints and force capabilities.

density(*)--The average number of mines per meter of minefield front.

density altitude(*)--An atmospheric density expressed in terms of the altitude which corresponds with that density in the standard atmosphere.

departmental intelligence--Intelligence that any department or agency of the Federal Government requires to execute its own mission.

Department of Defense Intelligence Information System--The aggregation of DOD personnel, procedures, equipment, computer programs, and supporting communications that support the timely and comprehensive preparation and presentation of intelligence and intelligence information to military commanders and national-level decisionmakers. Also called **DODIIS.**

Department of the Air Force--The executive part of the Department of the Air Force at the seat of government and all field headquarters, forces, reserve components, installations, activities, and functions under the control or supervision of the Secretary of the Air Force. See also **Military Department.**

Department of the Army--The executive part of the Department of the Army at the seat of government and all field headquarters, forces, reserve components, installations, activities, and functions under the control or supervision of the Secretary of the Army. See also **Military Department.**

Department of the Navy--The executive part of the Department of the Navy at the seat of government; the headquarters, US Marine Corps; the entire operating forces of the United States Navy, including naval aviation, and of the US Marine Corps, including the reserve components of such forces; all field activities, headquarters, forces, bases, installations, activities, and functions under the control or supervision of the Secretary of the Navy; and the US Coast Guard when operating as a part of the Navy pursuant to law. See also **Military Department.**

departure airfield--An airfield on which troops and/or materiel are enplaned for flight. See also **airfield.**

departure end(*)--That end of a runway nearest to the direction in which initial departure is made.

departure point(*)--1. A navigational check point used by aircraft as a marker for setting course. 2. In amphibious operations, an air control point at the seaward end of the helicopter approach lane system from which helicopter waves are dispatched along the selected helicopter approach lane to the initial point.

deployability posture--The state or stage of a unit's preparedness for deployment to participate in a military operation, defined in five levels as follows: **a. normal deployability posture.** The unit is conducting normal activities. Commanders are monitoring the situation in any area of tension and reviewing plans. No visible overt actions are being taken to increase deployability posture. Units not at home station report their scheduled closure time at home station or the time required to return to home station if ordered to return before scheduled time and desired mode of transportation are available. **b. increased deployability posture.** The unit is relieved from commitments not pertaining to the mission. Personnel are recalled from training areas, pass, and leave, as required, to meet the deployment schedule. Preparation for deployment of equipment and supplies is initiated. Pre-deployment personnel actions are completed. Essential equipment and supplies located at continental United States (CONUS) or overseas installations are identified. **c. advanced deployability posture.** All essential personnel, mobility equipment, and accompanying supplies are checked, packed, rigged for deployment,

and positioned with deploying unit. The unit remains at home station. Movement requirements are confirmed. Airlift, sealift, and intra-CONUS transportation resources are identified, and initial movement schedules are completed by the Transportation Component Commands. **d. marshaled deployability posture.** The first increment of deploying personnel, mobility equipment, and accompanying supplies is marshaled at designated ports of embarkation but not loaded. Sufficient aircraft or sealift assets are positioned at, or en route to, the port of embarkation, either to load the first increment or to sustain a flow, as required by the plan or directive being considered for execution. Supporting airlift control elements (ALCEs), stage crews (if required), and support personnel adequate to sustain the airlift flow at onload, en route, and offload locations will be positioned, as required. **e. loaded deployability posture.** All first increment equipment and accompanying supplies are loaded aboard ships and prepared for departure to the designated objective area. Personnel are prepared for loading on minimum notice. Follow-on increments of cargo and personnel are en route or available to meet projected ship loading schedules. Sufficient airlift is positioned and loaded at the port of embarkation to move the first increment or to initiate and sustain a flow, as required by the plan or directive being considered for execution. Supporting ALCEs, stage aircrews (if required), and support personnel adequate to sustain the airlift flow at onload, en route, and offload locations are positioned, as required.

deployed nuclear weapons--1. When used in connection with the transfer of weapons between the Department of Energy and the Department of Defense, this term describes those weapons transferred to and in the custody of the Department of Defense. 2. Those nuclear weapons specifically authorized by the Joint Chiefs of Staff to be transferred to the custody of the storage facilities, carrying or delivery units of the armed forces.

deployment--1. In naval usage, the change from a cruising approach or contact disposition to a disposition for battle. 2. The movement of forces within areas of operation. 3. The positioning of forces into a formation for battle. 4. The relocation of forces and materiel to desired areas of operations. Deployment encompasses all activities from origin or home station through destination, specifically including intra-continental United States, intertheater, and intratheater movement legs, staging, and holding areas.

deployment data base--The JOPES (Joint Operation Planning and Execution System) data base containing the necessary information on forces, materiel, and filler and replacement personnel movement requirements to support execution. The data base reflects information contained in the refined time-phased force and deployment data from the deliberate planning process or developed during the various phases of the crisis action planning process, and the movement schedules or tables developed by the transportation component commands to support the deployment of required forces, personnel, and materiel. See also **time-phased force and deployment data.**

deployment diagram--In the assault phase of an amphibious operation, a diagram showing the formation in which the boat group proceeds from the rendezvous area to the line of departure and the method of deployment into the landing formation.

deployment planning--Encompasses all activities from origin or home station through destination, specifically including intra-continental United States, intertheater, and intratheater movement legs, staging areas, and holding areas.

deployment preparation order--An order issued by competent authority to move forces or prepare forces for movement (e.g., increase deployability posture of units).

depot--**1. supply**--An activity for the receipt, classification, storage, accounting, issue, maintenance, procurement, manufacture, assembly, research, salvage, or disposal of material. **2. personnel**--An activity for the reception, processing, training, assignment, and forwarding of personnel replacements.

depot maintenance--That maintenance performed on materiel requiring major overhaul or a complete rebuild of parts, assemblies, subassemblies, and end-items, including the manufacture of parts, modifications, testing, and reclamation as required. Depot maintenance serves to support lower categories of maintenance by providing technical assistance and performing that maintenance beyond their responsibility. Depot maintenance provides stocks of serviceable equipment by using more extensive facilities for repair than are available in lower level maintenance activities.

depression angle--See **angle of depression.**

depth(*)--In maritime/hydrographic use, the vertical distance from the plane of the hydrographic datum to the bed of the sea, lake, or river.

depth contour(*)--A line connecting points of equal depth below the hydrographic datum. Also called **bathymetric contour or depth curve.**

depth curve--See **depth contour.**

description of target(*)--In artillery and naval gunfire support, an element in the call for fire in which the observer or spotter describes the installation, personnel, equipment, or activity to be taken under fire.

descriptive name(*)--Written indication on maps and charts, used to specify the nature of a feature (natural or artificial) shown by a general symbol.

desired appreciation--See **appreciations.**

desired effects--The damage or casualties to the enemy or material which a commander desires to achieve from a nuclear weapon detonation. Damage effects on material are classified as light, moderate, or severe. Casualty effects on personnel may be immediate, prompt, or delayed.

desired ground zero(*)--The point on the surface of the Earth at, or vertically below or above, the center of a planned nuclear detonation. Also called **DGZ.** See also **actual ground zero; ground zero.**

despatch route--See **dispatch route.**

destroy (beam)--In air intercept, a code meaning, "The interceptor will be vectored to a standard beam attack for interception and destruction of the target."

destroy (cutoff)--In air intercept, a code meaning, "Intercept and destroy. Command vectors will produce a cutoff attack."

destroyed--A condition of a target so damaged that it cannot function as intended nor be restored to a usable condition. In the case of a building, all vertical supports and spanning members are damaged to such an extent that nothing is salvageable. In the case of bridges, all spans must have dropped and all piers must require replacement.

destroyer--A high-speed warship designed to operate offensively with strike forces, with hunter-killer groups, and in support of amphibious assault operations. Destroyers also operate defensively to screen support forces and con-voys against submarine, air, and surface threats. Normal armament consists of 3-inch and 5-inch dual-purpose guns and various antisubmarine warfare weapons. Designated as **DD.**

destroy (frontal)--In air intercept, a command meaning, "The interceptor will be vectored to a standard frontal attack for interception and destruction of the target."

destroy (stern)--In air intercept a command meaning, "The interceptor will be vectored to a standard stern attack for interception and destruction of the target."

destruction--A type of adjustment for destroying a given target.

destruction area--An area in which it is planned to destroy or defeat the enemy airborne threat. The area may be further subdivided into air intercept, missile (long-, medium-, and short--range), or antiaircraft gun zones.

destruction fire--Fire delivered for the sole purpose of destroying material objects. See also **fire.**

destruction fire mission(*)--In artillery, fire delivered for the purpose of destroying a point target. See also **fire.**

destruction radius(*)--In mine warfare, the maximum distance from an exploding charge of stated size and type at which a mine will be destroyed by sympathetic detonation of the main charge, with a stated probability of destruction, regardless of orientation.

detachment(*)--1. A part of a unit separated from its main organization for duty elsewhere. 2. A temporary military or naval unit formed from other units or parts of units.

detailed photographic report(*)--A comprehensive, analytical, intelligence report written as a result of the interpretation of photography usually covering a single subject, a target, target complex, and of a detailed nature.

detained--See **missing.**

detainee--A term used to refer to any person captured or otherwise detained by an armed force.

detainee collecting point--A facility or other location where detainees are assembled for subsequent movement to a detainee processing station.

detainee processing station--A facility or other location where detainees are administratively processed and provided custodial care pending disposition and subsequent release, transfer, or movement to a prisoner-of-war or civilian internee camp.

detecting circuit(*)--The part of a mine firing circuit which responds to the influence of a target.

detection--1. In tactical operations, the perception of an object of possible military interest but unconfirmed by recognition. 2. In surveillance, the determination and transmission by a surveillance system that an event has occurred. 3. In arms control, the first step in the process of ascertaining the occurrence of a violation of an arms-control agreement.

deterioration limit(*)--A limit placed on a particular product characteristic to define the minimum acceptable quality requirement for the product to retain its NATO code number.

deterrence--The prevention from action by fear of the consequences. Deterrence is a state of mind brought about by the existence of a credible threat of unacceptable counteraction.

deterrent options--A course of action, developed on the best economic, diplomatic, political, and military judgment, designed to dissuade an adversary from a current course of action or contemplated operations. (In constructing an operation plan, a range of options should be presented to effect deterrence. Each option requiring deployment of forces should be a separate force module.)

detonating cord(*)--A flexible fabric tube containing a high explosive designed to transmit the detonation wave.

detonator(*)--A device containing a sensitive explosive intended to produce a detonation wave.

detour(*)--Deviation from those parts of a route where movement has become difficult or impossible to ensure continuity of movement to the destination. The modified part of the route is known as a "detour."

deviation(*)--1. The distance by which a point of impact or burst misses the target. See also **circular error probable; delivery error; dispersion error; horizontal error.** 2. The angular difference between magnetic and compass headings.

DEW--See **directed-energy warfare.**

diapositive(*)--A positive photograph on a transparent medium. See also **transparency.**

died of wounds received in action--A casualty category applicable to a hostile casualty, other than the victim of a terrorist activity, who dies of wounds or other injuries received in action after having reached a medical treatment facility. Also called **DWRIA.** See also **casualty category.**

differential ballistic wind(*)--In bombing, a hypothetical wind equal to the difference in velocity between the ballistic wind and the actual wind at a release altitude.

diffraction loading(*)--The total force which is exerted on the sides of a structure by the advancing shock front of a nuclear explosion.

dip(*)--In naval mine warfare, the amount by which a moored mine is carried beneath its set depth by a current or tidal stream acting on the mine casing and mooring.

diplomatic authorization(*)--Authority for overflight or landing obtained at government-to--government level through diplomatic channels.

dip needle circuit(*)--In naval mine warfare, a mechanism which responds to a change in the magnitude of the vertical component of the total magnetic field.

direct action--Short-duration strikes and other small-scale offensive actions by special operations forces to seize, destroy, capture, recover, or inflict damage on designated personnel or materiel. In the conduct of these operations, special operations forces may employ raid, ambush, or direct assault tactics; emplace mines and other munitions; conduct standoff attacks by fire from air, ground, or maritime platforms; provide terminal guidance for precision-guided munitions; and conduct independent sabotage. Also called **DA.**

direct action fuze--See **impact action fuze; proximity fuze; self-destroying fuse; time fuze.**

direct air support center--A subordinate operational component of a tactical air control system designed for control and direction of close air support and other tactical air support operations, and normally collocated with fire-support coordination elements. See also **direct air support center (airborne).**

direct air support center (airborne)--An airborne aircraft equipped with the necessary staff personnel, communications, and operations facilities to function as a direct air support center. See also **direct air support center.**

directed energy--An umbrella term covering technologies that relate to the production of a beam of concentrated electromagnetic energy or atomic or subatomic particles. Also called **DE.** See also **directed-energy device; directed-energy weapon.**

directed-energy device--A system using directed energy primarily for a purpose other than as a weapon. Directed-energy devices may produce effects that could allow the device to be used as a weapon against certain threats, for example, laser rangefinders and designators used against sensors that are sensitive to light. See also **directed energy; directed-energy weapon.**

directed-energy protective measures--That division of directed-energy warfare involving actions taken to protect friendly equipment, facilities, and personnel to ensure friendly effective uses of the electromagnetic spectrum that are threatened by hostile directed-energy weapons and devices.

directed-energy warfare--Military action involving the use of directed-energy weapons, devices, and countermeasures to either cause direct damage or destruction of enemy equipment, facilities, and personnel, or to determine, exploit, reduce, or prevent hostile use of the electromagnetic spectrum through damage, destruction, and disruption. It also includes actions taken to protect friendly equipment, facilities, and personnel and retain friendly use of the electromagnetic spectrum. Also called **DEW.** See also **directed energy; directed-energy device; directed-energy weapon; electromagnetic spectrum; electronic warfare.**

directed-energy weapon--A system using directed energy primarily as a direct means to damage or destroy enemy equipment, facilities, and personnel. See also **directed energy; directed-energy device.**

directed exercise--See **JCS-directed exercise.**

direct exchange--A supply method of issuing serviceable materiel in exchange for unserviceable materiel on an item-for-item basis.

direct fire--Gunfire delivered on a target, using the target itself as a point of aim for either the gun or the director.

direct illumination(*)--Illumination provided by direct light from pyrotechnics or searchlights.

directing staff--See **exercise directing staff.**

direction(*)--1. In artillery and naval gunfire support, a term used by a spotter/observer in a call for fire to indicate the bearing of the spotting line. 2. See intelligence cycle.

directional gyro indicator--An azimuth gyro with a direct display and means for setting the datum to a specified compass heading.

direction finding--A procedure for obtaining bearings of radio frequency emitters by using a highly directional antenna and a display unit on an intercept receiver or ancillary equipment.

direction of attack--A specific direction or route that the main attack or center of mass of the unit will follow. The unit is restricted, required to attack as indicated, and is not normally allowed to bypass the enemy. The direction of attack is used primarily in counterattacks or to ensure that supporting attacks make maximal contribution to the main attack.

directive(*)--1. A military communication in which policy is established or a specific action

is ordered. 2. A plan issued with a view to putting it into effect when so directed, or in the event that a stated 'contingency arises. 3. Broadly speaking, any communication which initiates or governs action, conduct, or procedure.

direct laying--Laying in which the sights of weapons are aligned directly on the target. See also **lay.**

direct support--A mission requiring a force to support another specific force and authorizing it to answer directly the supported force's request for assistance. See also **close support; general support; mutual support; support.**

direct support artillery(*)--Artillery whose primary task is to provide fire requested by the supported unit.

direct supporting fire(*)--Fire delivered in support of part of a force, as opposed to general supporting fire which is delivered in support of the force as a whole. See also **close supporting fire; deep supporting fire; supporting fire.**

disaffected person--A person who is alienated or estranged from those in authority or lacks loyalty to the government; a state of mind.

disarmament--The reduction of a military establishment to some level set by international agreement. See also **arms control; arms control agreement; arms control measure.**

disarmed mine(*)--A mine for which the arming procedure has been reversed, rendering the mine inoperative. It is safe to handle and transport and can be rearmed by simple action.

disaster control--Measures taken before, during, or after hostile action or natural or manmade disasters to reduce the probability of damage, minimize its effects, and initiate recovery. See also **area damage control; damage control.**

discriminating circuit(*)--That part of the operating circuit of a sea mine which distinguishes between the response of the detecting circuit to the passage of a ship and the response to other disturbances (e.g., influence sweep, countermining, etc.)

disembarkation schedule--See **debarkation schedule.**

disengagement--In arms control, a general term for proposals that would result in the geographic separation of opposing nonindigenous forces without directly affecting indigenous military forces.

dispatch route(*)--In road traffic, a roadway over which full control, both as to priorities of use and the regulation of movement of traffic in time and space is exercised. Movement authorization is required for its use, even by a single vehicle. See also **route.**

dispensary--See **clinic.**

dispenser(*)--In air armament, a container or device which is used to carry and release submunitions. See also **cluster bomb unit.**

dispersal--Relocation of forces for the purpose of increasing survivability. See also **dispersion.**

dispersal airfield--An airfield, military or civil, to which aircraft might move before H-hour on either a temporary duty or permanent change of station basis and be able to conduct operations. See also **airfield.**

dispersed movement pattern(*)--A pattern for ship-to-shore movement which provides additional separation of landing craft both laterally and in depth. This pattern is used when nuclear weapon threat is a factor.

dispersed site(*)--A site selected to reduce concentration and vulnerability by its separation from other military targets or a recognized threat area.

dispersion(*)--1. A scattered pattern of hits around the mean point of impact of bombs and projectiles dropped or fired under identical conditions. 2. In antiaircraft gunnery, the scattering of shots in range and deflection about the mean point of explosion. 3. The spreading or separating of troops, materiel, establishments, or activities which are usually concentrated in limited areas to reduce vulnerability. 4. In chemical and biological operations, the dissemination of agents in liquid or aerosol form. 5. In airdrop operations, the scatter of personnel and/or cargo on the drop zone. 6. In naval control of shipping, the reberthing of a ship in the periphery of the port area or in the vicinity of the port for its own protection in order to minimize the risk of damage from attack. See also **convoy dispersal point.** See also **circular error probable; delivery error; deviation; dispersion error; horizontal error.**

dispersion error(*)--The distance from the point of impact or burst of a round to the mean point of impact or burst.

dispersion pattern(*)--The distribution of a series of rounds fired from one weapon or group of weapons under conditions as nearly identical as possible the points of bursts or impact being dispersed about a point called the mean point of impact.

displaced person(*)--A civilian who is involuntarily outside the national boundaries of his or her country. See also **evacuee; evacuees; refugee; refugees.**

displacement--In air intercept, separation between target and interceptor tracks established to position the interceptor in such a manner as to provide sufficient maneuvering and acquisition space.

disposition(*)--1. Distribution of the elements of a command within an area, usually the exact location of each unit headquarters and the deployment of the forces subordinate to it. 2. A prescribed arrangement of the stations to be occupied by the several formations and single ships of a fleet, or major subdivisions of a fleet, for any purpose, such as cruising, approach, maintaining contact, or battle. 3. A prescribed arrangement of all the tactical units composing a flight or group of aircraft. See also **deployment; dispersion.** (DOD) 4. The removal of a patient from a medical treatment facility by reason of return to duty, transfer to another treatment facility, death, or other termination of medical case.

disruptive pattern(*)--In surveillance, an arrangement of suitably colored irregular shapes which, when applied to the surface of an object, is intended to enhance its camouflage.

dissemination--See **intelligence cycle.**

distance--1. The space between adjacent individual ships or boats measured in any direction between foremasts. 2. The space between adjacent men, animals, vehicles, or units in a formation measured from front to rear. 3. The space between known reference points or a ground observer and a target, measured in meters (artillery), in yards (naval gunfire), or in units specified by the observer. See also **interval.**

distributed fire(*)--Fire so dispersed as to engage most effectively an area target. See also **fire.**

distribution--1. The arrangement of troops for any purpose, such as a battle, march, or maneuver. 2. A planned pattern of projectiles about a point. 3. A planned spread of fire to cover a desired frontage or depth. 4. An

official delivery of anything, such as orders or supplies. 5. That functional phase of military logistics that embraces the act of dispensing materiel, facilities, and services. 6. The process of assigning military personnel to activities, units, or billets.

distribution point(*)--A point at which supplies and/or ammunition, obtained from supporting supply points by a division or other unit, are broken down for distribution to subordinate units. Distribution points usually carry no stocks; items drawn are issued completely as soon as possible.

distribution system--That complex of facilities, installations, methods, and procedures designed to receive, store, maintain, distribute, and control the flow of military materiel between the point of receipt into the military system and the point of issue to using activities and units.

DISUM--See **daily intelligence summary.**

ditching--Controlled landing of a distressed aircraft on water.

diversion--1. The act of drawing the attention and forces of an enemy from the point of the principal operation; an attack, alarm, or feint that diverts attention. 2. A change made in a prescribed route for operational or tactical reasons. A diversion order will not constitute a change of destination. 3. A rerouting of cargo or passengers to a new transshipment point or destination or on a different mode of transportation prior to arrival at ultimate destination. 4. In naval mine warfare, a route or channel bypassing a dangerous area. A diversion may connect one channel to another or it may branch from a channel and rejoin it on the other side of the danger. See also **demonstration.**

diversionary attack(*)--An attack wherein a force attacks, or threatens to attack, a target other than the main target for the purpose of drawing enemy defenses away from the main effort. See also **demonstration.**

diversionary landing--An operation in which troops are actually landed for the purpose of diverting enemy reaction away from the main landing.

divert--1. "Proceed to divert field or carrier as specified." 2. To change the target, mission, or destination of an airborne flight.

diving chamber--See **hyperbaric chamber.**

division(*)--1. A tactical unit/formation as follows: a. A major administrative and tactical unit/formation which combines in itself the necessary arms and services required for sustained combat, larger than a regiment/brigade and smaller than a corps. b. A number of naval vessels of similar type grouped together for operational and administrative command, or a tactical unit of a naval aircraft squadron, consisting of two or more sections. c. An air division is an air combat organization normally consisting of two or more wings with appropriate service units. The combat wings of an air division will normally contain similar type units. 2. An organizational part of a headquarters that handles military matters of a particular nature, such as personnel, intelligence, plans, and training, or supply and evacuation. 3. A number of personnel of a ship's complement grouped together for operational and administrative command.

division artillery--Artillery that is permanently an integral part of a division. For tactical purposes, all artillery placed under the command of a division commander is considered division artillery.

division slice--See **slice.**

dock landing ship--A naval ship designed to transport and launch loaded amphibious craft and vehicles with their crews and embarked personnel in amphibious assault, and to render limited docking and repair service to small ships and craft; and one that is capable of acting as a control ship in an amphibious assault. Designated LSD.

doctrine--Fundamental principles by which the military forces or elements thereof guide their actions in support of national objectives. It is authoritative but requires judgment in application. See also **combined doctrine; joint doctrine; multi-Service doctrine.**

DOD Internal Audit Organizations--The Army Audit Agency; Naval Audit Service; Air Force Audit Agency; and the Office of the Assistant Inspector General for Auditing, Office of the Inspector General, DOD.

dolly--Airborne data link equipment.

dome--See **spray dome.**

domestic air traffic--Air traffic within the continental United States.

domestic emergencies--Emergencies affecting the public welfare and occurring within the 50 states, District of Columbia, Commonwealth of Puerto Rico, US possessions and territories, or any political subdivision thereof, as a result of enemy attack, insurrection, civil disturbance, earthquake, fire, flood, or other public disasters or equivalent emergencies that endanger life and property or disrupt the usual process of government. The term domestic emergency includes any or all of the emergency conditions defined below: **a. civil defense emergency**--A domestic emergency disaster situation resulting from devastation created by an enemy attack and requiring emergency operations during and following that attack. It may be proclaimed by appropriate authority in anticipation of an attack. **b. civil disturbances**--Riots, acts of violence, insurrections, unlawful obstructions or assemblages, or other disorders prejudicial to public law and order. The term civil disturbance includes all domestic conditions requiring or likely to require the use of Federal Armed Forces pursuant to the provisions of Chapter 15 of Title 10, United States Code. **c. major disaster**--Any flood, fire, hurricane, tornado, earthquake, or other catastrophe which, in the determination of the President, is or threatens to be of sufficient severity and magnitude to warrant disaster assistance by the Federal Government under Public Law 606, 91st Congress (42 United States Code 58) to supplement the efforts and available resources of State and local governments in alleviating the damage, hardship, or suffering caused thereby. **d. natural disaster**--All domestic emergencies except those created as a result of enemy attack or civil disturbance.

domestic intelligence--Intelligence relating to activities or conditions within the United States that threaten internal security and that might require the employment of troops; and intelligence relating to activities of individuals or agencies potentially or actually dangerous to the security of the Department of Defense.

dominant user concept--The concept that the Service which is the principal consumer will have the responsibility for performance of a support workload for all using Services.

doppler effect(*)--The phenomenon evidenced by the change in the observed frequency of a sound or radio wave caused by a time rate of change in the effective length of the path of travel between the source and the point of observation.

doppler radar--A radar system that differentiates between fixed and moving targets by detecting the apparent change in frequency of the reflect-ed wave due to motion of target or the observer.

dormant(*)--In mine warfare, the state of a mine during which a time delay feature in a mine prevents it from being actuated.

dose rate contour line(*)--A line on a map, diagram, or overlay joining all points at which the radiation dose rate at a given time is the same.

dosimetry(*)--The measurement of radiation doses. It applies to both the devices used (dosimeters) and to the techniques.

double agent--Agent in contact with two opposing intelligence services, only one of which is aware of the double contact or quasi-intelligence services.

double flow route(*)--A route of at least two lanes allowing two columns of vehicles to proceed simultaneously, either in the same direction or in opposite directions. See also **limited access route; single flow route.**

doubtful(*)--In artillery and naval gunfire support, a term used by an observer or spotter to indicate that he was unable to determine the difference in range between the target and a round or rounds.

down(*)--In artillery and naval gunfire support: 1. A term used in a call for fire to indicate that the target is at a lower altitude than the reference point used in identifying the target. 2. A correction used by an observer/spotter in time fire to indicate that a decrease in height of burst is desired.

downgrade--To determine that classified information requires, in the interests of national security, a lower degree of protection against unauthorized disclosure than currently provided, coupled with a changing of the classification designation to reflect such lower degree.

down lock(*)--A device for locking retractable landing gear in the down or extended position.

DPG--See **Defense Planning Guidance.**

drafter--A person who actually composes the message for release by the originator or the releasing officer. See also **originator.**

draft plan(*)--A plan for which a draft plan has been coordinated and agreed with the other military headquarters and is ready for coordination with the nations involved, that is those nations who would be required to take national actions to support the plan. It may be used for future planning and exercises and may form the basis for an operation order to be implemented in time of emergency. See also **initial draft plan; coordinated draft plan; final plan; operation plan.**

drag--Force of aerodynamic resistance caused by the violent currents behind the shock front.

drag loading(*)--The force on an object or structure due to transient winds accompanying the passage of a blast wave. The drag pressure is the product of the dynamic pressure and the drag coefficient which is dependent upon the shape (or geometry) of the structure or object. See also **dynamic pressure.**

Dragon--A manportable medium antitank weapon, consisting of a round (missile and launcher) and a tracker that provides antitank/assault fire of infantry platoon level for employment against tanks and hard point targets such as emplaced weapons or fortifications. Designated as **M-47.**

drift(*)--In ballistics, a shift in projectile direction due to gyroscopic action which results from

gravitational and atmospherically induced torques on the spinning projectile.

drift angle(*)--The angle measured in degrees between the heading of an aircraft or ship and the track made good.

drifting mine(*)--A buoyant or neutrally buoyant mine, free to move under the influence of waves, wind, current or tide.

drill mine(*)--An inert filled mine or mine-like body, used in loading, laying or discharge practice and trials. See also **mine.**

drone--A land, sea, or air vehicle that is remotely or automatically controlled. See also **remotely piloted vehicle; unmanned aerial vehicle.**

droop stop(*)--A device to limit downward vertical motion of helicopter rotor blades upon rotor shutdown.

drop(*)--In artillery and naval gunfire support, a correction used by an observer/spotter to indicate that a decrease in range along a spotting line is desired.

drop altitude(*)--The altitude above mean sea level at which airdrop is executed. See also **altitude; drop height.**

drop height(*)--The vertical distance between the drop zone and the aircraft. See also **altitude; drop altitude.**

dropmaster--1. An individual qualified to prepare, perform acceptance inspection, load, lash, and eject material for airdrop. 2. An aircrew member who, during parachute operations, will relay any required information between pilot and jumpmaster. See also **air dispatcher (cargo).**

drop message(*)--A message dropped from an aircraft to a ground or surface unit.

drop track--In air intercept, the unit having reporting responsibility for a particular track is dropping that track and will no longer report it. Other units holding an interest in that track may continue to report it.

drop zone(*)--A specific area upon which airborne troops, equipment, or supplies are airdropped.

dry deck shelter--A shelter module that attaches to the hull of a specially configured submarine to provide the submarine with the capability to launch and recover special operations personnel, vehicles and equipment while submerged. The dry deck shelter provides a working environment at one atmosphere for the special operations element during transit and has structural integrity to the collapse depth of the host submarine. Also called **DDS.**

DSN--See **Defense Switched Network.**

D-to-P assets required on D-day--As applied to the D-to-P concept, this asset requirement represents those stocks that must be physically available on D-day to meet initial allowance requirements, to fill the wartime pipeline between the producers and users (even if P-day and D-day occur simultaneously), and to provide any required D-to-P consumption/production differential stockage. The D-to-P assets required on D-day are also represented as the difference between the D-to-P Materiel Readiness Gross Requirements and the cumulative sum of all production deliveries during the D-to-P period. See also **D-to-P concept.**

D-to-P concept--A logistic planning concept by which the gross materiel readiness requirement in support of approved forces at planned wartime rates for conflicts of indefinite duration will be satisfied by a balanced mix of assets on hand on D-day and assets to be gained from production through P-day when the planned rate of production deliveries to the users equals the planned wartime rate of expenditure (consumption). See also **D-day consumption/production differential assets; D-day pipeline assets; D-to-P assets required on D-day; D-to-P materiel readiness gross requirement.**

D-to-P materiel readiness gross requirement--As applied to the D-to-P concept, the gross requirement for all supplies/materiel needed to meet all initial pipeline and anticipated expenditure (consumption) requirements between D-day and P-day. Includes initial allowances, CONUS and overseas operating and safety levels, intransit levels of supply, and the cumulative sum of all items expended (consumed) during the D-to-P period. See also **D-to-P concept.**

dual agent--One who is simultaneously and independently employed by two or more intelligence agencies covering targets for both.

dual-capable forces--Forces capable of employing dual-capable weapons.

dual capable unit(*)--A nuclear certified delivery unit capable of executing both conventional and nuclear missions.

dual-firing circuit(*)--An assembly comprising two independent firing systems, both electric or both non-electric, so that the firing of either system will detonate all charges. See also **combination firing circuit.**

dual (multi)-capable weapons--1. Weapons, weapon systems, or vehicles capable of selective equipage with different types or mixes of armament or firepower. 2. Sometimes restricted to weapons capable of handling either nuclear or non-nuclear munitions.

dual (multi)-purpose weapons--Weapons which possess the capability for effective application in two or more basically different military functions and/or levels of conflict.

dual-purpose weapon--A weapon designed for delivering effective fire against air or surface targets.

dual warning phenomenology--Deriving warning information from two systems observing separate physical phenomena (e.g., radar/infrared or visible light/X-ray) associated with the same events to attain high credibility while being less susceptible to false reports or spoofing.

duck--In air intercept, a code meaning, "Trouble headed your way" (usually followed by "bogey, salvos," etc.).

dud(*)--Explosive munition which has not been armed as intended or which has failed to explode after being armed. See also **absolute dud; dwarf dud; flare dud; nuclear dud.**

dud probability--The expected percentage of failures in a given number of firings.

due in--Quantities of materiel scheduled to be received from vendors, repair facilities, assembly operation, interdepot transfers, and other sources.

dummy--See **decoy.**

dummy message(*)--A message sent for some purpose other than its content, which may consist of dummy groups or may have a meaningless text.

dummy minefield(*)--In naval mine warfare, a minefield containing no live mines and presenting only a psychological threat.

dummy run--Any simulated firing practice, particularly a dive bombing approach made without release of a bomb. Also called **dry run.**

dump(*)--A temporary storage area, usually in the open, for bombs, ammunition, equipment, or supplies.

duplicate negative(*)--A negative reproduced from negative or diapositive.

durable materiel--See **non-expendable supplies and materiel.**

Duster (antiaircraft weapon)--A self-propelled, twin 40-mm antiaircraft weapon for use against low-flying aircraft. Designated as **M42.**

DUSTWUN--See **duty status - whereabouts unknown.**

duty status - whereabouts unknown--A transitory casualty status, applicable only to military personnel, that is used when the responsible commander suspects the member may be a casualty whose absence is involuntary, but does not feel sufficient evidence currently exists to make a definite determination of missing or deceased. Also called **DUSTWUN.** See also **casualty status.**

dwarf dud--A nuclear weapon that, when launched at or emplaced on a target, fails to provide a yield within a reasonable range of that which could be anticipated with normal operation of the weapon. This constitutes a dud only in a relative sense.

dwell at/on(*)--In artillery and naval gunfire support, this term is used when fire is to continue for an indefinite period at specified time or on a particular target or targets.

DWRIA--See **died of wounds received in action.**

dynamic pressure(*)--Pressure resulting from some medium in motion, such as the air following the shock front of a blast wave.

E

E-2--See **Hawkeye.**

EA-6A--See **Intruder.**

EA-6B--See **Prowler.**

EAD--See **earliest arrival date.**

Eagle--A twin-engine supersonic, turbofan, all--weather tactical fighter, capable of employing a variety of air-launched weapons in the air-to--air role. The Eagle is air refuelable and is also capable of long-range air superiority missions. Designated as **F-15.**

EALT--See **earliest anticipated launch time.**

earliest anticipated launch time--The earliest time expected for a special operations tactical element and its supporting platform to depart the staging or marshalling area together en route to the operations area. Also called **EALT.**

earliest arrival date--A day, relative to C-day, that is specified by a planner as the earliest date when a unit, a resupply shipment, or replacement personnel can be accepted at a port of debarkation during a deployment. Used with the latest arrival data, it defines a delivery window for transportation planning. Also called **EAD.** See also **latest arrival date.**

Early Spring--An antireconnaissance satellite weapon system.

early time--See **span of detonation (atomic demolition munition employment).**

early warning(*)--Early notification of the launch or approach of unknown weapons or weapon carriers. See also **attack assessment; tactical warning.**

earmarking of stocks(*)--The arrangement whereby nations agree, normally in peacetime, to identify a proportion of selected items of their war reserve stocks to be called for by specified NATO commanders.

earthing(*)--The process of making a satisfactory electrical connection between the structure, including the metal skin, of an object or vehicle, and the mass of the Earth, to ensure a common potential with the Earth. See also **bonding; grounding.**

ease turn--Decrease rate of turn.

echelon(*)--1. A subdivision of a headquarters, i.e., forward echelon, rear echelon. 2. Separate level of command. As compared to a regiment, a division is a higher echelon, a battalion is a lower echelon. 3. A fraction of a command in the direction of depth to which a principal combat mission is assigned; i.e., attack echelon, support echelon, reserve echelon. 4. A formation in which its subdivisions are placed one behind another, with a lateral and even spacing to the same side.

echeloned displacement(*)--Movement of a unit from one position to another without discontinuing performance of its primary function. (DOD) Normally, the unit divides into two functional elements (base and advance); and, while the base continues to operate, the advance element displaces to a new site where, after it becomes operational, it is joined by the base element.

economic action--The planned use of economic measures designed to influence the policies or actions of another state, e.g., to impair the war-making potential of a hostile power or to generate economic stability within a friendly power.

economic mobilization(*)--The process of preparing for and carrying out such changes in the organization and functioning of the national economy as are necessary to provide for the most effective use of resources in a national emergency.

economic order quantity--That quantity derived from a mathematical technique used to determine the optimum (lowest) total variable costs required to order and hold inventory.

economic potential(*)--The total capacity of a nation to produce goods and services.

economic potential for war--That share of the total economic capacity of a nation that can be used for the purposes of war.

economic retention stock--That portion of the quantity of an item excess of the approved force retention level that has been determined will be more economical to retain for future peacetime issue in lieu of replacement of future issues by procurement. To warrant economic retention, items must have a reasonably predictable demand rate.

economic warfare--Aggressive use of economic means to achieve national objectives.

economy of force theater--Theater in which risk is accepted to allow a concentration of sufficient force in the theater of focus. See also **theater of focus.**

EEFI--See **essential elements of friendly information.**

EEI--See **essential elements of information.**

effective damage--That damage necessary to render a target element inoperative, unserviceable, nonproductive, or uninhabitable.

ejection(*)--1. Escape from an aircraft by means of an independently propelled seat or capsule. 2. In air armament, the process of forcefully separating an aircraft store from an aircraft to achieve satisfactory separation.

ejection systems(*)--a. **command ejection system**--A system in which the pilot of an aircraft or the occupant of the other ejection seat(s) initiates ejection resulting in the automatic ejection of all occupants; b. **command select ejection system**--A system permitting the optional transfer from one crew station to another of the control of a command ejection system for automatic ejection of all occupants; c. **independent ejection system**--An ejection system which operates independently of other ejection systems installed in one aircraft; d. **sequenced ejection system**--A system which ejects the aircraft crew in sequence to ensure a safe minimum total time of escape without collision.

electrode sweep(*)--In naval mine warfare, a magnetic cable sweep in which the water forms part of the electric circuit.

electro-explosive device--An explosive or pyrotechnic component that initiates an explosive, burning, electrical, or mechanical train and is activated by the application of electrical energy.

electromagnetic compatibility--The ability of systems, equipment, and devices that utilize the electromagnetic spectrum to operate in their intended operational environments without suffering unacceptable degradation or causing unintentional degradation because of electromagnetic radiation or response. It involves the application of sound electromagnetic spectrum management; system, equipment, and device design configuration that ensures interference--free operation; and clear concepts and doctrines that maximize operational effectiveness. Also called **EMC.** See also **electromagnetic spec-**

trum; electronic warfare; spectrum management.

electromagnetic deception--The deliberate radiation, reradiation, alteration, suppression, absorption, denial, enhancement, or reflection of electromagnetic energy in a manner intended to convey misleading information to an enemy or to enemy electromagnetic-dependent weapons, thereby degrading or neutralizing the enemy's combat capability. Among the types of electromagnetic deception are: **a. manipulative electromagnetic deception**--Actions to eliminate revealing, or convey misleading, electromagnetic telltale indicators that may be used by hostile forces. **b. simulative electromagnetic deception**--Actions to simulate friendly, notional, or actual capabilities to mislead hostile forces. **c. imitative electromagnetic deception**--The introduction of electromagnetic energy into enemy systems that imitates enemy emissions. See also **electronic warfare.**

electromagnetic environment--The resulting product of the power and time distribution, in various frequency ranges, of the radiated or conducted electromagnetic emission levels that may be encountered by a military force, system, or platform when performing its assigned mission in its intended operational environment. It is the sum of electromagnetic interference; electromagnetic pulse; hazards of electromagnetic radiation to personnel, ordnance, and volatile materials; and natural phenomena effects of lightning and p-static. Also called **EME.**

electromagnetic environmental effects--The impact of the electromagnetic environment upon the operational capability of military forces, equipment, systems, and platforms. It encompasses all electromagnetic disciplines, including electromagnetic compatibility/electromagnetic interference; electromagnetic vulnerability; electromagnetic pulse; electronic protection, hazards of electromagnetic radiation to personnel, ordnance, and volatile materials; and natural phenomena effects of lightning and p-static. Also called **E3.**

electromagnetic hardening--Action taken to protect personnel, facilities, and/or equipment by filtering, attenuating, grounding, bonding, and/or shielding against undesirable effects of electromagnetic energy. See also **electronic warfare.**

electromagnetic interference--Any electromagnetic disturbance that interrupts, obstructs, or otherwise degrades or limits the effective performance of electronics/electrical equipment. It can be induced intentionally, as in some forms of electronic warfare, or unintentionally, as a result of spurious emissions and responses, intermodulation products, and the like. Also called **EMI.**

electromagnetic intrusion--The intentional insertion of electromagnetic energy into transmission paths in any manner, with the objective of deceiving operators or of causing confusion. See also **electronic warfare.**

electromagnetic jamming--The deliberate radiation, reradiation, or reflection of electromagnetic energy for the purpose of preventing or reducing an enemy's effective use of the electromagnetic spectrum, and with the intent of degrading or neutralizing the enemy's combat capability. See also **electromagnetic spectrum; electronic warfare; spectrum management.**

electromagnetic pulse--The electromagnetic radiation from a nuclear explosion caused by Compton-recoil electrons and photoelectrons from photons scattered in the materials of the nuclear device or in a surrounding medium. The resulting electric and magnetic fields may couple with electrical/electronic systems to produce damaging current and voltage surges.

May also be caused by nonnuclear means. Also called **EMP.**

electromagnetic radiation--Radiation made up of oscillating electric and magnetic fields and propagated with the speed of light. Includes gamma radiation, X-rays, ultraviolet, visible, and infrared radiation, and radar and radio waves.

electromagnetic radiation hazards--Hazards caused by a transmitter/antenna installation that generates electromagnetic radiation in the vicinity of ordnance, personnel, or fueling operations in excess of established safe levels or increases the existing levels to a hazardous level; or a personnel, fueling, or ordnance installation located in an area that is illuminated by electromagnetic radiation at a level that is hazardous to the planned operations or occupancy. These hazards will exist when an electromagnetic field of sufficient intensity is generated to: a. Induce or otherwise couple currents and/or voltages of magnitudes large enough to initiate electroexplosive devices or other sensitive explosive components of weapon systems, ordnance, or explosive devices. b. Cause harmful or injurious effects to humans and wildlife. c. Create sparks having sufficient magnitude to ignite flammable mixtures of materials that must be handled in the affected area. Also called **EMR Hazards, RADHAZ, HERO.**

electromagnetic spectrum--The range of frequencies of electromagnetic radiation from zero to infinity. It is divided into 26 alphabetically designated bands. See also **electronic warfare.**

electromagnetic vulnerability--The characteristics of a system that cause it to suffer a definite degradation (incapability to perform the designated mission) as a result of having been subjected to a certain level of electromagnetic environmental effects. Also called **EMV.**

electronic attack--See **electronic warfare.**

electronic imagery dissemination--The transmission of imagery or imagery products by any electronic means. This includes the following four categories: **a. primary imagery dissemination system**--The equipment and procedures used in the electronic transmission and receipt of un-exploited original or near-original quality imagery in near-real time. **b. primary imagery dissemination**--The electronic transmission and receipt of unexploited original or near-original quality imagery in near-real time through a primary imagery dissemination system. **c. secondary imagery dissemination system**--The equipment and procedures used in the electronic transmission and receipt of exploited non--original quality imagery and imagery products in other than real or near-real time. **d. secondary imagery dissemination**--The electronic transmission and receipt of exploited nonoriginal quality imagery and imagery products in other than real or near-real time through a secondary imagery dissemination system.

electronic line of sight--The path traversed by electromagnetic waves that is not subject to reflection or refraction by the atmosphere.

electronic masking(*)--The controlled radiation of electromagnetic energy on friendly frequencies in a manner to protect the emissions of friendly communications and electronic systems against enemy electronic warfare support measures/signals intelligence, without significantly degrading the operation of friendly systems.

electronic probing--Intentional radiation designed to be introduced into the devices or systems of potential enemies for the purpose of learning the functions and operational capabilities of the devices or systems.

electronic protection--See **electronic warfare.**

electronic reconnaissance--The detection, identification, evaluation, and location of foreign electromagnetic radiations emanating from other than nuclear detonations or radioactive sources.

electronics intelligence--Technical and geolocation intelligence derived from foreign non-communications electromagnetic radiations emanating from other than nuclear detonations or radioactive sources. Also called **ELINT**. See also **electronic warfare; intelligence; signals intelligence; telemetry intelligence.**

electronics security--The protection resulting from all measures designed to deny unauthorized persons information of value that might be derived from their interception and study of noncommunications electromagnetic radiations, e.g., radar.

electronic warfare--Any military action involving the use of electromagnetic and directed energy to control the electromagnetic spectrum or to attack the enemy. Also called **EW**. The three major subdivisions within electronic warfare are: electronic attack, electronic protection, and electronic warfare support. **a. electronic attack**--That division of electronic warfare involving the use of electromagnetic or directed energy to attack personnel, facilities, or equipment with the intent of degrading, neutralizing, or destroying enemy combat capability. Also called **EA**. EA includes: 1) actions taken to prevent or reduce an enemy's effective use of the electromagnetic spectrum, such as jamming and electromagnetic deception, and 2) employment of weapons that use either electromagnetic or directed energy as their primary destructive mechanism (lasers, radio frequency weapons, particle beams). **b. electronic protection**--That division of electronic warfare involving actions taken to protect personnel, facilities, and equipment from any effects of friendly or enemy employment of electronic warfare that degrade, neutralize, or destroy

friendly combat capability. Also called **EP. c. electronic warfare support**--That division of electronic warfare involving actions tasked by, or under direct control of, an operational commander to search for, intercept, identify, and locate sources of intentional and unintentional radiated electromagnetic energy for the purpose of immediate threat recognition. Thus, electronic warfare support provides information required for immediate decisions involving electronic warfare operations and other tactical actions such as threat avoidance, targeting, and homing. Also called **ES**. Electronic warfare support data can be used to produce signals intelligence (SIGINT), both communications intelligence (COMINT), and electronics intelligence (ELINT). See also **command and control warfare; communications intelligence; directed energy; directed-energy device; directed-energy warfare; directed-energy weapon; electromagnetic compatibility; electromagnetic deception; electromagnetic hardening; electromagnetic jamming; electromagnetic spectrum; electronics intelligence; frequency deconfliction; signals intelligence; spectrum management; suppression of enemy air defenses.**

electronic warfare support--See **electronic warfare.**

electro-optical intelligence--Intelligence other than signals intelligence derived from the optical monitoring of the electromagnetic spectrum from ultraviolet (0.01 micrometers) through far infrared (1,000 micrometers). Also called **ELECTRO-OPTINT**. See also **intelligence; laser intelligence.**

electro-optics(*)--The technology associated with those components, devices and systems which are designed to interact between the electromagnetic (optical) and the electric (electronic) state.

ELECTRO-OPTINT--See **electro-optical intelligence.**

element of resupply--See **early resupply; improvised (early) resupply; initial (early) resupply; planned resupply; resupply of Europe.**

elements of national power--All the means that are available for employment in the pursuit of national objectives.

elevation(*)--The vertical distance of a point or level on or affixed to the surface of the Earth measured from mean sea level. See also **altitude; height.**

elevation guidance(*)--Information which will enable the pilot or autopilot of an aircraft to follow the required glide path.

elevation tint--See **hypsometric tinting.**

elevator--In air intercept, a code meaning, "Take altitude indicated" (in thousands of feet, calling off each 5,000-foot increment passed through).

elicitation (intelligence)--Acquisition of information from a person or group in a manner that does not disclose the intent of the interview or conversation. A technique of human source intelligence collection, generally overt, unless the collector is other than he purports to be.

eligible traffic--Traffic for which movement requirements are submitted and space is assigned or allocated. Such traffic must meet eligibility requirements specified in Joint Travel Regulations for the Uniformed Services and publications of the Department of Defense and Military Departments governing eligibility for land, sea, and air transportation, and be in accordance with the guidance of the Joint Chiefs of Staff.

ELINT--See **electronics intelligence.**

embarkation(*)--The process of putting personnel and/or vehicles and their associated stores and equipment into ships and/or aircraft. See also **loading.**

embarkation and tonnage table--A consolidated table showing personnel and cargo, by troop or naval units, loaded aboard a combat-loaded ship.

embarkation area(*)--An area ashore, including a group of embarkation points, in which final preparations for embarkation are completed and through which assigned personnel and loads for craft and ships are called forward to embark. See also **mounting area.**

embarkation element (unit) (group)--A temporary administrative formation of personnel with supplies and equipment embarking or to be embarked (combat loaded) aboard the ships of one transport element (unit) (group). It is dissolved upon completion of the embarkation. An embarkation element normally consists of two or more embarkation teams: a unit, of two or more elements; and a group, of two or more units. See also **embarkation organization; embarkation team.**

embarkation officer--An officer on the staff of units of the landing force who advises the commander thereof on matters pertaining to embarkation planning and loading ships. See also **combat cargo officer.**

embarkation order(*)--An order specifying dates, times, routes, loading diagrams, and methods of movement to shipside or aircraft for troops and their equipment. See also **movement table.**

embarkation organization--A temporary administrative formation of personnel with supplies and equipment embarking or to be embarked (combat loaded) aboard amphibious shipping. See also **embarkation element (unit) (group); embarkation team.**

embarkation phase--In amphibious operations, the phase which encompasses the orderly assembly of personnel and materiel and their subsequent loading aboard ships and/or aircraft in a sequence designed to meet the requirements of the landing force concept of operations ashore.

embarkation plans--The plans prepared by the landing force and appropriate subordinate commanders containing instructions and information concerning the organization for embarkation, assignment to shipping, supplies and equipment to be embarked, location and assignment of embarkation areas, control and communication arrangements, movement schedules and embarkation sequence, and additional pertinent instructions relating to the embarkation of the landing force.

embarkation team--A temporary administrative formation of all personnel with supplies and equipment embarking or to be embarked (combat loaded) aboard one ship. See also **embarkation element (unit) (group); embarkation organization.**

EMC--See **electromagnetic compatibility.**

EMCON--See **emission control.**

emergency anchorage(*)--An anchorage, which may have a limited defense organization, for naval vessels, mobile support units, auxiliaries, or merchant ships. See also **advanced fleet anchorage; assembly anchorage; holding anchorage; working anchorage.**

emergency burial(*)--A burial, usually on the battlefield, when conditions do not permit either evacuation for interment in a cemetery or burial according to national or international legal regulations. See also **burial.**

emergency locator beacon(*)--A generic term for all radio beacons used for emergency locating purposes. See also **personal locator beacon; crash locator beacon.**

emergency priority--A category of immediate mission request that takes precedence over all other priorities, e.g., an enemy breakthrough. See also **immediate mission request; priority of immediate mission requests.**

emergency relocation site--A site located where practicable outside a prime target area to which all or portions of a civilian or military headquarters may be moved. As a minimum, it is manned to provide for the maintenance of the facility, communications, and data base. It should be capable of rapid activation, of supporting the initial requirements of the relocated headquarters for a predetermined period, and of expansion to meet wartime requirements of the relocated headquarters.

emergency resupply--A resupply mission that occurs based on a predetermined set of circumstances and time interval should radio contact not be established or, once established, is lost between a special operations tactical element and its base. See also **automatic resupply; on-call resupply.**

emergency risk (nuclear)--A degree of risk where anticipated effects may cause some temporary shock, casualties and may significantly reduce the unit's combat efficiency. See also **degree of risk; moderate risk (nuclear); negligible risk (nuclear).**

emergency scramble--In air intercept, a code meaning, "Carrier(s) addressed immediately launch all available fighter aircraft as combat air patrol." If all available are not required, numerals and/or type may be added.

emergency substitute(*)--A product which may be used, in an emergency only, in place of another product, but only on the advice of technically qualified personnel of the nation using the product, who will specify the limitations. See also **acceptable product; standardized product.**

emission control--The selective and controlled use of electromagnetic, acoustic, or other emitters to optimize command and control capabilities while minimizing, for operations security: a. detection by enemy sensors; b. minimize mutual interference among friendly systems; and/or c. execute a military deception plan. Also called **EMCON.** See also **electronic warfare.**

emission control orders--Orders, referred to as EMCON orders, used to authorize, control, or prohibit the use of electronic emission equipment. See also **control of electromagnetic radiation.**

emplacement(*)--1. A prepared position for one or more weapons or pieces of equipment, for protection against hostile fire or bombardment, and from which they can execute their tasks. 2. The act of fixing a gun in a prepared position from which it may be fired.

enabling mine countermeasures--Countermeasures designed to counter mines once they have been laid. This includes both passive and active mine countermeasures. See also **mine countermeasures.**

encipher--To convert plain text into unintelligible form by means of a cipher system.

encrypt--To convert plain text into unintelligible forms by means of a cryptosystem. (Note: The term encrypt covers the meanings of encipher and encode.) See also **cryptosystem.**

end item--A final combination of end products, component parts, and/or materials that is ready

for its intended use, e.g., ship, tank, mobile machine shop, aircraft.

end of mission(*)--In artillery and naval gunfire support, an order given to terminate firing on a specific target.

endurance(*)--The time an aircraft can continue flying, or a ground vehicle or ship can continue operating, under specified conditions, e.g., without refueling. See also **endurance distance.**

endurance distance(*)--Total distance that a ground vehicle or ship can be self-propelled at any specified endurance speed.

endurance loading--The stocking aboard ship for a period of time, normally covering the number of months between overhauls, of items with all of the following characteristics: a. low price; b. low weight and cube; c. a predictable usage rate; and d. nondeteriorative. See also **loading.**

enemy capabilities--Those courses of action of which the enemy is physically capable, and that, if adopted, will affect accomplishment of our mission. The term "capabilities" includes not only the general courses of action open to the enemy, such as attack, defense, or withdrawal, but also all the particular courses of action possible under each general course of action. "Enemy capabilities" are considered in the light of all known factors affecting military operations, including time, space, weather, terrain, and the strength and disposition of enemy forces. In strategic thinking, the capabilities of a nation represent the courses of action within the power of the nation for accomplishing its national objectives throughout the range of military operations.

engage(*)--In air defense, a fire control order used to direct or authorize units and/or weapon systems to fire on a designated target. See also **cease engagement; hold fire.**

engagement--In air defense, an attack with guns or air-to-air missiles by an interceptor aircraft, or the launch of an air defense missile by air defense artillery and the missile's subsequent travel to intercept.

engagement control(*)--In air defense, that degree of control exercised over the operational functions of an air defense unit that are related to detection, identification, engagement, and destruction of hostile targets.

envelopment(*)--An offensive maneuver in which the main attacking force passes around or over the enemy's principal defensive positions to secure objectives to the enemy's rear. See also **turning movement.**

environmental services--The various combinations of scientific, technical, and advisory activities (including modification processes, i.e., the influence of manmade and natural factors) required to acquire, produce, and supply information on the past, present, and future states of space, atmospheric, oceanographic, and terrestrial surroundings for use in military planning and decisionmaking processes, or to modify those surroundings to enhance military operations.

ephemeris--A publication giving the computed places of the celestial bodies for each day of the year or for other regular intervals.

equipment--In logistics, all nonexpendable items needed to outfit or equip an individual or organization. See also **assembly; component; part; subassembly; supplies.**

equipment operationally ready--The status of an item of equipment in the possession of an operating unit that indicates it is capable of fulfilling its intended mission and in a system configuration that offers a high assurance of an effective, reliable, and safe performance.

equivalent focal length(*)--The distance measured along the optical axis of the lens from the rear nodal point to the plane of best average definition over the entire field used in a camera. See also **focal length.**

escalation--An increase in scope or violence of a conflict, deliberate or unpremeditated.

escapee--Any person who has been physically captured by the enemy and succeeds in getting free. See also **evasion and escape.**

escape line--A planned route to allow personnel engaged in clandestine activity to depart from a site or area when possibility of compromise or apprehension exists.

escape route--See **evasion and escape route.**

escort(*)--1. A combatant unit(s) assigned to accompany and protect another force or convoy. 2. Aircraft assigned to protect other aircraft during a mission. 3. An armed guard that accompanies a convoy, a train, prisoners, etc. 4. An armed guard accompanying persons as a mark of honor. (DOD) 5. To convoy. 6. A member of the Armed Forces assigned to accompany, assist, or guide an individual or group, e.g., an escort officer.

escort forces--Combat forces of various types provided to protect other forces against enemy attack.

espionage--Actions directed toward the acquisition of information through clandestine operations.

espionage against the United States--Overt, covert, or clandestine activity designed to obtain information relating to the national defense with intent or reason to believe that it will be used to the injury of the United States or to the advantage of a foreign nation. For espionage crimes see Chapter 37 of Title 18, United States Code.

essential communications traffic--Transmissions (record/voice) of any precedence which must be sent electrically in order for the command or activity concerned to avoid a serious impact on mission accomplishment or safety or life.

essential elements of friendly information--Key questions likely to be asked by adversary officials and intelligence systems about specific friendly intentions, capabilities, and activities, so they can obtain answers critical to their operational effectiveness. Also called **EEFI.**

essential elements of information--The critical items of information regarding the enemy and the environment needed by the commander by a particular time to relate with other available information and intelligence in order to assist in reaching a logical decision. Also called **EEI.**

essential industry--Any industry necessary to the needs of a civilian or war economy. The term includes the basic industries as well as the necessary portions of those other industries that transform the crude basic raw materials into useful intermediate or end products, e.g., the iron and steel industry, the food industry, and the chemical industry.

essential secrecy--The condition achieved from the denial of critical information to adversaries.

establishment(*)--1. An installation, together with its personnel and equipment, organized as an operating entity. **(NATO)** 2. The table setting out the authorized numbers of men and major equipment in a unit/formations; sometimes called "table of organization" or "table of organization and equipment." See also **activity; base; equipment.**

estimate--1. An analysis of a foreign situation, development, or trend that identifies its major elements, interprets the significance, and appraises the future possibilities and the prospective results of the various actions that might be taken. 2. An appraisal of the capabilities, vulnerabilities, and potential courses of action of a foreign nation or combination of nations in consequence of a specific national plan, policy, decision, or contemplated course of action. 3. An analysis of an actual or contemplated clandestine operation in relation to the situation in which it is or would be conducted in order to identify and appraise such factors as available and needed assets and potential obstacles, accomplishments, and consequences. See also **intelligence estimate.** 4. In air intercept, a code meaning, "Provide a quick estimate of the height/depth/range/size of designated contact," or "I estimate height/depth/range/size of designated contact is _____."

evacuation--1. The process of moving any person who is wounded, injured, or ill to and/or between medical treatment facilities. 2. The clearance of personnel, animals, or materiel from a given locality. 3. The controlled process of collecting, classifying, and shipping unserviceable or abandoned materiel, United States and foreign, to appropriate reclamation, maintenance, technical intelligence, or disposal facilities.

evacuation control ship(*)--In an amphibious operation, a ship designated as **a control point for landing craft, amphibious vehicles, and helicopters evacuating casualties from the beaches.** Medical personnel embarked in the evacuation control ship effect distribution of casualties throughout the attack force in accordance with ship's casualty capacities and specialized medical facilities available, and also perform emergency surgery.

evacuation convoy(*)--A convoy which is used for evacuation of dangerously exposed waters.

See also **evacuation of dangerously exposed waters.**

evacuation of dangerously exposed waters(*)--The movement of merchant ships under naval control from severely threatened coastlines and dangerously exposed waters to safer localities. See also **dangerously exposed waters; severely threatened coastline.**

evacuation policy--1. Command decision indicating the length in days of the maximum period of noneffectiveness that patients may be held within the command for treatment. Patients who, in the opinion of responsible medical officers, cannot be returned to duty status within the period prescribed are evacuated by the first available means, provided the travel involved will not aggravate their disabilities. 2. A command decision concerning the movement of civilians from the proximity of military operations for security and safety reasons and involving the need to arrange for movement, reception, care, and control of such individuals. 3. Command policy concerning the evacuation of unserviceable or abandoned materiel and including designation of channels and destinations for evacuated materiel, the establishment of controls and procedures, and the dissemination of condition standards and disposition instructions.

evacuee--A civilian removed from a place of residence by military direction for reasons of personal security or the requirements of the military situation. See also **displaced person; expellee; refugee.**

evader--Any person isolated in hostile or unfriendly territory who eludes capture.

evaluation--In intelligence usage, appraisal of an item of information in terms of credibility, reliability, pertinency, and accuracy. Appraisal is accomplished at several stages within the intelligence cycle with progressively different contexts. Initial evaluations, made by case officers and report officers, are focused upon the reliability of the source and the accuracy of the information as judged by data available at or close to their operational levels. Later evaluations, by intelligence analysts, are primarily concerned with verifying accuracy of information and may, in effect, convert information into intelligence. Appraisal or evaluation of items of information or intelligence is indicated by a standard letter-number system. The evaluation of the reliability of sources is designated by a letter from A through F, and the accuracy of the information is designated by numeral 1 through 6. These are two entirely independent appraisals, and these separate appraisals are indicated in accordance with the system indicated below. Thus, information adjudged to be "probably true" received from a "usually reliable source" is designated "B-2" or "B2," while information of which the "truth cannot be judged" received from a "usually reliable source" is designated "B-6" or "B6."

Reliability of Source	Accuracy of Information
A--Completely reliable	1--Confirmed by other sources
B--Usually reliable	2--Probably true
C--Fairly reliable	3--Possibly true
D--Not usually reliable	4--Doubtful
E--Unreliable	5--Improbable
F--Reliability cannot be judged	6--Truth cannot be judged

See also **intelligence cycle; operational evaluation; technical evaluation.**

evaluation agent--That command or agency designated in the program directive to be responsible for the planning, coordination, and conduct of the required evaluation. The evaluation agent, normally the Joint Doctrine Center, J-7, identifies evaluation criteria and the media to be used, develops a proposed evaluation directive, coordinates exercise-related evaluation requirements with the sponsoring commands, and provides required evaluation reports to the Director, J-7. See also **joint**

doctrine; joint tactics, techniques, and procedures; joint test publication.

evasion and escape(*)--The procedures and operations whereby military personnel and other selected individuals are enabled to emerge from an enemy-held or hostile area to areas under friendly control.

evasion and escape intelligence--Processed information prepared to assist personnel to escape if captured by the enemy or to evade capture if lost in enemy-dominated territory.

evasion and escape net--The organization within enemy-held or hostile areas that operates to receive, move, and exfiltrate military personnel or selected individuals to friendly control. See also **unconventional warfare.**

evasion and escape route--A course of travel, preplanned or not, that an escapee or evader uses in an attempt to depart enemy territory in order to return to friendly lines.

exaggerated stereoscopy--See **hyperstereoscopy.**

exceptional transport(*)--In railway terminology, a load whose size, weight, or preparation entail special difficulties vis-a-vis the facilities or equipment of even one of the railway systems to be used.

excess property--The quantity of property in possession of any component of the Department of Defense that exceeds the quantity required or authorized for retention by that component.

executing commander (nuclear weapons)--A commander to whom nuclear weapons are released for delivery against specific targets or in accordance with approved plans. See also **commander(s); releasing commander (nuclear weapons).**

execution planning--The phase of the crisis action planning process in which an approved operation plan or other National Command Authorities-designated course of action is adjusted, refined, and translated into an operation order. Execution planning can proceed on the basis of prior deliberate planning, or it can take place under a no plan situation.

exercise(*)--A military maneuver or simulated wartime operation involving planning, preparation, and execution. It is carried out for the purpose of training and evaluation. It may be a combined, joint, or single-Service exercise, depending on participating organizations. See also **command post exercise; field exercise; maneuver.**

exercise directing staff(*)--A group of officers who by virtue of experience, qualifications, and a thorough knowledge of the exercise instructions, are selected to direct or control an exercise.

exercise filled mine(*)--In naval mine warfare, a mine containing an inert filling and an indicating device. See also **explosive filled mine; fitted mine; mine.**

exercise incident(*)--An occurrence injected by directing staffs into the exercise which will have an effect on the forces being exercised, or their facilities, and which will require action by the appropriate commander and/or staff being exercised.

exercise mine(*)--In naval mine warfare, a mine suitable for use in mine warfare exercises, fitted with visible or audible indicating devices to show where and when it would normally fire. See also **drill mine; mine; practice mine.**

exercise specifications(*)--The fundamental requirements for an exercise, providing in advance an outline of the concept, form, scope, setting, aim, objectives, force requirements, political implications, analysis arrangements, and costs.

exercise sponsor(*)--The commander who conceives a particular exercise and orders that it be planned and executed either by the commander's staff or by a subordinate headquarters.

exercise study(*)--An activity which may take the form of a map exercise, a war game, a series of lectures, a discussion group, or an operational analysis.

exercise term--A combination of two words, normally unclassified, used exclusively to designate a test, drill, or exercise. An exercise term is employed to preclude the possibility of confusing exercise directives with actual operations directives.

exfiltration--The removal of personnel or units from areas under enemy control.

existence load--Consists of items other than those in the fighting load that are required to sustain or protect the combat soldier. These items may be necessary for increased personal and environmental protection and are not normally carried by the individual. See also **fighting load.**

exoatmosphere--See **nuclear exoatmospheric burst.**

expedition--A military operation conducted by an armed force to accomplish a specific objective in a foreign country.

expeditionary force--An armed force organized to accomplish a specific objective in a foreign country.

expellee--A civilian outside the boundaries of the country of his or her nationality or ethnic origin who is being forcibly repatriated to that country or to a third country for political or other purposes. See also **displaced person; evacuee; refugee.**

expendable property--Property that may be consumed in use or loses its identity in use and may be dropped from stock record accounts when it is issued or used.

expendable supplies and material--Supplies which are consumed in use, such as ammunition, paint, fuel, cleaning and preserving materials, surgical dressings, drugs, medicines, etc., or which lose their identity, such as spare parts, etc. Also called **consumable supplies and material.**

exploder(*)--A device designed to generate an electric current in a firing circuit after deliberate action by the user in order to initiate an explosive charge or charges.

exploitation(*)--1. Taking full advantage of success in battle and following up initial gains. 2. Taking full advantage of any information that has come to hand for tactical, operational, or strategic purposes. 3. An offensive operation that usually follows a successful attack and is designed to disorganize the enemy in depth.

exploratory hunting(*)--In naval mine warfare, a parallel operation to search sweeping, in which a sample of the route or area is subjected to minehunting procedures to determine the presence or absence of mines.

explosive filled mine(*)--In mine warfare, a mine containing an explosive charge but not necessarily the firing train needed to detonate it. See also **exercise filled mine; fitted mine.**

explosive ordnance(*)--All munitions containing explosives, nuclear fission or fusion materials and biological and chemical agents. This includes bombs and warheads; guided and ballistic missiles; artillery, mortar, rocket, and small arms ammunition; all mines, torpedoes, and depth charges; demolition charges; pyrotechnics; clusters and dispensers; cartridge and propellant actuated devices; electro-explosive devices; clandestine and improvised explosive devices; and all similar or related items or components explosive in nature.

explosive ordnance disposal(*)--The detection, identification, on-site evaluation, rendering safe, recovery, and final disposal of unexploded explosive ordnance. It may also include explosive ordnance which has become hazardous by damage or deterioration.

explosive ordnance disposal incident(*)--The suspected or detected presence of unexploded explosive ordnance, or damaged explosive ordnance, which constitutes a hazard to operations, installations, personnel or material. Not included in this definition are the accidental arming or other conditions that develop during the manufacture of high explosive material, technical service assembly operations or the laying of mines and demolition charges.

explosive ordnance disposal procedures(*)--Those particular courses or modes of action taken by explosive ordnance disposal personnel for access to, diagnosis, rendering safe, recovery, and final disposal of explosive ordnance or any hazardous material associated with an explosive ordnance disposal incident. **a. access procedures**--Those actions taken to locate exactly and gain access to unexploded explosive ordnance. **b. diagnostic procedures**--Those actions taken to identify and evaluate unexploded explosive ordnance. **c. render safe procedures**--The portion of the explosive ordnance disposal procedures involving the application of special explosive ordnance disposal methods and tools to provide for the interruption of functions or separation of essential components of unexploded explosive ordnance to prevent an unacceptable detonation. **d. recovery procedures**--Those actions

taken to recover unexploded explosive ordnance. **e. final disposal procedures**--The final disposal of explosive ordnance which may include demolition or burning in place, removal to a disposal area, or other appropriate means.

explosive ordnance disposal unit--Personnel with special training and equipment who render explosive ordnance safe (such as bombs, mines, projectiles, and booby traps), make intelligence reports on such ordnance, and supervise the safe removal thereof.

explosive train(*)--A succession of initiating and igniting elements arranged to cause a charge to function.

exposure dose(*)--The exposure dose at a given point is a measurement of radiation in relation to its ability to produce ionization. The unit of measurement of the exposure dose is the roentgen.

exposure station--See **air station (photogrammetry).**

extended communications search--In search and rescue operations, consists of contacting all possible sources of information on the missing craft, including physically checking possible locations such as harbors, marinas, and airport ramps. An extended communications search is normally conducted after a preliminary communications search has yielded no results and when the mission is upgraded to the alert phase. Also called **EXCOM.** See also **preliminary communications search; search and rescue incident classification, Subpart b.**

extent of a military exercise(*)--The scope of an exercise in relation to the involvement of NATO and/or national commands. See also **inter-command exercise; intra-command exercise; NATO-wide exercise; scale of an exercise.**

extent of damage--The visible plan area of damage to a target element, usually expressed in units of 1,000 square feet, in detailed damage analysis and in approximate percentages in immediate-type damage assessment reports; e.g., 50 percent structural damage.

external audience--All people who are not part of the internal audience of US military members and civilian employees and their immediate families. Part of the concept of "Publics." Includes many varied subsets that may be referred to as "Audiences" or "Publics." See also **internal audience; public.**

external reinforcing force(*)--A reinforcing force which is principally stationed in peacetime outside its intended Major NATO Command area of operations.

extraction parachute(*)--An auxiliary parachute designed to release and extract and deploy cargo from aircraft in flight and deploy cargo parachutes. See also **gravity extraction.**

extraction zone(*)--A specified drop zone used for the delivery of supplies and/or equipment by means of an extraction technique from an aircraft flying very close to the ground.

F

F-4--See **Phantom II.**

F-14--See **Tomcat.**

F-15--See **Eagle.**

F-16--See **Fighting Falcon.**

F-111--A twin-engine, supersonic, turbofan, all-weather tactical fighter. It is capable of employing nuclear and nonnuclear weapons. It also has the capability for operating from very short, relatively unprepared air strips.

FA--See **feasibility assessment.**

F/A-18--See **Hornet.**

fabricator--Individuals or groups who, without genuine resources, invent information or inflate or embroider over news for personal gain or for political purposes.

facility--A real property entity consisting of one or more of the following: a building, a structure, a utility system, pavement, and underlying land. See also **base.**

facsimile(*)--A system of telecommunication for the transmission of fixed images with a view to their reception in a permanent form.

faded--In air intercept, a code meaning, "Contact has disappeared from reporting station's scope, and any position information given is estimated."

faker--A friendly aircraft simulating a hostile in an air defense exercise.

fallout(*)--The precipitation to Earth of radioactive particulate matter from a nuclear cloud; also applied to the particulate matter itself.

fallout contours(*)--Lines joining points which have the same radiation intensity that define a fallout pattern, represented in terms of roentgens per hour.

fallout pattern(*)--The distribution of fallout as portrayed by fallout contours.

fallout prediction--An estimate, made before and immediately after a nuclear detonation, of the location and intensity of militarily significant quantities of radioactive fallout.

fallout safe height of burst--The height of burst at or above which no militarily significant fallout will be reproduced as a result of a nuclear weapon detonation. See also **types of burst.**

fallout wind vector plot(*)--A wind vector diagram based on the wind structure from the surface of the Earth to the highest altitude of interest.

false origin(*)--A fixed point to the south and west of a grid zone from which grid distances are measured eastward and northward.

famished--In air intercept, a code meaning, "Have you any instructions for me?"

fan camera photography(*)--Photography taken simultaneously by an assembly of three or more cameras systematically installed at fixed angles relative to each other so as to provide wide lateral coverage with overlapping images. See also **tri-camera photography.**

fan cameras(*)--An assembly of three or more cameras systematically disposed at fixed angles relative to each other so as to provide wide lateral coverage with overlapping images. See also **split cameras.**

fan marker beacon(*)--A type of radio beacon, the emissions of which radiate in a vertical, fan-shaped pattern. The signal can be keyed for identification purposes. See also **radio beacon; Z marker beacon.**

farm gate type operations--Operational assistance and specialized tactical training provided a friendly foreign air force by the United States Armed Forces to include, under certain specified conditions, the flying of operational missions in combat by combined United States/foreign aircrews as a part of the training being given when such missions are beyond the capability of the foreign air force.

feasibility--Operation plan review criterion. The determination of whether the assigned tasks could be accomplished by using available resources. See also **acceptability; adequacy; completeness; suitability.**

feasibility assessment--A basic target analysis that provides an initial determination of the viability of a proposed target for special operations forces employment. Also called **FA.**

feasibility test--An operation plan review criteria to determine whether or not a plan is within the capacity of the resources that can be made available. See also **logistic implications test.**

feature(*)--In cartography, any object or configuration of ground or water represented on the face of the map or chart.

feature line overlap(*)--A series of overlapping air photographs which follow the line of a ground feature, e.g., river, road, railway, etc.

FEBA--See **forward edge of the battle area.**

Federal Modal Agencies--See **transportation operating agencies.**

Federal Stock Number--The Federal Stock Number of an item of supply consists of the applicable 4-digit class code number from the Federal Supply Classification plus a sequentially assigned 7-digit Federal Item Identification Number. The number shall be arranged as follows: 4210-196-5439. See also **National Stock Number.** Note: Federal Stock Numbers were replaced by National Stock Numbers effective 30 September 1974.

federal supply class management--Those functions of materiel management that can best be accomplished by Federal Supply Classification, such as cataloging, characteristic screening, standardization, interchangeability and substitution grouping, multi-item specification management, and engineering support of the foregoing.

Federal Transport Agencies--See **transportation operating agencies.**

feet dry--1. In air operations, a code meaning, "I am, or contact designated is, over land." 2. In landing craft air cushion (LCAC) operations, a code meaning operations over land.

feet wet--1. In air operations, a code meaning, "I am, or contact designated is, over water." 2. In landing craft air cushion (LCAC) operations, a code meaning operations over water.

ferret--An aircraft, ship, or vehicle especially equipped for the detection, location, recording, and analyzing of electromagnetic radiation.

few (raid size)--In air intercept usage, seven or fewer aircraft. See also **many (raid size).**

FEZ--See **fighter engagement zone.**

FF--See **frigate.**

FFG--See **guided missile frigate.**

F-hour--See **times.**

FID--See **foreign internal defense.**

field army--Administrative and tactical organization composed of a headquarters, certain organic Army troops, service support troops, a variable number of corps, and a variable number of divisions. See also **Army corps.**

field artillery--Equipment, supplies, ammunition, and personnel involved in the use of cannon, rocket, or surface-to-surface missile launchers. Field artillery cannons are classified according to caliber as:
light--120mm and less.
medium--121-160mm.
heavy--161-210mm.
very heavy--greater than 210mm.
See also **direct support artillery; general support artillery.**

field artillery observer--A person who watches the effects of artillery fire, adjusts the center of impact of that fire onto a target, and reports the results to the firing agency. See also **naval gunfire spotting team; spotter.**

field control(*)--A series of points whose relative positions and elevations are known. These positions are used in basic data in mapping and charting. Normally, these positions are established by survey methods, and are sometimes referred to as "trig control" or "trigonometrical net(work)." See also **common control (artillery); control point; ground control.**

field exercise(*)--An exercise conducted in the field under simulated war conditions in which troops and armament of one side are actually present, while those of the other side may be imaginary or in outline. See also **command post exercise.**

field fortifications(*)--An emplacement or shelter of a temporary nature which can be constructed with reasonable facility by units requiring no more than minor engineer supervisory and equipment participation.

field headquarters--See **command post.**

field of fire(*)--The area which a weapon or a group of weapons may cover effectively with fire from a given position.

field of view(*)--1. In photography, the angle between two rays passing through the perspective center (rear nodal point) of a camera lens to the two opposite sides of the format. Not to be confused with "angle of view." See also **angle of view.** 2. The total solid angle available to the gunner when looking through the gunsight.

field of vision(*)--The total solid angle available to the gunner from his or her normal position. See also **field of view.**

field press censorship--The security review of news material subject to the jurisdiction of the Armed Forces of the United States, including all information or material intended for dissemination to the public. See also **censorship.**

fighter controller--See **air controller.**

fighter cover(*)--The maintenance of a number of fighter aircraft over a specified area or force for the purpose of repelling hostile air activities. See also **airborne alert; cover.**

fighter direction aircraft(*)--An aircraft equipped and manned for directing fighter aircraft. See also **combat information ship.**

fighter direction ship(*)--A ship equipped and manned for directing fighter aircraft operations. See also **combat information ship.**

fighter engagement zone--See **weapon engagement zone.**

fighter interceptor--See **interceptor.**

fighter sweep(*)--An offensive mission by fighter aircraft to seek out and destroy enemy aircraft or targets of opportunity in an allotted area of operations.

Fighting Falcon--A single engine, supersonic, turbofan, all-weather multipurpose tactical fighter/bomber. It is capable of employing nuclear/nonnuclear weapons. Air superiority is its primary mission with air interdiction and close air support as secondary. An air refueling capability increases its flexibility. Designated as **F-16.**

fighting load--Consists of items of individual clothing, equipment, weapons, and ammunition that are carried by, and are essential to, the effectiveness of the combat soldier and the accomplishment of the immediate mission of the unit when the soldier is on foot. See also **existence load.**

filler--A substance carried in an ammunition container such as a projectile, mine, bomb, or grenade. A filler may be an explosive, chemical, or inert substance.

filler personnel--Individuals of suitable grade and skill initially required to bring a unit or organization to its authorized strength.

film badge(*)--A photographic film packet to be carried by personnel, in the form of a badge, for measuring and permanently recording (usually) gamma-ray dosage.

filter(*)--In electronics, a device which transmits only part of the incident energy and may thereby change the spectral distribution of energy: a. High pass filters transmit energy above a certain frequency; b. Low pass filters transmit energy below a certain frequency; c. Band pass filters transmit energy of a certain band-width; d. Band stop filters transmit energy outside a specific frequency band.

filter center--The location in an aircraft control and warning system at which information from observation posts is filtered for further dissemination to air defense control centers and air defense direction centers.

FIM-43--See **Redeye.**

FIM-92A--See **Stinger.**

final approach(*)--That part of an instrument approach procedure in which alignment and descent for landing are accomplished. a. In a non-precision approach it normally begins at the final approach fix or point and ends at the missed approach point or fix. b. In a precision approach the final approach commences at the glide path intercept point and ends at the decision height/altitude.

final destination(*)--In naval control of shipping, the final destination of a convoy or of an individual ship (whether in convoy or independent) irrespective of whether or not routing instructions have been issued.

final disposal procedures--See **explosive ordnance disposal procedures.**

final plan(*)--A plan for which drafts have been coordinated and approved and which has been signed by or on behalf of a competent authority. See also **operation plan.**

final protective fire(*)--An immediately available prearranged barrier of fire designed to impede enemy movement across defensive lines or areas.

financial property accounting--The establishment and maintenance of property accounts in monetary terms; the rendition of property reports in monetary terms.

fire(*)--1. The command given to discharge a weapon(s). 2. To detonate the main explosive charge by means of a firing system. 3. See also **barrage fire; call fire; close supporting fire; concentrated fire; counterfire; counter-preparation fire; covering fire; deep support-ing fire; destruction fire; direct fire; direct supporting fire; distributed fire; grazing fire; harassing fire; indirect fire; interdiction fire; neutralization fire; observed fire; prepara-tion fire; radar fire; registration fire; sched-uled fire; searching fire; supporting fire; suppressive fire; unobserved fire; zone fire.**

fireball(*)--The luminous sphere of hot gases which forms a few millionths of a second after detonation of a nuclear weapon and immediate-ly starts expanding and cooling.

fire barrage (specify)--An order to deliver a prearranged barrier of fire. Specification of the particular barrage may be by code name, numbering system, unit assignment, or other designated means.

Firebee--A remotely controlled target drone pow-ered by a turbojet engine. It achieves high subsonic speeds and is designed to be ground launched or air launched. It is used to test, train, and evaluate weapon systems employing surface-to-air and air-to-air missiles. Designat-ed as **BQM-34.**

fire capabilities chart(*)--A chart, usually in the form of an overlay, showing the areas which can be reached by the fire of the bulk of the weapons of a unit.

fire control(*)--The control of all operations in connection with the application of fire on a target.

fire control radar(*)--Radar used to provide target information inputs to a weapon fire control system.

fire control system(*)--A group of interrelated fire control equipments and/or instruments designed for use with a weapon or group of weapons.

fire coordination--See **fire support coordination.**

fire coordination area(*)--An area with specified restraints into which fires in excess of those restraints will not be delivered without approval of the authority establishing the restraints.

fire direction center(*)--That element of a com-mand post, consisting of gunnery and commu-nication personnel and equipment, by means of which the commander exercises fire direction and/or fire control. The fire direction center receives target intelligence and requests for fire, and translates them into appropriate fire direction.

fire for effect(*)--1. Fire which is delivered after the mean point of impact or burst is within the desired distance of the target or adjust-ing/ranging point. 2. Term in a call for fire to indicate the adjustment/ranging is satisfactory and fire for effect is desired.

fire message--See **call for fire.**

fire mission(*)--1. Specific assignment given to a fire unit as part of a definite plan. 2. Order used to alert the weapon/battery area and indicate that the message following is a call for fire.

fire plan(*)--A tactical plan for using the weapons of a unit or formation so that their fire will be coordinated.

firepower(*)--1. The amount of fire which may be delivered by a position, unit, or weapon system. 2. Ability to deliver fire.

firepower umbrella(*)--An area of specified dimensions defining the boundaries of the airspace over a naval force at sea within which the fire of ships' antiaircraft weapons can endanger aircraft, and within which special procedures have been established for the identification and operation of friendly aircraft. See also **air defense operations area.**

fire storm(*)--Stationary mass fire, generally in built-up urban areas, generating strong, inrushing winds from all sides; the winds keep the fires from spreading while adding fresh oxygen to increase their intensity.

fire support area(*)--An appropriate maneuver area assigned to fire support ships from which to deliver gunfire support of an amphibious operation. See also **naval support area.**

fire support coordinating measure--A measure employed by land or amphibious commanders to facilitate the rapid engagement of targets and simultaneously provide safeguards for friendly forces. See also **fire support coordination.**

fire support coordination(*)--The planning and executing of fire so that targets are adequately covered by a suitable weapon or group of weapons.

fire support coordination center(*)--A single location in which are centralized communications facilities and personnel incident to the coordination of all forms of fire support. See also **supporting arms coordination center.**

fire support coordination line(*)--A line established by the appropriate ground commander to ensure coordination of fire not under the commander's control but which may affect current tactical operations. The fire support coordination line is used to coordinate fires of air, ground, or sea weapons systems using any type of ammunition against surface targets. The fire support coordination line should follow well--defined terrain features. The establishment of the fire support coordination line must be coordinated with the appropriate tactical air commander and other supporting elements. Supporting elements may attack targets forward of the fire support coordination line without prior coordination with the ground force commander provided the attack will not produce adverse surface effects on or to the rear of the line. Attacks against surface targets behind this line must be coordinated with the appropriate ground force commander. Also called **FSCL.**

fire support group(*)--A temporary grouping of ships under a single commander charged with supporting troop operations ashore by naval gunfire. A fire support group may be further subdivided into fire support units and fire support elements.

fire support station--An exact location at sea within a fire support area from which a fire support ship delivers fire.

fire task--See **fire mission.**

fire time--See **span of detonation (atomic demolition munition employment).**

firing area(*)--In a sweeper-sweep combination it is the horizontal area at the depth of a particular mine in which the mine will detonate. The firing area has exactly the same dimensions as the interception area but will lie astern of it unless the mine detonates immediately when actuated.

firing chart--Map, photo map, or grid sheet showing the relative horizontal and vertical positions of batteries, base points, base point lines, check points, targets, and other details needed in preparing firing data.

firing circuit(*)--1. In land operations, an electrical circuit and/or pyrotechnic loop designed to detonate connected charges from a firing point. 2. In naval mine warfare, that part of a mine circuit which either completes the detonator circuit or operates a ship counter.

firing mechanism--See **firing circuit.**

firing point(*)--That point in the firing circuit where the device employed to initiate the detonation of the charges is located.

firing system(*)--In demolition, a system composed of elements designed to fire the main charge or charges.

first generation negative--See **generation (photography).**

first generation positive--See **generation (photography).**

first light--The beginning of morning nautical twilight; i.e., when the center of the morning sun is 12 degrees below the horizon.

first salvo at--In naval gunfire support, a portion of a ship's message to an observer or spotter to indicate that because of proximity to troops, the ship will not fire at the target but offset the first salvo a specific distance from the target.

first strike--The first offensive move of a war. (Generally associated with nuclear operations.)

FISINT--See **foreign instrumentation signals intelligence.**

fission(*)--The process whereby the nucleus of a heavy element splits into (generally) two nuclei of lighter elements, with the release of substantial amounts of energy.

fission products(*)--A general term for the complex mixture of substances produced as a result of nuclear fission.

fission to yield ratio(*)--The ratio of the yield derived from nuclear fission to the total yield; it is frequently expressed in percent.

fitted mine(*)--In naval mine warfare, a mine containing an explosive charge, a primer, detonator, and firing system. See also **exercise filled mine; explosive filled mine.**

fix(*)--A position determined from terrestrial, electronic, or astronomical data.

fixed ammunition(*)--Ammunition in which the cartridge case is permanently attached to the projectile. See also **ammunition.**

fixed capital property--1. Assets of a permanent character having continuing value. 2. As used in military establishments, includes real estate and equipment installed or in use, either in productive plants or in field operations. Synonymous with fixed assets.

fixed medical treatment facility(*)--A medical treatment facility which is designed to operate for an extended period of time at a specific site.

fixed port--Water terminals with an improved network of cargo-handling facilities designed for the transfer of oceangoing freight. See also **port; water terminal.**

fixed price incentive contract--A fixed price type of contract with provision for the adjustment of profit and price by a formula based on the relationship that final negotiated total cost bears to negotiated target cost as adjusted by approved changes.

fixed price type contract--A type of contract that generally provides for a firm price or, under appropriate circumstances, may provide for an adjustable price for the supplies or services being procured. Fixed price contracts are of several types so designed as to facilitate proper pricing under varying circumstances.

fixed station patrol(*)--One in which each scout maintains station relative to an assigned point on a barrier line while searching the surrounding area. Scouts are not stationary but remain underway and patrol near the center of their assigned stations. A scout is a surface ship, submarine, or aircraft.

fixer network(*)--A combination of radio or radar direction-finding installations which, operating in conjunction, are capable of plotting the position relative to the ground of an aircraft in flight.

fixer system--See **fixer network.**

flag days (red or green)--Red flag days are those during which movement requirements cannot be met; green flag days are those during which the requisite amount or a surplus of transportation capability exists.

flag officer--A term applied to an officer holding the rank of general, lieutenant general, major general, or brigadier general in the US Army, Air Force or Marine Corps or admiral, vice admiral, rear admiral or commodore in the US Navy or Coast Guard.

flame field expedients--Simple, handmade devices used to produce flame or illumination.

flame thrower(*)--A weapon that projects incendiary fuel and has provision for ignition of this fuel.

flammable cargo--See **inflammable cargo.**

flank guard(*)--A security element operating to the flank of a moving or stationary force to protect it from enemy ground observation, direct fire, and surprise attack.

flanking attack(*)--An offensive maneuver directed at the flank of an enemy. See also **frontal attack.**

flare(*)--The change in the flight path of an aircraft so as to reduce the rate of descent for touchdown.

flare dud--A nuclear weapon that when launched at a target, detonates with anticipated yield but at an altitude appreciably greater than intended. This is not a dud insofar as yield is concerned, but it is a dud with respect to the effects on the target and the normal operation of the weapon.

flash blindness(*)--Impairment of vision resulting from an intense flash of light. It includes temporary or permanent loss of visual functions and may be associated with retinal burns. See also **dazzle.**

flash burn(*)--A burn caused by excessive exposure (of bare skin) to thermal radiation.

flash message--A category of precedence reserved for initial enemy contact messages or operational combat messages of extreme urgency. Brevity is mandatory. See also **precedence.**

flash ranging--Finding the position of the burst of a projectile or of an enemy gun by observing its flash.

flash report--Not to be used. See **inflight report.**

flash suppressor(*)--Device attached to the muzzle of the weapon which reduces the amount of visible light or flash created by burning propellant gases.

flash-to-bang time(*)--The time from light being first observed until the sound of the nuclear detonation is heard.

flat(*)--In photography, lacking in contrast.

flatted cargo--Cargo placed in the bottom of the holds, covered with planks and dunnage, and held for future use. Flatted cargo usually has room left above it for the loading of vehicles that may be moved without interfering with the flatted cargo. Frequently, flatted cargo serves in lieu of ballast. Sometimes called understowed cargo. See also **cargo.**

fleet--An organization of ships, aircraft, Marine forces, and shore-based fleet activities all under the command of a commander or commander in chief who may exercise operational as well as administrative control. See also **major fleet; numbered fleet.**

fleet ballistic missile submarine--A nuclear-powered submarine designed to deliver ballistic missile attacks against assigned targets from either a submerged or surfaced condition. Designated as **SSBN.**

fleet in being--A fleet (force) that avoids decisive action, but, because of its strength and location, causes or necessitates counter-concentrations and so reduces the number of opposing units available for operations elsewhere.

Fleet Marine Force--A balanced force of combined arms comprising land, air, and service elements of the US Marine Corps. A Fleet Marine Force is an integral part of a US Fleet and has the status of a type command.

flexible response--The capability of military forces for effective reaction to any enemy threat or attack with actions appropriate and adaptable to the circumstances existing.

flight--1. In Navy and Marine Corps usage, a specified group of aircraft usually engaged in a common mission. 2. The basic tactical unit in the Air Force, consisting of four or more aircraft in two or more elements. 3. A single aircraft airborne on a nonoperational mission.

flight advisory--A message dispatched to aircraft in flight or to interested stations to advise of any deviation or irregularity.

flight deck--1. In certain airplanes, an elevated compartment occupied by the crew for operating the airplane in flight. 2. The upper deck of an aircraft carrier that serves as a runway.

flight following(*)--The task of maintaining contact with specified aircraft for the purpose of determining en route progress and/or flight termination.

flight information center(*)--A unit established to provide flight information service and alerting service.

flight information region(*)--An airspace of defined dimensions within which flight information service and alerting service are provided. See also **air traffic control center; area control center.**

flight information service(*)--A service provided for the purpose of giving advice and information useful for the safe and efficient conduct of flights.

flight levels(*)--Surfaces of constant atmospheric pressure which are related to a specific pressure datum, 1013.2 mb (29.92 in), and are separated by specific pressure intervals. (Flight levels are expressed in three digits that represent hundreds of feet; e.g., flight level 250 represents a barometric altimeter indication of 25,000 feet and flight level 255 is an indication of 25,500 feet.)

flight operations center--The element of the tactical Army air traffic regulation system which provides for aircraft flight following, separation of aircraft under instrument conditions, and identification of friendly aircraft to friendly air defense agencies.

flight path(*)--The line connecting the successive positions occupied, or to be occupied, by an aircraft, missile, or space vehicle as it moves through air or space.

flight plan(*)--Specified information provided to air traffic services units relative to an intended flight or portion of a flight of an aircraft.

flight plan correlation--A means of identifying aircraft by association with known flight plans.

flight profile(*)--The flight path of an aircraft expressed in terms of altitude, speed, range, and maneuver.

flight quarters--A ship configuration that assigns and stations personnel at critical positions to conduct safe flight operations.

flight readiness firing--A missile system test of short duration conducted with the propulsion system operating while the missile is secured to the launcher. Such a test is performed to determine the readiness of the missile system and launch facilities prior to flight test.

flight surgeon(*)--A physician specially trained in aviator medical practice whose primary duty is the medical examination and medical care of aircrew.

flight test(*)--Test of an aircraft, rocket, missile, or other vehicle by actual flight or launching. Flight tests are planned to achieve specific test objectives and gain operational information.

flight visibility--The average forward horizontal distance from the cockpit of an aircraft in flight at which prominent unlighted objects may be seen and identified by day and prominent lighted objects may be seen and identified by night.

floating base support(*)--A form of logistic support in which supplies, repairs, mainte-nance, and other services are provided in harbor or at an anchorage for operating forces from ships.

floating dump--Emergency supplies preloaded in landing craft, amphibious vehicles, or in landing ships. Floating dumps are located in the vicinity of the appropriate control officer who directs their landing as requested by the troop commander concerned.

floating mine(*)--In naval mine warfare, a mine visible on the surface. See also **drifting mine; free mine; watching mine; mine.**

floating reserve(*)--In an amphibious operation, reserve troops which remain embarked until needed. See also **general reserve.**

flooder(*)--In naval mine warfare, a device fitted to a buoyant mine which, on operation after a preset time, floods the mine case and causes it to sink to the bottom.

FLOT--See **forward line of own troops.**

flotation(*)--The capability of a vehicle to float in water.

FLS--See **Naval Forward Logistics Site.**

fly(ing) at speed--In air intercept, a code meaning, "Fly at (mach__/__) indicated air speed," or, "My indicated air speed is (__ knots/mach __)."

FM--See **force module.**

FMC--See **full mission capable.**

FMP--See **force module package.**

FMS--See **foreign military sales.**

foam path--A path of fire extinguisher foam laid on a runway to assist aircraft in an emergency landing.

FOB--See **forward operations base.**

focal length--See **calibrated focal length; equivalent focal length; nominal focal length.**

focal plane(*)--The plane, perpendicular to the optical axis of the lens, in which images of points in the object field of the lens are focused.

FOD--See **foreign object damage.**

folded optics(*)--Any optical system containing reflecting components for the purpose of reducing the physical length of the system or for the purpose of changing the path of the optical axis.

follow-up--In amphibious operations, the landing of reinforcements and stores after the assault and assault follow-on echelons have been landed. See also **assault; assault follow-on echelon.**

follow-up echelon(*)--In air transport operations, elements moved into the objective area after the assault echelon.

follow-up shipping--Ships not originally a part of the amphibious task force but which deliver troops and supplies to the objective area after the assault phase has begun.

force--1. An aggregation of military personnel, weapon systems, vehicles and necessary support, or combination thereof. 2. A major subdivision of a fleet.

force closure--The point in time when a supported commander determines that sufficient personnel and equipment resources are in the assigned area of operations to carry out assigned tasks.

force combat air patrol--A patrol of fighters maintained over the task force to destroy enemy aircraft that threaten the force. See also **combat air patrol.**

force list--A total list of forces required by an operation plan, including assigned forces, augmentation forces, and other forces to be employed in support of the plan.

force module--A grouping of combat, combat support, and combat service support forces, with their accompanying supplies and the required nonunit resupply and personnel necessary to sustain forces for a minimum of 30 days. The elements of force modules are linked together or are uniquely identified so that they may be extracted from or adjusted as an entity in the Joint Operation Planning and Execution System data bases to enhance flexibility and usefulness of the operation plan during a crisis. Also called **FM.** See also **force module package.**

force module package--A force module with a specific functional orientation (e.g. air superiority, close air support, reconnaissance, ground defense) that include combat, associated combat support, and combat service support forces. Additionally, force module packages will contain sustainment in accordance with logistic policy contained in Joint Strategic Capabilities Plan Annex B. Also called **FMP.** See also **force module.**

force multiplier--A capability that, when added to and employed by a combat force, significantly increases the combat potential of that force and thus enhances the probability of successful mission accomplishment.

force protection--Security program designed to protect soldiers, civilian employees, family members, facilities, and equipment, in all locations and situations, accomplished through planned and integrated application of combatting terrorism, physical security, operations security, personal protective services, and supported by intelligence, counterintelligence, and other security programs.

force rendezvous(*)--A checkpoint at which formations of aircraft or ships join and become part of the main force. Also called **group rendezvous.**

force requirement number--An alphanumeric code used to uniquely identify force entries in a given operation plan time-phased force and deployment data. Also called **FRN.**

force(s)--See **airborne force; air transported force; armed forces; balanced collective forces; black forces; blue forces; combined force; covering force; forces allocated to NATO; garrison force; national forces for the defense of the NATO area; NATO assigned forces; NATO command forces; NATO earmarked forces; Navy Cargo Handling Force; orange forces; other forces for NATO; purple forces; task force; underway replenishment force; white forces.**

forces in being(*)--Forces classified as being in state of readiness "A" or "B" as prescribed in the appropriate Military Committee document.

force sourcing--The identification of the actual units, their origins, ports of embarkation, and movement characteristics to satisfy the time-phased force requirements of a supported commander.

force structure--See **military capability.**

force tabs--With reference to war plans, the statement of time-phased deployments of major combat units by major commands and geographical areas.

force tracking--The identification of units and their specific modes of transport during movement to an objective area.

fordability--See **deep fording; shallow fording.**

foreign instrumentation signals intelligence---Technical information and intelligence information derived from the intercept of foreign instrumentation signals by other than the intended recipients. Foreign instrumentation signals intelligence is a category of signals intelligence. Note: Foreign instrumentation signals include but are not limited to signals from telemetry, beaconry, electronic interrogators, tracking/fusing/arming/firing command systems, and video data links. Also called **FISINT.** See also **telemetry intelligence; signals intelligence.**

foreign intelligence--Information relating to capabilities, intentions, and activities of foreign powers, organizations, or persons, but not including counterintelligence, except for information on international terrorist activities. See also **intelligence.**

foreign internal defense--Participation by civilian and military agencies of a government in any of the action programs taken by another government to free and protect its society from subversion, lawlessness, and insurgency. Also called **FID.** See also **internal defense.**

foreign military sales--That portion of United States security assistance authorized by the Foreign Assistance Act of 1961, as amended, and the Arms Export Control Act of 1976, as amended. This assistance differs from the Military Assistance Program and the International Military Education and Training Program in that the recipient provides reimbursement for defense articles and services transferred. Also called **FMS.**

foreign military sales trainees--Foreign nationals receiving training conducted by the Department of Defense on a reimbursable basis, at the country's request.

foreign object damage--Rags, pieces of paper, line, articles of clothing, nuts, bolts, or tools that, when misplaced or caught by air currents normally found around aircraft operations (jet blast, rotor or prop wash, engine intake), cause damage to aircraft systems or weapons or injury to personnel. Also called **FOD.**

format(*)--1. In photography, the size and/or shape of a negative or of the print therefrom. 2. In cartography, the shape and size of a map or chart.

formation(*)--1. An ordered arrangement of troops and/or vehicles for a specific purpose. 2. An ordered arrangement of two or more ships, units, or aircraft proceeding together under a commander.

formatted message text(*)--A message text composed of several sets ordered in a specified sequence, each set characterized by an identifier and containing information of a specified type, coded and arranged in an ordered sequence of character fields in accordance with the NATO message text formatting rules. It is designed to permit both manual and automated handling and processing. See also **free form message text; structured message text.**

formerly restricted data--Information removed from the Restricted Data category upon a joint determination by the Department of Energy (or antecedent agencies) and Department of Defense that such information relates primarily to the military utilization of atomic weapons and that such information can be adequately safeguarded as classified defense information. (Section 142d, Atomic Energy Act of 1954, as amended.) See also **restricted data.**

form lines(*)--Lines resembling contours, but representing no actual elevations, which have been sketched from visual observation or from inadequate or unreliable map sources, to show collectively the configuration of the terrain.

forward aeromedical evacuation(*)--That phase of evacuation which provides airlift for patients between points within the battlefield, from the battlefield to the initial point of treatment, and to subsequent points of treatment within the combat zone.

forward air controller--An officer (aviator/pilot) member of the tactical air control party who, from a forward ground or airborne position, controls aircraft in close air support of ground troops.

forward air control post--A highly mobile US Air Force tactical air control system radar facility subordinate to the control and reporting center and/or post used to extend radar coverage and control in the forward combat area.

forward area--An area in proximity to combat.

forward arming and refueling point--A temporary facility, organized, equipped, and deployed by an aviation commander, and normally located in the main battle area closer to the area of operation than the aviation unit's combat service area, to provide fuel and ammunition necessary for the employment of aviation maneuver units in combat. The forward arming and refueling point permits combat aircraft to rapidly refuel and rearm simultaneously. Also called **FARP.**

forward edge of the battle area(*)--The foremost limits of a series of areas in which ground combat units are deployed, excluding the areas in which the covering or screening forces are operating, designated to coordinate fire support, the positioning of forces, or the maneuver of units. Also called **FEBA.**

forward line of own troops--A line which indicates the most forward positions of friendly forces in any kind of military operation at a specific time. The forward line of own troops normally identifies the forward location of covering and screening forces. Also called **FLOT.**

forward motion compensation--See **image motion compensation.**

forward oblique air photograph--Oblique photography of the terrain ahead of the aircraft.

forward observer--An observer operating with front line troops and trained to adjust ground or naval gunfire and pass back battlefield information. In the absence of a forward air controller, the observer may control close air support strikes. See also **spotter.**

forward operations base--In special operations, a base usually located in friendly territory or afloat that is established to extend command and control or communications or to provide support for training and tactical operations. Facilities may be established for temporary or longer duration operations and may include an airfield or an unimproved airstrip, an anchorage, or a pier. A forward operations base may be the location of special operations component headquarters or a smaller unit that is controlled and/or supported by a main operations base. Also called **FOB.** See also **advanced operations base; main operations base.**

forward overlap--See **overlap.**

forward recovery mission profile--A mission profile that involves the recovery of an aircraft at a neutral/friendly forward area airfield or landing site.

forward slope(*)--Any slope which descends towards the enemy.

forward tell(*)--The transfer of information to a higher level of command. See also **track telling.**

four-round illumination diamond(*)--A method of distributing the fire of illumination shells which, by a combination of lateral spread and range spread, provides illumination of a large area.

fox away--In air intercept, a code meaning, "Missile has fired or been released from aircraft."

fragmentary order--An abbreviated form of an operation order, usually issued on a day-to-day basis, that eliminates the need for restating information contained in a basic operation order. It may be issued in sections.

frame(*)--In photography, any single exposure contained within a continuous sequence of photographs.

freak--In air intercept usage, a word meaning frequency in megacycles.

freddie--In air intercept usage, a controlling unit.

free air anomaly--The difference between observed gravity and theoretical gravity that has been computed for latitude and corrected for elevation of the station above or below the geoid, by application of the normal rate of change of gravity for change of elevation, as in free air.

free air overpressure(*)--The unreflected pressure, in excess of the ambient atmospheric pressure, created in the air by the blast wave from an explosion. See also **overpressure.**

free drop(*)--The dropping of equipment or supplies from an aircraft without the use of parachutes. See also **airdrop; air movement; free fall; high velocity drop; low velocity drop.**

free fall--A parachute maneuver in which the parachute is manually activated at the discretion of the jumper or automatically at a preset altitude. See also **airdrop; air movement; free drop; high velocity drop; low velocity drop.**

free field overpressure--See **free air overpressure.**

free form message text(*)--A message text without prescribed format arrangements. It is intended for fast drafting as well as manual handling and processing. See also **formatted message text; structured message text.**

free issue--Materiel provided for use or consumption without charge to the fund or fund subdivision that finances the activity to which issued.

free lance--In air intercept, a code meaning, "Self-control of aircraft is being employed."

free mine(*)--In naval mine warfare, a moored mine whose mooring has parted or been cut.

free play exercise(*)--An exercise to test the capabilities of forces under simulated contingency and/or wartime conditions, limited only by those artificialities or restrictions required by peacetime safety regulations. See also **controlled exercise.**

free rocket(*)--A rocket not subject to guidance or control in flight.

freight consolidating activity--A transportation activity that receives less than carload/truckload shipments of materiel for the purpose of assembling them into carload/truckload lots for onward movement to the ultimate consignee or to a freight distributing activity or other break bulk point. See also **freight distributing activity.**

freight distributing activity--A transportation activity that receives and unloads consolidated carloads/truckloads of less than carload/truckload shipments of material and forwards the individual shipments to the ultimate consignee. See also **freight consolidating activity.**

frequency deconfliction--A systematic management procedure to coordinate the use of the electromagnetic spectrum for operations, communications, and intelligence functions. Frequency deconfliction is one element of electromagnetic spectrum management. See also **electromagnetic spectrum; electronic warfare; spectrum management.**

fresh target--A request or command sent by the observer or spotter to the firing ship to indicate that fire will be shifted from the original target to a new target by spots (corrections) applied to the computer solution being generated.

friendly--A contact positively identified as friendly. See also **bogey; hostile.**

friendly fire--In casualty reporting, a casualty circumstance applicable to persons killed in action or wounded in action mistakenly or accidentally by friendly forces actively engaged with the enemy, who are directing fire at a hostile force or what is thought to be a hostile force. See also **casualty.**

frigate--A warship designed to operate independently, or with strike, antisubmarine warfare, or amphibious forces against submarine, air, and surface threats. (Normal armament consists of 3-inch and 5-inch dual-purpose guns and advanced antisubmarine warfare weapons.) Designated as **FF.** See also **guided missile frigate.**

FRN--See **force requirement number.**

front(*)--1. The lateral space occupied by an element measured from the extremity of one flank to the extremity of the other flank. 2. The direction of the enemy. 3. The line of contact of two opposing forces. 4. When a combat situation does not exist or is not assumed, the direction toward which the command is faced.

frontal attack(*)--1. An offensive maneuver in which the main action is directed against the front of the enemy forces. (DOD) 2. In air intercept, an attack by an interceptor aircraft that terminates with a heading crossing angle greater than 135 degrees.

frustrated cargo--Any shipment of supplies and/or equipment which while en route to destination is stopped prior to receipt and for which further disposition instructions must be obtained.

full beam spread--See **indirect illumination.**

full charge--The larger of the two propelling charges available for naval guns.

full mission capable--Material condition of an aircraft or training device indicating that it can perform all of its missions. Also called **FMC.** See also **mission capable; partial mission capable; partial mission capable, maintenance; partial mission capable, supply.**

full mobilization--See **mobilization.**

functional component command--A command normally, but not necessarily, composed of forces of two or more Services which may be established in peacetime or war to perform particular operational missions that may be of short duration or may extend over a period of time.

functions--The appropriate or assigned duties, responsibilities, missions, or tasks of an individual, office, or organization. As defined in the National Security Act of 1947, as amended, the term "function" includes functions, powers, and duties (5 United States Code 171n (a)).

fusion--1. The process whereby the nuclei of light elements combine to form the nucleus of a heavier element, with the release of tremendous amounts of energy. 2. In intelligence usage, the process of examining all sources of intelligence and information to derive a complete assessment of activity.

fusion center--In intelligence usage, a physical location to accomplish fusion. It normally has sufficient intelligence automated data processing capability to assist in the process.

fuze(*)--A device which initiates an explosive train. See also **base fuze; boresafe fuze; impact action fuze; proximity fuze; self-destroying fuze; shuttered fuze; time fuze.**

fuze cavity(*)--A recess in a charge for receiving a fuze.

fuze (specify)--In artillery and naval gunfire support, a command or request to indicate the type of fuze action desired; i.e., delay, quick, time, proximity.

G

gadget--Radar equipment. (Type of equipment may be indicated by a letter as listed in operation orders.) May be followed by a color to indicate state of jamming. Colors will be used as follows: **a. green**--Clear of jamming. **b. amber**--Sector partially jammed. **c. red**--Sector completely jammed. **d. blue**--Completely jammed.

Galaxy--A large cargo transport aircraft powered by four turbofan engines, capable of carrying a very large payload (including outsize cargo and personnel) into forward area air fields. It further is capable of refueling in flight. Designated C-5.

gamma rays--High energy electromagnetic radiation emitted from atomic nuclei during a nuclear reaction. Gamma rays and very high energy X-rays differ only in origin. X-rays do not originate from atomic nuclei but are produced in other ways.

gap(*)--An area within a minefield or obstacle belt, free of live mines or obstacles, whose width and direction will allow a friendly force to pass through in tactical formation. See also **phoney minefield.**

gap filler radar(*)--A radar used to supplement the coverage of the principal radar in areas where coverage is inadequate.

gap (imagery)--Any space where imagery fails to meet minimum coverage requirements. This might be a space not covered by imagery or a space where the minimum specified overlap was not obtained. See also **holiday.**

gap marker(*)--In landmine warfare, markers used to indicate a minefield gap. Gap markers at the entrance to, and exit from, the gap will be referenced to a landmark or intermediate marker. See also **marker.**

garble--An error in transmission, reception, encryption, or decryption that changes the text of a message or any portion thereof in such a manner that it is incorrect or undecryptable.

garnishing(*)--In surveillance, natural or artificial material applied to an object to achieve or assist camouflage.

garrison force(*)--All units assigned to a base or area for defense, development, operation, and maintenance of facilities. See also **force(s).**

gate--In air intercept, a code meaning, "Fly at maximum possible speed (or power)." (To be maintained for a limited time only, depending on type of aircraft. Use of afterburners, rockets, etc., in accordance with local doctrine.)

general agency agreement--A contract between the Maritime Administration and a steamship company which, as general agent, exercises administrative control over a government--owned ship for employment by the Military Sealift Command. See also **Military Sealift Command.**

general air cargo(*)--Cargo without hazardous or dangerous properties and not requiring extra precautions for air transport.

general and complete disarmament--Reductions of armed forces and armaments by all states to levels required for internal security and for an international peace force. Connotation is "total disarmament" by all states.

general cargo--Cargo which is susceptible for loading in general, nonspecialized stowage areas; e.g., boxes, barrels, bales, crates, packages, bundles, and pallets.

general map--A map of small scale used for general planning purposes. See also **map.**

general military intelligence--Intelligence concerning the (1) military capabilities of foreign countries or organizations or (2) topics affecting potential US or allied military operations, relating to the following subjects: armed forces capabilities, including order of battle, organization, training, tactics, doctrine, strategy, and other factors bearing on military strength and effectiveness; area and terrain intelligence, including urban areas, coasts and landing beaches, and meteorological, oceanographic, and geological intelligence; transportation in all modes; military materiel production and support industries; military and civilian C3 systems; military economics, including foreign military assistance; insurgency and terrorism; military-political-sociological intelligence; location, identification, and description of military-related installations; government control; escape and evasion; and threats and forecasts. (Excludes scientific and technical intelligence.) Also called **GMI.** See also **intelligence; military intelligence.**

general orders--1. Permanent instructions, issued in order form, that apply to all members of a command, as compared with special orders, which affect only individuals or small groups. General orders are usually concerned with matters of policy or administration. 2. A series of permanent guard orders that govern the duties of a sentry on post.

general purchasing agents--Agents who have been appointed in the principal overseas areas of operations to supervise, control, coordinate, negotiate, and develop the local procurement of supplies, services, and facilities by United States Armed Forces, in order that the most

effective utilization may be made of local resources and production.

general quarters--A condition of readiness when naval action is imminent. All battle stations are fully manned and alert; ammunition is ready for instant loading; guns and guided missile launchers may be loaded.

general reserve(*)--Reserve of troops under the control of the overall commander. See also **floating reserve.**

general staff--A group of officers in the headquarters of Army or Marine divisions, Marine brigades and aircraft wings, or similar or larger units that assist their commanders in planning, coordinating, and supervising operations. A general staff may consist of four or more principal functional sections: personnel (G-1), military intelligence (G-2), operations and training (G-3), logistics (G-4), and (in Army organizations) civil affairs/military government (G-5). (A particular section may be added or eliminated by the commander, dependent upon the need that has been demonstrated.) The comparable Air Force staff is found in the wing and larger units, with sections designated Personnel, Operations, etc. G-2 Air and G-3 Air are Army officers assigned to G-2 or G-3 at division, corps, and Army headquarters level who assist in planning and coordinating joint operations of ground and air units. Naval staffs ordinarily are not organized on these lines, but when they are, they are designated N-1, N-2, etc. Similarly, a joint staff may be designated J-1, J-2, etc. In Army brigades and smaller units and in Marine Corps units smaller than a brigade or aircraft wing, staff sections are designated S-1, S-2, etc., with corresponding duties; referred to as a unit staff in the Army and as an executive staff in the Marine Corps. See also **staff.**

general stopping power(*)--The percentage of a group of vehicles in battle formation likely to be stopped by mines when attempting to cross a minefield.

general support(*)--That support which is given to the supported force as a whole and not to any particular subdivision thereof. See also **close support; direct support; mutual support; support.**

general support artillery(*)--Artillery which executes the fire directed by the commander of the unit to which it organically belongs or is attached. It fires in support of the operation as a whole rather than in support of a specific subordinate unit.

general support-reinforcing--A tactical artillery mission. General support-reinforcing artillery has the mission of supporting the force as a whole and of providing reinforcing fires for another artillery unit.

general support rocket system--A multiple rocket launcher system that supplements cannon artillery by delivery of large quantities of firepower in a short time against critical, time-sensitive targets.

general unloading period(*)--In amphibious operations, that part of the ship-to-shore movement in which unloading is primarily logistic in character, and emphasizes speed and volume of unloading operations. It encompasses the unloading of units and cargo from the ships as rapidly as facilities on the beach permit. It proceeds without regard to class, type, or priority of cargo, as permitted by cargo handling facilities ashore. See also **initial unloading period.**

general war--Armed conflict between major powers in which the total resources of the belligerents are employed, and the national survival of a major belligerent is in jeopardy.

generation (photography)--The preparation of successive positive/negative reproductions from an original negative/positive (first-generation). For example, the first positive produced from an original negative is a second-generation product; the negative made from this positive is a third-generation product; and the next positive or print from that negative is a fourth-generation product.

geodetic datum--See **datum (geodetic).**

geographic coordinates(*)--The quantities of latitude and longitude which define the position of a point on the surface of the Earth with respect to the reference spheroid. See also **coordinates.**

geographic reference points--A means of indicating position, usually expressed either as double letters or as code words that are established in operation orders or by other means.

georef(*)--A worldwide position reference system that may be applied to any map or chart graduated in latitude and longitude regardless of projection. It is a method of expressing latitude and longitude in a form suitable for rapid reporting and plotting. (This term is derived from the words "The World Geographic Reference System.")

glide bomb--A bomb fitted with airfoils to provide lift and which is carried and released in the direction of a target by an airplane.

glide mode--In a flight control system, a control mode in which an aircraft is automatically positioned to the center of the glide slope course.

GMI--See **general military intelligence.**

go around mode--In an automatic flight control system, a control mode which terminates an aircraft approach and programs a climb. See also **overshoot.**

goldie--The term, peculiar to air support radar team operations, indicating that the aircraft automatic flight-control system and ground-control bombing system are engaged and awaiting electronic ground control commands.

goldie lock--The term, peculiar to air support radar team operations, indicating ground controller has electronic control of the aircraft.

go no-go(*)--The condition or state of operability of a component or system: "go," functioning properly; or "no-go," not functioning properly.

government-owned, contract-operated ships---Those ships to which the US Government holds title and which the Military Sealift Command operates under a contract (i.e., nongovernment--manned). These ships are designated United States Naval Ships and use the prefix "USNS" with the ship name and the letter "T" as a prefix to the ship classification (e.g., T-AKR). See also **Military Sealift Command; United States Naval Ship.**

government-owned, Military Sealift Command-operated ships--Those ships to which the US Government holds title and which the Military Sealift Command operates with US Government (civil service) employees. These ships are designated United States Naval Ships and use the prefix "USNS" with the ship name and the letter "T" as a prefix to the ship classification (e.g., T-AKR). See also **Military Sealift Command; United States Naval Ship.**

gradient circuit(*)--In mine warfare, a circuit which is actuated when the rate of change, with time, of the magnitude of the influence is within predetermined limits.

grand slam--All enemy aircraft originally sighted are shot down.

graphic(*)--Any and all products of the cartographic and photogrammetric art. A graphic may be a map, chart, or mosaic or even a film strip that was produced using cartographic techniques.

graphic scale(*)--A graduated line by means of which distances on the map, chart, or photograph may be measured in terms of ground distance. See also **scale.**

grapnel(*)--In naval mine warfare, a device fitted to a mine mooring designed to grapple the sweep wire when the mooring is cut.

graticule(*)--1. In cartography, a network of lines representing the Earth's parallels of latitude and meridians of longitude. 2. In imagery interpretation, see **reticle**.

graticule ticks(*)--In cartography, short lines indicating where selected meridians and parallels intersect.

graves registration--Supervision and execution of matters pertaining to the identification, removal, and burial of the dead, and collection and processing of their effects. See also **burial.**

gravity extraction(*)--The extraction of cargoes from the aircraft by influence of their own weight. See also **extraction parachute.**

graze(*)--In artillery and naval gunfire support, a spotting, or an observation, by a spotter or an observer to indicate that all bursts occurred on impact.

grazing fire(*)--Fire approximately parallel to the ground where the center of the cone of fire does not rise above one meter from the ground. See also **fire.**

Greenwich Mean Time--See **Universal Time.** Also called **GMT.** (NATO: See ZULU time.)

grey propaganda--Propaganda that does not specifically identify any source. See also **propaganda.**

grid--1. Two sets of parallel lines intersecting at right angles and forming squares; the grid is superimposed on maps, charts, and other similar representations of the Earth's surface in an accurate and consistent manner to permit identification of ground locations with respect to other locations and the computation of direction and distance to other points. 2. A term used in giving the location of a geographic point by grid coordinates. See also **military grid; military grid reference system.**

grid bearing--Bearing measured from grid north.

grid convergence--The horizontal angle at a place between true north and grid north. It is proportional to the longitude difference between the place and the central meridian. See also **convergence.**

grid convergence factor(*)--The ratio of the grid convergence angle to the longitude difference. In the Lambert Conical Orthomorphic projection, this ratio is constant for all charts based on the same two standard parallels. See also **constant of the cone; convergence; grid convergence.**

grid coordinates(*)--Coordinates of a grid coordinate system to which numbers and letters are assigned for use in designating a point on a gridded map, photograph, or chart. See also **coordinates.**

grid coordinate system(*)--A plane-rectangular coordinate system usually based on, and mathematically adjusted to, a map projection in order that geographic positions (latitudes and longitudes) may be readily transformed into plane coordinates and the computations relating to them may be made by the ordinary method of plane surveying. See also **coordinates.**

grid interval(*)--The distance represented between the lines of a grid.

grid magnetic angle(*)--Angular difference in direction between grid north and magnetic north. It is measured east or west from grid north. Grid magnetic angle is sometimes called grivation and/or grid variation.

grid navigation(*)--A method of navigation using a grid overlay for direction reference. See also **navigational grid.**

grid north(*)--The northerly or zero direction indicated by the grid datum of directional reference.

grid ticks(*)--Small marks on the neatline of a map or chart indicating additional grid reference systems included on that sheet. Grid ticks are sometimes shown on the interior grid lines of some maps for ease of referencing.

grid variation--See **grid magnetic angle.**

grivation--See **grid magnetic angle.**

gross error--A nuclear weapon detonation at such a distance from the desired ground zero as to cause no nuclear damage to the target.

grossly transportation feasible--A determination made by the supported commander that a draft operation plan can be supported with the apportioned transportation assets. This determination is made by using a transportation feasibility estimator to simulate movement of personnel and cargo from port of embarkation to port of debarkation within a specified time frame.

gross weight(*)--1. Weight of a vehicle, fully equipped and serviced for operation, including the weight of the fuel, lubricants, coolant, vehicle tools and spares, crew, personal equipment, and load. 2. Weight of a container or pallet including freight and binding. See also **net weight.**

ground alert(*)--That status in which aircraft on the ground/deck are fully serviced and armed, with combat crews in readiness to take off within a specified short period of time (usually 15 minutes) after receipt of a mission order. See also **airborne alert; alert.**

ground combat element--See **Marine air-ground task force.**

ground control(*)--A system of accurate measurements used to determine the distances and directions or differences in elevation between points on the Earth. See also **common control (artillery); control point; field control; traverse.**

ground control (geodetic)--See **ground control.**

ground-controlled approach procedure(*)--The technique for talking down, through the use of both surveillance and precision approach radar, an aircraft during its approach so as to place it in a position for landing. See also **automatic approach and landing.**

ground controlled interception(*)--A technique which permits control of friendly aircraft or guided missiles for the purpose of effecting interception. See also **air interception.**

ground fire--Small arms ground-to-air fire directed against aircraft.

grounding(*)--The bonding of an equipment case, frame or chassis, to an object or vehicle structure to ensure a common potential. See also **bonding; earthing.**

ground liaison officer--An officer trained in offensive air support activities. Ground liaison officers are normally organized into parties under the control of the appropriate Army commander to provide liaison to Air Force and naval units engaged in training and combat operations.

ground liaison party--An army unit consisting of a variable number of personnel responsible for liaison with a tactical air support agency.

ground liaison section--An army unit consisting of a variable number of army officers, other ranks, and vehicles responsible for army/air liaison, under control of army headquarters.

ground mine--See **bottom mine.**

ground nadir(*)--The point on the ground vertically beneath the perspective center of the camera lens. On a true vertical photograph this coincides with the principal point.

ground observer center--A center to which ground observer teams report and which in turn will pass information to the appropriate control and/or reporting agency.

ground observer team--Small units or detachments deployed to provide information of aircraft movements over a defended area, obtained either by aural or visual means.

ground position(*)--The position on the Earth vertically below an aircraft.

ground readiness--That status wherein aircraft can be armed and serviced and personnel alerted to take off within a specified length of time after receiving orders.

ground return(*)--The reflection from the terrain as displayed and/or recorded as an image.

ground signals(*)--A visual signal displayed on an airfield to give local air traffic rules information to flight crews in the air. See also **signal area.**

ground speed(*)--The horizontal component of the speed of an aircraft relative to the Earth's surface.

ground speed mode--In a flight control system, a control mode in which the ground speed of an aircraft is automatically controlled to a computed value.

ground visibility--Prevailing horizontal visibility near the Earth's surface as reported by an accredited observer.

ground zero(*)--The point on the surface of the Earth at, or vertically below or above, the center of a planned or actual nuclear detonation. See also **actual ground zero; desired ground zero.**

group--1. A flexible administrative and tactical unit composed of either two or more battalions or two or more squadrons. The term also applies to combat support and combat service support units. 2. A number of ships and/or aircraft, normally a subdivision of a force, assigned for a specific purpose.

group burial--A burial in a common grave of two or more individually unidentified remains. See also **burial.**

group of targets(*)--Two or more targets on which fire is desired simultaneously. A group of targets is designated by a letter/number combination or a nickname.

group rendezvous--A check point at which formations of the same type will join before proceeding. See also **force rendezvous.**

guard(*)--A security element whose primary task is to protect the main force by fighting to gain time, while also observing and reporting information. See also **flank guard; screen.**

guerrilla--A combat participant in guerrilla warfare. See also **unconventional warfare.**

guerrilla force--A group of irregular, predominantly indigenous personnel organized along military lines to conduct military and paramilitary operations in enemy-held, hostile, or denied territory.

guerrilla warfare(*)--Military and paramilitary operations conducted in enemy-held or hostile territory by irregular, predominantly indigenous forces. See also **unconventional warfare.**

guidance--1. Policy, direction, decision, or instruction having the effect of an order when promulgated by a higher echelon. 2. The entire process by which target intelligence information received by the guided missile is used to effect proper flight control to cause timely direction changes for effective target interception. See also **active homing guidance; celestial guidance; command guidance; homing guidance; inertial guidance; midcourse guidance; passive homing guidance; preset guidance; semiactive homing guidance; stellar guidance; terminal guidance; terrestrial reference guidance.**

guidance coverage(*)--That volume of space in which guidance information (azimuth and/or elevation and/or distance) is provided to aircraft to the specified performance and accuracy. This may be specified either with relation to airfield/airstrip geometry, making assumptions about deployment of ground equipment, or with relation to the coverage provided by individual ground units.

guidance station equipment(*)--The ground-based portion of the missile guidance system necessary to provide guidance during missile flight.

guided missile--An unmanned vehicle moving above the surface of the Earth whose trajectory or flight path is capable of being altered by an external or internal mechanism. See also **aerodynamic missile; ballistic missile.**

guided missile cruiser--A warship designed to operate with strike and amphibious forces against air, surface, and subsurface threats. Normal armaments consist of 5-inch guns, an advanced area-defense antiair-warfare missile system, and antisubmarine-warfare weapons. Designated as **CG.**

guided missile destroyer--For mission, see destroyer. This destroyer type is equipped with Standard guided missiles, naval guns, long--range sonar, and antisubmarine-warfare weapons, including ASROC. Designated as **DDG.**

guided missile equipment carrier--A self-propelled, full-tracked, amphibious, air transportable, unarmored carrier for various guided missile systems and their equipment.

guided missile frigate--Equipped with Standard missile launchers, 5"/54 or 76-mm gun battery, torpedoes, embarked helicopters, and towed array sonar. Designated as **FFG.** See also **frigate.**

guided missile submarine--A submarine designed to have an additional capability to launch guided missile attacks. Designated as **SSG and SSGN.** The SSGN is nuclear powered.

guide specification(*)--Minimum requirements to be used as a basis for the evaluation of a national specification covering a fuel, lubricant or associated product proposed for standardization action.

guinea-pig(*)--In naval mine warfare, a ship used to determine whether an area can be considered safe from influence mines under certain conditions, or, specifically, to detonate pressure mines.

gull(*)--In electronic warfare, a floating radar reflector used to simulate a surface target at sea for deceptive purposes.

gun--1. A cannon with relatively long barrel, operating with relatively low angle of fire, and having a high muzzle velocity. 2. A cannon with tube length 30 calibers or more. See also **howitzer; mortar.**

gun carriage(*)--A mobile or fixed support for a gun. It sometimes includes the elevating and traversing mechanisms. Also called **carriage.**

guns/weapons free--In air intercept, means fire may be opened on all aircraft not recognized as friendly.

guns/weapons tight--In air intercept, means do not open fire, or cease firing on any aircraft (or on bogey specified, or in section indicated) unless target(s) known to be hostile.

gun-target line(*)--An imaginary straight line from the gun(s) to the target. See also **spotting line.**

gun-type weapon--A device in which two or more pieces of fissionable material, each less than a critical mass, are brought together very rapidly so as to form a supercritical mass that can explode as the result of a rapidly expanding fission chain.

gyro-magnetic compass--A directional gyroscope whose azimuth scale is maintained in alignment with the magnetic meridian by a magnetic detector unit.

H

H-2--See Sea Sprite.

H-3--See Sea King.

H-46--See Sea Knight.

hachuring(*)--A method of representing relief upon a map or chart by shading in short disconnected lines drawn in the direction of the slopes.

half-life(*)--The time required for the activity of a given radioactive species to decrease to half of its initial value due to radioactive decay. The half-life is a characteristic property of each radioactive species and is independent of its amount or condition. The effective half-life of a given isotope is the time in which the quantity in the body will decrease to half as a result of both radioactive decay and biological elimination.

half-residence time(*)--As applied to delayed fallout, it is the time required for the amount of weapon debris deposited in a particular part of the atmosphere to decrease to half of its initial value.

half thickness(*)--Thickness of absorbing material necessary to reduce by one-half the intensity of radiation which passes through it.

halftone(*)--Any photomechanical printing surface or the impression therefrom in which detail and tone values are represented by a series of evenly spaced dots in varying size and shape, varying in direct proportion to the intensity of the tones they represent. See also **halftone screen.**

halftone screen(*)--A series of regular spaced opaque lines on glass, crossing at right angles, producing transparent apertures between inter

sections. Used in a process camera to break up a solid or continuous tone image into a pattern of small dots. See also **halftone.**

handover--The passing of control authority of an aircraft from one control agency to another control agency. Handover action may be accomplished between control agencies of separate Services when conducting joint operations or between control agencies within a single command and control system. Handover action is complete when the receiving controller acknowledges assumption of control authority.

handover line(*)--A control feature, preferably following easily defined terrain features, at which responsibility for the conduct of combat operations is passed from one force to another.

hang fire(*)--An undesired delay in the functioning of a firing system.

harassing (air)--The attack of any target within the area of land battle not connected with interdiction or close air support. It is designed to reduce the enemy's combat effectiveness.

harassing fire(*)--Fire designed to disturb the rest of the enemy troops, to curtail movement, and, by threat of losses, to lower morale. See also **fire.**

harassment--An incident in which the primary objective is to disrupt the activities of a unit, installation, or ship, rather than to inflict serious casualties or damage.

harbor--A restricted body of water, an anchorage, or other limited coastal water area and its mineable water approaches, from which shipping operations are projected or supported. Generally, a harbor is part of a base, in which case the harbor defense force forms a component element of the base defense force established for the local defense of the base and its included harbor.

harbor defense(*)--The defense of a harbor or anchorage and its water approaches against external threats such as: a. submarine, submarine-borne, or small surface craft attack; b. enemy minelaying operations; and c. sabotage. The defense of a harbor from guided or dropped missiles while such missiles are airborne is considered to be a part of air defense. See also **port security.**

hard beach--A portion of a beach especially prepared with a hard surface extending into the water, employed for the purpose of loading or unloading directly into or from landing ships or landing craft.

hardened site(*)--A site, normally constructed under rock or concrete cover, designed to provide protection against the effects of conventional weapons. It may also be equipped to provide protection against the side effects of a nuclear attack and against a chemical or a biological attack.

hard missile base(*)--A launching base that is protected against a nuclear explosion.

hard port--Alter heading to magnetic heading indicated, turning to the port in a tight turn (three digit group), or alter heading indicated number of degrees to the port in a tight turn (one or two digit group with word "degrees").

hardstand(*)--1. A paved or stabilized area where vehicles are parked. 2. Open ground area having a prepared surface and used for the storage of materiel.

hard starboard--Alter heading to magnetic heading indicated, turning to the starboard in a tight turn (three digit group), or alter heading indicated number of degrees to the starboard in a tight turn (one or two digit group with word "degrees").

hardware--1. The generic term dealing with physical items as distinguished from its capability or function such as equipment, tools, implements, instruments, devices, sets, fittings, trimmings, assemblies, subassemblies, components, and parts. The term is often used in regard to the stage of development, as in the passage of a device or component from the design stage into the hardware stage as the finished object. 2. In data automation, the physical equipment or devices forming a computer and peripheral components. See also **software.**

harmful appreciations--See **appreciations.**

harmonization--The process and/or results of adjusting differences or inconsistencies to bring significant features into agreement.

Harpoon--An all-weather, anti-ship cruise missile capable of being employed from surface ships (RGM-84), aircraft (AGM-84A) or submarines (UGM-84). The missile is turbojet powered and employs a low level cruise trajectory. Terminal guidance is active radar. A 500--pound conventional warhead is employed.

Harrier--A single-engine, vectored thrust, turbojet, vertical and/or short takeoff and landing light attack aircraft, designed to operate from land bases and naval vessels in a close air support role. Capable of carrying a variety of conventional and/or nuclear weapons. Designated as **AV-8.**

hasty attack(*)--In land operations, an attack in which preparation time is traded for speed in order to exploit an opportunity. See also **deliberate attack.**

hasty breaching(*)--The rapid creation of a route through a minefield, barrier, or fortification by any expedient method.

hasty breaching (land mine warfare)--The creation of lanes through enemy minefields by expedient methods such as blasting with demolitions, pushing rollers or disabled vehicles through the minefields when the time factor does not permit detailed reconnaissance, deliberate breaching, or bypassing the obstacle. See also **breaching.**

hasty crossing(*)--The crossing of an inland water obstacle using the crossing means at hand or those readily available, and made without pausing for elaborate preparations. See also **deliberate crossing.**

hasty defense(*)--A defense normally organized while in contact with the enemy or when contact is imminent and time available for the organization is limited. It is characterized by improvement of the natural defensive strength of the terrain by utilization of foxholes, emplacements, and obstacles. See also **deliberate defense.**

hatch list--A list showing, for each hold section of a cargo ship, a description of the items stowed, their volume and weight, the consignee of each, and the total volume and weight of materiel in the hold.

havens (moving)--See **moving havens.**

Hawk--A mobile air defense artillery, surface-to--air missile system that provides non-nuclear, low to medium altitude air defense coverage for ground forces. Designated as **MIM-23.**

Hawkeye--A twin turboprop, multicrew airborne early warning and interceptor control aircraft designed to operate from aircraft carriers. It carries a long-range radar and integrated computer system for the detection and tracking of airborne targets at all altitudes. Designated as **E-2.**

HC-130--See **Hercules.**

heading--In air intercept, a code meaning, "My, or bogey's, magnetic course is _____."

heading crossing angle--In air intercept, the angular difference between interceptor heading and target heading at the time of intercept.

heading hold mode--In a flight control system, a control mode which automatically maintains an aircraft heading that exists at the instant of completion of a maneuver.

heading indicator(*)--An instrument which displays heading transmitted electrically from a remote compass system.

heading select feature--A flight control system feature which permits selection or preselection of desired automatically controlled heading or headings of an aircraft.

heads up--In air intercept, a code meaning, "Enemy got through (part or all)," or, "I am not in position to engage target."

head-up display(*)--A display of flight, navigation, attack, or other information superimposed upon the pilot's forward field of view. See also **horizontal situation display.**

heavy antitank weapon--A weapon capable of operating from ground or vehicle, used to defeat armor and other material targets.

heavy artillery--See **field artillery.**

heavy drop--A system of delivery of heavy supplies and equipment by parachute.

heavy-lift cargo--1. Any single cargo lift, weighing over 5 long tons, and to be handled aboard ship. 2. In Marine Corps usage, individual units of cargo that exceed 800 pounds in weight or 100 cubic feet in volume. See also **cargo.**

heavy-lift ship(*)--A ship specially designed and capable of loading and unloading heavy and bulky items. It has booms of sufficient capacity to accommodate a single lift of 100 tons.

height--The vertical distance of an object, point, or level above the ground or other established reference plane. Height may be indicated as follows:
very low--Less than 500 feet.
low--500 to 2,000 feet (above ground level).
medium--2,000 to 25,000 feet.
high--25,000 to 50,000 feet.
very high--More than 50,000 feet.

height datum--See **altitude datum.**

height delay--See **altitude delay..**

height hole--See **altitude hole.**

height of burst(*)--The vertical distance from the Earth's surface or target to the point of burst. See also **optimum height of burst; safe burst height; types of burst.**

helicopter approach route(*)--The track or series of tracks along which helicopters move to a specific landing site or landing zone. See also **helicopter lane; helicopter retirement route.**

helicopter assault force(*)--A task organization combining helicopters, supporting units, and helicopter-borne troop units for use in helicopter-borne assault operations.

helicopter break-up point--A control point at which helicopters returning from a landing zone break formation and are released to return to base or are dispatched for other employment.

helicopter departure point--See **departure point.**

helicopter direction center(*)--In amphibious operations, the primary direct control agency for the helicopter group/unit commander operating under the overall control of the tactical air control center.

helicopter drop point--A designated point within a landing zone where helicopters are unable to land because of the terrain, but in which they can discharge cargo or troops while hovering.

helicopter external air transport certification categories--Three categories of equipment suitability for helicopter external air transport (HEAT). The Services assign individual items of equipment to the appropriate categories, which are as follows: **a. certified**--Those items of equipment and their associated rigging procedures which have completed a certification process comprised of engineering analysis, static load testing (in accordance with applicable military standards), validating rigging procedures, evaluating successful flight tests, and the issuance of a certification statement. **b. suitable**--Those items of equipment and their associated rigging procedures that have not been certified, but have demonstrated acceptable static lift and flight characteristics during flight testing by the US Army Airborne and Special Operations Test Board. **c. prohibited**--Those items of equipment and their associated rigging procedures that are prohibited from helicopter external air transport as determined by each Service.

helicopter landing site--A designated subdivision of a helicopter landing zone in which a single flight or wave of assault helicopters land to embark or disembark troops and/or cargo.

helicopter landing zone--A specified ground area for landing assault helicopters to embark or disembark troops and/or cargo. A landing zone may contain one or more landing sites.

helicopter lane(*)--A safety air corridor in which helicopters fly to or from their destination during helicopter operations. See also **helicopter approach route; helicopter retirement route.**

helicopter retirement route(*)--The track or series of tracks along which helicopters move from a specific landing site or landing zone. See also **helicopter approach route; helicopter lane.**

helicopter support team(*)--A task organization formed and equipped for employment in a landing zone to facilitate the landing and movement of helicopter-borne troops, equipment and supplies, and to evacuate selected casualties and enemy prisoners of war.

helicopter team--The combat-equipped troops lifted in one helicopter at one time.

helicopter transport area--Areas to the seaward and on the flanks of the outer transport and landing ship areas, but preferably inside the area screen, for launching and/or recovering helicopters.

helicopter wave--See **wave.**

helipad(*)--A prepared area designated and used for takeoff and landing of helicopters. (Includes touchdown or hover point.)

heliport(*)--A facility designated for operating, basing, servicing, and maintaining helicopters.

herbicide--A chemical compound that will kill or damage plants. See also **anticrop agent; antiplant agent.**

Hercules--A medium range troop and cargo transport designed for air-drop or airland delivery into a combat zone as well as conventional airlift. This aircraft is equipped with four turboprop engines, and integral ramp and cargo

door. The D model is ski equipped. The E model has additional fuel capacity for extended range. Designated as **C130.** The inflight tanker configurations are designated KC-130 and HC-130, which is also used for the aerial rescue mission. The gunship version is designated AC-130.

H-hour--See **times.**

HIDACZ--See **high-density airspace control zone.**

high--A height between 25,000 and 50,000 feet.

high airburst--The fallout safe height of burst for a nuclear weapon that increases damage to or casualties on soft targets, or reduces induced radiation contamination at actual ground zero. See also **types of burst.**

high altitude(*)--Conventionally, an altitude above 10,000 meters (33,000 feet). See also **altitude.**

high altitude bombing--Horizontal bombing with the height of release over 15,000 feet.

high altitude burst(*)--The explosion of a nuclear weapon which takes place at a height in excess of 100,000 feet (30,000 meters). See also **types of burst.**

high-altitude missile engagement zone--See **weapon engagement zone.**

high angle(*)--In artillery and naval gunfire support, an order or request to obtain high angle fire.

high angle fire(*)--Fire delivered at angles of elevation greater than the elevation that corresponds to the maximum range of the gun and ammunition concerned; fire, the range of which decreases as the angle of elevation is increased.

high-density airspace control zone--Airspace designated in an airspace control plan or airspace control order, in which there is a concentrated employment of numerous and varied weapons and airspace users. A high-density airspace control zone has defined dimensions, which usually coincide with geographical features or navigational aids. Access to a high-density airspace control zone is normally controlled by the maneuver commander. The maneuver commander can also direct a more restrictive weapons status within the high-density airspace control zone. Also called **HIDACZ.**

high explosive cargo--Cargo such as artillery ammunition, bombs, depth charges, demolition material, rockets, and missiles. See also **cargo.**

high oblique--See **oblique air photograph.**

high-risk personnel--Personnel who, by their grade, assignment, symbolic value, or relative isolation, are likely to be attractive or accessible terrorist targets. See also **antiterrorism.**

high speed submarine--A submarine capable of submerged speeds of 20 knots or more.

high value asset control items--Items of supply identified for intensive management control under approved inventory management techniques designed to maintain an optimum inventory level of high investment items. Also called **hi-value asset control items.**

high velocity drop(*)--A drop procedure in which the drop velocity is greater than 30 feet per second (low velocity drop) and lower than free drop velocity. See also **airdrop.**

high-water mark--Properly, a mark left on a beach by wave wash at the preceding high water. It does not necessarily correspond to the high-water line. Because it can be deter-

mined by simple observation, it is frequently used in place of the high-water line, which can be determined only by a survey. When so used, it is called the high-water line.

hill shading(*)--A method of representing relief on a map by depicting the shadows that would be cast by high ground if light were shining from a certain direction.

HIMEZ--See **high-altitude missile engagement zone.**

hoist(*)--In helicopters, the mechanism by which external loads may be raised or lowered vertically.

hold(*)--1. A cargo stowage compartment aboard ship. 2. To maintain or retain possession of by force, as a position or an area. 3. In an attack, to exert sufficient pressure to prevent movement or redisposition of enemy forces. 4. As applied to air traffic, to keep an aircraft within a specified space or location which is identified by visual or other means in accordance with Air Traffic Control instructions.

holdee--See **transient.**

hold fire(*)--In air defense, an emergency order to stop firing. Missiles already in flight must be prevented from intercepting, if technically possible.

holding anchorage(*)--An anchorage where ships may lie: a. if the assembly or working anchorage, or port, to which they have been assigned is full; b. when delayed by enemy threats or other factors from proceeding immediately on their next voyage; c. when dispersed from a port to avoid the effects of a nuclear attack. See also **assembly anchorage; emergency anchorage; working anchorage.**

holding attack(*)--An attack designed to hold the enemy in position, to deceive him as to where the main attack is being made, to prevent him from reinforcing the elements opposing the main attack, and/or to cause him to commit his reserves prematurely at an indecisive location.

holding pattern mode--Automatic control of an aircraft to fly the programmed holding pattern.

holding point--A geographically or electronically defined location used in stationing aircraft in flight in a predetermined pattern in accordance with air traffic control clearance. See also **orbit point.**

holding position(*)--A specified location on the airfield, close to the active runway and identified by visual means, at which the position of a taxiing aircraft is maintained in accordance with air traffic control instructions.

holiday--An unintentional omission in imagery coverage of an area. See also **gap (imagery).**

hollow charge(*)--A shaped charge producing a deep cylindrical hole of relatively small diameter in the direction of its axis of rotation.

home recovery mission profile--A mission profile that involves the recovery of an aircraft at its permanent or temporarily assigned operating base.

homing(*)--The technique whereby a mobile station directs itself, or is directed, towards a source of primary or reflected energy, or to a specified point.

homing adaptor(*)--A device which, when used with an aircraft radio receiver, produces aural and/or visual signals which indicate the direction of a transmitting radio station with respect to the heading of the aircraft.

homing guidance(*)--A system by which a missile steers itself towards a target by means of a self-contained mechanism which is activated by some distinguishing characteristics of the target. See also **active homing guidance; guidance; passive homing guidance; semi-- active homing guidance.**

homing mine(*)--In naval mine warfare, a mine fitted with propulsion equipment that homes on to a target. See also **mine.**

hook--A procedure used by an air controller to electronically direct the data processing equipment of a semi-automatic command and control system to take a specified action on a specific radar blip or symbol.

horizon--In general, the apparent or visible junction of the Earth and sky, as seen from any specific position. Also called **the apparent, visible, or local horizon.** A horizontal plane passing through a point of vision or perspective center. The apparent or visible horizon approximates the true horizon only when the point of vision is very close to sea level.

horizontal action mine(*)--In land mine warfare, a mine designed to produce a destructive effect in a plane approximately parallel to the ground.

horizontal error(*)--The error in range, deflection, or in radius, which a weapon may be expected to exceed as often as not. Horizontal error of weapons making a nearly vertical approach to the target is described in terms of circular error probable. Horizontal error of weapons producing elliptical dispersion pattern is expressed in terms of probable error. See also **circular error probable; delivery error; deviation; dispersion error.**

horizontal loading(*)--Loading of items of like character in horizontal layers throughout the holds of a ship. See also **loading.**

horizontal situation display(*)--An electronically generated display on which navigation information and stored mission and procedural data can be presented. Radar information and television picture can also be displayed either as a map overlay or as a separate image. See also **head-up display.**

horizontal situation indicator(*)--An instrument which may display bearing and distance to a navigation aid, magnetic heading, track/course and track/course deviation.

horizontal stowage--The lateral distribution of unit equipment or categories of supplies so that they can be unloaded simultaneously from two or more holds.

horn(*)--In naval mine warfare, a projection from the mine shell of some contact mines which, when broken or bent by contact, causes the mine to fire.

Hornet--A twin-engine supersonic, fighter/attack aircraft. The C (single-seat) and D (dual-seat) models have an all-weather intercept, identify and destroy capability. The Hornet is equipped with an electronic self-protection jammer, and is both air-to-air and air-to-ground capable. It is designed to operate from both land bases and aircraft carriers. Designated as **F/A-18.**

hospital--A medical treatment facility capable of providing inpatient care. It is appropriately staffed and equipped to provide diagnostic and therapeutic services, as well as the necessary supporting services required to perform its assigned mission and functions. A hospital may, in addition, discharge the functions of a clinic.

hostage--A person held as a pledge that certain terms or agreements will be kept. (The taking of hostages is forbidden under the Geneva Conventions, 1949.)

host country--A nation in which representatives or organizations of another state are present because of government invitation and/or international agreement.

hostile--A contact positively identified as enemy. See also **bogey; friendly.**

hostile acts--Basic rules established by higher authority for defining and recognizing hostile acts by aircraft, submarines, surface units, and ground forces will be promulgated by the commanders of unified or specified commands, and by other appropriate commanders when so authorized.

hostile casualty--A person who is the victim of a terrorist activity or who becomes a casualty "in action." "In action" characterizes the casualty as having been the direct result of hostile action, sustained in combat or relating thereto, or sustained going to or returning from a combat mission provided that the occurrence was directly related to hostile action. Included are persons killed or wounded mistakenly or accidentally by friendly fire directed at a hostile force or what is thought to be a hostile force. However, not to be considered as sustained in action and not to be interpreted as hostile casualties are injuries or death due to the elements, self-inflicted wounds, combat fatigue, and except in unusual cases, wounds or death inflicted by a friendly force while the individual is in an absent-without-leave, deserter, or dropped-from-rolls status or is voluntarily absent from a place of duty. See also **casualty; casualty type; nonhostile casualty.**

hostile environment--See **operational environment.**

hostile track(*)--The classification assigned to a track which, based upon established criteria, is determined to be an enemy threat.

host nation(*)--A nation which receives the forces and/or supplies of allied nations and/or NATO organizations to be located on, to operate in, or to transit through its territory.

host nation assistance--See **host nation support.**

host nation support--Civil and/or military assistance rendered by a nation to foreign forces within its territory during peacetime, crises or emergencies, or war based on agreements mutually concluded between nations.

hot photo interpretation report--A preliminary unformatted report of significant information from tactical reconnaissance imagery dispatched prior to compilation of the Initial Photo Interpretation Report. It should pertain to a single objective, event, or activity of significant interest to justify immediate reporting. Also called **HOTPHOTOREP.**

HOTPHOTOREP--See **hot photo interpretation report.**

hot report--Not to be used. See **Joint Tactical Air Reconnaissance/Surveillance Mission Report.**

hot spot(*)--Region in a contaminated area in which the level of radioactive contamination is considerably greater than in neighboring regions in the area.

Hound Dog--A turbojet-propelled, air-to-surface missile designed to be carried externally on the B-52. It is equipped with a nuclear warhead and can be launched for either high or low altitude attacks against enemy targets, supplementing the internally carried firepower of the B-52. Designated as **AGM-28B.**

hovering(*)--A self-sustaining maneuver whereby a fixed, or nearly fixed, position is maintained relative to a spot on the surface of the Earth or underwater.

hovering ceiling(*)--The highest altitude at which the helicopter is capable of hovering in standard atmosphere. It is usually stated in two figures: hovering in ground effect and hovering out of ground effect.

howitzer--1. A cannon which combines certain characteristics of guns and mortars. The howitzer delivers projectiles with medium velocities, either by low or high trajectories. 2. Normally a cannon with a tube length of 20 to 30 calibers; however, the tube length can exceed 30 calibers and still be considered a howitzer when the high angle fire zoning solution permits range overlap between charges. See also **gun; mortar.**

Huey--See **Iroquois.**

human intelligence(*)--A category of intelligence derived from information collected and provided by human sources. Also called **HUMINT.**

humanitarian and civic assistance--Assistance to the local populace provided by predominantly US forces in conjunction with military operations and exercises. This assistance is specifically authorized by title 10, United States Code, section 401, and funded under separate authorities. Assistance provided under these provisions is limited to (1) medical, dental, and veterinary care provided in rural areas of a country; (2) construction of rudimentary surface transportation systems; (3) well drilling and construction of basic sanitation facilities; and (4) rudimentary construction and repair of public facilities. Assistance must fulfill unit training requirements that incidentally create humanitarian benefit to the local populace. See also **humanitarian assistance.**

humanitarian assistance--Programs conducted to relieve or reduce the results of natural or manmade disasters or other endemic conditions such as human pain, disease, hunger, or privation that might present a serious threat to life or that can result in great damage to or lóss of property. Humanitarian assistance provided by US forces is limited in scope and duration. The assistance provided is designed to supplement or complement the efforts of the host nation civil authorities or agencies that may have the primary responsibility for providing humanitarian assistance.

human resources intelligence--The intelligence information derived from the intelligence collection discipline that uses human beings as both sources and collectors, and where the human being is the primary collection instrument. Also called **HUMINT**.

HUMINT--See **human resources intelligence.**

hung weapons--Those weapons or stores on an aircraft that the pilot has attempted to drop or fire but could not because of a malfunction of the weapon, rack or launcher, or aircraft release and control system.

hunter track(*)--In naval mine warfare, the track to be followed by the hunter (or sweeper) to ensure that the hunting (or sweeping) gear passes over the lap track.

hydrofoil patrol craft--A patrol combatant, missile, fast surface patrol craft, capable of quick reaction and offensive operations against major enemy surface combatants. Designated as **PHM.**

hydrogen bomb--See **thermonuclear weapon.**

hydrographic chart(*)--A nautical chart showing depths of water, nature of bottom, contours of bottom and coastline, and tides and currents in a given sea or sea and land area.

hydrographic reconnaissance--Reconnaissance of an area of water to determine depths, beach gradients, the nature of the bottom, and the location of coral reefs, rocks, shoals, and manmade obstacles.

hydrographic section (beach party)--A section of a beach party whose duties are to clear the beach of damaged boats, conduct hydrographic reconnaissance, assist in removing underwater obstructions, act as stretcher bearers, and furnish relief boat crews.

hydrography(*)--The science which deals with the measurements and description of the physical features of the oceans, seas, lakes, rivers, and their adjoining coastal areas, with particular reference to their use for navigational purposes.

hyperbaric chamber(*)--A chamber used to induce an increase in ambient pressure as would occur in descending below sea level, in a water or air environment. It is the only type of chamber suitable for use in the treatment of decompression sickness in flying or diving. Also called **compression chamber; diving chamber; recompression chamber.**

hyperbolic navigation system(*)--A radio navigation system which enables the position of an aircraft equipped with a suitable receiver to be fixed by two or more intersecting hyperbolic position lines. The system employs either a time difference measurement of pulse transmissions or a phase difference measurement of phase-locked continuous wave transmissions. See also **Decca; loran.**

hyperfocal distance(*)--The distance from the lens to the nearest object in focus when the lens is focused at infinity.

hypergolic fuel(*)--Fuel which will spontaneously ignite with an oxidizer, such as aniline with fuming nitric acid. It is used as the propulsion agent in certain missile systems.

hypersonic(*)--Of or pertaining to speeds equal to, or in excess of, five times the speed of sound. See also **speed of sound.**

hyperstereoscopy(*)--Stereoscopic viewing in which the relief effect is noticeably exaggerated, caused by the extension of the camera base. Also called **exaggerated stereoscopy.**

hypobaric chamber(*)--A chamber used to induce a decrease in ambient pressure as would occur in ascending to altitude. This type of chamber is primarily used for training and experimental purposes. Also called **altitude chamber; decompression chamber.**

hypsometric tinting(*)--A method of showing relief on maps and charts by coloring in different shades those parts which lie between selected levels. Sometimes referred to as **elevation tint; altitude tint; layer tint.**

I

IA--See **initial assessment**.

IDAD--See **internal defense and development**.

identification--1. The process of determining the friendly or hostile character of an unknown detected contact. 2. In arms control, the process of determining which nation is responsible for the detected violations of any arms control measure. 3. In ground combat operations, discrimination between recognizable objects as being friendly or enemy, or the name that belongs to the object as a member of a class.

identification, friend or foe(*)--A system using electromagnetic transmissions to which equipment carried by friendly forces automatically responds, for example, by emitting pulses, thereby distinguishing themselves from enemy forces. Also called **IFF**.

Identification Friend or Foe personal identifier--The discrete Identification Friend or Foe code assigned to a particular aircraft, ship, or other vehicle for identification by electronic means.

identification, friend or foe/selective identification feature procedures--The directives that govern the use of identification, friend or foe selective identification feature equipment. See also **identification, friend or foe**.

identification maneuver--A maneuver performed for identification purposes.

identify--A code meaning, "Identify the contact designated by any means at your disposal." See also **identification, recognition**.

identity--See **identification; recognition**.

IDT--See **inactive duty training**.

IFF--See **identification, friend or foe**.

igloo space--Area in an earth-covered structure of concrete and/or steel designed for the storage of ammunition and explosives. See also **storage**.

igniter(*)--A device designed to produce a flame or flash which is used to initiate an explosive train.

ignition system--See **firing system**.

I go--A code meaning, "I am leaving my patrol/mission in _____ minutes." See also **I stay**.

III--See **incapacitating illness or injury**.

illumination by diffusion--See **indirect illumination**.

illumination by reflection--See **indirect illumination**.

image format--Actual size of negative, scope, or other medium on which image is produced.

image motion compensation(*)--Movement intentionally imparted to film at such a rate as to compensate for the forward motion of an air or space vehicle when photographing ground objects.

imagery(*)--Collectively, the representations of objects reproduced electronically or by optical means on film, electronic display devices, or other media.

imagery collateral(*)--The reference materials which support the imagery interpretation function.

181

imagery correlation(*)--The mutual relationship between the different signatures on imagery from different types of sensors in terms of position and the physical characteristics signified.

imagery data recording(*)--The transposing of information relating to the airborne vehicle and sensor, such as speed, height, tilt, position, and time, to the matrix block on the sensor record at the moment of image acquisition.

imagery exploitation(*)--The cycle of processing and printing imagery to the positive or negative state, assembly into imagery packs, identification, interpretation, mensuration, information extraction, the preparation of reports, and the dissemination of information.

imagery intelligence--Intelligence derived from the exploitation of collection by visual photography, infrared sensors, lasers, electro-optics, and radar sensors such as synthetic aperture radar wherein images of objects are reproduced optically or electronically on film, electronic display devices, or other media. Also called **IMINT.** See also **intelligence; photographic intelligence.**

imagery interpretation(*)--1. The process of location, recognition, identification, and description of objects, activities, and terrain represented on imagery. **(NATO)** 2. The extraction of information from photographs or other recorded images.

imagery interpretation key(*)--Any diagram, chart, table, list, or set of examples, etc., which is used to aid imagery interpreters in the rapid identification of objects visible on imagery.

imagery pack(*)--An assembly of the records from different imagery sensors covering a common target area.

imagery sortie(*)--One flight by one aircraft for the purpose of recording air imagery.

IMC--See **instrument meteorological conditions.**

IMINT--See **imagery intelligence.**

imitative electromagnetic deception--See **electromagnetic deception.**

immediate air support(*)--Air support to meet specific requests which arise during the course of a battle and which by their nature cannot be planned in advance. See also **air support.**

immediate decontamination(*)--Decontamination carried out by individuals upon coming contaminated, to save life and minimize casualties. This may include decontamination of some personal clothing and/or equipment. See also **decontamination, operational decontamination, thorough decontamination.**

immediate destination(*)--The next destination of a ship or convoy, irrespective of whether or not onward routing instructions have been issued to it.

immediately vital cargo(*)--A cargo already loaded which the consignee country regards as immediately vital for the prosecution of the war or for national survival, notwithstanding the risk to the ship. If the cargo is carried in a ship of another nation, then that nation must agree to the delivery of the cargo. The use of this term is limited to the period of implementation of the shipping movement policy. See also **cargo.**

immediate message--A category of precedence reserved for messages relating to situations that gravely affect the security of national/allied forces or populace and that require immediate delivery to the addressee(s). See also **precedence.**

immediate mission request--A request for an air strike on a target which, by its nature, could not be identified sufficiently in advance to permit detailed mission coordination and planning. See also **preplanned mission request.**

immediate mission request (reconnaissance)--A request for a mission on a target which, by its nature, could not be identified sufficiently in advance to permit detailed mission coordination and planning.

immediate nuclear support--Nuclear support to meet specific requests which arise during the course of a battle, and which, by their nature, cannot be planned in advance. See also **preplanned nuclear support; nuclear support.**

immediate operational readiness--Those operations directly related to the assumption of an alert or quick-reaction posture. Typical operations include strip alert, airborne alert/indoctrination, no-notice launch of an alert force, and the maintenance of missiles in an alert configuration. See also **nuclear weapon exercise; nuclear weapon maneuver.**

impact action fuze(*)--A fuze that is set in action by the striking of a projectile or bomb against an object, e.g., percussion fuze, contact fuze. Synonymous with direct action fuze. See also **fuze.**

impact area--An area having designated boundaries within the limits of which all ordnance will detonate or impact.

impact point--See **point of impact.**

impact pressure(*)--The difference between pitot pressure and static pressure.

implementation--Procedures governing the mobilization of the force and the deployment, employment, and sustainment of military operations in response to execution orders issued by the National Command Authorities.

implementation planning--Operational planning associated with the conduct of a continuing operation, campaign, or war to attain defined objectives. At the national level, it includes the development of strategy and the assignment of strategic tasks to the combatant commanders. At the theater level, it includes the development of campaign plans to attain assigned objectives and the preparation of operation plans and operation orders to prosecute the campaign. At lower levels, implementation planning prepares for the execution of assigned tasks or logistic missions. See also **joint operation planning.**

implosion weapon--A weapon in which a quantity of fissionable material, less than a critical mass at ordinary pressure, has its volume suddenly reduced by compression (a step accomplished by using chemical explosives) so that it becomes supercritical, producing a nuclear explosion.

imprest fund--A cash fund of a fixed amount established through an advance of funds, without appropriation change, to an authorized imprest fund cashier to effect immediate cash payments of relatively small amounts for authorized purchases of supplies and nonpersonal services.

imprint(*)--Brief note in the margin of a map giving all or some of the following: date of publication, printing, name of publisher, printer, place of publication, number of copies printed, and related information.

improved conventional munitions--Munitions characterized by the delivery of two or more antipersonnel or antimateriel and/or antiarmor submunitions by an artillery warhead or projectile.

improvised early resupply(*)--The onward movement of commodities which are available on land and which can be readily loaded into ships. See also **element of resupply.**

improvised explosive device(*)--A device placed or fabricated in an improvised manner incorporating destructive, lethal, noxious, pyrotechnic, or incendiary chemicals and designed to destroy, incapacitate, harass, or distract. It may incorporate military stores, but is normally devised from nonmilitary components.

improvised mine--A mine fabricated from available materials at or near its point of use.

improvised nuclear device--A device incorporating radioactive materials designed to result in the dispersal of radioactive material or in the formation of nuclear-yield reaction. Such devices may be fabricated in a completely improvised manner or may be an improvised modification to a US or foreign nuclear weapon. Also called **IND.**

inactive aircraft inventory--Aircraft in storage or bailment and/or government-furnished equipment on loan or lease outside of the Defense establishment or otherwise not available to the Military Services.

inactive duty training--Authorized training performed by a member of a Reserve component not on active duty or active duty for training and consisting of regularly scheduled unit training assemblies, additional training assemblies, periods of appropriate duty or equivalent training, and any special additional duties authorized for Reserve component personnel by the Secretary concerned, and performed by them in connection with the prescribed activities of the organization in which they are assigned with or without pay. Does not include work or study associated with correspondence courses. Also called **IDT.** See also **active duty for training.**

inactive status--Status of reserve members on an inactive status list of a reserve component or assigned to the Inactive Army National Guard. Those in an inactive status may not train for points or pay, and may not be considered for promotion.

inbound traffic--Traffic originating in an area outside continental United States destined for or moving in the general direction of continental United States.

incapacitating agent--An agent that produces temporary physiological or mental effects, or both, which will render individuals incapable of concerted effort in the performance of their assigned duties.

incapacitating illness or injury--The casualty status of a person whose illness or injury requires hospitalization but medical authority does not classify as very seriously ill or injured or seriously ill or injured and the illness or injury makes the person physically or mentally unable to communicate with the next of kin. Also called **III.** See also **casualty status.**

incentive type contract--A contract that may be of either a fixed price or cost reimbursement nature, with a special provision for adjustment of the fixed price or fee. It provides for a tentative target price and a maximum price or maximum fee, with price or fee adjustment after completion of the contract for the purpose of establishing a final price or fee based on the contractor's actual costs plus a sliding scale of profit or fee that varies inversely with the cost but which in no event shall permit the final price or fee to exceed the maximum price or fee stated in the contract. See also **cost contract; fixed price type contract.**

incident classification--See **search and rescue incident classification.**

incident control point--A designated point close to a terrorist incident where crisis management forces will rendezvous and establish control capability before initiating a tactical reaction. See also **antiterrorism**.

incidents--Brief clashes or other military disturbances generally of a transitory nature and not involving protracted hostilities.

inclination angle--See **pitch angle**.

IND--See **improvised nuclear device**.

indefinite call sign(*)--A call sign which does not represent a specific facility, command, authority, activity, or unit, but which may represent any one or any group of these. See also **call sign**.

indefinite delivery type contract--A type of contract used for procurements where the exact time of delivery is not known at time of contracting.

independent(*)--A merchant ship under naval control sailed singly and unescorted by a warship. See also **military independent**.

independent ejection system--See **ejection systems**.

independent mine(*)--A mine which is not controlled by the user after laying. See also **mine**.

independent review--In computer modeling and simulation, a review performed by competent, objective reviewers who are independent of the model developer. Independent review includes either (a) a detailed verification and/or validation of the model or simulation; or (b) an examination of the verification and/or validation performed by the model or simulation developer. See also **accreditation; configuration nagement; validation; verification**.

indicated airspeed--See **airspeed**.

indicated airspeed hold mode--In a flight control system, a control mode in which desired indicated airspeed of an aircraft is maintained automatically.

indicating--In air intercept, a code meaning, "Contact speed, by plot, is _____."

indications and warning--Those intelligence activities intended to detect and report time--sensitive intelligence information on foreign developments that could involve a threat to the United States or allied military, political, or economic interests or to US citizens abroad. It includes forewarning of enemy actions or intentions; the imminence of hostilities; insurgency; nuclear/non-nuclear attack on the United States, its overseas forces, or allied nations; hostile reactions to United States reconnaissance activities; terrorists' attacks; and other similar events.

indications (intelligence)--Information in various degrees of evaluation, all of which bears on the intention of a potential enemy to adopt or reject a course of action.

indicator(*)--In intelligence usage, an item of information which reflects the intention or capability of a potential enemy to adopt or reject a course of action.

indirect air support--All forms of air support provided to land or naval forces which do not immediately assist those forces in the tactical battle.

indirect fire--Fire delivered on a target that is not itself used as a point of aim for the weapons or the director.

indirect illumination(*)--Battlefield illumination provided by employing searchlight or pyrotechnic illuminants using diffusion or reflection. **a.**

Illumination by diffusion: Illumination of an area beneath and to the flanks of a slightly elevated searchlight or of pyrotechnic illuminants, by the light scattered from atmospheric particles. **b. Illumination by reflection:** Illumination of an area by reflecting light from low cloud. Either or both of these effects are present when a searchlight is used in defilade or with its beam spread to maximum width.

indirect laying(*)--Aiming a gun either by sighting at a fixed object, called the aiming point, instead of the target or by using a means of pointing other than a sight, such as a gun director, when the target cannot be seen from the gun position.

individual equipment--Referring to method of use, signifies personal clothing and equipment, for the personal use of the individual. See also **equipment.**

individual mobilization augmentee--An individual reservist attending drills who receives training and is preassigned to an active component organization, a Selective Service System, or a Federal Emergency Management Agency billet that must be filled on, or shortly after, mobilization. Individual mobilization augmentees train on a part-time basis with these organizations to prepare for mobilization. Inactive duty training for individual mobilization augmentees is decided by component policy and can vary from 0 to 48 drills a year.

individual mobilization augmentee detachment---An administrative unit organized to train and manage individual mobilization augmentees.

individual protective equipment(*)--In nuclear, biological and chemical warfare, the personal clothing and equipment required to protect an individual from biological and chemical hazards and some nuclear effects.

individual ready reservist--A member of the Ready Reserve not assigned to the Selected Reserve and not on active duty.

individual reserves--The supplies carried on a soldier, animal, or vehicle for individual use in an emergency. See also **reserve supplies.**

individual sponsored dependent--A dependent not entitled to travel to the overseas command at Government expense or who enters the command without endorsement of the appropriate overseas commander.

induced environment--Any manmade or equipment-made environment which directly or indirectly affects the performance of man or materiel.

induced precession(*)--A precession resulting from a torque deliberately applied to a gyro.

induced radiation(*)--Radiation produced as a result of exposure to radioactive materials, particularly the capture of neutrons. See also **contamination; initial radiation; residual radiation; residual radioactivity.**

induction circuit(*)--In naval mine warfare, a circuit actuated by the rate of change in a magnetic field due to the movement of the ship or the changing current in the sweep.

industrial mobilization--The transformation of industry from its peacetime activity to the industrial program necessary to support the national military objectives. It includes the mobilization of materials, labor, capital, production facilities, and contributory items and services essential to the industrial program. See also **mobilization.**

industrial preparedness--The state of preparedness of industry to produce essential materiel to support the national military objectives.

industrial preparedness program--Plans, actions, or measures for the transformation of the industrial base, both government-owned and civilian-owned, from its peacetime activity to the emergency program necessary to support the national military objectives. It includes industrial preparedness measures such as modernization, expansion, and preservation of the production facilities and contributory items and services for planning with industry.

industrial property--As distinguished from military property, means any contractor acquired or government-furnished property, including materials, special tooling, and industrial facilities, furnished or acquired in the performance of a contract or subcontract.

industrial readiness--See **industrial preparedness.**

inert filling(*)--A prepared non-explosive filling of the same weight as the explosive filling.

inertial guidance--A guidance system designed to project a missile over a predetermined path, wherein the path of the missile is adjusted after launching by devices wholly within the missile and independent of outside information. The system measures and converts accelerations experienced to distance traveled in a certain direction. See also **guidance.**

inertial navigation system(*)--A self-contained navigation system using inertial detectors, which automatically provides vehicle position, heading, and velocity.

inert mine(*)--A mine or replica of a mine incapable of producing an explosion.

in extremis--A situation of such exceptional urgency that immediate action must be taken to minimize imminent loss of life or catastrophic degradation of the political or military situation.

infiltration--1. The movement through or into an area or territory occupied by either friendly or enemy troops or organizations. The movement is made, either by small groups or by individuals, at extended or irregular intervals. When used in connection with the enemy, it infers that contact is avoided. 2. In intelligence usage, placing an agent or other person in a target area in hostile territory. Usually involves crossing a frontier or other guarded line. Methods of infiltration are: black (clandestine); grey (through legal crossing point but under false documentation); white (legal).

inflammable cargo--Cargo such as drummed gasoline and oils. See also **cargo.**

inflight phase--The flight of a missile or space vehicle from launch to detonation or impact.

inflight report--The transmission from the airborne system of information obtained both at the target and en route.

influence field(*)--The distribution in space of the influence of a ship or minesweeping equipment.

influence mine--A mine actuated by the effect of a target on some physical condition in the vicinity of the mine or on radiations emanating from the mine; includes acoustic, magnetic, pressure, seismic, and underwater potential. See also **mine.**

influence release sinker(*)--A sinker which holds a moored or rising mine at the sea-bed and releases it when actuated by a suitable ship influence.

influence sweep(*)--A sweep designed to produce an influence similar to that produced by a ship and thus actuate mines.

informant--1. A person who, wittingly or unwittingly, provides information to an agent, a clandestine service, or the police. 2. In reporting, a person who has provided specific information and is cited as a source.

information(*)--1. Unprocessed data of every description which may be used in the production of intelligence. (DOD) 2. The meaning that a human assigns to data by means of the known conventions used in their representation. See also **combat information; intelligence cycle.**

information box(*)--A space on an annotated overlay, mosaic, map, etc., which is used for identification, reference, and scale information. See also **reference box.**

information processing--See **intelligence cycle.**

information report--Report used to forward raw information collected to fulfill intelligence requirements.

information requirements(*)--Those items of information regarding the enemy and his environment which need to be collected and processed in order to meet the intelligence requirements of a commander. See also **priority intelligence requirements.**

informer--Person who intentionally discloses to police or to a security service information about persons or activities considered suspect, usually for a financial reward.

infrared film--Film carrying an emulsion especially sensitive to "near-infrared." Used to photograph through haze, because of the penetrating power of infrared light; and in camouflage detection to distinguish between living vegetation and dead vegetation or artificial green pigment.

infrared imagery--That imagery produced as a result of sensing electromagnetic radiations emitted or reflected from a given target surface in the infrared position of the electromagnetic

spectrum (approximately 0.72 to 1,000 microns).

infrared linescan system(*)--A passive airborne infrared recording system which scans across the ground beneath the flight path, adding successive lines to the record as the vehicle advances along the flight path.

infrared photography--Photography employing an optical system and direct image recording on film sensitive to near-infrared wavelength (infrared film). Note: Not to be confused with infrared imagery.

infrared radiation(*)--Radiation emitted or reflected in the infrared portion of the electromagnetic spectrum.

infrastructure(*)--A term generally applicable to all fixed and permanent installations, fabrications, or facilities for the support and control of military forces. See also **bilateral infrastructure; common infrastructure; national infrastructure.**

initial active duty for training--Basic military training and technical skill training required for all accessions. For nonprior service male enlistees, between the ages of 18 1/2 and 26, initial active duty for training shall be not less than 12 weeks and start insofar as practical within 270 days after enlistment. Initial active duty for training for all other enlistees and inductees shall be prescribed by the Secretary concerned and start insofar as practical within 360 days of entry into the Service, except in time of war or national emergency declared by Congress or the President when basic training shall be not less than 12 weeks or its equivalent. Reservists may not be assigned to active duty on land outside the United States or its territories and possessions until basic training has been completed.

initial approach(*)--a. That part of an instrument approach procedure in which the aircraft has departed an initial approach fix or point and is maneuvering to enter the intermediate or final approach. It ends at the intermediate fix or point or, where no intermediate segment is established, at the final approach fix or point. b. That part of a visual approach of an aircraft immediately prior to arrival over the airfield of destination, or over the reporting point from which the final approach to the airfield is commenced.

initial approach area(*)--An area of defined width lying between the last preceding navigational fix or dead reckoning position and either the facility to be used for making an instrument approach or a point associated with such a facility that is used for demarcating the termination of initial approach.

initial assessment--An assessment that provides a basic determination of the viability of the infiltration and exfiltration portion of a proposed special operations forces mission. Also called **IA.**

initial contact report--See **contact report.**

initial draft plan(*)--A plan which has been drafted and coordinated by the originating headquarters, and is ready for external coordination with other military headquarters. It cannot be directly implemented by the issuing commander, but it may form the basis for an operation order issued by the commander in the event of an emergency. See also **draft plan; coordinated draft plan; final plan; operation plan.**

initial early resupply(*)--The onward movement of ships which are already loaded with cargoes which will serve the requirements after D-day. This includes such shipping evacuation from major ports/major water terminals and subsequently dispersed to secondary ports/alternate water terminals and anchorages. See also **element of resupply.**

initial entry into Military Service--Entry for the first time into military status (active duty or reserve) by induction, enlistment, or appointment in any Service of the Armed Forces of the United States. Appointment may be as a commissioned or warrant officer; as a cadet or midshipman at the Service academy of one of the armed forces; or as a midshipman, US Naval Reserve, for US Naval Reserve Officers' Training Corps training at a civilian institution.

initial issues--The issue of materiel not previously furnished to an individual or organization, including new inductees and newly activated organizations, and the issue of newly authorized items of materiel.

initial operational capability--The first attainment of the capability to employ effectively a weapon, item of equipment, or system of approved specific characteristics, and which is manned or operated by an adequately trained, equipped, and supported military unit or force. Also called **IOC.**

initial path sweeping(*)--In naval mine warfare, initial sweeping to clear a path through a mined area dangerous to the following mine sweepers. See also **precursor sweeping.**

initial photo interpretation report--A first-phase interpretation report, subsequent to the Joint Tactical Air Reconnaissance/Surveillance Mission Report, presenting the results of the initial readout of new imagery to answer the specific requirements for which the mission was requested.

initial point--1. The first point at which a moving target is located on a plotting board. 2. A well-defined point, easily distinguishable visually and/or electronically, used as a starting point for the bomb run to the target. **3. air-**

borne--A point close to the landing area where serials (troop carrier air formations) make final alterations in course to pass over individual drop or landing zones. **4. helicopter**--An air control point in the vicinity of the landing zone from which individual flights of helicopters are directed to their prescribed landing sites. **5.** Any designated place at which a column or element thereof is formed by the successive arrival of its various subdivisions, and comes under the control of the commander ordering the move. See also **target approach point.**

initial programmed interpretation report(*)--A standardized imagery interpretation report providing information on programmed mission objectives or other vital intelligence information which can be readily identified near these objectives, and which has not been reported elsewhere. Also called **IPIR.**

initial provisioning--The process of determining the range and quantity of items (i.e., spares and repair parts, special tools, test equipment, and support equipment) required to support and maintain an item for an initial period of service. Its phases include the identification of items of supply, the establishment of data for catalog, technical manual, and allowance list preparation, and the preparation of instructions to assure delivery of necessary support items with related end articles.

initial radiation(*)--The radiation, essentially neutrons and gamma rays, resulting from a nuclear burst and emitted from the fireball within one minute after burst. See also **induced radiation; residual radiation.**

initial reserves--In an amphibious operations, those supplies which normally are unloaded immediately following the assault waves; usually the supplies for the use of the beach organization, battalion landing teams, and other elements of regimental combat teams for the purpose of initiating and sustaining combat

until higher supply installations are established. See also **reserve supplies.**

initial response force--The first unit, usually military police, on the scene of a terrorist incident. See also **antiterrorism.**

initial unloading period(*)--In amphibious operations, that part of the ship-to-shore movement in which unloading is primarily tactical in character and must be instantly responsive to landing force requirements. All elements intended to land during this period are serialized. See also **general unloading period.**

initial vector--The initial command heading to be assumed by an interceptor after it has been committed to intercept an airborne object.

initial velocity--See **muzzle velocity.**

initiating directive--An order to the commander, amphibious task force, to conduct an amphibious operation. It is issued by the unified commander, subunified commander, Service component commander, or joint force commander delegated overall responsibility for the operation.

initiation of procurement action--That point in time when the approved document requesting procurement and citing funds is forwarded to the procuring activity. See also **procurement lead time.**

injury--A term comprising such conditions as fractures, wounds, sprains, strains, dislocations, concussions, and compressions. In addition, it includes conditions resulting from extremes of temperature or prolonged exposure. Acute poisonings, except those due to contaminated food, resulting from exposure to a toxic or poisonous substance are also classed as injuries. See also **battle casualty; casualty; nonbattle casualty; wounded.**

inland search and rescue region--The inland areas of continental United States, except waters under the jurisdiction of the United States. See also **search and rescue region.**

inner transport area--In amphibious operations, an area as close to the landing beach as depth of water, navigational hazards, boat traffic, and enemy action permit, to which transports may move to expedite unloading. See also **outer transport area; transport area.**

in-place force--1. A NATO assigned force which, in peacetime, is principally stationed in the designated combat zone of the NATO command to which it is committed. 2. Force within a combatant commander's area of responsibility and under the combatant commander's combatant command (command authority).

inshore patrol(*)--A naval defense patrol operating generally within a naval defense coastal area and comprising all elements of harbor defenses, the coastal lookout system, patrol craft supporting bases, aircraft, and Coast Guard stations.

inspection--In arms control, physical process of determining compliance with arms control measures.

installation--A grouping of facilities, located in the same vicinity, which support particular functions. Installations may be elements of a base. See also **base; base complex.**

installation commander--The individual responsible for all operations performed by an installation. See also **antiterrorism; base commander; installation.**

installation complex--In the Air Force, a combination of land and facilities comprised of a main installation and its noncontiguous properties (auxiliary air fields, annexes and missile fields) which provide direct support to or are supported by that installation. Installation complexes may comprise two or more properties, e.g., a major installation, a minor installation, or a support site, each with its associated annex(es) or support property(ies). See also **major installation; minor installation; support site.**

instructional mine(*)--An inert mine used for instruction and normally sectionalized for this purpose. See also **inert mine.**

instrument approach procedure(*)--A series of predetermined maneuvers for the orderly transfer of an aircraft under instrument flight conditions from the beginning of the initial approach to a landing or to a point from which a landing may be made visually or the missed approach procedure is initiated.

instrument flight(*)--Flight in which the path and attitude of the aircraft are controlled solely by reference to instruments.

instrument landing system(*)--A system of radio navigation intended to assist aircraft in landing which provides lateral and vertical guidance, which may include indications of distance from the optimum point of landing.

instrument meteorological conditions--Meteorological conditions expressed in terms of visibility, distance from cloud, and ceiling; less than minimums specified for visual meteorological conditions. Also called **IMC.** See also **visual meteorological conditions.**

in support(*)--An expression used to denote the task of providing artillery supporting fire to a formation or unit. Liaison and observation are not normally provided. See also **at priority call; direct support.**

in support of(*)--Assisting or protecting another formation, unit, or organization while remaining under original control.

insurgency(*)--An organized movement aimed at the overthrow of a constituted government through use of subversion and armed conflict.

insurgent--Member of a political party who rebels against established leadership. See also **anti-terrorism; counterinsurgency; insurgency.**

integrated fire control system--A system which performs the functions of target acquisition, tracking, data computation, and engagement control, primarily using electronic means assisted by electromechanical devices.

integrated logistics support--A composite of all the support considerations necessary to assure the effective and economical support of a system for its life cycle. It is an integral part of all other aspects of system acquisition and operation. Also called **ILS.**

integrated material management--The exercise of total Department of Defense management responsibility for a Federal Supply Group/Class, commodity, or item by a single agency. It normally includes computation of requirements, funding, budgeting, storing, issuing, cataloging, standardizing, and procuring functions.

integrated priority list--A list of a combatant commander's highest priority requirements, prioritized across Service and functional lines, defining shortfalls in key programs that, in the judgment of the combatant commander, adversely affect the capability of the combatant commander's forces to accomplish their assigned mission. The integrated priority list provides the combatant commander's recommendations for programming funds in the Planning, Programming, and Budgeting System process. Also called **IPL.**

integrated staff(*)--A staff in which one officer only is appointed to each post on the establishment of the headquarters, irrespective of nationality and Service. See also **combined staff; joint staff; parallel staff; staff.**

integrated tactical warning--See **tactical warning.**

integrated warfare--The conduct of military operations in any combat environment wherein opposing forces employ non-conventional weapons in combination with conventional weapons.

integrating circuit(*)--A circuit whose actuation is dependent on the time integral of a function of the influence.

integration--1. A stage in the intelligence cycle in which a pattern is formed through the selection and combination of evaluated information. 2. In photography, a process by which the average radar picture seen on several scans of the time base may be obtained on a print, or the process by which several photographic images are combined into a single image.

intelligence--1. The product resulting from the collection, processing, integration, analysis, evaluation, and interpretation of available information concerning foreign countries or areas. 2. Information and knowledge about an adversary obtained through observation, investigation, analysis, or understanding. See also **acoustic intelligence; all-source intelligence; basic intelligence; civil defense intelligence; combat intelligence; communications intelligence; critical intelligence; current intelligence; departmental intelligence; domestic intelligence; electronics intelligence; electro-optical intelligence; escape and evasion intelligence; foreign intelligence; foreign instrumentation signals intelligence; general military intelligence; human resources intelligence; imagery intelligence; joint intelligence;**

laser intelligence; measurement and signature intelligence; medical intelligence; merchant intelligence; military intelligence; national intelligence; nuclear intelligence; open source intelligence; operational intelligence; photographic intelligence; political intelligence; radar intelligence; radiation intelligence; scientific and technical intelligence; security intelligence; strategic intelligence; tactical intelligence; target intelligence; technical intelligence; technical operational intelligence; telemetry intelligence; terrain intelligence; unintentional radiation intelligence.

intelligence annex--A supporting document of an operation plan or order that provides detailed information on the enemy situation, assignment of intelligence tasks, and intelligence administrative procedures.

intelligence collection plan--A plan for gathering information from all available sources to meet an intelligence requirement. Specifically, a logical plan for transforming the essential elements of information into orders or requests to sources within a required time limit. See also **intelligence cycle.**

intelligence contingency funds--Appropriated funds to be used for intelligence activities when the use of other funds is not applicable or would either jeopardize or impede the mission of the intelligence unit.

intelligence cycle--The steps by which information is converted into intelligence and made available to users. There are five steps in the cycle: **a. planning and direction**--Determination of intelligence requirements, preparation of a collection plan, issuance of orders and requests to information collection agencies, and a continuous check on the productivity of collection agencies. **b. collection**--Acquisition of information and the provision of this information to processing and/or production elements. **c.**

processing--Conversion of collected information into a form suitable to the production of intelligence. **d. production**--Conversion of information into intelligence through the integration, analysis, evaluation, and interpretation of all source data and the preparation of intelligence products in support of known or anticipated user requirements. **e. dissemination**--Conveyance of intelligence to users in a suitable form.

intelligence data base--The sum of holdings of intelligence data and finished intelligence products at a given organization.

intelligence data handling systems--Information systems that process and manipulate raw information and intelligence data as required. They are characterized by the application of general purpose computers, peripheral equipment, and automated storage and retrieval equipment for documents and photographs. While automation is a distinguishing characteristic of intelligence data handling systems, individual system components may be either automated or manually operated.

intelligence doctrine--Fundamental principles that guide the preparation and subsequent provision of intelligence to a commander and staff to aid in planning and conducting military operations. See also **doctrine; joint doctrine; joint intelligence doctrine.**

intelligence estimate(*)--The appraisal, expressed in writing or orally, of available intelligence relating to a specific situation or condition with a view to determining the courses of action open to the enemy or potential enemy and the order of probability of their adoption.

intelligence journal--A chronological log of intelligence activities covering a stated period, usually 24 hours. It is an index of reports and messages that have been received and transmitted, and of important events that have occurred, and actions taken. The journal is a permanent and official record.

intelligence operations--The variety of intelligence tasks that are carried out by various intelligence organizations and activities. Predominantly, it refers to either intelligence collection or intelligence production activities. When used in the context of intelligence collection activities, intelligence operations refer to collection, processing, exploitation, and reporting of information. When used in the context of intelligence production activities, it refers to collation, integration, interpretation, and analysis, leading to the dissemination of a finished product.

intelligence preparation of the battlespace--An analytical methodology employed to reduce uncertainties concerning the enemy, environment, and terrain for all types of operations. Intelligence preparation of the battlespace builds an extensive data base for each potential area in which a unit may be required to operate. The data base is then analyzed in detail to determine the impact of the enemy, environment, and terrain on operations and presents it in graphic form. Intelligence preparation of the battlespace is a continuing process. Also called **IPB.**

intelligence-related activities--1. Those activities outside the consolidated defense intelligence program which: a. Respond to operational commanders' tasking for time-sensitive information on foreign entities; b. Respond to national intelligence community tasking of systems whose primary mission is support to operating forces; c. Train personnel for intelligence duties; d. Provide an intelligence reserve; or e. Are devoted to research and development of intelligence or related capabilities. 2. Specifically excluded are programs which are so closely integrated with a weapon system that their primary function is to provide immediate-use targeting data.

intelligence report--A specific report of information, usually on a single item, made at any level of command in tactical operations and disseminated as rapidly as possible in keeping with the timeliness of the information. Also called INTREP.

intelligence reporting--The preparation and conveyance of information by any means. More commonly, the term is restricted to reports as they are prepared by the collector and as they are transmitted by the collector to the latter's headquarters and by this component of the intelligence structure to one or more intelligence-producing components. Thus, even in this limited sense, reporting embraces both collection and dissemination. The term is applied to normal and specialist intelligence reports. See also **normal intelligence reports; specialist intelligence reports.**

intelligence requirement--Any subject, general or specific, upon which there is a need for the collection of information, or the production of intelligence. See also **essential elements of information; priority intelligence requirements.**

intelligence subject code--A system of subject and area references to index the information contained in intelligence reports as required by a general intelligence document reference service.

intelligence summary--A specific report providing a summary of items of intelligence at frequent intervals. See also **intelligence.**

intelligence system--Any formal or informal system to manage data gathering, to obtain and process the data, to interpret the data, and to provide reasoned judgments to decisionmakers as a basis for action. The term is not limited to intelligence organizations or services but includes any system, in all its parts, that accomplishes the listed tasks.

intensity factor(*)--A multiplying factor used in planning activities to evaluate the foreseeable intensity or the specific nature of an operation in a given area for a given period of time. It is applied to the standard day of supply in order to calculate the combat day of supply.

intensity mine circuit(*)--A circuit whose actuation is dependent on the field strength reaching a level differing by some pre-set minimum from that experienced by the mine when no ships are in the vicinity.

intensive management--The continuous process by which the supported and supporting commanders, the Services, transportation component commands, and appropriate Defense agencies ensure that movement data in the Joint Operation Planning and Execution System time-phased force and deployment data for the initial days of deployment and/or mobilization are current to support immediate execution.

intention--An aim or design (as distinct from capability) to execute a specified course of action.

intercepting search(*)--A type of search designed to intercept an enemy whose previous position is known and the limits of whose subsequent course and speed can be assumed.

interceptor(*)--A manned aircraft utilized for identification and/or engagement of airborne objects.

intercept point(*)--The point to which an airborne vehicle is vectored or guided to complete an interception.

intercept receiver(*)--A receiver designed to detect and provide visual and/or aural indication of electromagnetic emissions occurring within the particular portion of the electro-magnetic spectrum to which it is tuned.

interchangeability(*)--A condition which exists when two or more items possess such functional and physical characteristics as to be equivalent in performance and durability, and are capable of being exchanged one for the other without alteration of the items themselves, or of adjoining items, except for adjustment, and without selection for fit and performance. See also **compatibility.**

inter-chart relationship diagram(*)--A diagram on a map or chart showing names and/or numbers of adjacent sheets in the same or related series. Also called **index to adjoining sheets.** See also **map index.**

intercoastal traffic--Sea traffic between Atlantic, Gulf, and Great Lakes continental United States ports and Pacific continental United States ports.

intercom--A telephone apparatus by means of which personnel can talk to each other within an aircraft, tank, ship, or activity.

interconnection--The linking together of interoperable systems.

intercontinental ballistic missile--A ballistic missile with a range capability from about 3,000 to 8,000 nautical miles.

intercount dormant period(*)--In naval mine warfare, the period after the actuation of a ship counter before it is ready to receive another actuation.

interdepartmental/agency support--Provision of logistic and/or administrative support in services or materiel by one or more Military Services to one or more departments or agencies of the United States Government (other than military) with or without reimbursement. See also **international logistic support; inter-Service support; support.**

interdepartmental intelligence--Integrated departmental intelligence that is required by departments and agencies of the United States Government for the execution of their missions but which transcends the exclusive competence of a single department or agency to produce.

interdiction--An action to divert, disrupt, delay, or destroy the enemy's surface military potential before it can be used effectively against friendly forces. See also **air interdiction.**

interface--A boundary or point common to two or more similar or dissimilar command and control systems, sub-systems, or other entities against which or at which necessary information flow takes place.

interim financing--Advance payments, partial payments, loans, discounts, advances, and commitments in connection therewith; and guarantees of loans, discounts, advances, and commitments in connection therewith; and any other type of financing necessary for both performance and termination of contracts.

interim JTIDS message specification--See **tactical digital information link.**

interim overhaul--An availability for the accomplishment of necessary repairs and urgent alterations at a naval shipyard or other shore--based repair activity, normally scheduled halfway through the established regular overhaul cycle.

inter-look dormant period(*)--In mine warfare, the time interval after each look in a multi-look mine, during which the firing mechanism will not register.

intermediate approach(*)--That part of an instrument approach procedure in which aircraft configuration, speed and positioning adjustments are made. It blends the initial approach segment into the final approach segment. It

begins at the intermediate fix or point and ends at the final approach fix or point.

Intermediate Force Planning Level--The force level established during Planning Force development to depict the buildup from the Current Force to the Planning Force. The Intermediate Force Planning Level is insufficient to carry out strategy with a reasonable assurance of success and consequently cannot be referred to as the Planning Force. See also **Current Force; force; Programmed Forces.**

intermediate maintenance (field)--That maintenance which is the responsibility of and performed by designated maintenance activities for direct support of using organizations. Its phases normally consist of: a. calibration, repair, or replacement of damaged or unserviceable parts, components, or assemblies; b. the emergency manufacture of nonavailable parts; and c. providing technical assistance to using organizations.

intermediate marker (land mine warfare)(*)--A marker, natural, artificial or specially installed, which is used as a point of reference between the landmark and the minefield. See also **marker (land mine warfare).**

intermediate objective(*)--In land warfare, an area or feature between the line of departure and an objective which must be seized and/or held.

intermediate-range ballistic missile--A ballistic missile with a range capability from about 1,500 to 3,000 nautical miles.

intermediate-range bomber aircraft--A bomber designed for a tactical operating radius of between 1,000 to 2,500 nautical miles at design gross weight and design bomb load.

intermittent arming device(*)--A device included in a mine so that it will be armed only at set times.

intermittent illumination(*)--A type of fire in which illuminating projectiles are fired at irregular intervals.

internal audience--US military members and civilian employees and their immediate families. One of the audiences comprising the concept of "Publics." See also **external audience; public.**

internal defense and development--The full range of measures taken by a nation to promote its growth and to protect itself from subversion, lawlessness, and insurgency. It focuses on building viable institutions (political, economic, social, and military) that respond to the needs of society. Also called **IDAD.** See also **foreign internal defense.**

internal radiation(*)--Nuclear radiation (alpha and beta particles and gamma radiation) resulting from radioactive substances in the body.

internal security--The state of law and order prevailing within a nation.

international arms control organization--An appropriately constituted organization established to supervise and verify the implementation of arms control measures.

International Atomic Time--The time reference scale established by the Bureau International des Poids et Mesures on the basis of atomic clock readings from various laboratories around the world. Also called **TAI.**

international call sign(*)--A call sign assigned in accordance with the provisions of the International Telecommunications Union to identify a radio station. The nationality of the radio station is identified by the first or the first two characters. (When used in visual signaling, international call signs are referred to as "signal letters.") See also **call sign.**

international cooperative logistics(*)--Cooperation and mutual support in the field of logistics through the coordination of policies, plans, procedures, development activities, and the common supply and exchange of goods and services arranged on the basis of bilateral and multilateral agreements with appropriate cost reimbursement provisions.

international date line(*)--The line coinciding approximately with the anti-meridian of Greenwich, modified to avoid certain habitable land. In crossing this line there is a date change of one day. Also called **date line.**

international identification code(*)--In railway terminology, a code which identifies a military train from point of origin to final destination. The code consists of a series of figures, letters, or symbols indicating the priority, country of origin, day of departure, national identification code number, and country of destination of the train.

international loading gauge (GIC)(*)--The loading gauge upon which international railway agreements are based. A load whose dimensions fall within the limits of this gauge may move without restriction on most of the railways of Continental Western Europe. GIC is an abbreviation for "gabarit international de chargement," formerly called PPI.

international logistics--The negotiating, planning, and implementation of supporting logistics arrangements between nations, their forces, and agencies. It includes furnishing logistic support (major end items, materiel, and/or services) to, or receiving logistic support from, one or more friendly foreign governments, international organizations, or military forces, with or without reimbursement. It also includes planning and actions related to the intermeshing of a significant element, activity, or component of the military logistics systems or procedures of the United States with those of one or more foreign governments, international organizations, or military forces on a temporary or permanent basis. It includes planning and actions related to the utilization of United States logistics policies, systems, and/or procedures to meet requirements of one or more foreign governments, international organizations, or forces.

international logistic support--The provision of military logistic support by one participating nation to one or more participating nations, either with or without reimbursement. See also **interdepartmental/agency support; inter-Service support; support.**

international military education and training--- Formal or informal instruction provided to foreign military students, units, and forces on a nonreimbursable (grant) basis by offices or employees of the United States, contract technicians, and contractors. Instruction may include correspondence courses; technical, educational or informational publications; and media of all kinds. See also **United States Military Service Funded Foreign Training.**

International Peace Force--An appropriately constituted organization established for the purpose of preserving world peace.

interned--See **missing.**

interocular distance--The distance between the centers of rotation of the eyeballs of an individual or between the oculars of optical instruments.

interoperability(*)--1. The ability of systems, units or forces to provide services to and accept services from other systems, units, or forces and to use the services so exchanged to enable

them to operate effectively together. (DOD) 2. The condition achieved among communications-electronics systems or items of communications-electronics equipment when information or services can be exchanged directly and satisfactorily between them and/or their users. The degree of interoperability should be defined when referring to specific cases.

interoperation--The use of interoperable systems, units, or forces.

interphone--See **intercom.**

interpretability(*)--Suitability of imagery for interpretation with respect to answering adequately requirements on a given type of target in terms of quality and scale. **a. poor**--Imagery is unsuitable for interpretation to answer adequately requirements on a given type of target. **b. fair**--Imagery is suitable for interpretation to answer requirements on a given type of target but with only average detail. **c. good**--Imagery is suitable for interpretation to answer requirements on a given type of target in considerable detail. **d. excellent**--Imagery is suitable for interpretation to answer requirements on a given type of target in complete detail.

interpretation--A stage in the intelligence cycle in which the significance of information is judged in relation to the current body of knowledge.

interrogation (intelligence)--Systematic effort to procure information by direct questioning of a person under the control of the questioner.

inter-Service education--Military education which is provided by one Service to members of another Service. See also **military education; military training.**

inter-Service support--Action by one Military Service or element thereof to provide logistic and/or administrative support to another Military Service or element thereof. Such action can be recurring or nonrecurring in character on an installation, area, or worldwide basis. See also **interdepartmental/agency support; international logistic support; support.**

inter-Service training--Military training which is provided by one Service to members of another Service. See also **military education; military training.**

intertheater--Between theaters or between the continental United States and theaters. See also **intertheater traffic.**

intertheater traffic--Traffic between theaters exclusive of that between the continental United States and theaters.

interval(*)--1. The space between adjacent groups of ships or boats measured in any direction between the corresponding ships or boats in each group. 2. The space between adjacent individuals, ground vehicles, or units in a formation that are placed side by side, measured abreast. 3. The space between adjacent aircraft measured from front to rear in units of time or distance. 4. The time lapse between photographic exposures. 5. At battery right or left, an interval ordered in seconds is the time between one gun firing and the next gun firing. Five seconds is the standard interval. 6. At rounds of fire for effect the interval is the time in seconds between successive rounds from each gun.

interview (intelligence)--To gather information from a person who is aware that information is being given although there is ignorance of the true connection and purposes of the interviewer. Generally overt unless the collector is other than purported to be.

in the dark--In air intercept, a code meaning, "Not visible on my scope."

intracoastal sealift--Shipping used primarily for the carriage of personnel and/or cargo along a coast or into river ports to support operations within a given area.

intra-command exercise(*)--An exercise which involves an identified part of one Major NATO Command or subordinate command.

intransit aeromedical evacuation facility--A medical facility, on or in the vicinity of an air base, that provides limited medical care for intransit patients awaiting air transportation. This type of medical facility is provided to obtain effective utilization of transport airlift within operating schedules. It includes "remain overnight" facilities, intransit facilities at aerial ports of embarkation and debarkation, and casualty staging facilities in an overseas combat area. See also **aeromedical evacuation unit.**

intransit inventory--That materiel in the military distribution system that is in the process of movement from point of receipt from procurement and production (either contractor's plant or first destination, depending upon point of delivery) and between points of storage and distribution.

intransit stock--See **intransit inventory.**

in-transit visibility--The capability provided to a theater combatant commander to have visibility of units, personnel, and cargo while in transit through the Defense Transportation System.

intratheater--Within a theater. See also **intratheater traffic.**

intratheater traffic--Traffic within a theater.

Intruder--A twin-engine, turbojet, two-place, long-range, all-weather, aircraft carrier-based, low-altitude attack aircraft, possessing an integrated attack-navigation and central digital computer system to locate, track, and destroy small moving targets, and large fixed targets. The armament system consists of an assortment of nuclear and/or non-nuclear weapons, Sidewinder, Harpoon, napalm, and all standard Navy rockets. This aircraft can be air refueled. Designated as **A-6.**

intruder(*)--An individual, unit, or weapon system, in or near an operational or exercise area, which presents the threat of intelligence gathering or disruptive activity.

intruder operation(*)--An offensive operation by day or night over enemy territory with the primary object of destroying enemy aircraft in the vicinity of their bases.

intrusion--See **electromagnetic intrusion.**

invasion currency--See **military currency.**

inventory control(*)--That phase of military logistics which includes managing, cataloging, requirements determinations, procurement, distribution, overhaul, and disposal of materiel. Synonymous with materiel control, materiel management, inventory management, and supply management.

inventory control point--An organizational unit or activity within a DOD supply system that is assigned the primary responsibility for the materiel management of a group of items either for a particular Service or for the Defense Department as a whole. Materiel inventory management includes cataloging direction, requirements computation, procurement direction, distribution management, disposal direction, and, generally, rebuild direction.

inventory management--See **inventory control.**

inventory managers--See **inventory control point.**

inverter(*)--In electrical engineering, a device for converting direct current into alternating current. See also **rectifier.**

investigation--A duly authorized, systematized, detailed examination or inquiry to uncover facts and determine the truth of a matter. This may include collecting, processing, reporting, storing, recording, analyzing, evaluating, producing, and disseminating the authorized information.

investment costs--Those program costs required beyond the development phase to introduce into operational use a new capability; to procure initial, additional, or replacement equipment for operational forces; or to provide for major modifications of an existing capability. They exclude research, development, test and evaluation, military personnel, and Operation and Maintenance appropriation costs.

ionization(*)--The process of producing ions by the removal of electrons from, or the addition of electrons to, atoms or molecules.

ionosphere--That part of the atmosphere, extending from about 70 to 500 kilometers, in which ions and free electrons exist in sufficient quantities to reflect electromagnetic waves.

IPB--See **intelligence preparation of the battlespace.**

IPIR--See **initial programmed interpretation report.**

IPL--See **integrated priority list.**

Iroquois--A light single-rotor helicopter used for cargo/personnel transport and attack helicopter support. Some versions are armed with machine guns and light air-to-ground rockets. Designated as **UH-1.** Also called **Huey.**

irregular forces--Armed individuals or groups who are not members of the regular armed forces, police, or other internal security forces.

irregular outer edge(*)--In land mine warfare, short mine rows or strips laid in an irregular manner in front of a minefield facing the enemy to deceive the enemy as to the type or extent of the minefield. Generally, the irregular outer edge will only be used in minefields with buried mines.

isodose rate line--See **dose rate contour line.**

isotopes--Forms of the same element having identical chemical properties but differing in their atomic masses due to different numbers of neutrons in their respective nuclei and in their nuclear properties.

issue priority designator--See **priority designator.**

I stay--In air intercept, a code meaning, "Am remaining with you on patrol/mission _____ hours." See also **I go.**

item manager--An individual within the organization of an inventory control point or other such organization assigned management responsibility for one or more specific items of materiel.

J

jamming--See **barrage jamming; electronic attack; electromagnetic jamming; selective jamming; spot jamming.**

j-axis--A vertical axis in a system of rectangular coordinates; that line on which distances above or below (north or south) the reference line are marked, especially on a map, chart or graph.

JCMEC--See **joint captured materiel exploitation center.**

JDA--See **joint duty assignment.**

JDAL--See **Joint Duty Assignment List.**

JDISS--See **joint deployable intelligence support system.**

jet advisory service--The service provided certain civil aircraft while operating within radar and nonradar jet advisory areas. Within radar jet advisory areas, civil aircraft receiving this service are provided radar flight following, radar traffic information, and vectors around observed traffic. In nonradar jet advisory areas, civil aircraft receiving this service are afforded standard instrument flight rules separation from all other aircraft known to air traffic control to be operating within these areas.

jet conventional low-altitude bombing system--A maneuver used by jet aircraft to loft conventional ordnance by means of a low-altitude bombing system.

jet propulsion--Reaction propulsion in which the propulsion unit obtains oxygen from the air, as distinguished from rocket propulsion in which the unit carries its own oxygen-producing material. In connection with aircraft propulsion, the term refers to a gasoline or other fuel turbine jet unit that discharges hot gas through a tail pipe and a nozzle which provides a thrust that propels the aircraft. See also **rocket propulsion.**

jet stream--A narrow band of high velocity wind in the upper troposphere or in the stratosphere.

jettison--The selective release of stores from an aircraft other than normal attack.

jettisoned mines(*)--Mines which are laid as quickly as possible in order to empty the minelayer of mines, without regard to their condition or relative positions.

JEZ--See **joint engagement zone.**

JFACC--See **joint force air component commander.**

JFC--See **joint force commander.**

JFLCC--See **joint force land component commander.**

JFMCC--See **joint force maritime component commander.**

JFSOCC--See **joint force special operations component commander.**

JIC--See **joint intelligence center.**

JIF--See **joint interrogation vehicle.**

JMEM-SO--See **Joint Munitions Effectiveness Manual-Special Operations.**

JMFU--See **joint force meteorological and oceanographic forecast unit.**

JMO--See **joint force meteorological and ocean-ographic officer.**

JMP--See **joint manpower program.**

joiner(*)--An independent merchant ship sailed to join a convoy. See also **joiner convoy; joiner section.**

joiner convoy(*)--A convoy sailed to join the main convoy. See also **joiner; joiner section.**

joiner section(*)--A joiner or joiner convoy, after rendezvous, and while maneuvering to integrate with the main convoy.

joint(*)--Connotes activities, operations, organizations, etc., in which elements of more than one Service of the same nation participate. (When all Services are not involved, the participating Services shall be identified, e.g., Joint Army--Navy.) See also **combined.**

joint activities reporting to the Chairman, JCS--Activities or agencies that receive direct guidance from the Chairman of the Joint Chiefs of Staff or his designated Joint Staff agent, and which have joint manpower programs separate from the Joint Staff and unified command organizations. See also **activity; joint; joint staff.**

joint administrative publication--Publication of joint interest dealing with administrative matters prepared under the cognizance of Joint Staff directorates and applicable to the Military Departments, combatant commands, and other authorized agencies. It is authenticated by the Secretary of the Joint Staff "For the Chairman of the Joint Chiefs of Staff" and distributed through Service channels. A joint administrative reference category administrative publication will be approved by the Director of the Joint Staff and is applicable only to the Joint Staff. All other joint administrative publications will be approved by the Chairman of the Joint Chiefs of Staff. See also **joint publication.**

joint airborne training--Training operations or exercises involving airborne and appropriate troop carrier units. This training includes: a. air delivery of personnel and equipment; b. assault operations by airborne troops and/or air transportable units; c. loading exercises and local orientation fights of short duration; and d. maneuvers/exercises as agreed upon by Services concerned and/or as authorized by the Joint Chiefs of Staff.

joint amphibious operation(*)--An amphibious operation conducted by significant elements of two or more Services.

joint amphibious task force--A temporary grouping of units of two or more Services under a single commander, organized for the purpose of engaging in an amphibious landing for assault on hostile shores.

joint base--For purposes of base defense operations, a joint base is a locality from which operations of two or more of the Armed Forces of the Department of Defense are projected or supported and which is manned by significant elements of two or more Services or in which significant elements of two or more Services are located.

joint captured materiel exploitation center--Physical location for deriving intelligence information from captured enemy materiel. It is normally subordinate to the joint force/J-2. Also called **JCMEC.**

joint communications network--The aggregation of all the joint communications systems in a theater. The joint communications network includes the Joint Multi-channel Trunking and Switching System and the Joint Command and Control Communications System(s).

joint deployable intelligence support system--A transportable workstation and communications suite that electronically extends a joint intelligence center to a joint task force or other tactical user. Also called **JDISS.**

joint deployment community--Those headquarters, commands, and agencies involved in the training, preparation, movement, reception, employment, support, and sustainment of military forces assigned or committed to a theater of operations or objective area. The joint deployment community usually consists of the Joint Staff, Services, certain Service major commands (including the Service wholesale logistic commands), unified and specified commands (and their Service component commands), transportation operating agencies, joint task forces (as applicable), Defense Logistics Agency, and other Defense agencies (e.g., Defense Intelligence Agency) as may be appropriate to a given scenario. Also called **JDC.**

joint deployment system--A system that consists of personnel, procedures, directives, communications systems, and electronic data processing systems to directly support time-sensitive planning and execution, and to complement peacetime deliberate planning. Also called **JDS.**

joint doctrine--Fundamental principles that guide the employment of forces of two or more Services in coordinated action toward a common objective. It will be promulgated by the Chairman of the Joint Chiefs of Staff, in coordination with the combatant commands, Services, and Joint Staff. See also **Chairman of the Joint Chiefs of Staff Instruction; combined doctrine; doctrine; guidance; joint publication; joint tactics, techniques, and procedures; joint test publication; multi-Service doctrine.**

Joint Doctrine Working Party--A forum to include representatives of the Services and combatant commands with the purpose of systematic addressal of joint doctrine and joint tactics, techniques, and procedures (JTTP) issues such as project proposal examination, project scope development, project validation, and lead agent recommendation. The Joint Doctrine Working Party meets under the sponsorship of the Director, Operational Plans and Interoperability (J-7). See also **joint doctrine; joint publication; joint tactics, techniques, and procedures; joint test publication.**

joint duty assignment--An assignment to a designated position in a multi-Service, joint or multinational command or activity that is involved in the integrated employment or support of the land, sea, and air forces of at least two of the three Military Departments. Such involvement includes, but is not limited to, matters relating to national military strategy, joint doctrine and policy, strategic planning, contingency planning, and command and control of combat operations under a unified or specified command. Also called **JDA.**

Joint Duty Assignment List--Positions designated as **joint duty assignments are reflected in a list approved by the Secretary of Defense and maintained by the Joint Staff.** The Joint Duty Assignment List is reflected in the Joint Duty Assignment Management Information System. Also called **JDAL.**

joint engagement zone--See **weapon engagement zone.**

joint force--A general term applied to a force composed of significant elements, assigned or attached, of the Army, the Navy or the Marine Corps, and the Air Force, or two or more of these Services, operating under a single commander authorized to exercise operational control. See also **joint force commander.**

joint force air component commander--The joint force air component commander derives authority from the joint force commander who has the authority to exercise operational control, assign missions, direct coordination among subordinate commanders, redirect and organize forces to ensure unity of effort in the accomplishment of the overall mission. The joint force commander will normally designate a joint force air component commander. The joint force air component commander's responsibilities will be assigned by the joint force commander (normally these would include, but not be limited to, planning, coordination, allocation, and tasking based on the joint force commander's apportionment decision). Using the joint force commander's guidance and authority, and in coordination with other Service component commanders and other assigned or supporting commanders, the joint force air component commander will recommend to the joint force commander apportionment of air sorties to various missions or geographic areas. Also called **JFACC.** See also **joint force commander.**

joint force commander--A general term applied to a commander authorized to exercise combatant command (command authority) or operational control over a joint force. Also called **JFC.** See also **joint force.**

joint force land component commander--The commander within a unified command, subordinate unified command, or joint task force responsible to the establishing commander for making recommendations on the proper employment of land forces, planning and coordinating land operations, or accomplishing such operational missions as may be assigned. The joint force land component commander is given the authority necessary to accomplish missions and tasks assigned by the establishing commander. The joint force land component commander will normally be the commander with the preponderance of land forces and the requisite command and control capabilities. Also called **JFLCC.**

joint force maritime component commander--- The commander within a unified command, subordinate unified command, or joint task force responsible to the establishing commander for making recommendations on the proper employment of maritime forces and assets, planning and coordinating maritime operations, or accomplishing such operational missions as may be assigned. The joint force maritime component commander is given the authority necessary to accomplish missions and tasks assigned by the establishing commander. The joint force maritime component commander will normally be the commander with the preponderance of maritime forces and the requisite command and control capabilities. Also called **JFMCC.**

joint force meteorological and oceanographic forecast unit--A flexible, transportable, jointly supported collective of meteorological and oceanographic personnel and equipment formed to provide the joint task force commander and joint force meteorological and oceanographic officer with full meteorological and oceanographic services. Also called **JMFU.**

joint force meteorological and oceanographic officer--Officer designated to provide direct meteorological and oceanographic support to the joint task force commander. Also called **JMO.**

joint force special operations component commander--The commander within a unified command, subordinate unified command, or joint task force responsible to the establishing commander for making recommendations on the proper employment of special operations forces and assets, planning and coordinating special operations, or accomplishing such operational missions as may be assigned. The joint force special operations component commander is given the authority necessary to accomplish missions and tasks assigned by the establishing commander. The joint force special operations component commander will normally be the commander with the preponderance of special operations forces and the requisite command and control capabilities. Also called **JFSOCC.**

joint intelligence--Intelligence produced by elements of more than one Service of the same nation.

joint intelligence architecture--A dynamic, flexible structure that consists of the National Military Joint Intelligence Center, the theater joint intelligence centers, and subordinate joint force joint intelligence centers. This architecture encompasses automated data processing equipment capabilities, communications and information flow requirements, and responsibilities to provide theater and tactical commanders with the full range of intelligence required for planning and conducting operations.

joint intelligence center--The intelligence center of the joint force headquarters. The joint intelligence center is responsible for providing and producing the intelligence required to support the joint force commander and staff, components, task forces and elements, and the national intelligence community. Also called **JIC.** See also **joint intelligence architecture.**

joint intelligence doctrine--Fundamental principles that guide the preparation of intelligence and the subsequent provision of intelligence to support military forces of two or more Services employed in coordinated action. See also **intelligence doctrine.**

joint intelligence liaison element--A liaison element provided by the Central Intelligence Agency in support of a unified command or joint task force.

joint interrogation facility--Physical location for systematic interrogation of enemy prisoners of war to derive tactical intelligence in support of the joint force commander. It is normally subordinate to the joint intelligence center. Also called **JIF.**

joint logistics--The art and science of planning and carrying out, by a joint force commander and staff, logistic operations to support the protection, movement, maneuver, firepower, and sustainment of operating forces of two or more Services of the same nation.

joint manpower program--The document which reflects an activity's mission, functions, organization, current and projected manpower needs, and, when applicable, its required mobilization augmentation. A recommended joint manpower program also identifies and justifies any changes proposed by the commander/director of a joint activity for the next five fiscal years. Also called **JMP.**

joint matters--Matters relating to the integrated employment of land, sea, and air forces, including matters relating to national military strategy, strategic and contingency planning, and command and control of combat operations under a unified command.

joint movement center--The center established to coordinate the employment of all means of transportation (including that provided by allies or host nations) to support the concept of operations. This coordination is accomplished through establishment of transportation policies within the assigned area of responsibility, consistent with relative urgency of need, port and terminal capabilities, transportation asset availability, and priorities set by a joint force commander.

joint multi-channel trunking and switching system--That composite multi-channel trunking and switching system formed from assets of the Services, the Defense Communications System, other available systems, and/or assets controlled by the Joint Chiefs of Staff to provide an operationally responsive, survivable communication system, preferably in a mobile/transportable/recoverableconfiguration,for the joint force commander in an area of operations.

Joint Munitions Effectiveness Manual-Special Operations--A publication providing a single, comprehensive source of information covering weapon effectiveness, selection, and requirements for special operations munitions. In addition, the closely related fields of weapon characteristics and effects, target characteristics, and target vulnerability are treated in limited detail required by the mission planner. Although emphasis is placed on weapons that are currently in the inventory, information is also included for some weapons not immediately available but projected for the near future. Also called **JMEM-SO.**

joint nuclear accident coordinating center--A combined Defense Nuclear Agency and Department of Energy centralized agency for exchanging and maintaining information concerned with radiological assistance capabilities and coordinating assistance activities, when called upon, in connection with accidents involving radioactive materials.

joint operational intelligence agency--An intelligence agency in which the efforts of two or more Services are integrated to furnish that operational intelligence essential to the commander of a joint force and to supplement that available to subordinate forces of the command. The agency may or may not be part of such joint force commander's staff.

joint operational planning process--A coordinated Joint Staff procedure used by a commander to determine the best method of accomplishing assigned tasks and to direct the action necessary to accomplish the mission.

joint operation planning--Joint operation planning activities exclusively associated with the preparation of operation plans, operation plans in concept format, and operation orders (other than the Single Integrated Operation Plan) for the conduct of military operations by the combatant commanders in response to requirements established by the Chairman of the Joint Chiefs of Staff. As such, joint operation planning includes contingency planning, execution planning, and implementation planning. Joint operation planning is performed in accordance with formally established planning and execution procedures. See also **contingency planning; execution planning; implementation planning; Joint Operation Planning and Execution System.**

Joint Operation Planning and Execution System--A continuously evolving system that is being developed through the integration and enhancement of earlier planning and execution systems: Joint Operation Planning System and Joint Deployment System. It provides the foundation for conventional command and control by national- and theater-level commanders and their staffs. It is designed to satisfy their information needs in the conduct of joint planning and operations. Joint Operation Planning and Execution System (JOPES) includes joint operation planning policies, procedures, and reporting structures supported by communications and automated data processing systems. JOPES is used to monitor, plan, and execute mobilization, deployment, employment, and sustainment activities associated with joint operations. Also called **JOPES.** See also **joint operation planning.**

joint operations area--That portion of an area of conflict in which a joint force commander conducts military operations pursuant to an assigned mission and the administration incident to such military operations. Also called **JOA.**

joint operations center--A jointly manned facility of a joint force commander's headquarters established for planning, monitoring, and guiding the execution of the commander's decisions. Also called **JOC.**

joint publication--Publication of joint interest prepared under the cognizance of Joint Staff directorates and applicable to the Military Departments, combatant commands, and other authorized agencies. It is approved by the Chairman of the Joint Chiefs of Staff, in coordination with the combatant commands, Services, and Joint Staff. Also called **JP.** See also **Chairman of the Joint Chiefs of Staff Instruction; guidance; joint administrative publication; joint doctrine; joint tactics, techniques, and procedures; joint test publication.**

joint purchase--A method of purchase whereby purchases of a particular commodity for two or more departments are made by an activity established, staffed, and financed by them jointly for that purpose. See also **purchase.**

joint rear area--A specific land area within a joint force commander's area of operations designated to facilitate protection and operation of installations and forces supporting the joint force.

joint rear area coordinator--The officer with responsibility for coordinating the overall security of the joint rear area in accordance with joint force commander directives and priorities in order to assist in providing a secure environment to facilitate sustainment, host nation support, infrastructure development, and movements of the joint force. The joint rear area coordinator also coordinates intelligence support and ensures that area management is practiced with due consideration for security requirements. Also called **JRAC.**

joint rear area operations--Those operations in the unified and joint rear area that facilitate protection or support of the joint force.

joint rear tactical operations center--A joint operations cell tailored to assist the joint rear area coordinator in meeting mission responsibilities. Also called **JRTOC.**

joint rescue coordination center--See **rescue coordination center.**

joint servicing--That function performed by a jointly staffed and financed activity in support of two or more Military Services. See also **servicing.**

joint special operations air component commander--The commander within the joint force special operations command responsible for planning and executing joint special air operations and for coordinating and deconflicting such operations with conventional nonspecial operations air activities. The joint special operations air component commander normally will be the commander with the preponderance of assets and/or greatest ability to plan, coordinate, allocate, task, control, and support the assigned joint special operations aviation assets. The joint special operations air component commander may be directly subordinate to the joint force special operations component commander or to any nonspecial operations component or joint force commander as directed. Also called **JSOACC.** See also **special operations component commander.**

joint special operations area--A restricted area of land, sea, and airspace assigned by a joint force commander to the commander of a joint special operations force to conduct special operations activities. The commander of joint special operations forces may further assign a specific area or sector within the joint special operations area to a subordinate commander for mission execution. The scope and duration of the special operations forces' mission, friendly and hostile situation, and politico-military considerations all influence the number, composition, and sequencing of special operations forces deployed into a joint special operations area. It may be limited in size to accommodate a discrete direct action mission or may be extensive enough to allow a continuing broad range of unconventional warfare operations. Also called **JSOA.**

joint special operations task force--A joint task force composed of special operations units from more than one Service, formed to carry out a specific special operation or prosecute special operations in support of a theater campaign or other operations. The joint special operations task force may have conventional nonspecial operations units assigned or attached to support the conduct of specific missions. Also called **JSOTF.**

Joint Specialty Officer/joint specialist--An officer on the active duty list who is particularly trained in, and oriented toward, joint matters. Also called **JSO.**

Joint Specialty Officer nominee--An officer who has completed a program of Joint Professional Military Education (JPME), or an officer who has a critical occupational specialty tour. In either instance, the officer has been designated as **a Joint Specialty Officer nominee by the Military Department concerned.**

joint staff--1. The staff of a commander of a unified or specified command, or of a joint task force, which includes members from the several Services comprising the force. These members should be assigned in such a manner as to ensure that the commander understands the tactics, techniques, capabilities, needs, and limitations of the component parts of the force. Positions on the staff should be divided so that Service representation and influence generally reflect the Service composition of the force. **2.**

Joint Staff. The staff under the Chairman of the Joint Chiefs of Staff as provided for in the National Security Act of 1947, as amended by the DOD Reorganization Act of 1986. The Joint Staff assists the Chairman, and subject to the authority, direction, and control of the Chairman, the other members of the Joint Chiefs of Staff and the Vice Chairman in carrying out their responsibilities. See also **staff.**

Joint Staff doctrine sponsor--The sponsor for a joint doctrine or joint tactics, techniques, and procedures (JTTP) project. Each joint doctrine or JTTP project will be assigned a Joint Staff doctrine sponsor. The Joint Staff doctrine sponsor will assist the lead agent and primary review authority as requested and directed. The Joint Staff doctrine sponsor will coordinate the draft document with the Joint Staff and provide Joint Staff comments and recommendations to the primary review authority. See also **joint doctrine; joint tactics, techniques, and procedures.**

Joint Strategic Planning System--The primary means by which the Chairman of the Joint Chiefs of Staff, in consultation with the other members of the Joint Chiefs of Staff and the combatant commanders, carries out his statutory responsibilities to assist the President and Secretary of Defense in providing strategic direction to the Armed Forces; prepares strategic plans; prepares and reviews contingency plans; advises the President and Secretary of Defense on requirements, programs, and budgets; and provides net assessment on the capabilities of the Armed Forces of the United States and its allies as compared with those of their potential adversaries. Also called **JSPS.**

joint suppression of enemy air defenses--A broad term that includes all suppression of enemy air defenses activities provided by one component of the joint force in support of another. Also called **J-SEAD.** See also **air defense suppression; suppression of enemy air defenses.**

joint table of allowances--A document which authorizes end-items of materiel for units operated jointly by two or more military assistance advisory groups and missions. Also called **JTA.**

joint table of distribution--A manpower document which identifies the positions and enumerates the spaces that have been approved for each organizational element of a joint activity for a specific fiscal year (authorization year), and those spaces which have been accepted for planning and programming purposes for the four subsequent fiscal years (program years). Also called **JTD.** See also **joint manpower program.**

Joint Tactical Air Reconnaissance/Surveillance Mission Report--A preliminary report of information from tactical reconnaissance aircrews rendered by designated debriefing personnel immediately after landing and dispatched prior to compilation of the Initial Photo Interpretation Report. It provides a summary of the route conditions, observations, and aircrew actions and identifies sensor products. Also called **MISREP.**

joint tactics, techniques, and procedures--The actions and methods which implement joint doctrine and describe how forces will be employed in joint operations. They will be promulgated by the Chairman of the Joint Chiefs of Staff, in coordination with the combatant commands, Services, and Joint Staff. Also called **JTTP.** See also **Chairman of the Joint Chiefs of Staff Instruction; guidance; joint administrative publication; joint doctrine; joint test publication.**

joint targeting coordination board--A group formed by the joint force commander to accomplish broad targeting oversight functions that may include but are not limited to coordinating targeting information, providing targeting guidance and priorities, and preparing and/or refining joint target lists. The board is normally comprised of representatives from the joint force staff, all components, and if required, component subordinate units. Also called **JTCB.** See also **joint target list.**

joint target list--A consolidated list of selected targets considered to have military significance in the joint operations area.

joint task force--A force composed of assigned or attached elements of the Army, the Navy or the Marine Corps, and the Air Force, or two or more of these Services, which is constituted and so designated by the Secretary of Defense or by the commander of a unified command, a specified command, or an existing joint task force.

Joint Technical Coordinating Group for Munitions Effectiveness--A Joint Staff level organization tasked to produce generic target vulnerability and weaponeering studies. The special operations working group is a subordinate organization specializing in studies for special operations. Also called **JTCG-ME.**

joint test publication--A proposed version of a joint doctrine or joint tactics, techniques, and procedures publication that normally contains contentious issues and is nominated for a test publication and evaluation stage. Joint test publications are approved for evaluation by the Director, Operational Plans and Interoperability (J-7), Joint Staff. Publication of a test publication does not constitute Chairman of the Joint Chiefs of Staff approval of the publication. Prior to final approval as joint doctrine, test publications are expected to be further refined based upon evaluation results. Test publications are automatically superseded upon completion of the evaluation and promulgation of the proposed publication. See also **Chairman**

of the Joint Chiefs of Staff Instruction; guidance; joint doctrine; joint publication; joint tactics, techniques, and procedures.

Joint Worldwide Intelligence Communications System--The sensitive compartmented information portion of the Defense Information System Network. It incorporates advanced networking technologies that permit point-to-point or multipoint information exchange involving voice, text, graphics, data, and video teleconferencing. Also called **JWICS.**

joint zone (air, land, or sea)--An area established for the purpose of permitting friendly surface, air, and subsurface forces to operate simultaneously.

join up(*)--To form separate aircraft or groups of aircraft into a specific formation. See also **rendezvous.**

JOPES--See **Joint Operation Planning and Execution System.**

JP--See **joint publication.**

JRAC--See **joint rear area coordinator.**

JRTOC--See **joint rear tactical operations center.**

J-SEAD--See **joint suppression of enemy air defenses.**

JSO--See **Joint Specialty Officer/joint specialist.**

JSOA--See **joint special operations area.**

JSOACC--See **joint special operations air component commander.**

JSOTF--See **joint special operations task force.**

JSPS--See **Joint Strategic Planning System.**

JTCB--See **joint targeting coordination board.**

JTCG-ME--See **Joint Technical Coordinating Group for Munitions Effectiveness.**

JTTP--See **joint tactics, techniques, and procedures.**

judy--In air intercept, a code meaning, "I have contact and am taking over the intercept."

jumpmaster--The assigned airborne-qualified individual who controls parachutists from the time they enter the aircraft until they exit. See also **stick commander (air transport).**

jump speed(*)--The airspeed at which parachute troops can jump with comparative safety from an aircraft.

JWICS--See **Joint Worldwide Intelligence Communications System.**

K

KA-6--See **Intruder.**

KC-135--See **Stratotanker.**

K-day--The basic date for the introduction of a convoy system on any particular convoy lane. See also **D-day; M-day.**

key employee--Any Reservist identified by his or her employer, private or public, as filling a key position.

key facilities list--A register of selected command installations and industrial facilities of primary importance to the support of military operations or military production programs. It is prepared under the policy direction of the Joint Chiefs of Staff.

key point(*)--A concentrated site or installation, the destruction or capture of which would seriously affect the war effort or the success of operations.

key position--A civilian position, public or private (designated by the employer and approved by the Secretary concerned), that cannot be vacated during war or national emergency.

key terrain(*)--Any locality, or area, the seizure or retention of which affords a marked advantage to either combatant. See also **vital ground.**

KIA--See **killed in action.**

killed in action--A casualty category applicable to a hostile casualty, other than the victim of a terrorist activity, who is killed outright or who dies as a result of wounds or other injuries before reaching a medical treatment facility. Also called **KIA.** See also **casualty category.**

killing zone--An area in which a commander plans to force the enemy to concentrate so as to destroy him with conventional weapons or the tactical employment of nuclear weapons.

kill probability(*)--A measure of the probability of destroying a target.

kiloton weapon(*)--A nuclear weapon, the yield of which is measured in terms of thousands of tons of trinitrotoluene explosive equivalents, producing yields from 1 to 999 kilotons. See also **megaton weapon; nominal weapon; subkiloton weapon.**

kite(*)--In naval mine warfare, a device which when towed, submerges and planes at a predetermined level without sideways displacement.

L

LAD--See **latest arrival date.**

Lance--A mobile, storable, liquid propellant, surface-to-surface guided missile, with nuclear and nonnuclear capability; designed to support the Army corps with long-range fires. Designated as **XMGM-52.**

land arm mode(*)--A mode of operation in which automatic sequence is used to engage and disengage appropriate modes of an aircraft automatic flight control system in order to execute the various flight phases in the terminal area necessary for completing an automatic approach and landing.

land control operations--The employment of ground forces, supported by naval and air forces, as appropriate, to achieve military objectives in vital land areas. Such operations include destruction of opposing ground forces, securing key terrain, protection of vital land lines of communication, and establishment of local military superiority in areas of land operations. See also **sea control operations.**

land effect--See **coastal refraction.**

landing aid(*)--Any illuminating light, radio beacon, radar device, communicating device, or any system of such devices for aiding aircraft in an approach and landing.

landing approach(*)--The continuously changing position of an aircraft in space directed toward effecting a landing on a predetermined area.

landing area--1. The part of the objective area within which are conducted the landing operations of an amphibious force. It includes the beach, the approaches to the beach, the transport areas, the fire support areas, the air occupied by close supporting aircraft, and the land included in the advance inland to the initial objective. 2. (Airborne) The general area used for landing troops and materiel either by airdrop or air landing. This area includes one or more drop zones or landing strips. 3. Any specially prepared or selected surface of land, water, or deck designated or used for takeoff and landing of aircraft. See also **airfield.**

landing attack--An attack against enemy defenses by troops landed from ships, aircraft, boats, or amphibious vehicles. See also **assault.**

landing beach(*)--That portion of a shoreline usually required for the landing of a battalion landing team. However, it may also be that portion of a shoreline constituting a tactical locality (such as the shore of a bay) over which a force larger or smaller than a battalion landing team may be landed.

landing craft(*)--A craft employed in amphibious operations, specifically designed for carrying troops and equipment and for beaching, unloading, and retracting. Also used for logistic cargo resupply operations.

landing craft and amphibious vehicle assignment table--A table showing the assignment of personnel and materiel to each landing craft and amphibious vehicle and the assignment of the landing craft and amphibious vehicles to waves for the ship-to-shore movement.

landing craft availability table--A tabulation of the type and number of landing craft that will be available from each ship of the transport group. The table is the basis for the assignment of landing craft to the boat groups for the ship-to-shore movement.

213

landing diagram(*)--A graphic means of illustrating the plan for the ship-to-shore movement.

landing force(*)--A task organization of troop units, aviation and ground, assigned to an amphibious assault. It is the highest troop echelon in the amphibious operation. See also **amphibious force.**

landing force supplies--Those supplies remaining in assault shipping after initial combat supplies and floating dumps have been unloaded. They are landed selectively in accordance with the requirements of the landing force until the situation ashore permits the inception of general unloading.

landing force support party--The forward echelon of the combat service support element formed to facilitate the ship-to-shore movement. It may contain a surface assault support element (shore party) and a helicopter assault support element (helicopter support). The landing force support party is brought into existence by a formal activation order issued by the commander, landing force.

landing group--In amphibious operations, a subordinate task organization of the landing force capable of conducting landing operations, under a single tactical command, against a position or group of positions.

landing mat(*)--A prefabricated, portable mat so designed that any number of planks (sections) may be rapidly fastened together to form surfacing for emergency runways, landing beaches, etc.

landing plan--In amphibious operations, a collective term referring to all individually prepared naval and landing force documents that, taken together, present in detail all instructions for execution of the ship-to-shore movement.

landing point(*)--A point within a landing site where one helicopter or vertical takeoff and landing aircraft can land. See also **airfield.**

landing roll(*)--The movement of an aircraft from touchdown through deceleration to taxi speed or full stop.

landing schedule--In an amphibious operation, a schedule which shows the beach, hour, and priorities of landing of assault units, and which coordinates the movements of landing craft from the transports to the beach in order to execute the scheme of maneuver ashore.

landing sequence table--A document that incorporates the detailed plans for ship-to-shore movement of nonscheduled units.

landing ship(*)--An assault ship which is designed for long sea voyages and for rapid unloading over and on to a beach.

landing ship dock(*)--A ship designed to transport and launch loaded amphibious craft and/or amphibian vehicles with their crews and embarked personnel and/or equipment and to render limited docking and repair services to small ships and craft.

landing site(*)--1. A site within a landing zone containing one or more landing points. See also **airfield.** 2. In amphibious operations, a continuous segment of coastline over which troops, equipment and supplies can be landed by surface means.

landing threshold--The beginning of that portion of a runway usable for landing.

landing vehicle, tracked, engineer, model 1--A lightly armored amphibious vehicle designed for minefield and obstacle clearance in amphibious assaults and operations inland. Equipped with line charges for projection in advance of the vehicle and bulldozer-type blade with scarifier teeth. Designated as **LVTE-l.**

landing zone(*)--Any specified zone used for the landing of aircraft. See also **airfield.**

landing zone control--See **pathfinder drop zone control.**

landing zone control party(*)--Personnel specially trained and equipped to establish and operate communications devices from the ground for traffic control of aircraft/helicopters for a specific landing zone.

landmark(*)--A feature, either natural or artificial, that can be accurately determined on the ground from a grid reference.

land mine warfare--See **mine warfare.**

land projection operations--See **land, sea, or aerospace projection operations.**

land, sea, or aerospace projection operations--The employment of land, sea, or air forces, or appropriate combinations thereof, to project United States military power into areas controlled or threatened by enemy forces. Operations may include penetration of such areas by amphibious, airborne, or land-transported means, as well as air combat operations by land-based and/or carrier air.

land search--The search of terrain by Earth-bound personnel.

lane marker(*)--In land mine warfare, sign used to mark a minefield lane. Lane markers, at the entrance to and exit from the lane, may be referenced to a landmark or intermediate marker. See also **marker; minefield lane.**

lap(*)--In naval mine warfare, that section or strip of an area assigned to a single sweeper or formation of sweepers for a run through the area.

lap course(*)--In naval mine warfare, the true course desired to be made good during a run along a lap.

lap track(*)--In naval mine warfare, the center line of a lap; ideally, the track to be followed by the sweep or detecting gear.

lap turn(*)--In naval mine warfare, the maneuver a minesweeper carries out during the period between the completion of one run and the commencement of the run immediately following.

lap width(*)--In naval mine warfare, the swept path of the ship or formation divided by the percentage coverage being swept to.

large-lot storage--A quantity of material which will require four or more pallet columns stored to maximum height. Usually accepted as stock stored in carload or greater quantities. See also **storage.**

large-scale map--A map having a scale of 1:75-,000 or larger. See also **map.**

large spread--A report by an observer or a spotter to the ship to indicate that the distance between the bursts of a salvo is excessive.

laser designator(*)--A device that emits a beam of laser energy which is used to mark a specific place or object.

laser guidance unit(*)--A device which incorporates a laser seeker to provide guidance commands to the control system of a missile, projectile or bomb.

laser guided weapon(*)--A weapon which uses a seeker to detect laser energy reflected from a laser marked/designated target and through signal processing provides guidance commands to a control system which guides the weapon to the point from which the laser energy is being reflected.

laser illuminator(*)--A device for enhancing the illumination in a zone of action by irradiating with a laser beam.

laser intelligence--Technical and geo-location intelligence derived from laser systems; a subcategory of electro-optical intelligence. Also called **LASINT.** See also **electro-optical intelligence; intelligence.**

laser linescan system(*)--An active airborne imagery recording system which uses a laser as the primary source of illumination to scan the ground beneath the flight path, adding successive across-track lines to the record as the vehicle advances. See also **infrared linescan system.**

laser pulse duration(*)--The time during which the laser output pulse power remains continuously above half its maximum value.

laser rangefinder(*)--A device which uses laser energy for determining the distance from the device to a place or object.

laser seeker(*)--A device based on a direction sensitive receiver which detects the energy reflected from a laser designated target and defines the direction of the target relative to the receiver. See also **laser guided weapon.**

laser target designating system(*)--A system which is used to direct (aim or point) laser energy at a target. The system consists of the laser designator or laser target marker with its display and control components necessary to acquire the target and direct the beam of laser energy thereon.

laser target marker--See **laser designator.**

laser target marking system--See **laser target designating system.**

laser tracker(*)--A device which locks on to the reflected energy from a laser marked/designated target and defines the direction of the target relative to itself.

lashing(*)--See tie down. (DOD) See also **restraint of loads.**

lashing point--See **tie down point.**

LASINT--See **laser intelligence.**

late(*)--In artillery and naval gunfire support, a report made to the observer or spotter, whenever there is a delay in reporting "shot" by coupling a time in seconds with the report.

lateral gain(*)--The amount of new ground covered laterally by successive photographic runs over an area.

lateral route(*)--A route generally parallel to the forward edge of the battle area, which crosses, or feeds into, axial routes. See also **route.**

lateral spread--A technique used to place the mean point of impact of two or more units 100 meters apart on a line perpendicular to the gun-target line.

lateral tell--See **track telling.**

latest arrival date--A day, relative to C-day, that is specified by a planner as the latest date when a unit, a resupply shipment, or replacement personnel can arrive and complete unloading at the port of debarkation and support the concept of operations. Also called **LAD.** See also **earliest arrival date.**

late time--See **span of detonation (atomic demolition munition employment).**

latitude band(*)--Any latitudinal strip, designated by accepted units of linear or angular measurement, which circumscribes the Earth. Also called **latitudinal band.**

lattice(*)--A network of intersecting positional lines printed on a map or chart from which a fix may be obtained.

launch--The transition from static repose to dynamic flight of a missile.

launcher(*)--A structural device designed to support and hold a missile in position for firing.

launch pad(*)--A concrete or other hard surface area on which a missile launcher is positioned.

launch time--The time at which an aircraft or missile is scheduled to be airborne. See also **airborne order.**

launch under attack--Execution by National Command Authorities of Single Integrated Operational Plan forces subsequent to tactical warning of strategic nuclear attack against the United States and prior to first impact. Also called **LUA.**

law of armed conflict--See **law of war.**

law of war--That part of international law that regulates the conduct of armed hostilities. Also called **the law of armed conflict.** See also **rules of engagement.**

lay--1. Direct or adjust the aim of a weapon. 2. Setting of a weapon for a given range, or for a given direction, or both. 3. To drop one or more aerial bombs or aerial mines onto the surface from an aircraft. 4. To spread a smoke screen on the ground from an aircraft. 5. To calculate or project a course. 6. To lay on: a. to execute a bomber strike; b. to set up a mission.

laydown bombing(*)--A very low level bombing technique wherein delay fuzes and/or devices are used to allow the attacker to escape the effects of the bomb.

layer depth--The depth from the surface of the sea to the point above the first major negative thermocline at which sound velocity is maximum.

layer tint--See **hypsometric tinting.**

lay leader or lay reader--A volunteer ("lay leader" in Army and Air Force; "lay reader" in Navy and Marine Corps) appointed by the commanding officer and supervised and trained by the command chaplain to serve for a period of time to meet the needs of a particular religious faith group when their military chaplains are not available. The lay leader or lay reader may conduct services, but may not exercise any other activities usually reserved for the ordained clergy. See also **command chaplain; command chaplain of the combatant command; religious ministry support; religious ministry support plan; religious ministry support team; Service component command chaplain.**

lay reader--See **lay leader or lay reader.**

lay reference number(*)--In naval mine warfare, a number allocated to an individual mine by the minefield planning authority to provide a simple means of referring to it.

lazy--In air intercept, a code meaning, "Equipment indicated at standby."

LCC--See **amphibious command ship.**

lead agent--Individual Services, combatant commands, or Joint Staff directorates may be assigned as lead agents for developing and maintaining joint doctrine, joint tactics, techniques, and procedures (JTTP) publications, or joint administrative publications. The lead agent is responsible for developing, coordinating, reviewing, and maintaining an assigned doctrine, JTTP, or joint administrative publication. See also **coordinating review authority; joint administrative publication; joint doctrine; joint publication; joint tactics, techniques, and procedures; joint test publication; primary review authority.**

lead aircraft--1. The airborne aircraft designated to exercise command of other aircraft within the flight. 2. An aircraft in the van of two or more aircraft.

lead collision course(*)--A vector which, if maintained by an interceptor aircraft, will result in collision between the interceptor's fixed armament and the target.

lead pursuit(*)--An interceptor vector designed to maintain a course of flight at a predetermined point ahead of a target.

leapfrog(*)--Form of movement in which like supporting elements are moved successively through or by one another along the axis of movement of supported forces.

Leap Second--A second of time that is added to or removed from Coordinated Universal Time (UTC) to keep UTC within 0.9 seconds of UT1 (see Universal Time). Leap Seconds are normally introduced at the end of June or December if required. The decision to introduce a Leap Second is announced by the International Time Bureau (Bureau International de l'Heure, or BIH) approximately eight to ten weeks in advance. See also **Coordinated Universal Time.**

leaver(*)--A merchant ship which breaks off from a convoy to proceed to a different destination and becomes independent. Also called **convoy leaver.** See also **leaver convoy; leaver section.**

leaver convoy(*)--A convoy which has broken off from the main convoy and is proceeding to a different destination. See also **leaver; leaver section.**

leaver section(*)--A group of ships forming part of the main convoy which will subsequently break off to become leavers or a leaver convoy. See also **leaver; leaver convoy.**

left (or right)(*)--1. Terms used to establish the relative position of a body of troops. The person using the terms "left" or "right" is assumed to be facing in the direction of the enemy regardless of whether the troops are advancing towards or withdrawing from the enemy. 2. Correction used in adjusting fire to indicate that a lateral shift of the mean point of impact perpendicular to the reference line or spotting line is desired.

left (right) bank--That bank of a stream or river on the left (right) of the observer when he is facing in the direction of flow or downstream.

level--In air intercept, a word meaning, "Contact designated is at your angels."

level of detail--Within the current joint planning and execution systems, movement characteristics are described at five distinct levels of detail. These levels are: **a. level I.** aggregated level. Expressed as total number of passengers and total short tons, total measurement tons, total square feet and/or total hundreds of barrels by unit line number (ULN), cargo increment number (CIN), and personnel increment number (PIN). **b. level II.** summary level. Expressed as total number of passengers by ULN and PIN and short tons, measurement tons (including barrels), total square feet of bulk, oversize, outsize, and non-air-transportable cargo by ULN and CIN. **c. level III.** detail by cargo category. Expressed as total number of passengers by ULN and PIN and short tons, and/or measurement tons (including barrels), total square feet of cargo as identified by the ULN or CIN three-position cargo category code. **d. level IV.** detail expressed as number of passengers and individual dimensional data (expressed in length, width, and height in number of inches) of cargo by equipment type by ULN. **e. level V.** detail by priority of shipment. Expressed as total number of passengers by Service specialty code in deployment sequence by ULN individual weight (in pounds) and dimensional data (expressed in length, width, and height in number of inches) of equipment in deployment sequence by ULN.

level-of-effort munitions(*)--In stockpile planning, munitions stocked on the basis of expected daily expenditure rate, the number of combat days, and the attrition rate assumed, to counter targets the number of which is unknown. See also **threat-oriented munitions.**

level of effort-oriented items--Items for which requirements computations are based on such factors as equipment and personnel density and time and rate of use. See also **combination mission/level of effort-oriented items; mission-oriented items.**

level of supply(*)--The quantity of supplies or materiel authorized or directed to be held in anticipation of future demands. See also **operating level of supply; order and shipping time; procurement lead time; requisitioning objective; safety level of supply; stockage objective; strategic reserve.**

LGM-30--See **Minuteman.**

LHA--See **amphibious assault ship (general purpose).**

L-hour--See **times.**

liaison(*)--That contact or intercommunication maintained between elements of military forces to ensure mutual understanding and unity of purpose and action.

liberated territory(*)--Any area, domestic, neutral, or friendly, which, having been occupied by an enemy, is retaken by friendly forces.

LIC--See **low intensity conflict.**

licensed production--A direct commercial arrangement between a US company and a foreign government, international organization, or foreign company, providing for the transfer of production information which enables the foreign government, international organization, or commercial producer to manufacture, in whole or in part, an item of US defense equipment. A typical license production arrangement would include the functions of production engineering, controlling, quality assurance and determining of resource requirements. It may or may not include design engineering information and critical materials production and design information. A licensed production arrangement is accomplished under the provisions of a manufacturing license agreement per the US International Traffic in Arms Regulation (ITAR).

life cycle--The total phases through which an item passes from the time it is initially developed until the time it is either consumed in use or disposed of as being excess to all known materiel requirements.

lifeguard submarine(*)--A submarine employed for rescue in an area which cannot be adequately covered by air or surface rescue facilities because of enemy opposition, distance from friendly bases, or other reasons. It is stationed near the objective and sometimes along the route to be flown by the strike aircraft.

life support equipment--Equipment designed to sustain aircrew members and passengers throughout the flight environment, optimizing their mission effectiveness and affording a means of safe and reliable escape, descent, survival, and recovery in emergency situations.

light artillery--See **field artillery.**

light damage--See **nuclear damage (land warfare).**

lightening(*)--The operation (normally carried out at anchor) of transferring crude oil cargo from a large tanker to a smaller tanker, so reducing the draft of the larger tanker to enable it to enter port.

lighterage--A small craft designed to transport cargo or personnel from ship to shore. Lighterage includes amphibians, landing craft, discharge lighters, causeways, and barges.

light filter(*)--An optical element such as a sheet of glass, gelatine, or plastic dyed in a specific manner to absorb selectively light of certain colors.

light line(*)--A designated line forward of which vehicles are required to use black-out lights at night.

limited-access plan--The limited-access plan (like the close-hold plan) is an operation plan that has access restricted to individual Worldwide Military Command and Control System user Ids and terminal Ids. Unlike the close-hold plan, the limited-access plan can be distributed to more than one Joint Operation Planning and Execution System site. See also **close-hold plan.**

limited access route(*)--A one way route with one or more restrictions which preclude its use by the full range of military traffic. See also **double flow route; single flow route.**

limited denied war--Not to be used. No substitute recommended.

limited production type item--An item under development, commercially available or available from other Government agencies, for which an urgent operational requirement exists and for which no other existing item is substitutable. Such an item appears to fulfill an approved materiel requirement or other Military Department-approved requirements, and to be promising enough operationally to warrant initiating procurement and/or production for service issue prior to completion of development and/or test or adoption as a standard item.

limited standard item--An item of supply determined by standardization action as authorized for procurement only to support in-service military materiel requirements.

limited war--Armed conflict short of general war, exclusive of incidents, involving the overt engagement of the military forces of two or more nations.

limiting factor--A factor or condition that, either temporarily or permanently, impedes mission accomplishment. Illustrative examples are transportation network deficiencies, lack of in-place facilities, malpositioned forces or materiel, extreme climatic conditions, distance, transit or overflight rights, political conditions, etc.

limit of fire(*)--1. The boundary marking off the area on which gunfire can be delivered. 2. Safe angular limits for firing at aerial targets.

line(*)--In artillery and naval gunfire support, a spotting, or an observation, used by a spotter or an observer to indicate that a burst(s) occurred on the spotting line.

linear scale--See **graphic scale; scale.**

line of departure(*)--1. In land warfare, a line designated to coordinate the departure of attack elements. 2. In amphibious warfare, a suitably marked offshore coordinating line to assist assault craft to land on designated beaches at scheduled times.

line of position--In air intercept, a reference line which originates at a target and extends outward at a predetermined angle.

line overlap--See **overlap 1.**

liner--In air intercept, a code meaning, "Fly at speed giving maximum cruising range."

line-route map--A map or overlay for signal communications operations that shows the actual routes and types of construction of wire circuits in the field. It also gives the locations of switchboards and telegraph stations. See also **map.**

line search(*)--Reconnaissance along a specific line of communications, such as a road, railway or waterway, to detect fleeting targets and activities in general.

lines of communications--All the routes, land, water, and air, which connect an operating military force with a base of operations and along which supplies and military forces move. Also called **LOC.**

link(*)--1. In communications, a general term used to indicate the existence of communications facilities between two points. 2. A maritime route, other than a coastal or transit route, which links any two or more routes.

link encryption--The application of online crypto-operation to a link of a communications system so that all information passing over the link is encrypted in its entirety.

link-lift vehicle--The conveyance, together with its operating personnel, used to satisfy a movement requirement between nodes.

link-route segments--Route segments that connect nodes wherein link-lift vehicles perform the movement function.

liquid explosive(*)--Explosive which is fluid at normal temperatures.

liquid propellant--Any liquid combustible fed to the combustion chamber of a rocket engine.

listening watch--A continuous receiver watch established for the reception of traffic addressed to, or of interest to, the unit maintaining the watch, with complete log optional.

list of targets--A tabulation of confirmed or suspect targets maintained by any echelon for informational and fire support planning purposes. See also **target list.**

litter--A basket or frame utilized for the transport of injured persons.

litter patient--A patient requiring litter accommodations while in transit.

LKA--See **attack cargo ship.**

load(*)--The total weight of passengers and/or freight carried on board a ship, aircraft, train, road vehicle, or other means of conveyance. See also **airlift capability; airlift requirement; allowable load; combat load; standard load.**

load control group(*)--Personnel who are concerned with organization and control of loading within the pick-up zone.

loading(*)--The process of putting personnel, materiel, and supplies on board ships, aircraft, trains, road vehicles, or other means of conveyance. See also **embarkation.**

loading chart (aircraft)--Any one of a series of charts carried in an aircraft which shows the proper location for loads to be transported and which pertains to check-lists, balance records, and clearances for weight and balance.

loading plan(*)--All of the individually prepared documents which, taken together, present in detail all instructions for the arrangement of personnel, and the loading of equipment for one or more units or other special grouping of personnel or material moving by highway, water, rail, or air transportation. See also **ocean manifest.**

loading point(*)--A point where one aircraft can be loaded or unloaded.

loading site(*)--An area containing a number of loading points.

load spreader(*)--Material used to distribute the weight of a load over a given floor area to avoid exceeding designed stress.

localizer(*)--A directional radio beacon which provides to an aircraft an indication of its lateral position relative to a predetermined final approach course. See also **beacon; instrument landing system.**

localizer mode(*)--In a flight control system, a control mode in which an aircraft is automatically positioned to, and held at, the center of the localizer course.

local mean time(*)--The time interval elapsed since the mean sun's transit of the observer's anti-meridian.

local procurement--The process of obtaining personnel, services, supplies, and equipment from local or indigenous sources.

local purchase--The function of acquiring a decentralized item of supply from sources outside the Department of Defense.

local war--Not to be used. See **limited war.**

LOCAP--Low combat air patrol.

lock on(*)--Signifies that a tracking or target-seeking system is continuously and automatically tracking a target in one or more coordinates (e.g., range, bearing, elevation).

lodgment area--See **airhead; beachhead.**

loft bombing--A method of bombing in which the delivery plane approaches the target at a very low altitude, makes a definite pullup at a given point, releases the bomb at a predetermined point during the pullup, and tosses the bomb onto the target. See also **over-the-shoulder bombing; toss bombing.**

logair--Long-term contract airlift service within continental United States for the movement of cargo in support of the logistics systems of the Military Services (primarily the Army and Air Force) and Department of Defense agencies. See also **quicktrans.**

logistic assessment(*)--An evaluation of: a. The logistic support required to support particular military operations in a theater of operations, country, or area. b. The actual and/or potential logistics support available for the conduct of military operations either within the theater, country, or area, or located elsewhere.

logistic estimate of the situation--An appraisal resulting from an orderly examination of the logistic factors influencing contemplated courses of action to provide conclusions concerning the degree and manner of that influence. See also **estimate of the situation.**

logistic implications test--An analysis of the major logistic aspects of a joint strategic war plan and the consideration of the logistic implications resultant therefrom as they may limit the acceptability of the plan. The logistic analysis and consideration are conducted concurrently with the development of the strategic plan. The objective is to establish whether the logistic requirements generated by the plan are in balance with availabilities, and to set forth those logistic implications that should be weighed by the Joint Chiefs of Staff in their consideration of the plan. See also **feasibility test.**

logistic routes--See **lines of communications.**

logistics(*)--The science of planning and carrying out the movement and maintenance of forces. In its most comprehensive sense, those aspects of military operations which deal with: a. design and development, acquisition, storage, movement, distribution, maintenance, evacuation, and disposition of materiel; b. movement, evacuation, and hospitalization of personnel; c. acquisition or construction, maintenance, operation, and disposition of facilities; and d. acquisition or furnishing of services.

logistics-over-the-shore operations--The loading and unloading of ships without the benefit of fixed port facilities, in friendly or nondefended territory, and, in time of war, during phases of theater development in which there is no opposition by the enemy. Also called **LOTS.**

logistics sourcing--The identification of the origin and determination of the availability of the time-phased force and deployment data nonunit logistics requirements.

logistic support--Logistic support encompasses the logistic services, materiel, and transportation required to support the continental United States-based and worldwide deployed forces.

logistic support (medical)--Medical care, treatment, hospitalization, evacuation, furnishing of medical services, supplies, materiel, and adjuncts thereto.

LOMEZ--See **low-altitude missile engagement zone.**

long-range bomber aircraft--A bomber designed for a tactical operating radius over 2,500 nautical miles at design gross weight and design bomb load.

long-range transport aircraft--See **transport aircraft.**

look(*)--In mine warfare, a period during which a mine circuit is receptive of an influence.

loran(*)--A long-range radio navigation position fixing system using the time difference of reception of pulse type transmissions from two or more fixed stations. This term is derived from the words long-range electronic navigation.

lost(*)--In artillery and naval gunfire support, a spotting, or an observation used by a spotter or an observer to indicate that rounds fired by a gun or mortar were not observed.

lot--Specifically, a quantity of material all of which was manufactured under identical conditions and assigned an identifying lot number.

low--A height between five hundred and two thousand feet.

low airburst(*)--The fallout safe height of burst for a nuclear weapon which maximizes damage to or casualties on surface targets. See also **types of burst.**

low-altitude bombing--Horizontal bombing with the height of release between 900 and 8,000 feet.

low altitude bombing system mode--In a flight control system, a control mode in which the low altitude bombing maneuver of an aircraft is controlled automatically.

low-altitude missile engagement zone--See **weapon engagement zone.**

low angle(*)--In artillery and naval gunfire support, an order or request to obtain low angle fire.

low angle fire(*)--Fire delivered at angles of elevation below the elevation that corresponds to the maximum range of the gun and ammunition concerned.

low angle loft bombing(*)--Type of loft bombing of free fall bombs wherein weapon release occurs at an angle less than 35 degrees above the horizontal. See also **loft bombing.**

low dollar value item--An item which normally requires considerably less management effort than those in the other management intensity groupings.

low intensity conflict--Political-military confrontation between contending states or groups below conventional war and above the routine, peaceful competition among states. It frequently involves protracted struggles of competing principles and ideologies. Low intensity conflict ranges from subversion to the use of armed force. It is waged by a combination of means employing political, economic, informational, and military instruments. Low intensity conflicts are often localized, generally in the Third World, but contain regional and global security implications. Also called **LIC.**

low level flight--See **terrain flight.**

low level transit route(*)--A temporary corridor of defined dimensions established in the forward area to minimize the risk to friendly aircraft from friendly air defenses or surface forces.

low oblique--See **oblique air photograph.**

low velocity drop(*)--A drop procedure in which the drop velocity does not exceed 30 feet per second.

low visibility operations--Sensitive operations wherein the political-military restrictions inherent in covert and clandestine operations are either not necessary or not feasible; actions are taken as required to limit exposure of those involved and/or their activities. Execution of these operations is undertaken with the knowledge that the action and/or sponsorship of the operation may preclude plausible denial by the initiating power.

LPD--See **amphibious transport dock.**

LSD--See **dock landing ship.**

LST--See **tank landing ship.**

LVTE-l--See **landing vehicle, tracked, engineer, model 1.**

M

M-42--See **Duster (antiaircraft weapon)**.

M-47--See **Dragon**.

M48A3--See **tank, combat, full-tracked, 90-mm gun**.

M-60--See **tank, combat, full-tracked, 105-mm gun**.

M548--See **cargo carrier**.

M88A1--See **recovery vehicle**.

MAAG--See **military assistance advisory group**.

Mace--An improved version of the MGM-1C Matador missile, differing primarily in its improved guidance system, longer-range, low-level attack capability, and higher-yield warhead. The MGM-13A is guided by a self-contained radar guidance system. The MGM-13B is guided by an inertial guidance system. Designated as **MGM-13**.

mach front--See **mach stem**.

mach hold mode--In a flight control system, a control mode in which a desired flight (flying) speed of an aircraft expressed as a mach number is maintained automatically.

machmeter(*)--An instrument which displays the mach number of the aircraft derived from inputs of pilot and static pressures.

mach no/yes--In air intercept, a code meaning, "I have reached maximum speed and am not/am closing my target."

mach number--The ratio of the velocity of a body to that of sound in the surrounding medium.

mach number indicator--See **machmeter**.

mach stem(*)--The shock front formed by the fusion of the incident and reflected shock fronts from an explosion. The term is generally used with reference to a blast wave, propagated in the air, reflected at the surface of the Earth. In the ideal case, the mach stem is perpendicular to the reflecting surface and slightly convex (forward). Also called **mach front**.

mach trim compensator--In a flight control system, an automatic control sub-system which provides pitch trim of an aircraft as a function of mach number.

mach wave--See **mach stem**.

magnetic bearing--See **bearing**.

magnetic circuit--See **magnetic mine**.

magnetic compass(*)--An instrument containing a freely suspended magnetic element which displays the direction of the horizontal component of the Earth's magnetic field at the point of observation.

magnetic declination(*)--The angle between the magnetic and geographical meridians at any place, expressed in degrees east or west to indicate the direction of magnetic north from true north. In nautical and aeronautical navigation, the term magnetic variation is used instead of magnetic declination and the angle is termed variation of the compass or magnetic variation. Magnetic declination is not otherwise synonymous with magnetic variation which refers to regular or irregular change with time of the magnetic declination, dip, or intensity. See also **magnetic variation**.

magnetic equator(*)--A line drawn on a map or chart connecting all points at which the magnetic inclination (dip) is zero for a specified epoch. Also called **aclinic line.**

magnetic mine(*)--A mine which responds to the magnetic field of a target.

magnetic minehunting--The process of using magnetic detectors to determine the presence of mines or minelike objects.

magnetic north(*)--The direction indicated by the north seeking pole of a freely suspended magnetic needle, influenced only by the Earth's magnetic field.

magnetic tape--A tape or ribbon of any material impregnated or coated with magnetic or other material on which information may be placed in the form of magnetically polarized spots.

magnetic variation(*)--1. In navigation, at a given place and time, the horizontal angle between the true north and magnetic north measured east or west according to whether magnetic north lies east or west of true north. See also **magnetic declination.** 2. In cartography, the annual change in direction of the horizontal component of the Earth's magnetic field.

MAGTF--See **Marine air-ground task force.**

main airfield(*)--An airfield planned for permanent occupation in peacetime, at a location suitable for wartime utilization, and with operational facilities of a standard adequate to develop full use of its war combat potential. See also **alternative airfield; departure airfield; redeployment airfield.**

main armament--The request of the observer or spotter to obtain fire from the largest guns installed on the fire support ship.

main attack(*)--The principal attack or effort into which the commander throws the full weight of the offensive power at his disposal. An attack directed against the chief objective of the campaign or battle.

main battle area--That portion of the battlefield in which the decisive battle is fought to defeat the enemy. For any particular command, the main battle area extends rearward from the forward edge of the battle area to the rear boundary of the command's subordinate units.

main battle tank--See **tank, main battle.**

main convoy(*)--The convoy as a whole which sails from the convoy assembly port/anchorage to its destination. It may be supplemented by joiners or joiner convoys, and leavers or leaver convoys may break off.

main detonating line(*)--In demolition, a line of detonating cord used to transmit the detonation wave to two or more branches.

main line of resistance--A line at the forward edge of the battle position, designated for the purpose of coordinating the fire of all units and supporting weapons, including air and naval gunfire. It defines the forward limits of a series of mutually supporting defensive areas, but it does not include the areas occupied or used by covering or screening forces.

main operations base--In special operations, a base established by a joint force special operations component commander or a subordinate special operations component commander in friendly territory to provide sustained command and control, administration, and logistical support to special operations activities in designated areas. Also called **MOB.** See also **advanced operations base; forward operations base.**

main road--A road capable of serving as the principal ground line of communication to an area or locality. Usually it is wide enough and suitable for two-way, all-weather traffic at high speeds.

main supply route(*)--The route or routes designated within an area of operations upon which the bulk of traffic flows in support of military operations.

maintain--When used in the context of deliberate planning, the directed command will keep the referenced operation plan, operation plan in concept format, or concept summary, and any associated Joint Operation Planning and Execution System (JOPES) automated data processing files active in accordance with applicable tasking documents describing the type and level of update or maintenance to be performed. General guidance is contained in JOPES, Volumes I and II. See also **archive; retain.**

maintenance area--A general locality in which are grouped a number of maintenance activities for the purpose of retaining or restoring materiel to a serviceable condition.

maintenance engineering--The application of techniques, engineering skills, and effort, organized to ensure that the design and development of weapon systems and equipment provide adequately for their effective and economical maintenance.

maintenance (materiel)--1. All action taken to retain materiel in a serviceable condition or to restore it to serviceability. It includes inspection, testing, servicing, classification as to serviceability, repair, rebuilding, and reclamation. 2. All supply and repair action taken to keep a force in condition to carry out its mission. 3. The routine recurring work required to keep a facility (plant, building, structure, ground facility, utility system, or other real property) in such condition that it may be continuously used, at its original or designed capacity and efficiency for its intended purpose.

maintenance status--1. A nonoperating condition, deliberately imposed, with adequate personnel to maintain and preserve installations, materiel, and facilities in such a condition that they may be readily restored to operable condition in a minimum time by the assignment of additional personnel and without extensive repair or overhaul. 2. That condition of materiel which is in fact, or is administratively classified as, unserviceable, pending completion of required servicing or repairs.

major combat element--Those organizations and units described in the Joint Strategic Capabilities Plan that directly produce combat capability. The size of the element varies by Service, force capability, and the total number of such elements available. Examples are Army divisions and separate brigades, Air Force squadrons, Navy task forces, and Marine expeditionary forces. See also major force.

major disaster--See **domestic emergencies.**

major fleet--A principal, permanent subdivision of the operating forces of the Navy with certain supporting shore activities. Presently there are two such fleets: the Pacific Fleet and the Atlantic Fleet. See also **fleet.**

major force--A military organization comprised of major combat elements and associated combat support, combat service support, and sustainment increments. The major force is capable of sustained military operations in response to plan employment requirements. See also **major combat element.**

major installation--In the Air Force, a self-supporting center of operations for actions of importance to Air Force combat, combat support, or training. It is operated by an active, reserve, or Guard unit of group size or larger with all land, facilities and organic support needed to accomplish the unit mission. It must have real property accountability through ownership, lease, permit, or other written agreement for all real estate and facilities. Agreements with foreign governments which give the Air Force jurisdiction over real property meet this requirement. Shared use agreements (as opposed to joint use agreements where the Air Force owns the runway) do not meet the criteria to be major installations. This category includes Air Force bases; air bases; air reserve bases; and Air Guard bases. See also **installation complex; minor installation; other activity; support site.**

major nuclear power(*)--Any nation that possesses a nuclear striking force capable of posing a serious threat to every other nation.

major port(*)--Any port with two or more berths and facilities and equipment capable of discharging 100,000 tons of cargo per month from ocean-going ships. Such ports will be designated as **probable nuclear targets.** See also **port.**

major weapon system--One of a limited number of systems or subsystems which, for reasons of military urgency, criticality, or resource requirements, is determined by the Department of Defense as being vital to the national interest.

make safe--One or more actions necessary to prevent or interrupt complete function of the system (traditionally synonymous with "dearm," "disarm," and "disable"). Among the necessary actions are: (1) install (safety devices such as pins or locks); (2) disconnect (hoses, linkages, batteries); (3) bleed (accumulators, reservoirs); (4) remove (explosive devices such as initiators, fuzes, detonators); (5) intervene (as in welding, lockwiring).

management--A process of establishing and attaining objectives to carry out responsibilities. Management consists of those continuing actions of planning, organizing, directing, coordinating, controlling, and evaluating the use of men, money, materials, and facilities to accomplish missions and tasks. Management is inherent in command, but it does not include as extensive authority and responsibility as command.

management and control system (mobility)--Those elements of organizations and/or activities which are part of, or are closely related to, the mobility system, and which authorize requirements to be moved, to obtain and allocate lift resources, or to direct the operation of linklift vehicles.

maneuver(*)--1. A movement to place ships or aircraft in a position of advantage over the enemy. 2. A tactical exercise carried out at sea, in the air, on the ground, or on a map in imitation of war. 3. The operation of a ship, aircraft, or vehicle, to cause it to perform desired movements. 4. Employment of forces on the battlefield through movement in combination with fire, or fire potential, to achieve a position of advantage in respect to the enemy in order to accomplish the mission.

maneuverable reentry vehicle--A reentry vehicle capable of performing preplanned flight maneuvers during the reentry phase. See also **multiple independently targetable reentry vehicle; multiple reentry vehicle; reentry vehicle.**

maneuvering area--That part of an airfield used for takeoffs, landings, and associated maneuvers. See also **aircraft marshalling area.**

manifest--A document specifying in detail the passengers or items carried for a specific destination.

manipulative electromagnetic deception--See **electromagnetic deception.**

man portable(*)--Capable of being carried by one man. Specifically, the term may be used to qualify: 1. Items designed to be carried as an integral part of individual, crew-served, or team equipment of the dismounted soldier in conjunction with his assigned duties. Upper weight limit: approximately 14 kilograms (31 pounds.) 2. In land warfare, equipment which can be carried by one man over long distance without serious degradation of the performance of his normal duties.

manpower--See **manpower requirements; manpower resources.**

manpower management(*)--The means of manpower control to ensure the most efficient and economical use of available manpower.

manpower management survey(*)--Systematic evaluation of a functional area, utilizing expert knowledge, manpower scaling guides, experience, and other practical considerations in determining the validity and managerial efficiency of the function's present or proposed manpower establishment.

manpower requirements--Human resources needed to accomplish specified work loads of organizations.

manpower resources--Human resources available to the Services which can be applied against manpower requirements.

man space--The space and weight factor used to determine the combat capacity of vehicles, craft, and transport aircraft, based on the requirements of one person with individual equipment. The person is assumed to weigh between 222-250 pounds and to occupy 13.5 cubic feet of space. See also **boat space.**

man transportable--Items which are usually transported on wheeled, tracked, or air vehicles, but have integral provisions to allow periodic handling by one or more individuals for limited distances (100-500 meters). Upper weight limit: approximately 65 pounds per individual.

many (raid size)--In air intercept usage, 8 or more aircraft. See also **few (raid size).**

map(*)--A graphic representation, usually on a plane surface, and at an established scale, of natural or artificial features on the surface of a part or the whole of the Earth or other planetary body. The features are positioned relative to a coordinate reference system. See also **administrative map; battle map; chart index; chart series; chart sheet; controlled map; general map; large scale map; line route map; map chart; map index; map series; map sheet; medium-scale map; operation map; planimetric map; situation map; small-scale map; strategic map; tactical map; topographic map; traffic circulation map; weather map.**

map chart--A representation of a land-sea area, using the characteristics of a map to represent the land area and the characteristics of a chart to represent the sea area, with such special characteristics as to make the map-chart most useful in military operations, particularly amphibious operations. See also **map.**

map convergence(*)--The angle at which one meridian is inclined to another on a map or chart. See also **convergence.**

map exercise--An exercise in which a series of military situations is stated and solved on a map.

map index(*)--Graphic key primarily designed to give the relationship between sheets of a series, their coverage, availability, and further information on the series. See also **map.**

mapping camera--See **air cartographic camera.**

mapping, charting, and geodesy--Maps, charts, and other data used for military planning, operations, and training. These products and data support air, land, and sea navigation; weapon system guidance; target positioning; and other military activities. These data are presented in the forms of topographic, planimetric, imaged, or thematic maps and graphics; nautical and aeronautical charts and publications; and, in digital and textual formats, gazetteers, which contain geophysical and geodetic data and coordinate lists. Also called **MC&G.**

map reference(*)--A means of identifying a point on the surface of the Earth by relating it to information appearing on a map, generally the graticule or grid.

map reference code(*)--A code used primarily for encoding grid coordinates and other information pertaining to maps. This code may be used for other purposes where the encryption of numerals is required.

map series(*)--A group of maps or charts usually having the same scale and cartographic specifications, and with each sheet appropriately identified by producing agency as belonging to the same series.

map sheet(*)--An individual map or chart either complete in itself or part of a series. See also **map.**

margin(*)--In cartography, the area of a map or chart lying outside the border.

marginal data(*)--All explanatory information given in the margin of a map or chart which clarifies, defines, illustrates, and/or supplements the graphic portion of the sheet.

marginal information--See **marginal data.**

marginal weather--Weather which is sufficiently adverse to a military operation so as to require the imposition of procedural limitations. See also **adverse weather.**

Marine air command and control system--A US Marine Corps tactical air command and control system which provides the tactical air commander with the means to command, coordinate, and control all air operations within an assigned sector and to coordinate air operations with other Services. It is composed of command and control agencies with communications-electronics equipment that incorporates a capability from manual through semiautomatic control.

Marine air control squadron--The component of the Marine air control group which provides and operates ground facilities for the detection and interception of hostile aircraft and for the navigational direction of friendly aircraft in the conduct of support missions.

Marine air-ground task force--A task organization of Marine forces (division, aircraft wing, and service support groups) under a single command and structured to accomplish a specific mission. The Marine air-ground task force (MAGTF) components will normally include command, aviation combat, ground combat, and combat service support elements (including Navy Support Elements). Three types of Marine air-ground task forces which can be task organized are the Marine expeditionary unit, Marine expeditionary brigade, and Marine expeditionary force. The four elements of a Marine air-ground task force are: **a. command element (CE)**--The MAGTF headquarters. The CE is a permanent organization composed of the commander, general or executive and special staff sections, headquarters section, and requisite communications and service support facilities. The CE provides command, control, and coordination essential for effective planning and execution of operations by the other three elements of the MAGTF. There is only one CE in a MAGTF. **b. aviation combat element (ACE)**--The MAGTF element that is task organized to provide all or a portion of the functions of Marine Corps aviation in varying degrees based on the tactical situation and the MAGTF mission and size. These functions are air reconnaissance, antiair warfare, assault support, offensive air support, electronic warfare, and control of aircraft and missiles. The ACE is organized around an aviation headquarters and varies in size from a reinforced helicopter squadron to one or more Marine aircraft wing(s). It includes those aviation command (including air control agencies), combat, combat support, and combat service support units required by the situation. Normally, there is only one ACE in a MAGTF. **c. ground combat element (GCE)**--The MAGTF element that is task organized to conduct ground operations. The GCE is constructed around an infantry unit and varies in size from a reinforced infantry battalion to one or more reinforced Marine division(s). The GCE also includes appropriate combat support and combat service support units. Normally, there is only one GCE in a MAGTF. **d. combat service support element (CSSE)**--The MAGTF element that is task organized to provide the full range of combat service support necessary to accomplish the MAGTF mission. CSSE can provide supply, maintenance, transportation, deliberate engineer, health, postal, disbursing, enemy prisoner of war, automated information systems, exchange, utilities, legal, and graves registration services. The CSSE varies in size from a Marine expeditionary unit (MEU) service support group (MSSG) to a force service support group (FSSG). Normally, there is only one combat service support ele-

ment in a MAGTF. See also **combat service support elements; Marine expeditionary brigade; Marine expeditionary force; Marine expeditionary unit; task force.**

Marine air support squadron--The component of the Marine air control group which provides and operates facilities for the control of support aircraft operating in direct support of ground forces.

Marine base--A base for support of Marine ground forces, consisting of activities or facilities for which the Marine Corps has operating responsibilities, together with interior lines of communications and the minimum surrounding area necessary for local security. (Normally, not greater than an area of 20 square miles.) See also **base complex.**

Marine division/wing team--A Marine Corps air-ground team consisting of one division and one aircraft wing, together with their normal reinforcements.

marine environment--The oceans, seas, bays, estuaries, and other major water bodies, including their surface interface and interaction, with the atmosphere and with the land seaward of the mean high water mark.

Marine expeditionary brigade--A task organization which is normally built around a regimental landing team, a provisional Marine aircraft group, and a logistics support group. It is capable of conducting amphibious assault operations of a limited scope. During potential crisis situations, a Marine expeditionary brigade may be forward deployed afloat for an extended period in order to provide an immediate combat response. Also called **MEB.** See also **Marine air-ground task force.**

Marine expeditionary force--The Marine expeditionary force, the largest of the Marine air-ground task forces, is normally built around a division/wing team, but can include several divisions and aircraft wings, together with an appropriate combat service support organization. The Marine expeditionary force is capable of conducting a wide range of amphibious assault operations and sustained operations ashore. It can be tailored for a wide variety of combat missions in any geographic environment. Also called **MEF.** See also **Marine air-ground task force.**

Marine expeditionary unit--A task organization which is normally built around a battalion landing team, reinforced helicopter squadron, and logistic support unit. It fulfills routine forward afloat deployment requirements, provides an immediate reaction capability for crisis situations, and is capable of relatively limited combat operations. Also called **MEU.** See also **Marine air-ground task force.**

Marine expeditionary unit (special operations capable)--A forward-deployed, embarked US Marine Corps unit with enhanced capability to conduct special operations. The Marine expeditionary unit (special operations capable) is oriented toward amphibious raids, at night, under limited visibility, while employing emission control procedures. The Marine expeditionary unit (special operations capable) is not a Secretary of Defense-designated special operations force but, when directed by the National Command Authorities and/or the theater commander, may conduct hostage recovery or other special operations under in extremis circumstances when designated special operations forces are not available. Also called **MEU(SOC).**

maritime control area--An area generally similar to a defensive sea area in purpose except that it may be established any place on the high seas. Maritime control areas are normally established only in time of war. See also **defensive sea area.**

maritime defense sector(*)--One of the subdivisions of a coastal area.

maritime environment--The oceans, seas, bays, estuaries, islands, coastal areas, and the airspace above these, including amphibious objective areas.

maritime power projection--Power projection in and from the maritime environment, including a broad spectrum of offensive military operations to destroy enemy forces or logistic support or to prevent enemy forces from approaching within enemy weapons' range of friendly forces. Maritime power projection may be accomplished by amphibious assault operations, attack of targets ashore, or support of sea control operations.

maritime prepositioning ships--Civilian-crewed, Military Sealift Command-chartered ships which are organized into three squadrons and are usually forward-deployed. These ships are loaded with prepositioned equipment and 30 days of supplies to support three Marine expeditionary brigades. Also called **MPS.** See also **Navy Cargo Handling Battalion.**

maritime search and rescue region--The waters subject to the jurisdiction of the United States; the territories and possessions of the United States (except Canal Zone and the inland area of Alaska) and designated areas of the high seas. See also **search and rescue region.**

maritime special purpose force--A task-organized force formed from elements of a Marine expeditionary unit (special operations capable) and naval special warfare forces that can be quickly tailored to a specific mission. The maritime special purpose force can execute on short notice a wide variety of missions in a supporting, supported, or unilateral role. It focuses on operations in a maritime environment and is capable of operations in conjunction with or in support of special operations forces. The maritime special purpose force is integral to and directly relies upon the Marine expeditionary unit (special operations capable) for all combat and combat service support. Also called **MSPF.**

mark--1. In artillery and naval gunfire support: a. to call for fire on a specified location in order to orient the observer/spotter or to indicate targets; b. to report the instant of optimum light on the target produced by illumination shells. 2. In naval operations, to use a maritime unit to maintain an immediate offensive or obstructive capability against a specified target. See also **marker.**

marker(*)--1. A visual or electronic aid used to mark a designated point. 2. In land mine warfare: See **gap marker; intermediate marker; lane marker; row marker; strip marker.** 3. In naval operations, a maritime unit which maintains an immediate offensive or obstructive capability against a specified target. See also **beacon; shadower; mark.**

marker ship(*)--In an amphibious operation, a ship which takes accurate station on a designated control point. It may fly identifying flags by day and show lights to seaward by night.

marking error(*)--In naval mine warfare, the distance and bearing of a marker from a target.

marking fire(*)--Fire placed on a target for the purpose of identification.

marking panel(*)--A sheet of material displayed for visual communication, usually between friendly units. See also **panel code.**

mark mark--Command from ground controller for aircraft to release bombs; may indicate electronic ground-controlled release or voice command to aircrew.

married failure(*)--In naval mine warfare, a moored mine lying on the seabed connected to its sinker from which it has failed to release owing to defective mechanism.

MARS--See **Military Affiliate Radio System.**

marshalling(*)--1. The process by which units participating in an amphibious or airborne operation group together or assemble when feasible or move to temporary camps in the vicinity of embarkation points, complete preparations for combat, or prepare for loading. 2. The process of assembling, holding, and organizing supplies and/or equipment, especially vehicles of transportation, for onward movement. See also **stage; staging area.**

MASINT--See **measurement and signature intelligence.**

mass(*)--1. The concentration of combat power. 2. The military formation in which units are spaced at less than the normal distances and intervals.

mass casualty--Any large number of casualties produced in a relatively short period of time, usually as the result of a single incident such as a military aircraft accident, hurricane, flood, earthquake, or armed attack that exceeds local logistical support capabilities. See also **casualty.**

massed fire--1. The fire of the batteries of two or more ships directed against a single target. 2. Fire from a number of weapons directed at a single point or small area. See also **concentrated fire.**

master--The commanding officer of a United States Naval Ship, a commercial ship, or a government-owned general agency agreement ship operated for the Military Sealift Command by a civilian company to transport DOD cargo.

master film(*)--The earliest generation of imagery (negative or positive) from which subsequent copies are produced.

master force list--A file which contains the current status of each requirement for a given operation plan. The master force list is made available for file transfer service (FTS) transfer to other Worldwide Military Command and Control System activities from a file produced from the joint deployment system data base. Also called **MFL.**

master plot(*)--A portion of a map or overlay on which are drawn the outlines of the areas covered by an air photographic sortie. Latitude and longitude, map, and sortie information are shown. See also **sortie plot.**

materials handling(*)--The movement of materials (raw materials, scrap, semifinished, and finished) to, through, and from productive processes; in warehouses and storage; and in receiving and shipping areas.

materials handling equipment--Mechanical devices for handling of supplies with greater ease and economy.

materiel--All items (including ships, tanks, self--propelled weapons, aircraft, etc., and related spares, repair parts, and support equipment, but excluding real property, installations, and utilities) necessary to equip, operate, maintain, and support military activities without distinction as to its application for administrative or combat purposes. See also **equipment; personal property.**

materiel cognizance--Denotes responsibility for exercising supply management over items or categories of materiel.

materiel control--See **inventory control.**

materiel inventory objective--The quantity of an item required to be on hand and on order on M-day in order to equip, provide a materiel pipeline, and sustain the approved US force structure (active and reserve) and those Allied forces designated for US materiel support, through the period prescribed for war materiel planning purposes. It is the quantity by which the war materiel requirement exceeds the war materiel procurement capability and the war materiel requirement adjustment. It includes the M-day force materiel requirement and the war reserve materiel requirement.

materiel management--See **inventory control.**

materiel pipeline--The quantity of an item required in the worldwide supply system to maintain an uninterrupted replacement flow.

materiel planning--A subset of logistic planning and consists of a four-step process: **a. requirements definition.** Requirements for significant items must be calculated at item level detail (i.e., national stock number) to support sustainability planning and analysis. Requirements include unit roundout, consumption and attrition replacement, safety stock, and the needs of allies. **b. apportionment.** Items are apportioned to the combatant commanders based on a global scenario to avoid sourcing of items to multiple theaters. The basis for apportionment is the capability provided by unit stocks, host nation support, theater prepositioned war reserve stocks and industrial base, and continental United States Department of Defense stockpiles and available production. Item apportionment cannot exceed total capabilities. **c. sourcing.** Sourcing is the matching of available capabilities on a given date against item requirements to support sustainability analysis and the identification of locations to support transportation planning. Sourcing of any item is done within the combatant commander's apportionment. **d. documentation.** Sourced item requirements and corresponding shortfalls are major inputs to the combatant commander's sustainability analysis. Sourced item requirements are translated into movement requirements and documented in the Joint Operation Planning and Execution System data base for transportation feasibility analysis. Movement requirements for nonsignificant items are estimated in tonnage.

materiel readiness--The availability of materiel required by a military organization to support its wartime activities or contingencies, disaster relief (flood, earthquake, etc.), or other emergencies.

materiel release confirmation--A notification from a shipping/storage activity advising the originator of a materiel release order of the positive action taken on the order. It will also be used with appropriate shipment status document identifier codes as a reply to a followup initiated by the inventory control point.

materiel release order--An order issued by an accountable supply system manager (usually an inventory control point or accountable depot/stock point) directing a non-accountable activity (usually a storage site or materiel drop point) within the same supply distribution complex to release and ship materiel.

materiel requirements--Those quantities of items of equipment and supplies necessary to equip, provide a materiel pipeline, and sustain a Service, formation, organization, or unit in the fulfillment of its purposes or tasks during a specified period.

Maverick--An air-to-surface missile with launch and leave capability. It is designed for use against stationary or moving small, hard targets such as tanks, armored vehicles, and field fortifications. Designated as **AGM-65.**

maximum aircraft arresting hook load--The maximum load experienced by an aircraft arresting hook assembly during an arrestment.

maximum effective range(*)--The maximum distance at which a weapon may be expected to be accurate and achieve the desired result.

maximum elevation figure(*)--A figure, shown in each quadrangle bounded by ticked graticule lines on aeronautical charts, which represents the height in thousands and hundreds of feet, above mean sea level, of the highest known natural or manmade feature in that quadrangle, plus suitable factors to allow for inaccuracy and incompleteness of the topographical heighting information.

maximum landing weight(*)--The maximum gross weight due to design or operational limitations at which an aircraft is permitted to land.

maximum operating depth(*)--The depth which a submarine is not to exceed during operations. This depth is determined by the submarine's national naval authority. See also **test depth.**

maximum ordinate(*)--In artillery and naval gunfire support, the height of the highest point in the trajectory of a projectile above the horizontal plane passing through its origin. Also called **vertex height.**

maximum permissible concentration--See **radioactivity concentration guide.**

maximum permissible dose(*)--That radiation dose which a military commander or other appropriate authority may prescribe as the limiting cumulative radiation dose to be received over a specific period of time by members of the command, consistent with current operational military considerations.

maximum range(*)--The greatest distance a weapon can fire without consideration of dispersion.

maximum sustained speed(*)--In road transport, the highest speed at which a vehicle, with its rated payload, can be driven for an extended period on a level first-class highway without sustaining damage.

maximum takeoff weight(*)--The maximum gross weight due to design or operational limitations at which an aircraft is permitted to take off.

mayday--Distress call.

MC--See **mission capable.**

MC&G--See **mapping, charting, and geodesy.**

M-day--See **times.**

M-day force materiel requirement--The quantity of an item required to be on hand and on order (on M-day minus one day) to equip and provide a materiel pipeline for the approved peacetime US force structure, both active and reserve.

meaconing(*)--A system of receiving radio beacon signals and rebroadcasting them on the same frequency to confuse navigation. The meaconing stations cause inaccurate bearings to be obtained by aircraft or ground stations. See also **beacon.**

mean lethal dose(*)--1. The amount of nuclear irradiation of the whole body which would be fatal to 50 percent of the exposed personnel in a given period of time. 2. The dose of chemical agent that would kill 50 percent of exposed, unprotected and untreated personnel.

mean line of advance--In naval usage, the direction expected to be made good over a sustained period.

mean point of burst--See **mean point of impact.**

mean point of impact(*)--The point whose coordinates are the arithmetic means of the coordinates of the separate points of impact/burst of a finite number of projectiles fired or released at the same aiming point under a given set of conditions.

mean sea level--The average height of the surface of the sea for all stages of the tide; used as a reference for elevations.

means of transport--See **mode of transport.**

measurement and signature intelligence--Scientific and technical intelligence obtained by quantitative and qualitative analysis of data (metric, angle, spatial, wavelength, time dependence, modulation, plasma, and hydromagnetic) derived from specific technical sensors for the purpose of identifying any distinctive features associated with the source, emitter, or sender and to facilitate subsequent identification and/or measurement of the same. Also called **MASINT.** See also **intelligence.**

mechanical sweep(*)--In naval mine warfare, any sweep used with the object of physically contacting the mine or its appendages.

median incapacitating dose(*)--The amount or quantity of chemical agent which when introduced into the body will incapacitate 50 percent of exposed, unprotected personnel.

medical evacuees--Personnel who are wounded, injured, or ill and must be moved to or between medical facilities.

medical intelligence--That category of intelligence resulting from collection, evaluation, analysis, and interpretation of foreign medical, bio-scientific, and environmental information which is of interest to strategic planning and to military medical planning and operations for the conservation of the fighting strength of friendly forces and the formation of assessments of foreign medical capabilities in both military and civilian sectors.

medical officer(*)--Physician with officer rank.

medical treatment facility--A facility established for the purpose of furnishing medical and/or dental care to eligible individuals.

medium--As used in air intercept, a height between 2,000 and 25,000 feet.

medium-altitude bombing--Horizontal bombing with the height of release between 8,000 and 15,000 feet.

medium-angle loft bombing--Type of loft bombing wherein weapon release occurs at an angle between 35 and 75 degrees above the horizontal.

medium artillery--See **field artillery.**

medium atomic demolition munition--A low--yield, team-portable, atomic demolition munition which can be detonated either by remote control or a timer device.

medium-lot storage--Generally defined as a quantity of material which will require one to three pallet stacks, stored to maximum height. Thus, the term refers to relatively small lots as distinguished from definitely large or small lots. See also **storage.**

medium-range ballistic missile--A ballistic missile with a range capability from about 600 to 1,500 nautical miles.

medium-range bomber aircraft--A bomber designed for a tactical operating radius of under 1,000 nautical miles at design gross weight and design bomb load.

medium-range transport aircraft--See **transport aircraft.**

medium-scale map--A map having a scale larger than 1:600,000 and smaller than 1:75,000. See also **map.**

meeting engagement(*)--A combat action that occurs when a moving force, incompletely deployed for battle, engages an enemy at an unexpected time and place.

megaton weapon(*)--A nuclear weapon, the yield of which is measured in terms of millions of tons of trinitrotoluene explosive equivalents. See also **kiloton weapon; nominal weapon; subkiloton weapon.**

memory--See **storage.**

mercantile convoy(*)--A convoy consisting primarily of merchant ships controlled by the naval control of shipping organization.

merchant intelligence--In intelligence handling, communication instructions for reporting by merchant vessels of vital intelligence sightings. Also called **MERINT.**

merchant ship(*)--A vessel engaged in mercantile trade except river craft, estuarial craft, or craft which operate solely within harbor limits.

merchant ship casualty report--A report by message, or other means, of a casualty to a merchant ship at sea or in port. Merchant ship casualty reports are sent by the escort force commander or other appropriate authority to the operational control authority in whose area the casualty occurred.

merchant ship communications system(*)--A worldwide system of communications to and from merchant ships using the peacetime commercial organization as a basis but under Operational Control Authority, with the ability to employ the broadcast mode to ships when the situation makes radio silence necessary.

merchant ship control zone(*)--A defined area of sea or ocean inside which it may be necessary to offer guidance, control, and protection to allied shipping.

merchant ship reporting and control message system(*)--A worldwide message system for reporting the movements of and information relating to the control of merchant ships.

mercomms system--See **merchant ship communications system.**

merged--In air intercept, a code meaning, "Tracks have come together."

message--Any thought or idea expressed briefly in a plain or secret language and prepared in a form suitable for transmission by any means of communication.

message center--See **telecommunications center.**

message (telecommunications)--Record information expressed in plain or encrypted language and prepared in a format specified for intended transmission by a telecommunications system.

meteorological and oceanographic forecast center--Shore-based meteorological and oceanographic production activity. Also called **MFC.**

meteorological data--Meteorological facts pertaining to the atmosphere, such as wind, temperature, air density, and other phenomena which affect military operations.

metrology--The science of measurement, including the development of measurement standards and systems for absolute and relative measurements.

MEU(SOC)--See **Marine expeditionary unit (special operations capable).**

MFC--See **meteorological and oceanographic forecast center.**

MGM-13--See **Mace.**

MGM-29A--See **Sergeant.**

MGM-31A--See **Pershing.**

MGM-51--See **Shillelagh.**

MIA--See **missing in action.**

MIB--See **Military Intelligence Board.**

microform(*)--A generic term for any form, whether film, video tape, paper, or other medium, containing miniaturized or otherwise compressed images which cannot be read without special display devices.

midcourse guidance--The guidance applied to a missile between termination of the boost phase and the start of the terminal phase of flight. See also **guidance.**

midcourse phase--That portion of the trajectory of a ballistic missile between the boost phase and the reentry phase. See also **ballistic trajectory; boost phase; reentry phase; terminal phase.**

middleman--In air intercept, a code meaning, "Very high frequency or ultra-high frequency radio relay equipment."

midnight--In air intercept, a code meaning, "Changeover from close to broadcast control."

militarily significant fallout--Radioactive contamination capable of inflicting radiation doses on personnel which may result in a reduction of their combat effectiveness.

Military Affiliate Radio System--A program conducted by the Departments of the Army, Navy, and Air Force in which amateur radio stations and operators participate in and contribute to the mission of providing auxiliary and emergency communications on a local, national, or international basis as an adjunct to normal military communications. Also called **MARS.**

military assistance advisory group--A joint Service group, normally under the military command of a commander of a unified command and representing the Secretary of Defense, which primarily administers the US military assistance planning and programming in the host country. Also called **MAAG.**

Military Assistance Articles and Services List--A Department of Defense publication listing source, availability, and price of items and services for use by the unified commands and Military Departments in preparing military assistance plans and programs.

Military Assistance Program--That portion of the US security assistance authorized by the Foreign Assistance Act of 1961, as amended, which provides defense articles and services to recipients on a nonreimbursable (grant) basis.

Military Assistance Program Training--See **international military education and training.**

military capability--The ability to achieve a specified wartime objective (win a war or battle, destroy a target set). It includes four major components: force structure, modernization, readiness, and sustainability. **a. force structure**--Numbers, size, and composition of the units that comprise our Defense forces; e.g., divisions, ships, airwings. **b. modernization**--Technical sophistication of forces, units, weapon systems, and equipments. **c. readiness**--The ability of forces, units, weapon systems, or equipments to deliver the outputs for which they were designed (includes the ability to deploy and employ without unacceptable delays). **d. sustainability**--The ability to maintain the necessary level and duration of operational activity to achieve military objectives. Sustainability is a function of providing for and maintaining those levels of ready forces, materiel, and consumables necessary to support military effort.

military censorship--All types of censorship conducted by personnel of the Armed Forces of the United States, to include armed forces censorship, civil censorship, prisoner of war censorship, and field press censorship. See also **censorship.**

military characteristics--Those characteristics of equipment upon which depends its ability to perform desired military functions. Military characteristics include physical and operational characteristics but not technical characteristics.

military civic action--The use of preponderantly indigenous military forces on projects useful to the local population at all levels in such fields as education, training, public works, agriculture, transportation, communications, health, sanitation, and others contributing to economic and social development, which would also serve to improve the standing of the military forces with the population. (US forces may at times advise or engage in military civic actions in overseas areas.)

military convoy(*)--A land or maritime convoy that is controlled and reported as a military unit. A maritime convoy can consist of any combination of merchant ships, auxiliaries, or other military units.

military currency(*)--Currency prepared by a power and declared by its military commander to be legal tender for use by civilian and/or military personnel as prescribed in the areas occupied by its forces. It should be of distinctive design to distinguish it from the official currency of the countries concerned, but may be denominated in the monetary unit of either.

military damage assessment--An appraisal of the effects of an attack on a nation's military forces to determine residual military capability and to support planning for recovery and reconstitution. See also **damage assessment.**

military deception--Actions executed to mislead foreign decisionmakers, causing them to derive and accept desired appreciations of military capabilities, intentions, operations, or other activities that evoke foreign actions that contribute to the originator's objectives. There are three categories of military deception: **a. strategic military deception**--Military deception planned and executed to result in foreign national policies and actions which support the originator's national objectives, policies, and strategic military plans. **b. tactical military deception**--Military deception planned and executed by and in support of operational commanders against the pertinent threat to result in opposing operational actions favorable to the originator's plans and operations. **c. Department/Service military deception**--Military deception planned and executed by Military Services about military systems, doctrine, tactics, techniques, personnel or service operations, or other activities to result in foreign actions which increase or maintain the originator's capabilities relative to adversaries. See also **deception.**

Military Department--One of the departments within the Department of Defense created by the National Security Act of 1947, as amended. See also **Department of the Army; Department of the Navy; Department of the Air Force.**

military designed vehicle--A vehicle having military characteristics resulting from military research and development processes, designed primarily for use by forces in the field in direct connection with, or support of, combat or tactical operations.

military education--The systematic instruction of individuals in subjects which will enhance their knowledge of the science and art of war. See also **military training.**

military geographic documentation--Military geographic information which has been evaluated, processed, summarized, and published.

military geographic information--Comprises the information concerning physical aspects, resources, and artificial features which is necessary for planning and operations.

military geography--The specialized field of geography dealing with natural and manmade physical features that may affect the planning and conduct of military operations.

military government--See **civil affairs.**

military government ordinance--An enactment on the authority of a military governor promulgating laws or rules regulating the occupied territory under such control.

military governor(*)--The military commander or other designated person who, in an occupied territory, exercises supreme authority over the civil population subject to the laws and usages of war and to any directive received from the commander's government or superior.

military grid(*)--Two sets of parallel lines intersecting at right angles and forming squares; the grid is superimposed on maps, charts, and other similar representations of the surface of the Earth in an accurate and consistent manner to permit identification of ground locations with respect to other locations and the computation of direction and distance to other points. See also **military grid reference system.**

military grid reference system(*)--A system which uses a standard-scaled grid square, based on a point of origin on a map projection of the surface of the Earth in an accurate and consistent manner to permit either position referencing or the computation of direction and distance between grid positions. See a]so military grid.

military independent(*)--A merchant ship or auxiliary sailed singly but controlled and reported as a military unit. See also **independent.**

military intelligence--Intelligence on any foreign military or military-related situation or activity which is significant to military policy-making or the planning and conduct of military operations and activities.

Military Intelligence Board--A decisionmaking forum which formulates Defense intelligence policy and programming priorities. The Military Intelligence Board, chaired by the Director, Defense Intelligence Agency, who is dual-hatted as Director of Military Intelligence, consists of senior military and civilian intelligence officials of each Service, US Coast Guard, each combat support agency, the Joint Staff/J-2/J-6, Deputy Assistant Secretary of Defense (Intelligence), Intelligence Program Support Group, National Military Intelligence Production Center, National Military Intelligence Collection Center, National Military Intelligence Support Center, and the combatant command J-2s. Also called **MIB.**

military intervention--The deliberate act of a nation or a group of nations to introduce its military forces into the course of an existing controversy.

military land transportation resources--All military-owned transportation resources, desig-

nated for common-user, over the ground, point-to-point use.

military load classification(*)--A standard system in which a route, bridge, or raft is assigned class number(s) representing the load it can carry. Vehicles are also assigned number(s) indicating the minimum class of route, bridge, or raft they are authorized to use. See also **route classification.**

military necessity(*)--The principle whereby a belligerent has the right to apply any measures which are required to bring about the successful conclusion of a military operation and which are not forbidden by the laws of war.

military nuclear power(*)--A nation which has nuclear weapons and the capability for their employment. See also **nuclear power.**

military objectives--The derived set of military actions to be taken to implement National Command Authorities guidance in support of national objectives. Defines the results to be achieved by the military and assigns tasks to commanders. See also **national objectives.**

military occupation--A condition in which territory is under the effective control of a foreign armed force. See also **occupied territory; phases of military government.**

military options--A range of military force responses that can be projected to accomplish assigned tasks. Options include one or a combination of the following: civic action, humanitarian assistance, civil affairs, and other military activities to develop positive relationships with other countries; confidence building and other measures to reduce military tensions; military presence; activities to convey threats to adversaries and truth projections; military deceptions and psychological operations; quarantines, blockades, and harassment operations; raids; intervention campaigns; armed conflict involving air, land, maritime, and strategic warfare campaigns and operations; support for law enforcement authorities to counter international criminal activities (terrorism, narcotics trafficking, slavery, and piracy); support for law enforcement authorities to suppress domestic rebellion; and support for insurgencies, counterinsurgency, and civil war in foreign countries.

military platform--A side-loading platform generally at least 300 meters/1000 feet long for military trains.

military posture--The military disposition, strength, and condition of readiness as it affects capabilities.

military projection operations--See **land, sea, or aerospace projection operations.**

military requirement(*)--An established need justifying the timely allocation of resources to achieve a capability to accomplish approved military objectives, missions, or tasks. See also **objective force level.**

military resources--Military and civilian personnel, facilities, equipment, and supplies under the control of a DOD component.

Military Sealift Command--The US Transportation Command's component command responsible for designated sealift service. Also called **MSC.** See also **transportation component command.**

Military Service--A branch of the Armed Forces of the United States, established by act of Congress, in which persons are appointed, enlisted, or inducted for military service, and which operates and is administered within a military or executive department. The Military Services are: the United States Army, the United States Navy, the United States Air Force, the United States Marine Corps, and the United States Coast Guard.

military standard requisitioning and issue procedure--A uniform procedure established by the Department of Defense for use within the Department of Defense to govern requisition and issue of materiel within standardized priorities. Also called **MILSTRIP.**

military standard transportation and movement procedures--Uniform and standard transportation data, documentation, and control procedures applicable to all cargo movements in the Department of Defense transportation system. Also called **MILSTAMP.**

military strategy--The art and science of employing the armed forces of a nation to secure the objectives of national policy by the application of force or the threat of force. See also **strategy.**

military symbol(*)--A graphic sign used, usually on map, display or diagram, to represent a particular military unit, installation, activity, or other item of military interest.

military traffic--Department of Defense personnel, mail, and cargo to be, or being, transported.

Military Traffic Management Command--The US Transportation Command's component command responsible for military traffic, continental United States air and land transportation, and common-user water terminals. Also called **MTMC.** See also **transportation component command.**

military training--The instruction of personnel to enhance their capacity to perform specific military functions and tasks; the exercise of one or more military units conducted to enhance their combat readiness. See also **military education.**

MILVAN--Military-owned demountable container, conforming to United States and international standards, operated in a centrally controlled fleet for movement of military cargo.

MILVAN chassis--The compatible chassis to which the MILVAN is attached by coupling the lower four standard corner fittings of the container to compatible mounting blocks in the chassis to permit road movement.

MIM-23--See **Hawk.**

MIM-72--See **Chaparral.**

mine(*)--1. In land mine warfare, an explosive or other material, normally encased, designed to destroy or damage ground vehicles, boats, or aircraft, or designed to wound, kill, or otherwise incapacitate personnel. It may be detonated by the action of its victim, by the passage of time, or by controlled means. 2. In naval mine warfare, an explosive device laid in the water with the intention of damaging or sinking ships or of deterring shipping from entering an area. The term does not include devices attached to the bottoms of ships or to harbor installations by personnel operating underwater, nor does it include devices which explode immediately on expiration of a predetermined time after laying.

mineable waters(*)--Waters where naval mines of any given type may be effective against any given target.

mine clearance(*)--The process of removing all mines from a route or area.

mine countermeasures--All methods for preventing or reducing damage or danger from mines.

mined area(*)--An area declared dangerous due to the presence or suspected presence of mines.

mine defense(*)--The defense of a position, area, etc., by land or underwater mines. A mine defense system includes the personnel and equipment needed to plant, operate, maintain, and protect the minefields that are laid.

mine disposal(*)--The operation by suitably qualified personnel designed to render safe, neutralize, recover, remove, or destroy mines.

minefield(*)--1. In land warfare, an area of ground containing mines laid with or without a pattern. See also **defensive minefield; mixed minefield; nuisance minefield; phoney minefield; protective minefield.** 2. In naval warfare, an area of water containing mines laid with or without a pattern. See also **dummy minefield.**

minefield breaching(*)--In land mine warfare, the process of clearing a lane through a minefield under tactical conditions. See also **minefield lane.**

minefield density(*)--In land mine warfare, the average number of mines per meter of minefield front, or the average number of mines per square meter of minefield.

minefield lane(*)--A marked lane, unmined, or cleared of mines, leading through a minefield.

minefield marking--Visible marking of all points required in laying a minefield and indicating the extent of such minefields.

minefield record(*)--A complete written record of all pertinent information concerned on a minefield, submitted on a standard form by the officer in charge of the laying operations.

minefield report--An oral, electronic, or written communication concerning mining activities, friendly or enemy, submitted in a standard format by the fastest secure means available.

minehunting--Employment of sensor and neutralization systems, whether air, surface, or subsurface, to locate and dispose of individual mines. Minehunting is conducted to eliminate mines in a known field when sweeping is not feasible or desirable, or to verify the presence or absence of mines in a given area. See also **minesweeping.**

mine row(*)--A single row of mines or clusters. See also **mine strip.**

mine spotting(*)--In naval mine warfare, the process of visually observing a mine or minefield.

mine strip(*)--In land mine warfare, two parallel mine rows laid simultaneously six meters or six paces apart. See also **mine row.**

minesweeping--The technique of clearing mines using either mechanical, explosive, or influence sweep equipment. Mechanical sweeping removes, disturbs, or otherwise neutralizes the mine; explosive sweeping causes sympathetic detonations in, damages, or displaces the mine; and influence sweeping produces either the acoustic and/or magnetic influence required to detonate the mine. See also **minehunting.**

mine warfare--The strategic, operational, and tactical use of mines and mine countermeasures. Mine warfare is divided into two basic subdivisions: the laying of mines to degrade the enemy's capabilities to wage land, air, and maritime warfare; and the countering of enemy-laid mines to permit friendly maneuver or use of selected land or sea areas.

mine warfare chart(*)--A special naval chart, at a scale of 1:50,000 or larger (preferably 1:25,000 or larger) designed for planning and executing mine warfare operations, either based on an existing standard nautical chart, or produced to special specifications.

mine warfare forces (naval)--Navy forces charged with the strategic, operational, and tactical use of naval mines and their countermeasures. Such forces are capable of offensive and defensive measures in connection with laying and clearing mines.

mine warfare group(*)--A task organization of mine warfare units for the conduct of minelaying and/or mine countermeasures in maritime operations.

minewatching(*)--In naval mine warfare, the mine countermeasures procedure to detect, record, and, if possible, track potential minelayers and to detect, find the position of, and/or identify mines during the actual minelaying.

mine weapons(*)--The collective term for all weapons which may be used in mine warfare.

minimize--A condition wherein normal message and telephone traffic is drastically reduced in order that messages connected with an actual or simulated emergency shall not be delayed.

minimum aircraft operating surface(*)--The minimum surface on an airfield which is essential for the movement of aircraft. It includes the aircraft dispersal areas, the minimum operating strip, and the taxiways between them. See also **minimum operating strip.**

minimum-altitude bombing--Horizontal or glide bombing with the height of release under 900 feet. It includes masthead bombing, which is sometimes erroneously referred to as "skip bombing." See also **skip bombing.**

minimum attack altitude--The lowest altitude determined by the tactical use of weapons, terrain consideration, and weapons effects which permits the safe conduct of an air attack and/or minimizes effective enemy counteraction.

minimum crossing altitude--The lowest altitude at certain radio fixes at which an aircraft must cross when proceeding in the direction of a higher minimum en route instrument flight rules altitude.

minimum descent altitude(*)--The lowest altitude to which descent shall be authorized in procedures not using a glide slope, until the required visual reference has been established. See also **minimum descent height.**

minimum descent height(*)--The lowest height to which descent shall be authorized in procedures not using a glide slope, until the required visual reference has been established. See also **minimum descent altitude.**

minimum essential equipment--That part of authorized allowances of Army equipment, clothing, and supplies needed to preserve the integrity of a unit during movement without regard to the performance of its combat or service mission. Items common within this category will normally be carried by or accompany troops to the port and will be placed aboard the same ships with the troops. As used in movement directives, minimum essential equipment refers to specific items of both organizational and individual clothing and equipment.

minimum normal burst altitude--The altitude above terrain below which air defense nuclear warheads are not normally detonated.

minimum nuclear safe distance(*)--The sum of the radius of safety and the buffer distance.

minimum nuclear warning time(*)--The sum of system reaction time and personnel reaction time.

minimum obstruction clearance altitude--The specified altitude in effect between radio fixes on very high frequency omnirange airways, off-airway routes, or route segments, which meets obstruction clearance requirements for the entire route segment, and that assures acceptable navigational signal coverage only within 22 miles of a very high frequency omnirange.

minimum operating strip(*)--A runway which meets the minimum requirements for operating assigned and/or allocated aircraft types on a particular airfield at maximum or combat gross weight. See also **minimum aircraft operating surface.**

minimum range--1. Least range setting of a gun at which the projectile will clear an obstacle or friendly troops between the gun and the target. 2. Shortest distance to which a gun can fire from a given position.

minimum reception altitude--The lowest altitude required to receive adequate signals to determine specific very high frequency omnirange/tactical air navigation fixes.

minimum residual radioactivity weapon(*)--A nuclear weapon designed to have optimum reduction of unwanted effects from fallout, rainout, and burst site radioactivity. See also **salted weapon.**

minimum-risk level--A specific altitude or altitude block that allows homebound aircraft to return in a homebound direction without lateral restrictions. Also called **MRL.**

minimum-risk route--A temporary corridor of defined dimensions recommended for use by high-speed, fixed-wing aircraft that presents the minimum known hazards to low-flying aircraft transiting the combat zone. Also called **MRR.**

minimum safe altitude(*)--The altitude below which it is hazardous to fly owing to presence of high ground or other obstacles.

minor control--See **photogrammetric control.**

minor installation--In the Air Force, a facility operated by an active, reserve, or Guard unit of at least squadron size that does not otherwise satisfy all the criteria for a major installation. This category includes Air Force stations; air stations; Air Reserve stations; and Air Guard stations. Examples of minor installations are active, reserve, or Guard flying operations that are located at civilian-owned airports. See also **installation complex; major installation; other activity; support site.**

minor port(*)--A port having facilities for the discharge of cargo from coasters or lighters only. See also **port.**

Minuteman--A three-stage, solid propellant, ballistic missile which is guided to its target by an all-inertial guidance and control system. The missiles are equipped with nuclear warheads and designed for deployment in hardened and dispersed underground silos. With the improved third stage and the post-boost vehicle, the Minuteman III missile can deliver multiple independently targetable reentry vehicles and their penetration aids to multiple targets. Designated as **LGM-30.**

misfire(*)--1. Failure to fire or explode properly. 2. Failure of a primer or the propelling charge of a round or projectile to function wholly or in part.

MISREP--See **Joint Tactical Air Reconnaissance/Surveillance Mission Report.**

missed approach(*)--An approach which is not completed by landing.

missed approach procedure(*)--The procedures to be followed if, after an instrument approach, a landing is not effected and occurring normally: a. when the aircraft has descended to the decision height/altitude and has not established

visual contact; or b. when directed by air traffic control to pull up or to go around again.

missile assembly-checkout facility--A building, van, or other type structure located near the operational missile launching location and designed for the final assembly and checkout of the missile system.

missile control system(*)--A system that serves to maintain attitude stability and to correct deflections. See also **missile guidance system.**

missile destruct(*)--Intentional destruction of a missile or similar vehicle for safety or other reasons.

missile destruct system(*)--A system which, when operated by external command or preset internal means, destroys the missile or similar vehicle.

missile engagement zone--(NATO) See weapon engagement zone.

missile guidance system(*)--A system which evaluates flight information, correlates it with target data, determines the desired flight path of a missile, and communicates the necessary commands to the missile flight control system. See also **missile control system.**

missile intercept zone--That geographical division of the destruction area where surface-to-air missiles have primary responsibility for destruction of airborne objects. See also **destruction area.**

missile monitor--A mobile, electronic, air defense fire-distribution system for use at Army air defense group, battalion, and battery levels. It employs digital data to exchange information within the system and provides means for the Army air defense commander to monitor actions of the units and take corrective action when necessary. It automatically exchanges information with adjacent missile monitor systems when connected with them by data links.

missile release line--The line at which an attacking aircraft could launch an air-to-surface missile against a specific target.

missing--A casualty status for which the United States Code provides statutory guidance concerning missing members of the Military Services. Excluded are personnel who are in an absent without leave, deserter, or dropped--from-rolls status. A person declared missing is categorized as follows: **a. beleaguered**--The casualty is a member of an organized element that has been surrounded by a hostile force to prevent escape of its members. **b. besieged**--The casualty is a member of an organized element that has been surrounded by a hostile force for compelling it to surrender. **c. captured**--The casualty has been seized as the result of action of an unfriendly military or paramilitary force in a foreign country. **d. detained**--The casualty is prevented from proceeding or is restrained in custody for alleged violation of international law or other reason claimed by the government or group under which the person is being held. **e. interned**--The casualty is definitely known to have been taken into custody of a nonbelligerent foreign power as the result of and for reasons arising out of any armed conflict in which the Armed Forces of the United States are engaged. **f. missing**--The casualty is not present at his or her duty location due to apparent involuntary reasons and whose location is unknown. **g. missing in action**--The casualty is a hostile casualty, other than the victim of a terrorist activity, who is not present at his or her duty location due to apparent involuntary reasons and whose location is unknown. Also called **MIA**. See also **casualty category; casualty status**.

missing in action--See **missing**.

mission--1. The task, together with the purpose, that clearly indicates the action to be taken and the reason therefor. 2. In common usage, especially when applied to lower military units, a duty assigned to an individual or unit; a task. 3. The dispatching of one or more aircraft to accomplish one particular task.

mission capable--Material condition of an aircraft indicating it can perform at least one and potentially all of its designated missions. Mission capable is further defined as the sum of full mission capable and partial mission capable. Also called **MC**. See also **full mission capable; partial mission capable; partial mission capable, maintenance; partial mission capable, supply**.

mission cycle--The mission cycle, as it pertains to targeting, is a decisionmaking process used by commanders to employ forces. Within the cycle there are six general mission steps: detection, location, identification, decision, execution, and assessment.

mission-essential materiel--1. That materiel which is authorized and available to combat, combat support, combat service support, and combat readiness training forces to accomplish their assigned missions. 2. For the purpose of sizing organic industrial facilities, that Service--designated materiel authorized to combat, combat support, combat service support, and combat readiness training forces and activities, including Reserve and National Guard activities, which is required to support approved emergency and/or war plans, and where the materiel is used to: a. destroy the enemy or his capacity to continue war; b. provide battlefield protection of personnel; c. communicate under war conditions; d. detect, locate, or maintain surveillance over the enemy; e. provide combat transportation and support of men and materiel; and f. support training functions, but is suitable for employment under emergency plans to meet purposes enumerated above.

mission-oriented items--Items for which requirements computations are based upon the assessment of enemy capabilities expressed as a known or estimated quantity of total targets to be destroyed. See also **combination mission/level-of-effort-oriented items; level-of-effort-oriented items.**

mission report(*)--A standard report containing the results of a mission and significant sightings along the flight route.

mission review report (photographic interpretation)--An intelligence report containing information on all targets covered by one photographic sortie.

mission type order--1. Order issued to a lower unit that includes the accomplishment of the total mission assigned to the higher headquarters. 2. Order to a unit to perform a mission without specifying how it is to be accomplished.

mixed(*)--In artillery and naval gunfire support, a spotting, or an observation, by a spotter or an observer to indicate that the rounds fired resulted in an equal number of air and impact bursts.

mixed air(*)--In artillery and naval gunfire support, a spotting, or an observation, by a spotter or an observer to indicate that the rounds fired resulted in both air and impact bursts with a majority of the bursts being airbursts.

mixed bag(*)--In naval mine warfare, a collection of mines of various types, firing systems, sensitivities, arming delays and ship counter settings.

mixed graze(*)--In artillery and naval gunfire support, a spotting, or an observation, by a spotter or an observer to indicate that the rounds fired resulted in both air and impact bursts with a majority of the bursts being impact bursts.

mixed minefield(*)--A minefield containing both antitank and antipersonnel mines. See also **minefield.**

mixup, caution--In air intercept, a term meaning mixture of friendly and hostile aircraft.

MOB--See **main operations base.**

mobile defense--Defense of an area or position in which maneuver is used with organization of fire and utilization of terrain to seize the initiative from the enemy.

mobile mine(*)--In naval mine warfare, a mine designed to be propelled to its proposed laying position by propulsion equipment like a torpedo. It sinks at the end of its run and then operates like a mine. See also **mine.**

mobile support group (naval)--Provides logistic support to ships at an anchorage; in effect, a naval base afloat although certain of its supporting elements may be located ashore.

mobile training team--A team consisting of one or more US military or civilian personnel sent on temporary duty, often to a foreign nation, to give instruction. The mission of the team is to train indigenous personnel to operate, maintain, and employ weapons and support systems, or to develop a self-training capability in a particular skill. The National Command Authorities may direct a team to train either military or civilian indigenous personnel, depending upon host nation requests. Also called **MTT.**

mobility(*)--A quality or capability of military forces which permits them to move from place to place while retaining the ability to fulfill their primary mission.

mobility analysis--An in-depth examination of all aspects of transportation planning in support of operation plan and operation order development.

mobility echelon--A subordinate element of a unit that is scheduled for deployment separately from the parent unit.

mobility system support resources--Those resources that are required to: a. complement the airlift and sealift forces, and/or b. perform those work functions directly related to the origination, processing, or termination of a movement requirement.

mobilization--1. The act of assembling and organizing national resources to support national objectives in time of war or other emergencies. See also **industrial mobilization.** 2. The process by which the Armed Forces or part of them are brought to a state of readiness for war or other national emergency. This includes activating all or part of the Reserve components as well as assembling and organizing personnel, supplies, and materiel. Mobilization of the Armed Forces includes but is not limited to the following categories: **a. selective mobilization**--Expansion of the active Armed Forces resulting from action by Congress and/or the President to mobilize Reserve component units, individual ready reservists, and the resources needed for their support to meet the requirements of a domestic emergency that is not the result of an enemy attack. **b. partial mobilization**--Expansion of the active Armed Forces resulting from action by Congress (up to full mobilization) or by the President (not more than 1,000,000) to mobilize Ready Reserve component units, individual reservists, and the resources needed for their support to meet the requirements of a war or other national emergency involving an external threat to the national security. **c. full mobilization**--Expansion of the active Armed Forces resulting from action by Congress and the President to mobilize all Reserve component units in the existing approved force structure, all individual reservists, retired military personnel, and the resources needed for their support to meet the requirements of a war or other national emergency involving an external threat to the national security. **d. total mobilization**--Expansion of the active Armed Forces resulting from action by Congress and the President to organize and/or generate additional units or personnel, beyond the existing force structure, and the resources needed for their support, to meet the total requirements of a war or other national emergency involving an external threat to the national security.

mobilization base--The total of all resources available, or which can be made available, to meet foreseeable wartime needs. Such resources include the manpower and material resources and services required for the support of essential military, civilian, and survival activities, as well as the elements affecting their state of readiness, such as (but not limited to) the following: manning levels, state of training, modernization of equipment, mobilization materiel reserves and facilities, continuity of government, civil defense plans and preparedness measures, psychological preparedness of the people, international agreements, planning with industry, dispersion, and standby legislation and controls.

mobilization exercise--An exercise involving, either completely or in part, the implementation of mobilization plans.

mobilization reserves--Not to be used. See **war reserves.**

mock-up(*)--A model, built to scale, of a machine, apparatus, or weapon, used in studying the construction of, and in testing a new development, or in teaching personnel how to operate the actual machine, apparatus, or weapon.

mode (identification, friend or foe)--The number or letter referring to the specific pulse spacing of the signals transmitted by an interrogator.

mode of transport--The various modes used for a movement. For each mode, there are several means of transport. They are: a. inland surface transportation (rail, road, and inland waterway); b. sea transport (coastal and ocean); c. air transportation; and d. pipelines.

moderate damage--See **nuclear damage (land warfare).**

moderate risk (nuclear)--A degree of risk where anticipated effects are tolerable, or at worst a minor nuisance. See also **degree of risk; emergency risk (nuclear); negligible risk (nuclear).**

modernization--See **military capability.**

modification center--An installation consisting of an airfield and of facilities for modifying standard production aircraft to meet certain requirements which were not anticipated at the time of manufacture.

modify(*)--In artillery, an order by the person authorized to make modifications to a fire plan.

moment(*)--In air transport, the weight of a load multiplied by its distance from a reference point in the aircraft.

monitoring(*)--1. The act of listening, carrying out surveillance on, and/or recording the emissions of one's own or allied forces for the purposes of maintaining and improving procedural standards and security, or for reference, as applicable. 2. The act of listening, carrying out surveillance on, and/or recording of enemy emissions for intelligence purposes. 3. The act of detecting the presence of radiation and the measurement thereof with radiation measuring instruments. Also called **radiological monitoring.**

monitoring service--The general surveillance of known air traffic movements by reference to a radar scope presentation or other means, for the purpose of passing advisory information concerning conflicting traffic or providing navigational assistance. Direct supervision or control is not exercised, nor is positive separation provided.

moored mine(*)--A contact or influence-operated mine of positive buoyancy held below the surface by a mooring attached to a sinker or anchor on the bottom. See also **mine.**

mopping up(*)--The liquidation of remnants of enemy resistance in an area that has been surrounded or isolated, or through which other units have passed without eliminating all active resistance.

mortar--A muzzle-loading, indirect fire weapon with either a rifled or smooth bore. It usually has a shorter range than a howitzer, employs a higher angle of fire, and has a tube, length of 10 to 20 calibers. See also **gun; howitzer.**

mosaic(*)--An assembly of overlapping photographs that have been matched to form a continuous photographic representation of a portion of the surface of the Earth. See also **controlled mosaic; semicontrolled mosaic; uncontrolled mosaic.**

motorized unit(*)--A unit equipped with complete motor transportation that enables all of its personnel, weapons, and equipment to be moved at the same time without assistance from other sources.

mounting(*)--1. All preparations made in areas designated for the purpose, in anticipation of an operation. It includes the assembly in the mounting area, preparation and maintenance within the mounting area, movement to loading points, and subsequent embarkation into ships, craft, or aircraft if applicable. 2. A carriage or stand upon which a weapon is placed.

mounting area--A general locality where assigned forces of an amphibious or airborne operation, with their equipment, are assembled, prepared, and loaded in shipping and/or aircraft preparatory to an assault. See also **embarkation area.**

movement control--The planning, routing, scheduling, and control of personnel and cargo movements over lines of communications; also an organization responsible for these functions. See also **non-unit-related cargo; non-unit-related personnel.**

movement control post(*)--The post through which the control of movement is exercised by the commander, depending on operational requirements.

movement credit(*)--The allocation granted to one or more vehicles in order to move over a controlled route in a fixed time according to movement instructions.

movement directive--The basic document published by the Department of the Army or the Department of the Air Force, or jointly, which authorizes a command to take action to move a designated unit from one location to another.

movement group--Those ships and embarked units that load out and proceed to rendezvous in the objective area.

movement order--An order issued by a commander covering the details for a move of the command.

movement plan--In amphibious operations, the naval plan providing for the movement of the amphibious task force to the objective area. It includes information and instructions concern-

ing departure of ships from loading points, the passage at sea, and the approach to and arrival in assigned positions in the objective area.

movement report control center--The controlling agency for the entire movement report system. It has available all information relative to the movements of naval ships and other ships under naval control.

movement report system--A system established to collect and make available to certain commands vital information on the status, location, and movement of flag commands, commissioned fleet units, and ships under operational control of the Navy.

movement requirement--A stated movement mode and time-phased need for the transport of units, personnel, and/or materiel from a specified origin to a specified destination.

movement restriction(*)--A restriction temporarily placed on traffic into and/or out of areas to permit clearance or prevention of congestion.

movement schedule--A schedule developed to monitor or track a separate entity whether it is a force requirement, cargo or personnel increment, or lift asset. The schedule reflects the assignment of specific lift resources (such as an aircraft or ship) that will be used to move the personnel and cargo included in a specific movement increment. Arrival and departure times at ports of embarkation, etc., are detailed to show a flow and workload at each location. Movement schedules are detailed enough to support plan implementation.

movement table(*)--A table giving detailed instructions or data for a move. When necessary it will be qualified by the words road, rail, sea, air, etc., to signify the type of movement. Normally issued as an annex to a movement order or instruction.

moving havens--Restricted areas established to provide a measure of security to submarines and surface ships in transit through areas in which the existing attack restrictions would be inadequate to prevent attack by friendly forces. See also **moving submarine haven; moving surface ship haven.**

moving map display(*)--A display in which a symbol, representing the vehicle, remains stationary while the map or chart image moves beneath the symbol so that the display simulates the horizontal movement of the vehicle in which it is installed. Occasionally the design of the display is such that the map or chart image remains stationary while the symbol moves across a screen. See also **projected map display.**

moving mine(*)--The collective description of mines, such as drifting, oscillating, creeping, mobile, rising, homing, and bouquet mines.

moving submarine haven--Established by submarine notices, surrounding submarines in transit, extending 50 miles ahead, 100 miles behind, and 15 miles on each side of the estimated position of the submarine along its stated track. See also **moving havens.**

moving surface ship haven--Established by surface ship notices, and will normally be a circle with a specified radius centered on the estimated position of the ship or the guide of a group of ships. See also **moving havens.**

moving target indicator(*)--A radar presentation which shows only targets which are in motion. Signals from stationary targets are subtracted out of the return signal by the output of a suitable memory circuit.

MPS--See **maritime prepositioning ships.**

MRL--See **minimum-risk level.**

MRR--See **minimum-risk route.**

MSC--See **Military Sealift Command.**

MSC-controlled ships--Those ships assigned by the Military Sealift Command (MSC) for a specific operation. They may be MSC nucleus fleet ships, contract-operated MSC ships, MSC-controlled time or voyage-chartered commercial ships, or MSC-controlled ships allocated by the Maritime Administration to MSC to carry out DOD objectives.

MSPF--See **maritime special purpose force.**

MTMC--See **Military Traffic Management Command.**

MTT--See **mobile training team.**

multi-modal(*)--In transport operations, a term applied to the movement of passengers and cargo by more than one method of transport.

multiple drill--See **multiple unit training assemblies.**

multiple inactive duty training periods--Two scheduled inactive duty training periods performed in one calendar day, each at least four hours in duration. No more than two inactive duty training periods may be performed in one day.

multiple independently targetable reentry vehicle--A reentry vehicle carried by a delivery system which can place one or more reentry vehicles over each of several separate targets. See also **maneuverable reentry vehicle; multiple reentry vehicle; reentry vehicle.**

multiple reentry vehicle--The reentry vehicle of a delivery system which places more than one reentry vehicle over an individual target. See also **maneuverable reentry vehicle; multiple**

independently targetable reentry vehicle; reentry vehicle.

multiple unit training assemblies--Two or more unit training assemblies executed during one or more consecutive days. No more than two unit training assemblies may be performed in one calendar day.

multiple warning phenomenology--Deriving warning information from two or more systems observing separate physical phenomena associated with the same events to attain high credibility while being less susceptible to false reports or spoofing.

multi-Service doctrine--Fundamental principles that guide the employment of forces of two or more Services in coordinated action toward a common objective. It is ratified by two or more Services, and is promulgated in multi-Service publications that identify the participating Services, e.g., Army-Navy doctrine. See also **combined doctrine; joint doctrine; joint tactics, techniques, and procedures.**

multi-spectral imagery(*)--The image of an object obtained simultaneously in a number of discrete spectral bands.

munition(*)--A complete device charged with explosives, propellants, pyrotechnics, initiating composition, or nuclear, biological or chemical material for use in military operations, including demolitions. Certain suitably modified munitions can be used for training, ceremonial or nonoperational purposes. Also called **ammunition.** Note: In common usage, "munitions" (plural) can be military weapons, ammunition, and equipment. See also **explosive ordnance.**

music--In air intercept, a term meaning electromagnetic jamming.

mutual support(*)--That support which units render each other against an enemy, because of their assigned tasks, their position relative to each other and to the enemy, and their inherent capabilities. See also **close support; direct support; support.**

muzzle brake--A device attached to the muzzle of a weapon which utilizes escaping gas to reduce recoil.

muzzle compensator--A device attached to the muzzle of a weapon which utilizes escaping gas to control muzzle movement.

muzzle velocity--The velocity of a projectile with respect to the muzzle at the instant the projectile leaves the weapon.

N

napalm--1. Powdered aluminum soap or similar compound used to gelatinize oil or gasoline for use in napalm bombs or flame throwers. 2. The resultant gelatinized substance.

nap-of-the-earth flight--See **terrain flight.**

national censorship--The examination and control under civil authority of communications entering, leaving, or transiting the borders of the United States, its territories, or its possessions. See also **censorship.**

National Command Authorities--The President and the Secretary of Defense or their duly deputized alternates or successors. Also called **NCA.**

National Communications System--The telecommunications system that results from the technical and operational integration of the separate telecommunications systems of the several executive branch departments and agencies having a significant telecommunications capability. Also called **NCS.**

national defense area--An area established on non-Federal lands located within the United States or its possessions or territories for the purpose of safeguarding classified defense information or protecting DOD equipment and/or material. Establishment of a national defense area temporarily places such non-Federal lands under the effective control of the Department of Defense and results only from an emergency event. The senior DOD representative at the scene will define the boundary, mark it with a physical barrier, and post warning signs. The landowner's consent and cooperation will be obtained whenever possible; however, military necessity will dictate the final decision regarding location, shape, and size of the national defense area. Also called **NDA.**

National Defense Reserve Fleet--a. Including the Ready Reserve Force, a fleet composed of ships acquired and maintained by the Maritime Administration (MARAD) for use in mobilization or emergency. b. Less the Ready Reserve Force, a fleet composed of the older dry cargo ships, tankers, troop transports, and other assets in the MARAD's custody that are maintained at a relatively low level of readiness. They are acquired by MARAD from commercial ship operators under the provisions of the Merchant Marine Act of 1936; they are available only on mobilization or congressional declaration of an emergency. Because the ships are maintained in a state of minimum preservation, activation requires 30 to 90 days and extensive shipyard work for many. Also called **NDRF.** See also **Ready Reserve Force.**

national emergency--A condition declared by the President or the Congress by virtue of powers previously vested in them that authorize certain emergency actions to be undertaken in the national interest. Action to be taken may include partial, full, or total mobilization of national resources. See also **mobilization.**

national infrastructure(*)--Infrastructure provided and financed by a NATO member in its own territory solely for its own forces (including those forces assigned to or designated for NATO). See also **infrastructure.**

national intelligence--Integrated departmental intelligence that covers the broad aspects of national policy and national security, is of concern to more than one department or agency, and transcends the exclusive competence of a single department or agency.

national intelligence estimate--A strategic estimate of the capabilities, vulnerabilities, and probable courses of action of foreign nations which is produced at the national level as a composite of the views of the intelligence community.

national intelligence support team--A nationally sourced team composed of intelligence and communications experts from either Defense Intelligence Agency, Central Intelligence Agency, National Security Agency, or any combination of these agencies. Also called **NIST.**

national intelligence surveys--Basic intelligence studies produced on a coordinated interdepartmental basis and concerned with characteristics, basic resources, and relatively unchanging natural features of a foreign country or other area.

National Military Command System--The priority component of the Worldwide Military Command and Control System designed to support the National Command Authorities and Joint Chiefs of Staff in the exercise of their responsibilities. Also called **NMCS.**

national military strategy--The art and science of distributing and applying military power to attain national objectives in peace and war. See also **military strategy; national security strategy; strategy; theater strategy.**

national objectives--The aims, derived from national goals and interests, toward which a national policy or strategy is directed and efforts and resources of the nation are applied. See also **military objectives.**

national policy--A broad course of action or statements of guidance adopted by the government at the national level in pursuit of national objectives.

National Reconnaissance Office--A Department of Defense agency tasked to ensure that the United States has the technology and spaceborne and airborne assets needed to acquire intelligence worldwide, including support to such functions as monitoring of arms control agreements, indications and warning, and the planning and conducting of military operations. This mission is accomplished through research and development, acquisition, and operation of spaceborne and airborne intelligence data collection systems. Also called **NRO.**

national security--A collective term encompassing both national defense and foreign relations of the United States. Specifically, the condition provided by: a. a military or defense advantage over any foreign nation or group of nations, or b. a favorable foreign relations position, or c. a defense posture capable of successfully resisting hostile or destructive action from within or without, overt or covert. See also **security.**

National Security Council--A governmental body specifically designed to assist the President in integrating all spheres of national security policy. The President, Vice President, Secretary of State, and Secretary of Defense are statutory members. The Chairman of the Joint Chiefs of Staff; Director, Central Intelligence Agency; and the Assistant to the President for National Security Affairs serve as advisers. Also called **NSC.**

national security interests--The foundation for the development of valid national objectives that define US goals or purposes. National security interests include preserving US political identity, framework, and institutions; fostering economic well-being; and bolstering international order supporting the vital interests of the United States and its allies.

national security strategy--The art and science of developing, applying, and coordinating the instruments of national power (diplomatic, economic, military, and informational) to achieve objectives that contribute to national security. Also called **national strategy or grand strategy**. See also **military strategy; national military strategy; strategy; theater strategy**.

National Stock Number--The 13-digit stock number replacing the 11-digit Federal Stock Number. It consists of the 4-digit Federal Supply Classification code and the 9-digit National Item Identification Number. The National Item Identification Number consists of a 2-digit National Codification Bureau number designating the central cataloging office of the NATO or other friendly country which assigned the number and a 7-digit (xxxxxxx) nonsignificant number. The number shall be arranged as follows: 9999-00-999-9999. See also **Federal Stock Number**.

national strategy--The art and science of developing and using the political, economic, and psychological powers of a nation, together with its armed forces, during peace and war, to secure national objectives. See also **strategy**.

nation assistance--Civil and/or military assistance rendered to a nation by foreign forces within that nation's territory during peacetime, crises or emergencies, or war based on agreements mutually concluded between nations. Nation assistance programs include, but are not limited to, security assistance, foreign internal defense, other US Code Title 10 (DOD) programs, and activities performed on a reimbursable basis by Federal agencies or international organizations.

natural disaster--See **domestic emergencies**.

nautical chart--See **hydrographic chart**.

nautical mile--A measure of distance equal to one minute of arc on the Earth's surface. The United States has adopted the international

nautical mile equal to 1,852 meters or 6,076.11549 feet.

nautical plotting chart(*)--An outline chart, devoid of hydrographic information, of a specific scale and projection, usually portraying a graticule and compass rose, designed to be ancillary to standard nautical charts, and produced either as an individual chart or a part of a coordinated series.

naval advanced logistic support site--An overseas location used as the primary transshipment point in the theater of operations for logistic support. A naval advanced logistic support site possesses full capabilities for storage, consolidation, and transfer of supplies and for support of forward-deployed units (including replacements units) during major contingency and wartime periods. Naval advanced logistic support sites, with port and airfield facilities in close proximity, are located within the theater of operations but not near the main battle areas, and must possess the throughput capacity required to accommodate incoming and outgoing intertheater airlift and sealift. When fully activated, the naval advanced logistic support site should consist of facilities and services provided by the host nation, augmented by support personnel located in the theater of operations, or both. Also called **ALSS**. See also **naval forward logistic site**.

naval base--A naval base primarily for support of the forces afloat, contiguous to a port or anchorage, consisting of activities or facilities for which the Navy has operating responsibilities, together with interior lines of communication and the minimum surrounding area necessary for local security. (Normally, not greater than an area of 40 square miles.) See also **base complex**.

naval beach group(*)--A permanently organized naval command within an amphibious force comprised of a commander and staff, a beachmaster unit, an amphibious construction battalion, and an assault craft unit, designed to provide an administrative group from which required naval tactical components may be made available to the attack force commander and to the amphibious landing force commander to support the landing of one division (reinforced). See also **shore party.**

naval beach unit--See **naval beach group.**

naval campaign(*)--An operation or a connected series of operations conducted essentially by naval forces, including all surface, subsurface, air, and amphibious troops, for the purpose of gaining, extending, or maintaining control of the sea.

naval coastal warfare--Coastal sea control, harbor defense, and port security, executed both in coastal areas outside the United States in support of national policy and in the United States as part of this Nation's defense. Also called **NCW.**

naval coastal warfare area--An assigned geographic area of responsibility which includes offshore waters, harbor approaches, harbors, ports, waterfront facilities, and those internal waters and rivers which provide access to port facilities.

naval coastal warfare commander--An officer designated to conduct naval coastal warfare missions within a designated naval coastal geographic area. Also called **NCWC.**

naval construction force--The combined construction units of the Navy, including primarily the mobile construction battalions and the amphibious construction battalions. These units are part of the operating forces and represent the Navy's capability for advanced base construction.

naval control of shipping(*)--Control exercised by naval authorities of movement, routing, reporting, convoy organization, and tactical diversion of allied merchant shipping. It does not include the employment or active protection of such shipping.

naval control of shipping officer(*)--A naval officer appointed to form merchant convoys and control and coordinate the routing and movements of such convoys, independently sailed merchant ships, and hospital ships in and out of a port or base, subject to the directions of the operational control authority.

naval control of shipping organization--The organization within the Navy which carries out the specific responsibilities of the Chief of Naval Operations to provide for the control and protection of movements of merchant ships in time of war.

naval district--A geographically defined area in which one naval officer, designated commandant, is the direct representative of the Secretary of the Navy and the Chief of Naval Operations. The commandant has the responsibility for local naval defense and security and for the coordination of naval activities in the area.

naval forward logistic site--An overseas location, with port and airfield facilities nearby, which provides logistic support to naval forces within the theater of operations during major contingency and wartime periods. Naval forward logistic sites may be located in close proximity to main battle areas to permit forward staging of services, throughput of high priority cargo, advanced maintenance, and battle damage repair. Naval forward logistic sites are linked to in-theater naval advanced logistic support sites (ALSSs) by intratheater airlift and sealift, but may also serve as transshipment points for intertheater movement of high-priority cargo into areas of direct combat. In providing fleet logistic support, naval forward logistic site capabilities may range from very austere to near those of a naval advanced logistic support site. Also called **FLS**. See also **naval advanced logistic support site.**

naval gunfire liaison team(*)--Personnel and equipment required to coordinate and advise ground/landing forces on naval gunfire employment.

naval gunfire operations center(*)--The agency established in a ship to control the execution of plans for the employment of naval gunfire, process requests for naval gunfire support, and to allot ships to forward observers. Ideally located in the same ship as the Supporting Arms Coordination Center.

naval gunfire spotting team--The unit of a shore fire control party which designates targets; controls commencement, cessation, rate, and types of fire; and spots fire on the target. See also **field artillery observer; spotter.**

naval gunfire support--Fire provided by Navy surface gun systems in support of a unit or units tasked with achieving the commander's objectives. A subset of naval surface fire support. Also called **NGFS**. See also **naval surface fire support.**

naval mobile environmental team--A team of naval personnel organized, trained, and equipped to support maritime special operations by providing weather, oceanography, mapping, charting, and geodesy support. Also called **NMET.**

naval operation--1. A naval action, or the performance of a naval mission, which may be strategic, operational, tactical, logistic, or training. 2. The process of carrying on or training for naval combat to gain the objectives of any battle or campaign.

naval or Marine (air) base--An air base for support of naval or Marine air units, consisting of landing strips, seaplane alighting areas, and all components of related facilities for which the Navy or Marine Corps has operating responsibilities, together with interior lines of communication and the minimum surrounding area necessary for local security. (Normally, not greater than an area of 20 square miles.) See also **base complex.**

naval port control office--The authority established at a port or port complex to coordinate arrangements for logistic support and harbor services to ships under naval control and to otherwise support the naval control of shipping organization.

naval special warfare--A specific term describing a designated naval warfare specialty and covering operations generally accepted as being unconventional in nature and, in many cases, covert or clandestine in character. These operations include using specially trained forces assigned to conduct unconventional warfare, psychological operations, beach and coastal reconnaissance, operational deception operations, counterinsurgency operations, coastal and river interdiction, and certain special tactical intelligence collection operations that are in addition to those intelligence functions normally required for planning and conducting special operations in a hostile environment. Also called **NSW.**

naval special warfare forces--Those Active and Reserve component Navy forces designated by the Secretary of Defense that are specifically organized, trained, and equipped to conduct and support special operations. Also called **NSW forces or NAVSOF.**

naval special warfare group--A permanent Navy echelon III major command to which most naval special warfare forces are assigned for some operational and all administrative purposes. It consists of a group headquarters with command and control, communications, and support staff; sea-air-land teams; and sea-air-land team delivery vehicle teams. The group is the source of all deployed naval special warfare forces and administratively supports the naval special warfare units assigned to the theater combatant commanders. The group staff provides general operational direction and coordinates the activities of its subordinate units. A naval special warfare group is capable of task-organizing to meet a wide variety of requirements. Also called **NSWG.**

naval special warfare special operations component--The Navy special operations component of a unified or subordinate unified command or joint special operations task force. Also called **NAVSOC.**

naval special warfare task element--A provisional subordinate element of a naval special warfare task unit, employed to extend the command and control and support capabilities of its parent task unit. Also called **NSWTE.** See also **naval special warfare task unit.**

naval special warfare task group--A provisional naval special warfare organization that plans, conducts, and supports special operations in support of fleet commanders and joint force special operations component commanders. Also called **NSWTG.**

naval special warfare task unit--A provisional subordinate unit of a naval special warfare task group. Also called **NSWTU.** See also **naval special warfare task group.**

naval special warfare unit--A permanent Navy organization forward based to control and support attached naval special warfare forces. Also called **NSWU.**

naval stores(*)--Any articles or commodities used by a naval ship or station, such as equipment, consumable supplies, clothing, petroleum, oils, and lubricants, medical supplies, and ammunition.

naval support area(*)--A sea area assigned to naval ships detailed to support an amphibious operation. See also **fire support area.**

naval surface fire support--Fire provided by Navy surface gun, missile, and electronic warfare systems in support of a unit or units tasked with achieving the commander's objectives. Also called **NSFS.**

naval tactical data system--A complex of data inputs, user consoles, converters, adapters, and radio terminals interconnected with high-speed, general-purpose computers and its stored programs. Combat data is collected, processed, and composed into a picture of the overall tactical situation which enables the force commander to make rapid, accurate evaluations and decisions.

navigational grid(*)--A series of straight lines, superimposed over a conformal projection and indicating grid north, used as an aid to navigation. The interval of the grid lines is generally a multiple of 60 or 100 nautical miles. See also **military grid.**

navigation head(*)--A transshipment point on a waterway where loads are transferred between water carriers and land carriers. A navigation head is similar in function to a railhead or truckhead.

navigation mode--In a flight control system, a control mode in which the flight path of an aircraft is automatically maintained by signals from navigation equipment.

NAVSOC--See **naval special warfare special operations component.**

NAVSOF--See **naval special warfare forces.**

Navy Cargo Handling Battalion--A mobile logistics support unit capable of worldwide deployment in its entirety or in specialized detachments. It is organized, trained, and equipped to: a. load and off-load Navy and Marine Corps cargo carried in maritime pre-positioning ships and merchant breakbulk or container ships in all environments; b. to operate an associated temporary ocean cargo terminal; c. load and offload Navy and Marine Corps cargo carried in military-controlled aircraft; d. to operate an associated expeditionary air cargo terminal. Also called **CHB.** Three sources of Navy Cargo Handling Battalions are: a. Navy Cargo Handling and Port Group--The active duty, cargo handling, battalion-sized unit composed solely of active duty personnel. Also called **NAVCHAPGRU.** b. Naval Reserve Cargo Handling Training Battalion--The active duty, cargo handling training battalion composed of both active duty and reserve personnel. Also called **NRCHTB.** c. Naval Reserve Cargo Handling Battalion--A reserve cargo handling battalion composed solely of selected reserve personnel. Also called **NRCHB.** See also **maritime prepositioning ships.**

Navy Cargo Handling Force--The combined cargo handling units of the Navy, including primarily the Navy Cargo Handling and Port Group, the Naval Reserve Cargo Handling Training Battalion, and the Naval Reserve Cargo Handling Battalion. These units are part of the operating forces and represent the Navy's capability for open ocean cargo handling. See also **Navy Cargo Handling Battalion.**

Navy special operations component--The Navy component of a joint force special operations component. Also called **NAVSOC.** See also **Air Force special operations component; Army special operations component.**

Navy support element--The Maritime Prepositioning Force element that is composed of naval beach group staff and subordinate unit personnel, a detachment of Navy cargo handling force personnel, and other Navy components, as required. It is tasked with conducting the off-load and ship-to-shore movement of maritime prepositioned equipment/supplies.

NBC defense--Nuclear defense, biological defense, and chemical defense, collectively. The term may not be used in the context of US offensive operations.

NCA--See **National Command Authorities.**

NCS--See **National Communications System; net control station.**

NCW--See **naval coastal warfare.**

NCWC--See **naval coastal warfare commander.**

NDA--See **national defense area.**

N-day--See **times.**

NDRF--See **National Defense Reserve Fleet.**

near miss(*)--Any circumstances in flight when the degree of separation between two aircraft might constitute a hazardous situation. Also called **airmiss.**

near miss (aircraft)--Any circumstance in flight where the degree of separation between two aircraft is considered by either pilot to have constituted a hazardous situation involving potential risk of collision.

near real time(*)--Pertaining to the timeliness of data or information which has been delayed by the time required for electronic communication and automatic data processing. This implies that there are no significant delays. See also **real time.**

neatlines(*)--The lines that bound the body of a map, usually parallels and meridians. See also **graticule.**

need to know--A criterion used in security procedures which requires the custodians of classified information to establish, prior to disclosure, that the intended recipient must have access to the information to perform his or her official duties.

negative--As used in air intercept, means cancel or no.

negative phase of the shock wave--The period during which the pressure falls below ambient and then returns to the ambient value. See also **positive phase of the shock wave; shock wave.**

negative photo plane(*)--The plane in which a film or plate lies at the moment of exposure.

neglect(*)--In artillery and naval gunfire support, a report to the observer/spotter to indicate that the last round(s) was fired with incorrect data and that the round(s) will be fired again using correct data.

negligible risk (nuclear)--A degree of risk where personnel are reasonably safe, with the exceptions of dazzle or temporary loss of night vision. See also **degree of risk (nuclear); emergency risk (nuclear); moderate risk (nuclear).**

negotiations--A discussion between authorities and a barricaded offender or terrorist to effect hostage release and terrorist surrender. See also **antiterrorism.**

nerve agent(*)--A potentially lethal chemical agent which interferes with the transmission of nerve impulses.

net call sign(*)--A call sign which represents all stations within a net. See also **call sign.**

net, chain, cell system--Patterns of clandestine organization, especially for operational purposes. Net is the broadest of the three; it usually involves: a. a succession of echelons; and b. such functional specialists as may be required to accomplish its mission. When it consists largely or entirely of nonstaff employees, it may be called an agent net. Chain focuses attention upon the first of these elements; it is commonly defined as a series of agents and informants who receive instructions from and pass information to a principal agent by means of cutouts and couriers. Cell system emphasizes a variant of the first element of net; its distinctive feature is the grouping of personnel into small units that are relatively isolated and self-contained. In the interest of maximum security for the organization as a whole, each cell has contact with the rest of the organization only through an agent of the organization and a single member of the cell. Others in the cell do not know the agent, and nobody in the cell knows the identities or activities of members of other cells.

net (communications)--An organization of stations capable of direct communications on a common channel or frequency.

net control station--A communications station designated to control traffic and enforce circuit discipline within a given net. Also called **NCS.**

net inventory assets--That portion of the total materiel assets which is designated to meet the materiel inventory objective. It consists of the total materiel assets less the peacetime materiel consumption and losses through normal appropriation and procurement leadtime periods.

net sweep(*)--In naval mine warfare, a two-ship sweep, using a netlike device, designed to collect drifting mines or scoop them up from the sea bottom.

net weight--Weight of a ground vehicle without fuel, engine oil, coolant, on-vehicle materiel, cargo, or operating personnel.

neutrality--In international law, the attitude of impartiality, during periods of war, adopted by third states toward belligerent and recognized by the belligerent, which creates rights and duties between the impartial states and the belligerent. In a United Nations enforcement action, the rules of neutrality apply to impartial members of the United Nations except so far as they are excluded by the obligation of such members under the United Nations Charter.

neutralization(*)--In mine warfare, a mine is said to be neutralized when it has been rendered, by external means, incapable of firing on passage of a target, although it may remain dangerous to handle.

neutralization fire--Fire which is delivered to render the target ineffective or unusable. See also **fire.**

neutralize--As pertains to military operations, to render ineffective or unusable.

neutralize track--As used in air intercept, to render the target being tracked ineffective or unusable.

neutral state--In international law, a state which pursues a policy of neutrality during war. See also **neutrality.**

neutron induced activity(*)--Radioactivity induced in the ground or an object as a result of direct irradiation by neutrons.

NGFS--See **naval gunfire support.**

nickname--A combination of two separate unclassified words which is assigned an unclassified meaning and is employed only for unclassified administrative, morale, or public information purposes.

night cap--Night combat air patrol (written NCAP).

night effect(*)--An effect mainly caused by variations in the state of polarization of reflected waves, which sometimes result in errors in direction finding bearings. The effect is most frequent at nightfall.

NIST--See **national intelligence support team.**

NMCM--See **not mission capable, maintenance.**

NMCS--See **National Military Command System.**

NMCS--See **not mission capable, supply.**

NMET--See **naval mobile environmental team.**

node--A location in a mobility system where a movement requirement is originated, processed for onward movement, or terminated.

no-fire line(*)--A line short of which artillery or ships do not fire except on request or approval of the supported commander, but beyond which they may fire at any time without danger to friendly troops.

no joy--In air intercept, a code meaning, "I have been unsuccessful," or, "I have no information."

nominal filter(*)--A filter capable of cutting off a nominated minimum percentage by weight of solid particles greater than a stated micron size.

nominal focal length(*)--An approximate value of the focal length, rounded off to some standard figure, used for the classification of lenses, mirrors, or cameras.

nominal scale--See **principal scale; also scale.**

nominal weapon(*)--A nuclear weapon producing a yield of approximately 20 kilotons. See also **kiloton weapon; megaton weapon; subkiloton weapon.**

nonair transportable--That which is not transportable by air by virtue of dimension, weight, or special characteristics or restrictions.

nonaligned state--A state which pursues a policy of nonalignment.

nonalignment--The political attitude of a state which does not associate or identify itself with the political ideology or objective espoused by other states, groups of states, or international causes, or with the foreign policies stemming therefrom. It does not preclude involvement, but expresses the attitude of no precommitment to a particular state (or block) or policy before a situation arises.

nonappropriated funds--Funds generated by DOD military and civilian personnel and their dependents and used to augment funds appropriated by the Congress to provide a comprehensive, morale-building welfare, religious, educational, and recreational program, designed to improve the well-being of military and civilian personnel and their dependents.

noncontiguous facility--A facility for which the Service indicated has operating responsibility, but which is not located on, or in the immediate vicinity of, a base complex of that Service. Its area includes only that actually occupied by the facility, plus the minimum surrounding area necessary for close-in security. See also **base complex.**

nondeferrable issue demand--Issue demand related to specific periods of time which will not exist after the close of those periods, even though not satisfied during the period.

nondeployable account--An account where Reservists (officer and enlisted) either in units or individually are assigned to a reserve component category or a training/retired category when the individual has not completed initial active duty for training or its equivalent. Reservists in a nondeployable account are not considered as trained strength assigned to units or mobilization positions and are not deployable overseas on land with those units or mobilization positions. See also **training pipeline.**

noneffective sortie--Any aircraft dispatched which for any reason fails to carry out the purpose of the mission. Abortive sorties are included.

nonexpendable supplies and material--Supplies which are not consumed in use and which retain their original identity during the period of use, such as weapons, machines, tools, and equipment.

nonfixed medical treatment facility--A medical treatment facility designed to be moved from place to place, including medical treatment facilities afloat.

nonhostile casualty--A person who becomes a casualty due to circumstances not directly attributable to hostile action or terrorist activity. Casualties due to the elements, self-inflicted wounds, and combat fatigue are nonhostile casualties. See also **casualty; casualty type; hostile casualty.**

non-linear approach(*)--In approach and landing systems, a final approach in which the nominal flight path is not a straight line.

nonprecision approach--Radar-controlled approach or an approach flown by reference to navigation aids in which glide slope information is not available. See also **final approach; precision approach.**

nonprior service personnel--Individuals without any prior military service, who have not completed basic inactive duty training, and who receive a commission in or enlist directly into an Armed Force of the United States.

nonprogram aircraft--All aircraft, other than active and reserve categories, in the total aircraft inventory, including X-models; aircraft for which there is no longer a requirement either in the active or reserve category; and aircraft in the process of being dropped from the total aircraft inventory. See also **aircraft.**

nonrecurring demand--A request by an authorized requisitioner to satisfy a materiel requirement known to be a one-time occurrence. This materiel is required to provide initial stockage allowances, to meet planned program requirements, or to satisfy a one-time project or maintenance requirement. Nonrecurring demands normally will not be considered by the supporting supply system in the development of demand-based elements of the requirements computation.

non-registered publication(*)--A publication which bears no register number and for which periodic accounting is not required.

nonscheduled units--Units of the landing force held in readiness for landing during the initial unloading period, but not included in either scheduled or on-call waves. This category usually includes certain of the combat support units and most of the combat service support units with higher echelon (division and above) reserve units of the landing force. Their landing is directed when the need ashore can be predicted with a reasonable degree of accuracy.

nonstandard item--An item of supply determined by standardization action as not authorized for procurement.

nonstocked item--An item that does not meet the stockage criteria for a given activity and, therefore, is not stocked at the particular activity

non-submarine contact chart(*)--A special naval chart, at a scale of 1:100,000 to 1:1,000,000, showing bathymetry, bottom characteristics, wreck data, and non-submarine contact data for coastal and off-shore waters. It is designed for use in conducting submarine and anti-submarine warfare operations. Also called **non-sub contact chart.**

non-unit-related cargo--All equipment and supplies requiring transportation to an area of operations, other than those identified as the equipment or accompanying supplies of a specific unit (e.g., resupply, military support for allies, and support for nonmilitary programs, such as civil relief).

non-unit-related personnel--All personnel requiring transportation to or from an area of operations, other than those assigned to a specific unit (e.g., filler personnel; replacements; temporary duty/temporary additional duty personnel; civilians; medical evacuees; and retrograde personnel).

normal charge--Charge employing a standard amount of propellant to fire a gun under ordinary conditions, as compared with a reduced charge. See also **reduced charge.**

normal impact effect--See **cardinal point effect.**

normal intelligence reports--A category of reports used in the dissemination of intelligence, which is conventionally used for the immediate dissemination of individual items of intelligence. See also **intelligence reporting; specialist intelligence reports.**

normal lighting(*)--Lighting of vehicles as prescribed or authorized by the law of a given country without restrictions for military reasons. See also **reduced lighting.**

normal operations--Generally and collectively, the broad functions which a combatant commander undertakes when assigned responsibility for a given geographic or functional area. Except as otherwise qualified in certain unified command plan paragraphs which relate to particular commands, "normal operations" of a combatant commander include: planning for and execution of operations throughout the range of military operations; planning and conduct of cold war activities; planning for and administration of military assistance; and maintaining the relationships and exercising the directive or coordinating authority prescribed in Joint Pub 0-2, Admin. Pub 1.1, and Joint Pub 4-01.

no-strike target list--A list designated by a commander containing targets not to be destroyed. Destruction of targets on the list would interfere with or unduly hamper projected friendly military operations, or friendly relations with indigenous personnel or governments.

NOTAM--See **notice to airmen.**

notice to airmen(*)--A notice containing information concerning the establishment, condition, or change in any aeronautical facility, service, procedures, or hazard, the timely knowledge of which is essential to personnel concerned with flight operations. Also called **NOTAM.**

notional ship--A theoretical or average ship of any one category used in transportation planning (e.g., a Liberty ship for dry cargo; a T-2 tanker for bulk petroleum, oils, and lubricants; a personnel transport of 2,400 troop spaces.)

not mission capable, maintenance--Material condition indicating that systems and equipment are not capable of performing any of their assigned missions because of maintenance requirements. Also called **NMCM.** See also **not mission capable, supply.**

not mission capable, supply--Material condition indicating that systems and equipment are not capable of performing any of their assigned missions because of maintenance work stoppage due to a supply shortage. Also called **NMCS.** See also **not mission capable, maintenance.**

not seriously injured--The casualty status of a person whose injury may or may not require hospitalization; medical authority does not classify as very seriously injured, seriously injured, or incapacitating illness or injury; and the person can communicate with the next of kin. Also called **NSI.** See also **casualty status.**

no-wind position--See **air position.**

NRO--See **National Reconnaissance Office.**

NSC--See **National Security Council.**

NSFS--See **naval surface fire support.**

NSI--See **not seriously injured.**

NSW--See **naval special warfare.**

NSWG--See **naval special warfare group.**

NSWTE--See **naval special warfare task element.**

NSWTG--See **naval special warfare task group.**

NSWTG/TU--See **naval special warfare task group/unit.**

NSWTU--See **naval special warfare task unit.**

NSWU--See **naval special warfare unit.**

Nth country--A reference to additions to the group of powers possessing nuclear weapons---the next country of a series to acquire nuclear capabilities.

NUCINT--See **nuclear intelligence.**

nuclear accident--See **nuclear weapon(s) accident.**

nuclear airburst(*)--The explosion of a nuclear weapon in the air, at a height greater than the maximum radius of the fireball. See also **types of burst.**

nuclear bonus effects(*)--Desirable damage or casualties produced by the effects from friendly nuclear weapons that cannot be accurately calculated in targeting as the uncertainties involved preclude depending on them for a militarily significant result.

nuclear burst--See **types of burst.**

nuclear certifiable(*)--Indicates a unit or vehicle possessing the potential of passing functional tests and inspections of all normal and emergency systems affecting the nuclear weapons.

nuclear certified(*)--See nuclear certified delivery unit; nuclear certified delivery vehicle.

nuclear certified delivery unit(*)--Any level of organization and support elements which are capable of executing nuclear missions in accordance with appropriate bilateral arrangements and NATO directives. See also **nuclear delivery unit.**

nuclear certified delivery vehicle(*)--A delivery vehicle whose compatibility with a nuclear weapon has been certified by the applicable nuclear power through formal procedures. See also **nuclear delivery vehicle.**

nuclear cloud(*)--An all-inclusive term for the volume of hot gases, smoke, dust, and other particulate matter from the nuclear bomb itself and from its environment, which is carried aloft in conjunction with the rise of the fireball produced by the detonation of the nuclear weapon.

nuclear collateral damage(*)--Undesired damage or casualties produced by the effects from friendly nuclear weapons.

nuclear column(*)--A hollow cylinder of water and spray thrown up from an underwater burst of a nuclear weapon, through which the hot, high-pressure gases formed in the explosion are vented to the atmosphere. A somewhat similar column of dirt is formed in an underground explosion.

nuclear commitment(*)--A statement by a NATO member that specific forces have been committed or will be committed to NATO in a nuclear only or dual capable role.

nuclear coordination--A broad term encompassing all the actions involved with planning nuclear strikes, including liaison between commanders, for the purpose of satisfying support requirements or because of the extension of weapons effects into the territory of another.

nuclear damage(*)--1. Light Damage--Damage which does not prevent the immediate use of equipment or installations for which it was intended. Some repair by the user may be required to make full use of the equipment or installations. 2. Moderate Damage--Damage which prevents the use of equipment or installations until extensive repairs are made. 3. Severe Damage--Damage which prevents use of equipment or installations permanently.

nuclear damage assessment(*)--The determination of the damage effect to the population, forces, and resources resulting from actual nuclear attack. It is performed during and after an attack. The operational significance of the damage is not evaluated in this assessment.

nuclear defense(*)--The methods, plans, and procedures involved in establishing and exercising defensive measures against the effects of an attack by nuclear weapons or radiological warfare agents. It encompasses both the training for, and the implementation of, these methods, plans, and procedures. See also **NBC defense; radiological defense.**

nuclear delivery unit(*)--Any level of organization capable of employing a nuclear weapon system or systems when the weapon or weapons have been released by proper authority.

nuclear delivery vehicle(*)--That portion of the weapon system which provides the means of delivery of a nuclear weapon to the target.

nuclear detonation detection and reporting system(*)--A system deployed to provide surveillance coverage of critical friendly target areas, and indicate place, height of burst, yield, and ground zero of nuclear detonations.

nuclear dud--A nuclear weapon which, when launched at or emplaced on a target, fails to provide any explosion of that part of the weapon designed to produce the nuclear yield.

nuclear energy--All forms of energy released in the course of a nuclear fission or nuclear transformation.

nuclear equipoise--Not to be used. See **nuclear stalemate.**

nuclear exoatmospheric burst--The explosion of a nuclear weapon above the sensible atmosphere (above 120 kilometers) where atmospheric interaction is minimal. See also **types of burst.**

nuclear incident--An unexpected event involving a nuclear weapon, facility, or component, resulting in any of the following, but not constituting a nuclear weapon(s) accident: a. an increase in the possibility of explosion or radioactive contamination; b. errors committed in the assembly, testing, loading, or transportation of equipment, and/or the malfunctioning of equipment and materiel which could lead to an unintentional operation of all or part of the weapon arming and/or firing sequence, or which could lead to a substantial change in yield, or increased dud probability; and c. any act of God, unfavorable environment, or condition resulting in damage to the weapon, facility, or component.

nuclear intelligence--Intelligence derived from the collection and analysis of radiation and other effects resulting from radioactive sources. Also called **NUCINT.** See also **intelligence.**

nuclear logistic movement--The transport of nuclear weapons in connection with supply or maintenance operations. Under certain specified conditions, combat aircraft may be used for such movements.

nuclear nation(*)--Military nuclear powers and civil nuclear powers. See also **nuclear power.**

nuclear parity--A condition at a given point in time when opposing forces possess nuclear offensive and defensive systems approximately equal in overall combat effectiveness.

nuclear planning and execution--Worldwide Military Command and Control System application systems that support strategic and tactical nuclear planning, execution, termination, and reconstitution.

Nuclear Planning System--A system composed of personnel, directives, and electronic data processing systems to directly support theater nuclear combatant commanders in developing, maintaining, and disseminating nuclear operation plans.

nuclear power(*)--Not to be used without appropriate modifier. See also **civil nuclear power; major nuclear power; military nuclear power; nuclear nation; nuclear weapons state.**

nuclear radiation(*)--Particulate and electromagnetic radiation emitted from atomic nuclei in various nuclear processes. The important nuclear radiations, from the weapon standpoint, are alpha and beta particles, gamma rays, and neutrons. All nuclear radiations are ionizing radiations, but the reverse is not true; X-rays for example, are included among ionizing radiations, but they are not nuclear radiations since they do not originate from atomic nuclei.

nuclear reactor--A facility in which fissile material is used in a self-supporting chain reaction (nuclear fission) to produce heat and/or radiation for both practical application and research and development.

nuclear round--See **complete round.**

nuclear safety line(*)--A line selected, if possible, to follow well-defined topographical features and used to delineate levels of protective measures, degrees of damage or risk to friendly troops, and/or to prescribe limits to which the effects of friendly weapons may be permitted to extend.

nuclear stalemate--A concept which postulates a situation wherein the relative strength of opposing nuclear forces results in mutual deterrence against employment of nuclear forces.

nuclear strike warning(*)--A warning of impending friendly or suspected enemy nuclear attack.

nuclear support--The use of nuclear weapons against hostile forces in support of friendly air, land, and naval operations. See also **immedi-**

ate nuclear support; preplanned nuclear support.

nuclear surface burst(*)--An explosion of a nuclear weapon at the surface of land or water; or above the surface, at a height less than the maximum radius of the fireball. See also **types of burst.**

nuclear transmutation--Artificially induced modification (nuclear reaction) of the constituents of certain nuclei, thus giving rise to different nuclides.

nuclear underground burst(*)--The explosion of a nuclear weapon in which the center of the detonation lies at a point beneath the surface of the ground. See also **types of burst.**

nuclear underwater burst(*)--The explosion of a nuclear weapon in which the center of the detonation lies at a point beneath the surface of the water. See also **types of burst.**

nuclear vulnerability assessment(*)--The estimation of the probable effect on population, forces, and resources from a hypothetical nuclear attack. It is performed predominantly in the preattack period; however, it may be extended to the transattack or postattack periods.

nuclear warfare(*)--Warfare involving the employment of nuclear weapons. See also **postattack period; transattack period.**

nuclear warning message--A warning message which must be disseminated to all affected friendly forces any time a nuclear weapon is to be detonated if effects of the weapon will have impact upon those forces.

nuclear weapon(*)--A complete assembly (i.e., implosion type, gun type, or thermonuclear type), in its intended ultimate configuration which, upon completion of the prescribed arming, fusing, and firing sequence, is capable of producing the intended nuclear reaction and release of energy.

nuclear weapon degradation--The degeneration of a nuclear warhead to such an extent that the anticipated nuclear yield is lessened.

nuclear weapon employment time(*)--The time required for delivery of a nuclear weapon after the decision to fire has been made.

nuclear weapon exercise(*)--An operation not directly related to immediate operational readiness. It includes removal of a weapon from its normal storage location, preparing for use, delivery to an employment unit, and the movement in a ground training exercise, to include loading aboard an aircraft or missile and return to storage. It may include any or all of the operations listed above, but does not include launching or flying operations. Typical exercises include aircraft generation exercises, ground readiness exercises, ground tactical exercises, and various categories of inspections designed to evaluate the capability of the unit to perform its prescribed mission. See also **immediate operational readiness; nuclear weapon maneuver.**

nuclear weapon maneuver(*)--An operation not directly related to immediate operational readiness. It may consist of all those operations listed for a nuclear weapon exercise and is extended to include flyaway in combat aircraft, but does not include expenditure of the weapon. Typical maneuvers include nuclear operational readiness maneuvers and tactical air operations. See also **immediate operational readiness; nuclear weapon exercise.**

nuclear weapon(s) accident--An unexpected event involving nuclear weapons or radiological nuclear weapon components that results in any of the following; a. accidental or unauthorized launching, firing, or use by United States forces or United States supported allied forces, of a nuclear-capable weapon system which could create the risk of an outbreak of war; b. nuclear detonation; c. nonnuclear detonation or burning of a nuclear weapon or radiological nuclear weapon component; d. radioactive contamination; e. seizure, theft, loss, or destruction of a nuclear weapon or radiological nuclear weapon component, including jettisoning; f. public hazard, actual or implied.

nuclear weapons state--See **military nuclear power.**

nuclear weapons surety--Materiel, personnel, and procedures which contribute to the security, safety, and reliability of nuclear weapons and to the assurance that there will be no nuclear weapon accidents, incidents, unauthorized weapon detonations, or degradation in performance at the target.

nuclear yields--The energy released in the detonation of a nuclear weapon, measured in terms of the kilotons or megatons of trinitrotoluene required to produce the same energy release. Yields are categorized as:
very low--less than 1 kiloton.
low--1 kiloton to 10 kilotons.
medium--over 10 kilotons to 50 kilotons.
high--over 50 kilotons to 500 kilotons.
very high--over 500 kilotons. See also **nominal weapon; subkiloton weapon.**

nucleon--The common name for a constituent particle of the atomic nucleus. It is applied to protons and neutrons, but it is intended to include any other particle that is found to exist in the nucleus.

nuclide--All nuclear species, both stable (about 270) and unstable (about 500), of the chemical elements, as distinguished from the two or more nuclear species of a single chemical element which are called isotopes.

nudets--See **nuclear detonation detection and reporting system.**

nuisance minefield(*)--A minefield laid to delay and disorganize the enemy and to hinder the use of an area or route. See also **minefield.**

number . . . in (out)(*)--In artillery, term used to indicate a change in status of weapon number

_____ .

numbered beach--In amphibious operations, a subdivision of a colored beach, designated for the assault landing of a battalion landing team or similarly sized unit, when landed as part of a larger force. See also **colored beach.**

numbered fleet--A major tactical unit of the Navy immediately subordinate to a major fleet command and comprising various task· forces, elements, groups, and units for the purpose of prosecuting specific naval operations. See also **fleet.**

numbered wave--See **wave.**

numerical scale--See **scale.**

O

objective(*)--The physical object of the action taken, e.g., a definite tactical feature, the seizure and/or holding of which is essential to the commander's plan. See also **target.**

objective area(*)--1. A defined geographical area within which is located an objective to be captured or reached by the military forces. This area is defined by competent authority for purposes of command and control. (DOD) 2. The city or other geographical location where a civil disturbance is occurring or is anticipated, and where Federal Armed Forces are, or may be, employed.

objective force level--The level of military forces that needs to be attained within a finite time frame and resource level to accomplish approved military objectives, missions, or tasks. See also **military requirement.**

obligated reservist--An individual who has a statutory requirement imposed by the Military Selective Service Act of 1967 or Section 651, Title 10 United States Code to serve on active duty in the armed forces or to serve while not on active duty in a reserve component for a period not to exceed that prescribed by the applicable statute.

oblique air photograph(*)--An air photograph taken with the camera axis directed between the horizontal and vertical planes. Commonly referred to as an "oblique": **a. High Oblique.** One in which the apparent horizon appears, and **b. Low Oblique.** One in which the apparent horizon does not appear.

oblique air photograph strip--Photographic strip composed of oblique air photographs.

obliquity--The characteristic in wide-angle or oblique photography which portrays the terrain and objects at such an angle and range that details necessary for interpretation are seriously masked or are at a very small scale, rendering interpretation difficult or impossible.

observation helicopter(*)--Helicopter used primarily for observation and reconnaissance, but which may be used for other roles.

observation post(*)--A position from which military observations are made, or fire directed and adjusted, and which possesses appropriate communications; may be airborne.

observed fire(*)--Fire for which the point of impact or burst can be seen by an observer. The fire can be controlled and adjusted on the basis of observation. See also **fire.**

observed fire procedures(*)--A standardized procedure for use in adjusting indirect fire on a target.

observer identification(*)--In artillery and naval gunfire support, the first element of a call for fire to establish communication and to identify the observer/spotter.

observer-target line(*)--An imaginary straight line from the observer/spotter to the target. See also **spotting line.**

observer-target range--The distance along an imaginary straight line from the observer/spotter to the target.

obstacle--Any obstruction designed or employed to disrupt, fix, turn, or block the movement of an opposing force, and to impose additional losses in personnel, time, and equipment on the opposing force. Obstacles can exist naturally or can be manmade, or can be a combination of both.

obstacle belt--A brigade-level command and control measure, normally given graphically, to show where within an obstacle zone the ground tactical commander plans to limit friendly obstacle employment and focus the defense. It assigns an intent to the obstacle plan and provides the necessary guidance on the overall effect of obstacles within a belt. See also **obstacle.**

obstacle restricted areas--A command and control measure used to limit the type or number of obstacles within an area. See also **obstacle.**

obstacle zone--A division-level command and control measure, normally done graphically, to designate specific land areas where lower echelons are allowed to employ tactical obstacles. See also **obstacle.**

obstructor(*)--In naval mine warfare, a device laid with the sole object of obstructing or damaging mechanical minesweeping equipment.

occupation currency--See **military currency.**

occupied territory--Territory under the authority and effective control of a belligerent armed force. The term is not applicable to territory being administered pursuant to peace terms, treaty, or other agreement, express or implied, with the civil authority of the territory. See also **civil affairs agreement.**

ocean convoy(*)--A convoy whose voyage lies, in general, outside the continental shelf. See also **convoy.**

ocean manifest(*)--A detailed listing of the entire cargo loaded into any one ship showing all pertinent data which will readily identify such cargo and where and how the cargo is stowed.

oceanography--The study of the sea, embracing and integrating all knowledge pertaining to the sea and its physical boundaries, the chemistry and physics of seawater, and marine biology.

ocean station ship(*)--A ship assigned to operate within a specified area to provide several services, including search and rescue, meteorological information, navigational aid, and communications facilities.

offensive counter air operation(*)--An operation mounted to destroy, disrupt, or limit enemy air power as close to its source as possible.

offensive minefield(*)--In naval mine warfare, a minefield laid in enemy territorial water or waters under enemy control.

officer in tactical command(*)--In maritime usage, the senior officer present eligible to assume command, or the officer to whom he has delegated tactical command.

official information--Information which is owned by, produced for or by, or is subject to the control of the United States Government.

offset bombing(*)--Any bombing procedure which employs a reference or aiming point other than the actual target.

offset distance (nuclear)--The distance the desired ground zero or actual ground zero is offset from the center of an area target or from a point target.

offset point(*)--In air interception, a point in space relative to a target's flight path toward which an interceptor is vectored and from which the final or a preliminary turn to attack heading is made.

offshore patrol(*)--A naval defense patrol operating in the outer areas of navigable coastal waters. It is a part of the naval local defense forces consisting of naval ships and aircraft and operates outside those areas assigned to the inshore patrol.

off-the-shelf item--An item which has been developed and produced to military or commercial standards and specifications, is readily available for delivery from an industrial source, and may be procured without change to satisfy a military requirement.

oiler(*)--A naval or merchant tanker specially equipped and rigged for replenishing other ships at sea.

on berth--Said of a ship when it is properly moored to a quay, wharf, jetty, pier, or buoy or when it is at anchor and available for loading or discharging passengers and cargo.

on-call--1. A term used to signify that a prearranged concentration, air strike, or final protective fire may be called for. 2. Preplanned, identified force or materiel requirements without designated time-phase and destination information. Such requirements will be called forward upon order of competent authority. See also **call for fire; call mission.**

on-call resupply--A resupply mission planned before insertion of a special operations team into the operations area but not executed until requested by the operating team. See also **automatic resupply; emergency resupply.**

on-call target(*)--In artillery and naval gunfire support, a planned target other than a scheduled target on which fire is delivered when requested.

on-call target (nuclear)--A planned nuclear target other than a scheduled nuclear target for which a need can be anticipated but which will be delivered upon request rather than at a specific time. Coordination and warning of friendly troops and aircraft are mandatory.

on-call wave--See **wave.**

one day's supply(*)--A unit or quantity of supplies adopted as a standard of measurement, used in estimating the average daily expenditure under stated conditions. It may also be expressed in terms of a factor, e.g., rounds of ammunition per weapon per day. See also **standard day of supply; combat day of supply.**

one-look circuit(*)--A mine circuit which requires actuation by a given influence once only.

on hand--The quantity of an item that is physically available in a storage location and contained in the accountable property book records of an issuing activity.

on-scene commander--The person designated to coordinate the rescue efforts at the rescue site.

on station--1. In air intercept usage, a code meaning, "I have reached my assigned station." 2. In close air support and air interdiction, means airborne aircraft are in position to attack targets or to perform the mission designated by control agency.

on station time--The time an aircraft can remain on station. May be determined by endurance or orders.

on target--(DOD) In air intercept, a code meaning, "My fire control director(s)/system(s) have acquired the indicated contact and is (are) tracking successfully."

on the deck--At minimum altitude.

OPCON--See **operational control.**

OPDOC--See **operational documentation.**

open--Term used in a call for fire to indicate that the spotter or observer desires bursts to be separated by the maximum effective width of the burst of the shell fired.

open improved storage space--Open area which has been graded and hard surfaced or prepared with topping of some suitable material so as to permit effective material handling operations. See also **storage.**

open route(*)--A route not subject to traffic or movement control restrictions.

open sheaf--The lateral distribution of the fire of two or more pieces so that adjoining points of impact or points of burst are separated by the maximum effective width of burst of the type shell being used. See also **converged sheaf; parallel sheaf; sheaf; special sheaf.**

open source intelligence--Information of potential intelligence value that is available to the general public. Also called **OSINT.** See also **intelligence.**

open unimproved wet space--That water area specifically allotted to and usable for storage of floating equipment. See also **storage.**

operating forces--Those forces whose primary missions are to participate in combat and the integral supporting elements thereof. See also **combat forces; combat service support elements; combat support elements.**

operating level of supply--The quantities of materiel required to sustain operations in the interval between requisitions or the arrival of successive shipments. These quantities should be based on the established replenishment period (monthly, quarterly, etc.) See also **level of supply.**

operation(*)--A military action or the carrying out of a strategic, tactical, service, training, or administrative military mission; the process of carrying on combat, including movement,

supply, attack, defense and maneuvers needed to gain the objectives of any battle or campaign.

operational art--The employment of military forces to attain strategic and/or operational objectives through the design, organization, integration, and conduct of strategies, campaigns, major operations, and battles. Operational art translates the joint force commander's strategy into operational design, and, ultimately, tactical action, by integrating the key activities at all levels of war.

operational chain of command(*)--The chain of command established for a particular operation or series of continuing operations. See also **administrative chain of command; chain of command.**

operational characteristics--Those military characteristics which pertain primarily to the functions to be performed by equipment, either alone or in conjunction with other equipment; e.g., for electronic equipment, operational characteristics include such items as frequency coverage, channeling, type of modulation, and character of emission.

operational control--Transferable command authority that may be exercised by commanders at any echelon at or below the level of combatant command. Operational control is inherent in Combatant Command (command authority) and is the authority to perform those functions of command over subordinate forces involving organizing and employing commands and forces, assigning tasks, designating objectives, and giving authoritative direction necessary to accomplish the mission. Operational control includes authoritative direction over all aspects of military operations and joint training necessary to accomplish missions assigned to the command. Operational control should be exercised through the commanders of subordinate organizations; normally this authority is exercised through the Service component commanders. Operational control normally provides full authority to organize commands and forces and to employ those forces as the commander in operational control considers necessary to accomplish assigned missions. Operational control does not, in and of itself, include authoritative direction for logistics or matters of administration, discipline, internal organization, or unit training. Also called **OPCON.** See also **combatant command; combatant command (command authority); combatant commander; tactical control.**

operational control authority(*)--The naval commander responsible within a specified geographical area for the operational control of all maritime forces assigned to him and for the control of movement and protection of all merchant shipping under allied naval control.

operational decontamination(*)--Decontamination carried out by an individual and/or a unit, restricted to specific parts of operationally essential equipment, materiel and/or working areas, in order to minimize contact and transfer hazards and to sustain operations. This may include decontamination of the individual beyond the scope of immediate decontamination, as well as decontamination of mission-essential spares and limited terrain decontamination. See also **decontamination; immediate decontamination; thorough decontamination.**

operational documentation--Visual information documentation of activities to convey information about people, places, and things. It is general purpose documentation normally accomplished in peacetime. Also called **OPDOC.** See also **visual information documentation.**

operational environment--A composite of the conditions, circumstances, and influences which affect the employment of military forces and bear on the decisions of the unit commander.

Some examples are: **a. permissive environment**--operational environment in which host country military and law enforcement agencies have control and the intent and capability to assist operations that a unit intends to conduct. **b. uncertain environment**--operational environment in which host government forces, whether opposed to or receptive to operations that a unit intends to conduct, do not have totally effective control of the territory and population in the intended area of operations. **c. hostile environment**--operational environment in which hostile forces have control and the intent and capability to effectively oppose or react to the operations a unit intends to conduct.

operational evaluation--The test and analysis of a specific end item or system, insofar as practicable under Service operating conditions, in order to determine if quantity production is warranted considering: a. the increase in military effectiveness to be gained; and b. its effectiveness as compared with currently available items or systems, consideration being given to: (1) personnel capabilities to maintain and operate the equipment; (2) size, weight, and location considerations; and (3) enemy capabilities in the field. See also **technical evaluation.**

operational intelligence--Intelligence that is required for planning and conducting campaigns and major operations to accomplish strategic objectives within theaters or areas of operations. See also **intelligence; tactical intelligence; strategic intelligence.**

operational level of war--The level of war at which campaigns and major operations are planned, conducted, and sustained to accomplish strategic objectives within theaters or areas of operations. Activities at this level link tactics and strategy by establishing operational objectives needed to accomplish the strategic objectives, sequencing events to achieve the operational objectives, initiating actions, and applying resources to bring about and sustain these events. These activities imply a broader dimension of time or space than do tactics; they ensure the logistic and administrative support of tactical forces, and provide the means by which tactical successes are exploited to achieve strategic objectives. See also **strategic level of war; tactical level of war.**

operationally ready--1. As applied to a unit, ship, or weapon system--Capable of performing the missions or functions for which organized or designed. Incorporates both equipment readiness and personnel readiness. 2. As applied to personnel--Available and qualified to perform assigned missions or functions.

operational missile(*)--A missile which has been accepted by the using services for tactical and/or strategic use.

operational procedures(*)--The detailed methods by which headquarters and units carry out their operational tasks.

operational readiness(*)--The capability of a unit/formation, ship, weapon system or equipment to perform the missions or functions for which it is organized or designed. May be used in a general sense or to express a level or degree of readiness. See also **combat readiness.**

operational readiness evaluation(*)--An evaluation of the operational capability and effectiveness of a unit or any portion thereof.

operational requirement--See **military requirement.**

operational reserve(*)--An emergency reserve of men and/or material established for the support of a specific operation. See also **reserve supplies.**

operational route(*)--Land route allocated to a command for the conduct of a specific operation; derived from the corresponding basic military route network.

operational testing--A continuing process of evaluation which may be applied to either operational personnel or situations to determine their validity or reliability.

operational training(*)--Training that develops, maintains, or improves the operational readiness of individuals or units.

operation annexes--Those amplifying instructions which are of such a nature, or are so voluminous or technical, as to make their inclusion in the body of the plan or order undesirable.

operation exposure guide--The maximum amount of nuclear radiation which the commander considers a unit may be permitted to receive while performing a particular mission or missions.

operation map--A map showing the location and strength of friendly forces involved in an operation. It may indicate predicted movement and location of enemy forces. See also **map.**

operation order--A directive issued by a commander to subordinate commanders for the purpose of effecting the coordinated execution of an operation. Also called **OPORD.**

operation plan--Any plan, except for the Single Integrated Operation Plan, for the conduct of military operations. Plans are prepared by combatant commanders in response to requirements established by the Chairman of the Joint Chiefs of Staff and by commanders of subordinate commands in response to requirements tasked by the establishing unified commander. Operation plans (OPLANs) are prepared in either a complete format of an OPLAN or as a concept plan (CONPLAN). **a. OPLAN.** An operation plan for the conduct of joint operations that can be used as a basis for development of an operation order (OPORD). An OPLAN identifies the forces and supplies required to execute the CINC's Strategic Concept and a movement schedule of these resources to the theater of operations. The forces and supplies are identified in time-phased force deployment data (TPFDD) files. OPLANs will include all phases of the tasked operation. The plan is prepared with the appropriate annexes, appendixes, and TPFDD files as described in the Joint Operation Planning and Execution System manuals containing planning policies, procedures, and formats. Also called **OPLAN.** **b. CONPLAN.** An operation plan in an abbreviated format that would require considerable expansion or alteration to convert it into an OPLAN or OPORD. A CONPLAN contains the CINC's Strategic Concept and those annexes and appendixes deemed necessary by the combatant commander to complete planning. Generally, detailed support requirements are not calculated and TPFDD files are not prepared. Also called **CONPLAN.**

operations center--The facility or location on an installation, base, or facility used by the commander to command, control, and coordinate all crisis activities. See also **base defense operations center; command center.**

operations research--The analytical study of military problems undertaken to provide responsible commanders and staff agencies with a scientific basis for decision on action to improve military operations. Also known as **operational research; operations analysis.**

operations security--A process of identifying critical information and subsequently analyzing friendly actions attendant to military operations and other activities to: a. Identify those actions that can be observed by adversary intelligence systems. b. Determine indicators hostile intelligence systems might obtain that could be

interpreted or pieced together to derive critical information in time to be useful to adversaries. c. Select and execute measures that eliminate or reduce to an acceptable level the vulnerabilities of friendly actions to adversary exploitation. Also called **OPSEC.** See also **command and control warfare; operations security indicators; operations security measures; operations security planning guidance; operations security vulnerability.**

operations security indicators--Friendly detectable actions and open-source information that can be interpreted or pieced together by an adversary to derive critical information.

operations security measures--Methods and means to gain and maintain essential secrecy about critical information. The following categories apply: **a. action control**--The objective is to eliminate indicators or the vulnerability of actions to exploitation by adversary intelligence systems. Select what actions to undertake; decide whether or not to execute actions; and determine the "who," "when," "where," and "how" for actions necessary to accomplish tasks. **b. countermeasures**--The objective is to disrupt effective adversary information gathering or prevent their recognition of indicators when collected materials are processed. Use diversions, camouflage, concealment, jamming, threats, police powers, and force against adversary information gathering and processing capabilities. **c. counteranalysis**--The objective is to prevent accurate interpretations of indicators during adversary analysis of collected materials. This is done by confusing the adversary analyst through deception techniques such as covers.

operations security planning guidance--Guidance that serves as the blueprint for OPSEC planning by all functional elements throughout the organization. It defines the critical information that requires protection from adversary appreciations, taking into account friendly and adversary goals, estimated key adversary questions, probable adversary knowledge, desirable and harmful adversary appreciations, and pertinent intelligence system threats. It also should outline provisional operations security measures to ensure the requisite essential secrecy.

operations security vulnerability--A condition in which friendly actions provide operations security indicators that may be obtained and accurately evaluated by an adversary in time to provide a basis for effective adversary decisionmaking.

OPLAN--See **operation plan.**

OPORD--See **operation order.**

opportune lift--That portion of lift capability available for use after planned requirements have been met.

opportunity target--See **target of opportunity.**

opposite numbers--Officers (including foreign) having corresponding duty assignments within their respective Military Services or establishments.

OPSEC--See **operations security.**

optical axis(*)--In a lens element, the straight line which passes through the centers of curvature of the lens surfaces. In an optical system, the line formed by the coinciding principal axes of the series of optical elements.

optical landing system--A shipboard gyro-stabilized or shore-based device which indicates to the pilot displacement from a preselected glide path. See also **ground controlled approach procedure.**

optical minehunting(*)--The use of an optical system (e.g., television or towed diver) to detect and classify mines or minelike objects on or protruding from the seabed.

optimum height(*)--The height of an explosion which will produce the maximum effect against a given target.

optimum height of burst(*)--For nuclear weapons and for a particular target (or area), the height at which it is estimated a weapon of a specified energy yield will produce a certain desired effect over the maximum possible area.

orange commander(*)--The officer designated to exercise operational control over orange forces for a specific period during an exercise.

oranges (sour)--In air intercept, a code meaning, "Weather is unsuitable for aircraft mission."

oranges (sweet)--In air intercept, a code meaning, "Weather is suitable for aircraft mission."

orbital injection--The process of providing a space vehicle with sufficient velocity to establish an orbit.

orbit determination--The process of describing the past, present, or predicted position of a satellite in terms of orbital parameters.

orbiting--In air intercept, means circling, or circle and search.

orbit point(*)--A geographically or electronically defined location used in stationing aircraft in flight during tactical operations when a predetermined pattern is not established. See also **holding point.**

order(*)--A communication, written, oral, or by signal, which conveys instructions from a superior to a subordinate. **(DOD)** In a broad sense, the terms "order" and "command" are synonymous. However, an order implies discretion as to the details of execution whereas a command does not.

order and shipping time--The time elapsing between the initiation of stock replenishment action for a specific activity and the receipt by that activity of the materiel resulting from such action. Order and shipping time is applicable only to materiel within the supply system, and it is composed of the distinct elements, order time, and shipping time. See also **level of supply.**

order of battle(*)--The identification, strength, command structure, and disposition of the personnel, units, and equipment of any military force.

order time--1. The time elapsing between the initiation of stock replenishment action and submittal of requisition or order. 2. The time elapsing between the submittal of requisition or order and shipment of materiel by the supplying activity. See also **order and shipping time.**

ordinary priority--A category of immediate mission request which is lower than "urgent priority" but takes precedence over "search and attack priority," e.g., a target which is delaying a unit's advance but which is not causing casualties. See also **immediate mission request; priority of immediate mission requests.**

ordinary transport(*)--In railway terminology, a load whose size, weight or preparation does not entail special difficulties vis-a-vis the facilities or equipment of the railway systems to be used. See also **exceptional transport.**

ordnance--Explosives, chemicals, pyrotechnics, and similar stores, e.g., bombs, guns and ammunition, flares, smoke, napalm.

organic--Assigned to and forming an essential part of a military organization. Organic parts of a unit are those listed in its table of organization for the Army, Air Force, and Marine Corps, and are assigned to the administrative organizations of the operating forces for the Navy.

organizational equipment--Referring to method of use, signifies that equipment, other than individual equipment, which is used in furtherance of the common mission of an organization or unit. See also **equipment.**

organizational maintenance--That maintenance which is the responsibility of and performed by a using organization on its assigned equipment. Its phases normally consist of inspecting, servicing, lubricating, adjusting, and the replacing of parts, minor assemblies, and subassemblies.

organization for embarkation--In amphibious operations, the administrative grouping of the landing force for the overseas movement. It includes, in any vessel or embarkation group, the task organization that is established for landing as well as additional forces embarked for purposes of transport, labor, or for distribution to achieve a maximum of security.

organization for landing--In amphibious operations, the specific tactical grouping of the landing force for the assault.

organization of the ground(*)--The development of a defensive position by strengthening the natural defenses of the terrain and by assignment of the occupying troops to specific localities.

origin--Beginning point of a deployment where unit or non-unit-related cargo or personnel are located.

original destination(*)--In naval control of shipping, the original final destination of a convoy or an individual ship (whether in convoy or independent). This is particularly applicable to the original destination of a voyage begun in peacetime.

original negative--See **generation (photography).**

original positive--See **generation (photography).**

originating medical facility(*)--A medical facility that initially transfers a patient to another medical facility.

originator--The command by whose authority a message is sent. The responsibility of the originator includes the responsibility for the functions of the drafter and the releasing officer. See also **drafter; releasing officer.**

Orion--A four-engine, turboprop, all-weather, long-range, land-based antisubmarine aircraft. It is capable of carrying a varied assortment of search radar, nuclear depth charges, and homing torpedoes. It can be used for search, patrol, hunter-killer, and convoy escort operations. Designated as **P-3.** Electronic attack version is designated EP-3.

oropesa sweep(*)--In naval mine warfare, a form of sweep in which a length of sweep wire is towed by a single ship, lateral displacement being caused by an otter and depth being controlled at the ship end by a kite and at the other end by a float and float wire.

orthomorphic projection(*)--A projection in which the scale, although varying throughout the map, is the same in all directions at any point, so that very small areas are represented by correct shape and bearings are correct.

oscillating mine(*)--A mine, hydrostatically controlled, which maintains a pre-set depth below the surface of the water independently of the rise and fall of the tide. See also **mine.**

OSINT--See **open source intelligence.**

other activity--In the Air Force, a unit or activity that has little or no real property accountability over the real estate it occupies. Examples include active, Guard, or reserve Air Force units that are located on installations belonging to other Services or leased office space that supports recruiting detachments, Civil Air Patrol, etc. See also **installation complex; major installation; minor installation; support site.**

other war reserve materiel requirement--This level consists of the war reserve materiel requirement less the pre-positioned war reserve materiel requirement.

other war reserve materiel requirement, balance--That portion of the other war reserve materiel requirement which has not been acquired or funded. This level consists of the other war reserve materiel requirement less the other war reserve materiel requirement protectable.

other war reserve materiel requirement, protectable--The portion of the other war reserve materiel requirement which is protected for purposes of procurement, funding, and inventory management.

other war reserve stock--The quantity of an item acquired and placed in stock against the other war reserve materiel requirement.

otter(*)--In naval mine warfare, a device which, when towed, displaces itself sideways to a predetermined distance.

outbound traffic--Traffic originating in continental United States destined for overseas or overseas traffic moving in a general direction away from continental United States.

outer fix--A fix in the destination terminal area, other than the approach fix, to which aircraft are normally cleared by an air route traffic control center or a terminal area traffic control facility, and from which aircraft are cleared to the approach fix or final approach course.

outer landing ship areas--In amphibious operations, areas to which landing ships proceed initially after their arrival in the objective area. They are usually located on the flanks of the outer transport areas.

outer transport area--In amphibious operations, an area inside the antisubmarine screen to which assault transports proceed initially after arrival in the objective area. See also **inner transport area; transport area.**

outline map(*)--A map which represents just sufficient geographic information to permit the correlation of additional data placed upon it.

outline plan(*)--A preliminary plan which outlines the salient features or principles of a course of action prior to the initiation of detailed planning.

OV-10--See **Bronco.**

over(*)--In artillery and naval gunfire support, a spotting, or an observation, used by a spotter or an observer to indicate that a burst(s) occurred beyond the target in relation to the spotting line.

overhaul--The restoration of an item to a completely serviceable condition as prescribed by maintenance serviceability standards. See also **rebuild; repair.**

overhead clearance(*)--The vertical distance between the route surface and any obstruction above it.

overlap--1. In photography, the amount by which one photograph includes the same area covered by another, customarily expressed as a percentage. The overlap between successive air photographs on a flight line is called "forward overlap." The overlap between photographs in adjacent parallel flight lines is called "side overlap." 2. In cartography, that portion of a map or chart which overlaps the area covered by another of the same series. 3. In naval mine warfare, the width of that part of the swept path of a ship or formation which is also swept by an adjacent sweeper or formation or is reswept on the next adjacent lap.

overlap tell(*)--The transfer of information to an adjacent facility concerning tracks detected in the adjacent facility's area of responsibility. See also **track telling.**

overlap zone--A designated area on each side of a boundary between adjacent tactical air control systems wherein coordination and interaction between the systems is required.

overlay(*)--A printing or drawing on a transparent or semi-transparent medium at the same scale as a map, chart, etc., to show details not appearing or requiring special emphasis on the original.

overpressure(*)--The pressure resulting from the blast wave of an explosion. It is referred to as "positive" when it exceeds atmospheric pressure and "negative" during the passage of the wave when resulting pressures are less than atmospheric pressure.

overprint(*)--Information printed or stamped upon a map or chart, in addition to that originally printed, to show data of importance or special use.

overseas--All locations, including Alaska and Hawaii, outside the continental United States.

overseas search and rescue region--Overseas unified command areas (or portions thereof not included within the inland region or the maritime region). See also **search and rescue region.**

over the beach operations--See **logistics over the shore operations.**

over-the-horizon amphibious operations--An operational initiative launched from beyond visual and radar range of the shoreline.

over-the-horizon radar--A radar system that makes use of the atmospheric reflection and refraction phenomena to extend its range of detection beyond line of sight. Over-the-horizon radars may be either forward scatter or back scatter systems.

over-the-shoulder bombing--A special case of loft bombing where the bomb is released past the vertical in order that the bomb may be thrown back to the target. See also **loft bombing; toss bombing.**

overt operation--An operation conducted openly, without concealment. See also **clandestine operation; covert operation.**

P

P-3--See **Orion.**

pace(*)--For ground forces, the speed of a column or element regulated to maintain a prescribed average speed.

pace setter(*)--An individual, selected by the column commander, who travels in the lead vehicle or element to regulate the column speed and establish the pace necessary to meet the required movement order.

packaged forces--Forces of varying size and composition preselected for specific missions in order to facilitate planning and training.

packaged petroleum product--A petroleum product (generally a lubricant, oil, grease, or specialty item) normally packaged by a manufacturer and procured, stored, transported, and issued in containers having a fill capacity of 55 United States gallons (or 45 Imperial gallons, or 205 liters) or less.

padding--Extraneous text added to a message for the purpose of concealing its beginning, ending, or length.

pallet(*)--A flat base for combining stores or carrying a single item to form a unit load for handling, transportation, and storage by materials handling equipment.

palletized unit load(*)--Quantity of any item, packaged or unpackaged, which is arranged on a pallet in a specified manner and securely strapped or fastened thereto so that the whole is handled as a unit.

pan--In air intercept, a code meaning the calling station has a very urgent message to transmit concerning the safety of a ship, aircraft, or other vehicle, or of some person on board or within sight.

pancake--In air intercept, a code meaning, "Land," or, "I wish to land" (reason may be specified; e.g., "pancake ammo," "pancake fuel").

panel code(*)--A prearranged code designed for visual communications, usually between friendly units, by making use of marking panels. See also **marking panel.**

panoramic camera(*)--1. In aerial photography, a camera which, through a system of moving optics or mirrors, scans a wide area of the terrain, usually from horizon to horizon. The camera may be mounted vertically or obliquely within the aircraft, to scan across or along the line of flight. 2. In ground photography, a camera which photographs a wide expanse of terrain by rotating horizontally about the vertical axis through the center of the camera lens.

parachute deployment height(*)--The height above the intended impact point at which the parachute or parachutes are fully deployed.

paradrop(*)--Delivery by parachute of personnel or cargo from an aircraft in flight.

parallactic angle(*)--Angle formed by the optical axes of two instruments, for example, a telescope and its viewfinder seeing the same object. See also **angle of convergence.**

parallax difference(*)--The difference in displacement of the top of an object in relation to its base, as measured on the two images of the object on a stereo pair of photographs.

parallel chains of command--In amphibious operations, a parallel system of command, responding to the interrelationship of Navy, landing force, Air Force, and other major forces assigned, wherein corresponding commanders are established at each subordinate level of all components to facilitate coordinated planning for, and execution of, the amphibious operation.

parallel classification(*)--In railway terminology, the classification of ordinary transport military vehicles and equipment, based on a comparative study of the main characteristics of those vehicles and equipment and of those of the ordinary flat·wagons of a corresponding category onto which they can be loaded.

parallel sheaf--In artillery and naval gunfire support, a sheaf in which the planes (lines) of fire of all pieces are parallel. See also **converged sheaf; open sheaf; sheaf; special sheaf.**

parallel staff(*)--A staff in which one officer from each nation, or Service, working in parallel is appointed to each post. See also **combined staff; integrated staff; joint staff.**

paramilitary forces--Forces or groups which are distinct from the regular armed forces of any country, but resembling them in organization, equipment, training, or mission.

paraphrase--To change the phraseology of a message without changing its meaning.

pararescue team--Specially trained personnel qualified to penetrate to the site of an incident by land or parachute, render medical aid, accomplish survival methods, and rescue survivors.

parlimentaire--An agent employed by a commander of belligerent forces in the field to go in person within the enemy lines for the purpose of communicating or negotiating openly and directly with the enemy commander.

parrot--Identification Friend or Foe transponder equipment.

partial mission capable--Material condition of an aircraft or training device indicating that it can perform at least one but not all of its missions. Also called **PMC.** See also **full mission capable; mission capable; partial mission capable, maintenance; partial mission capable, supply.**

partial mission capable, maintenance--Material condition of an aircraft or training device indicating that it can perform at least one but not all of its missions because of maintenance requirements existing on the inoperable subsystem(s). Also called **PMCM.** See also **full mission capable; mission capable; partial mission capable; partial mission capable, supply.**

partial mission capable, supply--Material condition of an aircraft or training device indicating it can perform at least one but not all of its missions because maintenance required to clear the discrepancy cannot continue due to a supply shortage. Also called **PMCS.** See also **full mission capable; mission capable; partial mission capable; partial mission capable, maintenance.**

partial mobilization--See **mobilization.**

partial storage monitoring--A periodic inspection of major assemblies or components for nuclear weapons, consisting mainly of external observation of humidity, temperatures, and visual damage or deterioration during storage. This type of inspection is also conducted prior to and upon completion of a movement.

partisan warfare--Not to be used. See **guerrilla warfare.**

part number--A combination of numbers, letters, and symbols assigned by a designer, a manufacturer, or vendor to identify a specific part or item of materiel.

pass--1. A short tactical run or dive by an aircraft at a target. 2. A single sweep through or within firing range of an enemy air formation.

passage of lines(*)--An operation in which a force moves forward or rearward through another force's combat positions with the intention of moving into or out of contact with the enemy.

passenger mile--One passenger transported one mile. For air and ocean transport, use nautical miles; for rail, highway, and inland waterway transport in the Continental United States, use statute miles.

passive(*)--In surveillance, an adjective applied to actions or equipments which emit no energy capable of being detected.

passive air defense(*)--All measures, other than active air defense, taken to minimize the effectiveness of hostile air action. These measures include deception, dispersion, and the use of protective construction. See also **air defense.**

passive communications satellite--See **communications satellite.**

passive defense--Measures taken to reduce the probability of and to minimize the effects of damage caused by hostile action without the intention of taking the initiative. See also **active defense.**

passive homing guidance(*)--A system of homing guidance wherein the receiver in the missile utilizes radiation from the target. See also **guidance.**

passive mine(*)--1. A mine whose anticountermining device has been operated preventing the firing mechanism from being actuated. The mine will usually remain passive for a compar-

atively short time. 2. A mine which does not emit a signal to detect the presence of a target, in contrast to an active mine. See also **active mine.**

pass time(*)--In road transport, the time that elapses between the moment when the leading vehicle of a column passes a given point and the moment when the last vehicle passes the same point.

password(*)--A secret word or distinctive sound used to reply to a challenge. See also **challenge; countersign; reply.**

pathfinder drop zone control--The communication and operation center from which pathfinders exercise aircraft guidance.

pathfinder landing zone control--See **pathfinder drop zone control.**

pathfinders--1. Experienced aircraft crews who lead a formation to the drop zone, release point, or target. 2. Teams dropped or air landed at an objective to establish and operate navigational aids for the purpose of guiding aircraft to drop and landing zones. 3. A radar device used for navigating or homing to an objective when visibility precludes accurate visual navigation. 4. Teams air delivered into enemy territory for the purpose of determining the best approach and withdrawal lanes, landing zones, and sites for helicopterborne forces.

patient--A sick, injured, wounded, or other person requiring medical/dental care or treatment.

patrol(*)--A detachment of ground, sea, or air forces sent out for the purpose of gathering information or carrying out a destructive, harassing, mopping-up, or security mission. See also **combat air patrol; combat patrol; reconnaissance patrol; standing patrol.**

pattern bombing--The systematic covering of a target area with bombs uniformly distributed according to a plan.

pattern laying(*)--In land mine warfare, the laying of mines in a fixed relationship to each other.

payload(*)--1. The sum of the weight of passengers and cargo that an aircraft can carry. See also **load.** 2. The warhead, its container, and activating devices in a military missile. 3. The satellite or research vehicle of a space probe or research missile. 4. The load (expressed in tons of cargo or equipment, gallons of liquid, or number of passengers) which the vehicle is designed to transport under specified conditions of operation, in addition to its unladen weight.

payload build-up (missile and space)--The process by which the scientific instrumentation (sensors, detectors, etc.) and necessary mechanical and electronic subassemblies are assembled into a complete operational package capable of achieving the scientific objectives of the mission.

payload integration (missile and space)--The compatible installation of a complete payload package into the spacecraft and space vehicle.

payload (missile)--See **payload, Part 2.**

P-day--That point in time at which the rate of production of an item available for military consumption equals the rate at which the item is required by the Armed Forces.

peacetime force materiel assets--That portion of total materiel assets which is designated to meet the peacetime force materiel requirement. See also **war reserves.**

peacetime force materiel requirement--The quantity of an item required to equip, provide

a materiel pipeline, and sustain the United States force structure (active and reserve) and those allied forces designated for United States peacetime support in current Secretary of Defense guidance, including approved supply support arrangements with foreign military sales countries, and support the scheduled establishment through normal appropriation and procurement leadtime periods.

peacetime materiel consumption and losses--The quantity of an item consumed, lost, or worn-out beyond economical repair through normal appropriation and procurement leadtime periods.

peak overpressure(*)--The maximum value of overpressure at a given location which is generally experienced at the instant the shock (or blast) wave reaches that location. See also **shock wave.**

pecuniary liability--A personal, joint, or corporate monetary obligation to make good any lost, damaged, or destroyed property resulting from fault or neglect. It may also result under conditions stipulated in a contract or bond.

pencil beam(*)--A searchlight beam reduced to, or set at, its minimum width.

penetration(*)--In land operations, a form of offensive which seeks to break through the enemy's defense and disrupt the defensive system.

penetration aids--Techniques and/or devices employed by offensive aerospace weapon systems to increase the probability of penetration of enemy defenses.

penetration (air traffic control)--That portion of a published high altitude instrument approach procedure which prescribes a descent path from the fix on which the procedure is based to a fix or altitude from which an approach to the airport is made.

penetration (intelligence)--The recruitment of agents within or the infiltration of agents or technical monitoring devices in an organization or group for the purpose of acquiring information or of influencing its activities.

percentage clearance(*)--In mine warfare, the estimated percentage of mines of specified characteristics which have been cleared from an area or channel.

perception management--Actions to convey and/or deny selected information and indicators to foreign audiences to influence their emotions, motives, and objective reasoning; and to intelligence systems and leaders at all levels to influence official estimates, ultimately resulting in foreign behaviors and official actions favorable to the originator's objectives. In various ways, perception management combines truth projection, operations security, cover and deception, and psychological operations. See also **psychological operations.**

perimeter defense--A defense without an exposed flank, consisting of forces deployed along the perimeter of the defended area.

periodic intelligence summary--A report of the intelligence situation in a tactical operation, normally produced at corps level or its equivalent, and higher, usually at intervals of 24 hours, or as directed by the commander. Also called **PERINTSUM.**

peripheral war--Not to be used. See **limited war.**

perishable cargo--Cargo requiring refrigeration, such as meat, fruit, fresh vegetables, and medical department biologicals. See also **cargo.**

perishable target--A force or activity at a specific location whose value as a target can decrease substantially during a specified time. A significant decrease in value occurs when the target moves or the operational circumstances change to the extent that the target is no longer lucrative. See also **target.**

permafrost--Permanently frozen subsoil.

permanent echo(*)--Any dense and fixed radar return caused by reflection of energy from the Earth's surface. Distinguished from "ground clutter" by being from definable locations rather than large areas.

permissive action link--A device included in or attached to a nuclear weapon system to preclude arming and/or launching until the insertion of a prescribed discrete code or combination. It may include equipment and cabling external to the weapon or weapon system to activate components within the weapon or weapon system.

permissive environment--See **operational environment.**

Pershing--A mobile surface-to-surface inertially guided missile of a solid propellant type. It possesses a nuclear warhead capability and is designed to support the ground forces with the attack of long range ground targets. Designated as **MGM-31A.**

persistency(*)--In biological or chemical warfare, the characteristic of an agent which pertains to the duration of its effectiveness under determined conditions after its dispersal.

personal locator beacon(*)--An emergency radio locator beacon with a two-way speech facility carried by crew members, either on their person or in their survival equipment, and capable of providing homing signals to assist search and rescue operations. See also **crash locator beacon; emergency locator beacon.**

personal property--Property of any kind or any interest therein, except real property, records of the Federal Government, and naval vessels of the following categories: surface combatants, support ships, and submarines.

personnel--Those individuals required in either a military or civilian capacity to accomplish the assigned mission.

personnel increment number--A seven-character, alphanumeric field that uniquely describes a non-unit-related personnel entry (line) in a Joint Operation Planning and Execution System time-phased force and deployment data. Also called **PIN.**

personnel reaction time (nuclear)(*)--The time required by personnel to take prescribed protective measures after receipt of a nuclear strike warning.

personnel security investigation--An inquiry into the activities of an individual which is designed to develop pertinent information pertaining to trustworthiness and suitability for a position of trust as related to loyalty, character, emotional stability, and reliability.

perspective grid(*)--A network of lines, drawn or superimposed on a photograph, to represent the perspective of a systematic network of lines on the ground or datum plane.

petroleum intersectional service(*)--An intersectional or interzonal service in a theater of operations that operates pipe-lines and related facilities for the supply of bulk petroleum products to theater Army elements and other forces as directed.

petroleum, oils, and lubricants(*)--A broad term which includes all petroleum and associated products used by the Armed Forces. Also called POL.

Phalanx--A close-in weapons system providing automatic, autonomous terminal defense against the anti-ship cruise missile threat. The system includes self-contained search and track radars, weapons control, and a 20-mm M61 gun firing sub-caliber penetrators.

Phantom II--A twin-engine, supersonic, multipurpose, all-weather jet fighter/bomber. It operates from land and aircraft carriers and employs both air-to-air and air-to-surface weapons. The Phantom is a prime air interdiction/close air support and fleet defense vehicle. Special missions such as laser bombing, electronic bombing, and radar bombing are considered routine capabilities. It is capable of employing nuclear and nonnuclear weapons. Designated as **F-4.** RF-4 is the photo-reconnaissance version.

phantom order--A draft contract with an industrial establishment for wartime production of a specific product with provisions for necessary preplanning in time of peace and for immediate execution of the contract upon receipt of proper authority.

phase line(*)--A line utilized for control and coordination of military operations, usually a terrain feature extending across the zone of action. See also **report line.**

phases of military government--1. **assault**--That period which commences with first contact with civilians ashore and extends to the establishment of military government control ashore by the landing force. 2. **consolidation**--That period which commences with the establishment of military government control ashore by the landing force and extends to the establishment of control by occupation forces. 3. **occupation**--That period which commences when an area has been occupied in fact, and the military commander within that area is in a position to enforce public safety and order. See also **civil affairs; military occupation.**

Phoenix--A long-range air-to-air missile with electronic guidance/homing. Designated as **AIM-54A.**

phonetic alphabet--A list of standard words used to identify letters in a message transmitted by radio or telephone. The following are the authorized words, listed in order, for each letter in the alphabet: ALFA, BRAVO, CHAR-LIE, DELTA, ECHO, FOXTROT, GOLF, HOTEL, INDIA, JULIETT, KILO, LIMA, MIKE, NOVEMBER, OSCAR, PAPA, QUE-BEC, ROMEO, SIERRA, TANGO, UNI-FORM, VICTOR, WHISKEY, X-RAY, YAN-KEE, and ZULU.

phoney minefield(*)--An area free of live mines used to simulate a minefield, or section of a minefield, with the object of deceiving the enemy. See also **gap, minefield.**

PHOTINT--See **photographic intelligence.**

photoflash bomb(*)--A bomb designed to produce a brief and intense illumination for medium altitude night photography.

photoflash cartridge(*)--A pyrotechnic cartridge designed to produce a brief and intense illumination for low altitude night photography.

photogrammetric control(*)--Control established by photogrammetric methods as distinguished from control established by ground methods. Also called **minor control.**

photogrammetry(*)--The science or art of obtaining reliable measurements from photographic images.

photographic coverage--The extent to which an area is covered by photography from one

mission or a series of missions or in a period of time. Coverage, in this sense, conveys the idea of availability of photography and is not a synonym for the word "photography."

photographic intelligence--The collected products of photographic interpretation, classified and evaluated for intelligence use. Also called **PHOTINT.**

photographic interpretation--See **imagery interpretation.**

photographic panorama--A continuous photograph or an assemblage of overlapping oblique or ground photographs which have been matched and joined together to form a continuous photographic representation of the area.

photographic reading(*)--The simple recognition of natural or manmade features from photographs not involving imagery interpretation techniques.

photographic scale(*)--The ratio of a distance measured on a photograph or mosaic to the corresponding distance on the ground, classified as follows:
a. **very large scale**--1:4,999 and larger.
b. **large scale**--1:5,000 to 1:9,999.
c. **medium scale**--1:10,000 to 1:24,999.
d. **small scale**--1:25,000 to 1:49,999.
e. **very small scale**--1:50,000 and smaller.
See also **scale.**

photographic sortie--See **imagery sortie.**

photographic strip(*)--Series of successive overlapping photographs taken along a selected course or direction.

photo interpretation key--See **imagery interpretation key.**

photomap(*)--A reproduction of a photograph or photomosaic upon which the grid lines, marginal data, contours, place names, boundaries, and other data may be added.

photo nadir(*)--The point at which a vertical line through the perspective center of the camera lens intersects the photo plane.

physical characteristics--Those military characteristics of equipment which are primarily physical in nature, such as weight, shape, volume, water-proofing, and sturdiness.

physical security(*)--That part of security concerned with physical measures designed to safeguard personnel; to prevent unauthorized access to equipment, installations, material, and documents; and to safeguard them against espionage, sabotage, damage, and theft. See also **communications security; protective security; security.**

pictomap--A topographic map in which the photographic imagery of a standard mosaic has been converted into interpretable colors and symbols by means of a pictomap process.

pictorial symbolization(*)--The use of symbols which convey the visual character of the features they represent.

PID--See **plan identification number.**

pier--1. A structure extending into the water approximately perpendicular to a shore or a bank and providing berthing for ships and which may also provide cargo-handling facilities. 2. A structure extending into the water approximately perpendicular to a shore or bank and providing a promenade or place for other use, as a fishing pier. 3. A support for the spans of a bridge. See also **quay; wharf.**

pigeon--In air intercept, a code meaning, "The magnetic bearing and distance of base (or unit indicated) from you is _____ degrees _____ miles."

pillbox(*)--A small, low fortification that houses machine guns, antitank weapons, etc. A pillbox is usually made of concrete, steel, or filled sandbags.

pilot's trace(*)--A rough overlay to a map made by the pilot of a photographic reconnaissance aircraft during or immediately after a sortie. It shows the location, direction, number, and order of photographic runs made, together with the camera(s) used on each run.

PIN--See **personnel increment number.**

pinpoint(*)--1. A precisely identified point, especially on the ground, that locates a very small target, a reference point for rendezvous or for other purposes; the coordinates that define this point. 2. The ground position of aircraft determined by direct observation of the ground.

pinpoint photograph(*)--A single photograph or a stereo pair of a specific object or target.

pinpoint target(*)--In artillery and naval gunfire support, a target less than 50 meters in diameter.

pipeline(*)--In logistics, the channel of support or a specific portion thereof by means of which materiel or personnel flow from sources of procurement to their point of use.

pitch(*)--1. The rotation of an aircraft or ship about its lateral axis. 2. In air photography, the camera rotation about the transverse axis of the aircraft. Also called **tip.**

pitch angle(*)--The angle between the aircraft's longitudinal axis and the horizontal plane. Also called **inclination angle.**

plan for landing--In amphibious operations, a collective term referring to all individually prepared naval and landing force documents which, taken together, present in detail all instructions for execution of the ship-to-shore movement.

plan identification number--1. A command-unique four-digit number followed by a suffix indicating the Joint Strategic Capabilities Plan (JSCP) year for which the plan is written, e.g., "2220-95". 2. In the Joint Operation Planning and Execution System (JOPES) data base, a five-digit number representing the command-unique four-digit identifier, followed by a one character, alphabetic suffix indicating the operation plan option, or a one-digit numeric value indicating the JSCP year for which the plan is written. Also called **PID.**

planimetric map--A map representing only the horizontal position of features. Sometimes called a line map. See also **map.**

plan information capability--This capability allows a supported command to enter and update key elements of information in an operation plan stored in the Joint Operation Planning and Execution System.

planned target(*)--In artillery and naval gunfire support, a target on which fire is prearranged.

planned target (nuclear)--A nuclear target planned on an area or point in which a need is anticipated. A planned nuclear target may be scheduled or on call. Firing data for a planned nuclear target may or may not be determined in advance. Coordination and warning of friendly troops and aircraft are mandatory.

planning directive--In amphibious operations, the plan issued by commander, amphibious task force, following receipt of the initiating directive, to ensure that the planning process and interdependent plans developed by the amphibious task force headquarters and assigned major forces will be coordinated, the plan completed in the time allowed, and important aspects not overlooked.

planning factor(*)--A multiplier used in planning to estimate the amount and type of effort involved in a contemplated operation. Planning factors are often expressed as rates, ratios, or lengths of time.

planograph--A scale drawing of a storage area showing the approved layout of the area, location of bulk, bin, rack, and box pallet areas, aisles, assembly areas, walls, doorways, directions of storage, office space, wash rooms, and other support and operational areas.

plan position indicator(*)--A cathode ray tube on which radar returns are so displayed as to bear the same relationship to the transmitter as the objects giving rise to them.

plant equipment--Personal property of a capital nature, consisting of equipment, furniture, vehicles, machine tools, test equipment, and accessory and auxiliary items, but excluding special tooling and special test equipment, used or capable of use in the manufacture of supplies or for any administrative or general plant purpose.

plastic range(*)--The stress range in which a material will not fail when subjected to the action of a force, but will not recover completely so that a permanent deformation results when the force is removed.

plastic zone(*)--The region beyond the rupture zone associated with crater formation resulting from an explosion in which there is no visible rupture, but in which the soil is permanently deformed and compressed to a high density. See also **rupture zone.**

plate(*)--1. In cartography: a. a printing plate of zinc, aluminum, or engraved copper; b. collective term for all "states" of an engraved map reproduced from the same engraved printing plate; c. all detail to appear on a map or chart which will be reproduced from a single printing plate (e.g., the "blue plate" or the "contour plate"). 2. In photography, a transparent medium, usually glass, coated with a photographic emulsion. See also **diapositive; transparency.**

platform drop(*)--The airdrop of loaded platforms from rear loading aircraft with roller conveyors. See also **airdrop; airdrop platform.**

plot(*)--1. Map, chart, or graph representing data of any sort. 2. Representation on a diagram or chart of the position or course of a target in terms of angles and distances from positions; location of a position on a map or a chart. 3. The visual display of a single location of an airborne object at a particular instant of time. 4. A portion of a map or overlay on which are drawn the outlines of the areas covered by one or more photographs. See also **master plot.**

PMC--See **partial mission capable.**

PMCM--See **partial mission capable, maintenance.**

PMCS--See **partial mission capable, supply.**

POD--See **port of debarkation.**

POE--See **port of embarkation.**

pogo--In air intercept, a code meaning, "Switch to communications channel number preceding 'pogo.' If unable to establish communications, switch to channel number following 'pogo.' "

point defense--The defense or protection of special vital elements and installations; e.g., command and control facilities, air bases.

point designation grid(*)--A system of lines, having no relation to the actual scale, or orientation, drawn on a map, chart, or air photograph dividing it into squares so that points can be more readily located.

point of impact(*)--1. The point on the drop zone where the first parachutist or air dropped cargo item lands or is expected to land. 2. The point at which a projectile, bomb, or re-entry vehicle impacts or is expected to impact.

point of no return(*)--A point along an aircraft track beyond which its endurance will not permit return to its own or some other associated base on its own fuel supply.

point target--1. A target of such small dimension that it requires the accurate placement of ordnance in order to neutralize or destroy it. 2. **nuclear**--A target in which the ratio of radius of damage to target radius is equal to or greater than 5.

point target (nuclear)--See **point target, Part 2.**

point to point sealift--The movement of troops and/or cargo in Military Sealift Command nucleus or commercial shipping between established ports, in administrative landings or logistics over the shore operations. See also **administrative landing; administrative movement; logistics over the shore operations.**

poised mine(*)--A mine in which the ship counter setting has been run down to "one" and which is ready to detonate at the next actuation. See also **mine.**

POL--See **petroleum, oils, and lubricants.**

polar coordinates(*)--1. Coordinates derived from the distance and angular measurements from a fixed point (pole). 2. In artillery and naval gunfire support, the direction, distance, and vertical correction from the observer/spotter position to the target.

Polaris--An underwater/surface-launched, surface-to-surface, solid-propellant ballistic missile with inertial guidance and nuclear warhead. Designated as **UGM-27**.
UGM-27A--1,200 nautical mile range.
UGM-27B--1,500 nautical mile range.
UGM-27C--2,500 nautical mile range.

polar plot(*)--The method of locating a target or point on the map by means of polar coordinates.

political intelligence--Intelligence concerning foreign and domestic policies of governments and the activities of political movements.

political warfare--Aggressive use of political means to achieve national objectives.

politico-military gaming--Simulation of situations involving the interaction of political, military, sociological, psychological, economic, scientific, and other appropriate factors.

pool--1. To maintain and control a supply of resources or personnel upon which other activities may draw. The primary purpose of a pool is to promote maximum efficiency of use of the pooled resources or personnel, e.g., a petroleum pool, a labor and equipment pool. 2. Any combination of resources which serves a common purpose.

popeye--In air intercept, a code meaning, "In clouds or area of reduced visibility."

port--A place at which ships may discharge or receive their cargoes. It includes any port accessible to ships on the seacoast, navigable rivers or inland waterways. The term "ports" should not be used in conjunction with air facilities which are designated as **aerial ports, airports, etc.** See also **control port; indoctri-**

nation port; major port; minor port; secondary port; water terminal.

port capacity(*)--The estimated capacity of a port or an anchorage to clear cargo in 24 hours usually expressed in tons. See also **beach capacity; clearance capacity.**

port complex(*)--A port complex comprises one or more port areas of varying importance whose activities are geographically linked either because these areas are dependent on a common inland transport system or because they constitute a common initial destination for convoys.

port designator(*)--A group of letters identifying ports in convoy titles or messages.

port evacuation of cargoes(*)--The removal of cargoes from a threatened port to alternative storage sites.

port evacuation of shipping(*)--The movement of merchant ships from a threatened port for their own protection.

port of debarkation--The geographic point at which cargo or personnel are discharged. May be a seaport or aerial port of debarkation. For unit requirements, it may or may not coincide with the destination. Also called **POD.** See also **port of embarkation.**

port of embarkation--The geographic point in a routing scheme from which cargo or personnel depart. May be a seaport or aerial port from which personnel and equipment flow to port of debarkation. For unit and nonunit requirements, it may or may not coincide with the origin. Also called **POE.** See also **port of debarkation.**

port security(*)--The safeguarding of vessels, harbors, ports, waterfront facilities and cargo from internal threats such as: destruction, loss, or injury from sabotage or other subversive acts; accidents; thefts; or other causes of similar nature. See also **harbor defense; physical security; security.**

POS--See **primary operating stocks.**

Poseidon--A two-stage, solid propellant ballistic missile capable of being launched from a specially configured submarine operating in either its surface or submerged mode. The missile is equipped with inertial guidance, nuclear warheads, and a maneuverable bus that has the capability to carry up to 14 reentry bodies which can be directed to as many as 14 separate targets. Designated as **UGM-73A.**

positional defense--See **position defense.**

position defense(*)--The type of defense in which the bulk of the defending force is disposed in selected tactical localities where the decisive battle is to be fought. Principal reliance is placed on the ability of the forces in the defended localities to maintain their positions and to control the terrain between them. The reserve is used to add depth, to block, or restore the battle position by counterattack.

positive control--A method of airspace control which relies on positive identification, tracking, and direction of aircraft within an airspace, conducted with electronic means by an agency having the authority and responsibility therein.

positive identification and radar advisory zone--A specified area established for identification and flight following of aircraft in the vicinity of a fleet-defended area.

positive phase of the shock wave--The period during which the pressure rises very sharply to a value that is higher than ambient and then decreases rapidly to the ambient pressure. See also **negative phase of the shock wave; shock wave.**

possible--A term used to qualify a statement made under conditions wherein some evidence exists to support the statement. This evidence is sufficient to warrant mention, but insufficient to warrant assumption as true. See also **probable.**

postattack period--In nuclear warfare, that period which extends from the termination of the final attack until political authorities agree to terminate hostilities. See also **post-hostilities period; transattack period.**

post hostilities period--That period subsequent to the date of ratification by political authorities of agreements to terminate hostilities.

poststrike reconnaissance--Missions undertaken for the purpose of gathering information used to measure results of a strike.

pounce--In air intercept, a code meaning, "I am in position to engage target."

PPI gauge--See **international loading gauge.**

practice mine(*)--1. In land mine warfare, an inert mine to which is fitted a fuze and a device to indicate, in a non-lethal fashion, that the fuze has been activated. See also **mine.** 2. In naval mine warfare, an inert-filled mine but complete with assembly, suitable for instruction and for practice in preparation. See also **drill mine.**

prearranged fire(*)--Fire that is formally planned and executed against targets or target areas of known location. Such fire is usually planned well in advance and is executed at a predetermined time or during a predetermined period of time. See also **fire; on call; scheduled fire.**

preassault operation--In amphibious operations, an operation conducted in the amphibious objective area before the assault phase begins.

precautionary launch--The launching of nuclear loaded aircraft under imminent nuclear attack so as to preclude friendly aircraft destruction and loss of weapons on the ground/carrier.

precedence--**1. communications**--A designation assigned to a message by the originator to indicate to communications personnel the relative order of handling and to the addressee the order in which the message is to be noted. **2. reconnaissance**--A letter designation, assigned by a unit requesting several reconnaissance missions, to indicate the relative order of importance, within an established priority, of the mission requested. See also **flash message; immediate message; priority message; routine message.**

precession--See **apparent precession; induced precession; real precession.**

precise frequency--A frequency requirement accurate to within one part in 1,000,000,000.

precise time--A time requirement accurate to within 10 milliseconds.

precision approach--An approach in which range, azimuth, and glide slope information are provided to the pilot. See also **final approach; nonprecision approach.**

precision bombing--Bombing directed at a specific point target.

precursor front(*)--An air pressure wave which moves ahead of the main blast wave for some distance as a result of a nuclear explosion of appropriate yield and low burst height over a heat-absorbing (or dusty) surface. The pressure at the precursor front increases more gradually than in a true (or idea,) shock wave, so that the behavior in the precursor region is said to be nonideal.

precursor sweeping(*)--The sweeping of an area by relatively safe means in order to reduce the risk to mine countermeasures vessels in subsequent operations.

predicted fire(*)--Fire that is delivered without adjustment.

predominant height(*)--In air reconnaissance, the height of 51 percent or more of the structures within an area of similar surface material.

preemptive attack--An attack initiated on the basis of incontrovertible evidence that an enemy attack is imminent.

preemptive war--Not to be used. See **preemptive attack.**

preflight inspection--See **before-flight inspection.**

preinitiation--The initiation of the fission chain reaction in the active material of a nuclear weapon at any time earlier than that at which either the designed or the maximum compression or degree of assembly is attained.

prelanding operations--In amphibious operations, operations conducted between the commencement of the assault phase and the commencement of the ship-to-shore movement by the main body of the amphibious task force. They encompass similar preparations conducted by the advanced force but focus on the landing area, concentrating specifically on the landing beaches and the helicopter landing zones to be used by the main landing force. Prelanding operations also encompass final preparations for the ship-to-shore movement.

pre-launch survivability--The probability that a delivery and/or launch vehicle will survive an enemy attack under an established condition of warning.

preliminary communications search--In search and rescue operations, consists of contacting and checking major facilities within the areas where the craft might be or might have been seen. A preliminary communications search is normally conducted during the uncertainty phase. Also called **PRECOM.** See also **extended communications search; search and rescue incident classification, Subpart a.**

preliminary demolition target(*)--A target, other than a reserved demolition target, which is earmarked for demolition and which can be executed immediately after preparation, provided that prior authority has been granted. See also **demolition target; reserved demolition target.**

preliminary movement schedule--A projection of the routing of movement requirements reflected in the time-phased force and deployment data, from origin to destination, including identification of origins, ports of embarkation, ports of debarkation, and en route stops; associated time frames for arrival and departure at each location; type of lift assets required to accomplish the move; and cargo details by carrier. Schedules are sufficiently detailed to support comparative analysis of requirements against capabilities and to develop location workloads for reception and onward movement.

preload loading(*)--The loading of selected items aboard ship at one port prior to the main loading of the ship at another. See also **loading.**

premature dud--See **flare dud.**

preparation fire(*)--Fire delivered on a target preparatory to an assault. See also **fire.**

preplanned air support(*)--Air support in accordance with a program, planned in advance of operations. See also **air support.**

preplanned mission request--A request for an air strike on a target which can be anticipated sufficiently in advance to permit detailed mission coordination and planning.

preplanned nuclear support--Nuclear support planned in advance of operations. See also **immediate nuclear support; nuclear support.**

preposition(*)--To place military units, equipment, or supplies at or near the point of planned use or at a designated location to reduce reaction time, and to ensure timely support of a specific force during initial phases of an operation.

prepositioned war reserve materiel requirement, balance--That portion of the pre-positioned war reserve materiel requirement which has not been acquired or funded. This level consists of the pre-positioned war reserve materiel requirement, less the prepositioned war reserve requirement, protectable.

prepositioned war reserve materiel requirement, protectable--That portion of the pre-positioned war reserve materiel requirement which is protected for purposes of procurement, funding and inventory management.

prepositioned war reserve requirement--That portion of the war reserve materiel requirement which the current Secretary of Defense guidance dictates be reserved and positioned at or near the point of planned use or issue to the user prior to hostilities to reduce reaction time and to assure timely support of a specific force/project until replenishment can be effected.

prepositioned war reserve stock--The assets that are designated to satisfy the prepositioned war reserve materiel requirement.

prescribed nuclear load(*)--A specified quantity of nuclear weapons to be carried by a delivery unit. The establishment and replenishment of this load after each expenditure is a command decision and is dependent upon the tactical situation, the nuclear logistical situation, and the capability of the unit to transport and utilize the load. It may vary from day to day and among similar delivery units.

prescribed nuclear stockage(*)--A specified quantity of nuclear weapons, components of nuclear weapons, and warhead test equipment to be stocked in special ammunition supply points or other logistical installations. The establishment and replenishment of this stockage is a command decision and is dependent upon the tactical situation, the allocation, the capability of the logistical support unit to store and maintain the nuclear weapons, and the nuclear logistical situation. The prescribed stockage may vary from time to time and among similar logistical support units.

preset guidance--A technique of missile control wherein a predetermined flight path is set into the control mechanism and cannot be adjusted after launching. See also **guidance.**

Presidential Selected Reserve Callup Authority--Provision of a public law that provides the President a means to activate not more than 200,000 members of the Selected Reserve for 90 days to meet the support requirements of any operational mission without a declaration of a national emergency. It further grants the President authority to extend the original 90 days for an additional 90 days in the interest of national security. This authority has particular utility when used in circumstances in which the escalatory national or international signals of partial or full mobilization would be undesirable. Forces available under this authority can provide a tailored, limited-scope, deterrent, or operational response, or may be used as a precursor to any subsequent mobilization. See also **mobilization.**

pressure altimeter--See **barometric altimeter.**

pressure-altitude(*)--An atmospheric pressure expressed in terms of altitude which corresponds to that pressure in the standard atmosphere. See also **altitude.**

pressure breathing(*)--The technique of breathing which is required when oxygen is supplied direct to an individual at a pressure higher than the ambient barometric pressure.

pressure front--See **shock front.**

pressure mine(*)--1. In land mine warfare, a mine whose fuze responds to the direct pressure of a target. 2. In naval mine warfare, a mine whose circuit responds to the hydrodynamic pressure field of a target. See also **mine.**

pressure mine circuit--See **pressure mine.**

pressure suit(*)--1. Partial--A skin tight suit which does not completely enclose the body but which is capable of exerting pressure on the major portion of the body in order to counteract an increased intrapulmonary oxygen pressure. 2. Full--A suit which completely encloses the body and which a gas pressure, sufficiently above ambient pressure for maintenance of function, may be sustained.

pressurized cabin--The occupied space of an aircraft in which the air pressure has been increased above that of the ambient atmosphere by compression of the ambient atmosphere into the space.

prestrike reconnaissance--Missions undertaken for the purpose of obtaining complete information about known targets for use by the strike force.

prevention--The security procedures undertaken by the public and private sector in order to discourage terrorist acts. See also **antiterrorism.**

prevention of stripping equipment--See **antirecovery device.**

preventive maintenance--The care and servicing by personnel for the purpose of maintaining equipment and facilities in satisfactory operating condition by providing for systematic inspection, detection, and correction of incipient failures either before they occur or before they develop into major defects.

preventive war--A war initiated in the belief that military conflict, while not imminent, is inevitable, and that to delay would involve greater risk.

prewithdrawal demolition target--A target prepared for demolition preliminary to a withdrawal, the demolition of which can be executed as soon after preparation as convenient on the orders of the officer to whom the responsibility for such demolitions has been delegated. See also **demolition target.**

primary aircraft authorization--Aircraft authorized to a unit for performance of its operational mission. The primary authorization forms the basis for the allocation of operating resources to include manpower, support equipment, and flying-hour funds. See also **backup aircraft authorization.**

primary aircraft inventory--The aircraft assigned to meet the primary aircraft authorization. See also **backup aircraft inventory.**

primary censorship--Armed forces censorship performed by personnel of a company, battery, squadron, ship, station, base, or similar unit on the personal communications of persons assigned, attached, or otherwise under the jurisdiction of a unit. See also **censorship.**

primary control officer--In amphibious operations, the officer embarked in a primary control ship assigned to control the movement of landing craft, amphibious vehicles, and landing ships to and from a colored beach.

primary control ship--In amphibious operations, a ship of the task force designated to provide support for the primary control officer and a combat information center control team for a colored beach.

primary imagery dissemination--See **electronic imagery dissemination.**

primary imagery dissemination system--See **electronic imagery dissemination.**

primary interest--Principal, although not exclusive, interest and responsibility for accomplishment of a given mission, including responsibility for reconciling the activities of other agencies that possess collateral interest in the program.

primary operating stocks--Logistics resources on hand or on order necessary to support day-to-day operational requirements, and which, in part, can also be used to offset sustaining combat requirements. Also called **POS.**

primary review authority--The organization assigned by the lead agent to perform the actions and coordination necessary to develop and maintain the assigned joint publication under cognizance of the lead agent. See also **JCS publication; lead agent.**

primary target--An object of high publicity value to terrorists. See also **antiterrorism; secondary targets.**

primed charge(*)--A charge ready in all aspects for ignition.

prime mover--A vehicle, including heavy construction equipment, possessing military characteristics, designed primarily for towing heavy, wheeled weapons and frequently providing facilities for the transportation of the crew of, and ammunition for, the weapon.

principal items--End items and replacement assemblies of such importance that management techniques require centralized individual item management throughout the supply system, to include depot level, base level, and items in the hands of using units. These specifically include the items where, in the judgment of the Services, there is a need for central inventory control, including centralized computation of requirements, central procurement, central direction of distribution, and central knowledge and control of all assets owned by the Services.

principal operational interest--When used in connection with an established facility operated by one Service for joint use by two or more Services, the term indicates a requirement for the greatest use of, or the greatest need for, the services of that facility. The term may be applied to a Service, but is more applicable to a command.

principal parallel(*)--On an oblique photograph, a line parallel to the true horizon and passing through the principal point.

principal plane(*)--A vertical plane which contains the principal point of an oblique photograph, the perspective center of the lens, and the ground nadir.

principal point(*)--The foot of the perpendicular to the photo plane through the perspective center. Generally determined by intersection of the lines joining opposite collimating or fiducial marks.

principal scale(*)--In cartography, the scale of a reduced or generating globe representing the sphere or spheroid, defined by the fractional relation of their respective radii. Also called **nominal scale**. See also **scale.**

principal vertical(*)--On an oblique photograph, a line perpendicular to the true horizon and passing through the principal point.

printing size of a map or chart(*)--The dimensions of the smallest rectangle which will contain a map or chart, including all the printed material in its margin.

print reference(*)--A reference to an individual print in an air photographic sortie.

priority--With reference to operation plans and the tasks derived therefrom, an indication of relative importance rather than an exclusive and final designation of the order of accomplishment.

priority designator--A two-digit issue and priority code (01 through 15) placed in military standard requisitioning and issue procedure requisitions. It is based upon a combination of factors which relate the mission of the requisitioner and the urgency of need or the end use and is used to provide a means of assigning relative rankings to competing demands placed on the Department of Defense supply system.

priority intelligence requirements(*)--Those intelligence requirements for which a commander has an anticipated and stated priority in his task of planning and decisionmaking. See also **information requirements; intelligence cycle.**

priority message--A category of precedence reserved for messages that require expeditious action by the addressee(s) and/or furnish essential information for the conduct of operations in progress when routine precedence will not suffice. See also **precedence.**

priority national intelligence objectives--A guide for the coordination of intelligence collection and production in response to requirements relating to the formulation and execution of national security policy. They are compiled annually by the Washington Intelligence Community and flow directly from the intelligence mission as set forth by the National Security Council. They are specific enough to provide a basis for planning the allocation of collection and research resources, but not so specific as to constitute in themselves research and collection requirements.

priority of immediate mission requests--See **emergency priority; ordinary priority; search and attack priority; urgent priority.**

priority of preplanned mission requests--1. Targets capable of preventing the execution of the plan of action. 2. Targets capable of immediate serious interference with the plan of action. 3. Targets capable of ultimate serious interference with the execution of the plan of action. 4. Targets capable of limited interference with the execution of the plan of action.

priority system for mission requests for tactical reconnaissance--Priority I--Takes precedence over all other requests except previously assigned priorities I. The results of these requests are of paramount importance to the immediate battle situation or objective. **Priority II**--The results of these requirements are in support of the general battle situation and will be accomplished as soon as possible after priorities I. These are requests to gain current battle information. **Priority III**--The results of these requests update the intelligence data base but do not affect the immediate battle situation. **Priority IV**--The results of these requests are of a routine nature. These results will be fulfilled when the reconnaissance effort permits. See also **precedence.**

prior permission(*)--Permission granted by the appropriate authority prior to the commencement of a flight or a series of flights landing in or flying over the territory of the nation concerned.

prisoner of war--A detained person as defined in Articles 4 and 5 of the Geneva Convention Relative to the Treatment of Prisoners of War of August 12, 1949. In particular, one who, while engaged in combat under orders of his or her government, is captured by the armed forces of the enemy. As such, he or she is entitled to the combatant's privilege of immunity from the municipal law of the capturing state for warlike acts which do not amount to breaches of the law of armed conflict. For example, a prisoner of war may be, but is not limited to, any person belonging to one of the following categories who has fallen into the power of the enemy: a member of the armed forces, organized militia or volunteer corps; a person who accompanies the armed forces without actually being a member thereof; a member of a merchant marine or civilian aircraft crew not qualifying for more favorable treatment; or individuals who, on the approach of the enemy, spontaneously take up arms to resist the invading forces.

prisoner of war branch camp(*)--A subsidiary camp under the supervision and administration of a prisoner of war camp.

prisoner of war camp--An installation established for the internment and administration of prisoners of war.

prisoner of war censorship--The censorship of the communications to and from enemy prisoners of war and civilian internees held by the United States Armed Forces. See also **censorship.**

prisoner of war compound(*)--A subdivision of a prisoner of war enclosure.

prisoner of war enclosure(*)--A subdivision of a prisoner of war camp.

prisoner of war personnel record(*)--A form for recording the photograph, fingerprints, and other pertinent personal data concerning the prisoner of war, including that required by the Geneva Convention.

proactive measures--In antiterrorism, measures taken in the preventive stage of antiterrorism designed to harden targets and detect actions before they occur.

proactive mine countermeasures--Measures intended to prevent the enemy from successfully laying mines. See also **mine countermeasures.**

probability of damage(*)--The probability that damage will occur to a target expressed as a percentage or as a decimal.

probability of detection--The probability that the search object will be detected under given conditions if it is in the area searched.

probable--A term used to qualify a statement made under conditions wherein the available evidence indicates that the statement is factual until there is further evidence in confirmation or denial. See also **possible.**

probable error--See **horizontal error.**

probable error deflection--Error in deflection which is exceeded as often as not.

probable error height of burst--Error in height of burst which projectile/missile fuzes may be expected to exceed as often as not.

probable error range--Error in range which is exceeded as often as not.

probably destroyed(*)--In air operations, a damage assessment on an enemy aircraft seen to break off combat in circumstances which lead to the conclusion that it must be a loss although it is not actually seen to crash.

procedural control(*)--A method of airspace control which relies on a combination of previously agreed and promulgated orders and procedures.

procedure--A procedure begins with a specific, documentable event that causes an activity to occur. The activity must produce a product that normally affects another external organization. Frequently, that product will be the event that causes another procedure to occur. It is important to recognize that a procedure determines "what" an organization must do at critical periods but does not direct "how" it will be done.

procedure turn(*)--An aircraft maneuver in which a turn is made away from a designated track followed by a turn in the opposite direction, both turns being executed at a constant rate so as to permit the aircraft to intercept and proceed along the reciprocal of the designated track.

procedure word--A word or phrase limited to radio telephone procedure used to facilitate communication by conveying information in a condensed standard form. Also called **proword.**

processing(*)--1. In photography, the operations necessary to produce negatives, diapositives, or prints from exposed films, plates, or paper. 2. See intelligence cycle.

proclamation--A document published to the inhabitants of an area which sets forth the basis of authority and scope of activities of a commander in a given area and which defines the obligations, liabilities, duties, and rights of the population affected.

procurement--The process of obtaining personnel, services, supplies, and equipment. See also **central procurement.**

procurement lead time--The interval in months between the initiation of procurement action and receipt into the supply system of the production model (excludes prototypes) purchased as the result of such actions, and is composed of two elements, production lead time and administrative lead time. See also **administrative lead time; initiation of procurement action; level of supply; receipt into the supply system.**

production--The conversion of raw materials into products and/or components thereof, through a series of manufacturing processes. It includes functions of production engineering, controlling, quality assurance, and the determination of resources requirements.

production base--The total national industrial production capacity available for the manufacture of items to meet materiel requirements.

production lead time--The time interval between the placement of a contract and receipt into the supply system of materiel purchased. Two entries are provided: **a. initial**--The time interval if the item is not under production as of the date of contract placement. **b. reorder**--The time interval if the item is under production as of the date of contract placement. See also **procurement lead time.**

production logistics--That part of logistics concerning research, design, development, manufacture, and acceptance of materiel. In consequence, production logistics includes: standardization and interoperability, contracting, quality assurance, initial provisioning, transportability, reliability and defect analysis, safety standards,

specifications and production processes, trials and testing (including provision of necessary facilities), equipment documentation, configuration control, and modifications.

production loss appraisal--An estimate of damage inflicted on an industry in terms of quantities of finished products denied the enemy from the moment of attack through the period of reconstruction to the point when full production is resumed.

proficiency training aircraft--Aircraft required to maintain the proficiency of pilots and other aircrew members who are assigned to nonflying duties.

profile--See **flight profile.**

proforma(*)--A standard form. See also **standard NATO data message.**

program aircraft--The total of the active and reserve aircraft. See also **aircraft.**

program manager--See **system manager.**

Programmed Forces--The forces that exist for each year of the Future Years Defense Program. They contain the major combat and tactical support forces that are expected to execute the national strategy within manpower, fiscal, and other constraints. See also **Current Force; force; Intermediate Force Planning Level.**

program of nuclear cooperation(*)--Presidentially approved bilateral proposals for the United States to provide nuclear weapons, and specified support to user nations who desire to commit delivery units to NATO in nuclear only or dual capable roles. After presidential approval in principle, negotiations will be initiated with the user nation to develop detailed support arrangements.

progress payment--Payment made as work progresses under a contract, upon the basis of costs incurred, of percentage of completion accomplished, or of a particular stage of completion. The term does not include payments for partial deliveries accepted by the Government under a contract, or partial payments on contract termination claims.

prohibited area--A specified area within the land areas of a state or territorial waters adjacent thereto over which the flight of aircraft is prohibited. May also refer to land or sea areas to which access is prohibited. See also **closed area; danger area; restricted area.**

project--A planned undertaking of something to be accomplished, produced, or constructed, having a finite beginning and a finite ending.

projected map display(*)--The displayed image of a map or chart projected through an optical or electro-optical system onto a viewing surface.

projectile--An object projected by an applied exterior force and continuing in motion by virtue of its own inertia, as a bullet, shell, or grenade. Also applied to rockets and to guided missiles.

projection(*)--In cartography, any systematic arrangement of meridians and parallels portraying the curved surface of the sphere or spheroid upon a plane.

projection print--An enlarged or reduced photographic print made by projection of the image of a negative or a transparency onto a sensitized surface.

project manager--See **system manager.**

proliferation (nuclear weapons)--The process by which one nation after another comes into possession of, or into the right to determine the use of nuclear weapons, each potentially able to launch a nuclear attack upon another nation.

prompt radiation--The gamma rays produced in fission and as a result of other neutron reactions and nuclear excitation of the weapon materials appearing within a second or less after a nuclear explosion. The radiations from these sources are known either as prompt or instantaneous gamma rays. See also **induced radiation; initial radiation; residual radiation.**

pronto--As quickly as possible.

propaganda--Any form of communication in support of national objectives designed to influence the opinions, emotions, attitudes, or behavior of any group in order to benefit the sponsor, either directly or indirectly. See also **black propaganda; grey propaganda; white propaganda.**

propellant--That source which provides the energy required for propelling a projectile. Specifically, an explosive charge for propelling a projectile; also a fuel, either solid or liquid, for propelling a rocket or missile.

propelled mine--See **mobile mine.**

property--1. Anything that may be owned. 2. As used in the military establishment, this term is usually confined to tangible property, including real estate and materiel. 3. For special purposes and as used in certain statutes, this term may exclude such items as the public domain, certain lands, certain categories of naval vessels and records of the Federal Government.

property account--A formal record of property and property transactions in terms of quantity and/or cost, generally by item. An official record of Government property required to be maintained.

proportional navigation--A method of homing navigation in which the missile turn rate is directly proportional to the turn rate in space of the line of sight.

protected site(*)--A facility which is protected by the use of camouflage or concealment, selective siting, construction of facilities designed to prevent damage from fragments caused by conventional weapons, or a combination of such measures.

protective clothing(*)--Clothing especially designed, fabricated, or treated to protect personnel against hazards caused by extreme changes in physical environment, dangerous working conditions, or enemy action.

protective minefield(*)--1. In land mine warfare, a minefield employed to assist a unit in its local, close-in protection. 2. In naval mine warfare, a minefield laid in friendly territorial waters to protect ports, harbors, anchorages, coasts, and coastal routes. See also **minefield.**

prototype--A model suitable for evaluation of design, performance, and production potential.

provisioning--See **initial provisioning.**

Prowler--A twin turbojet engine, quadruple crew, all-weather, electronic attack aircraft designed to operate from aircraft carriers. It contains a wide assortment of integrated, computer-controlled, active and passive electronic attack equipment. Designated as **EA-6B.**

proword--See **procedure word.**

proximity fuze(*)--A fuze wherein primary initiation occurs by remotely sensing the presence, distance, and/or direction of a target or its associated environment by means of a signal generated by the fuze or emitted by the target, or by detecting a disturbance of a natural field surrounding the target. See also **fuze.**

prudent limit of endurance(*)--The time during which an aircraft can remain airborne and still retain a given safety margin of fuel.

prudent limit of patrol(*)--The time at which an aircraft must depart from its operational area in order to return to its base and arrive there with a given safety margin (usually 20 percent) of fuel reserve for bad weather diversions.

pseudopursuit navigation--A method of homing navigation in which the missile is directed toward the instantaneous target position in azimuth, while pursuit navigation in elevation is delayed until more favorable angle of attack on the target is achieved.

psychological consolidation activities(*)--Planned psychological activities in peace and war directed at the civilian population located in areas under friendly control in order to achieve a desired behavior which supports the military objectives and the operational freedom of the supported commanders.

psychological operations--Planned operations to convey selected information and indicators to foreign audiences to influence their emotions, motives, objective reasoning, and ultimately the behavior of foreign governments, organizations, groups, and individuals. The purpose of psychological operations is to induce or reinforce foreign attitudes and behavior favorable to the originator's objectives. Also called **PSYOP.** See also **perception management.**

psychological warfare--The planned use of propaganda and other psychological actions having the primary purpose of influencing the opinions, emotions, attitudes, and behavior of hostile foreign groups in such a way as to support the achievement of national objectives. Also called **PSYWAR.**

PSYOP--See **psychological operations.**

PSYWAR--See **psychological warfare.**

public--Concept that includes all audiences, both internal and external. See also **external audience; internal audience.**

public affairs--Those public information and community relations activities directed toward the general public by the various elements of the Department of Defense.

public information--Information of a military nature, the dissemination of which through public news media is not inconsistent with security, and the release of which is considered desirable or nonobjectionable to the responsible releasing agency.

pull-up point(*)--The point at which an aircraft must start to climb from a low-level approach in order to gain sufficient height from which to execute the attack or retirement. See also **contact point; turn-in point.**

pulse duration--In radar, measurement of pulse transmission time in microseconds, that is, the time the radar's transmitter is energized during each cycle. Also called **pulse length and pulse width.**

pulsejet(*)--A jet-propulsion engine containing neither compressor nor turbine. Equipped with valves in the front which open and shut, it takes in air to create thrust in rapid periodic bursts rather than continuously.

pulse repetition frequency--In radar, the number of pulses that occur each second. Not to be confused with transmission frequency which is determined by the rate at which cycles are repeated within the transmitted pulse.

pulsing(*)--In naval mine warfare, a method of operating magnetic and acoustic sweeps in which the sweep is energized by current which

varies or is intermittent in accordance with a predetermined schedule.

punch--In air intercept, a code meaning, "You should very soon be obtaining a contact on the aircraft that is being intercepted." (Use only with "air intercept" interceptions.)

purchase--To procure property or services for a price; includes obtaining by barter. See also **collaborative purchase; joint purchase; single department purchase.**

purchase description--A statement outlining the essential characteristics and functions of an item, service, or material required to meet the minimum needs of the Government. It is used when a specification is not available or when specific procurement specifications are not required by the individual Military Departments or the Department of Defense.

purchase notice agreements--Agreements concerning the purchase of brand-name items for resale purposes established by each military Service under the control of the Defense Logistics Agency.

purchasing office--Any installation or activity, or any division, office, branch, section, unit, or other organizational element of an installation or activity charged with the functions of procuring supplies or services.

purple--In air intercept, a code meaning, "The unit indicated is suspected of carrying nuclear weapons" (i.e., "purple VB").

pursuit(*)--An offensive operation designed to catch or cut off a hostile force attempting to escape, with the aim of destroying it.

pyrotechnic(*)--A mixture of chemicals which when ignited is capable of reacting exothermically to produce light, heat, smoke, sound or gas, and may also be used to introduce a delay into an explosive train because of its known burning time. The term excludes propellants and explosives.

pyrotechnic delay(*)--A pyrotechnic device added to a firing system which transmits the ignition flame after a predetermined delay.

Q

q-message(*)--A classified message relating to navigational dangers, navigational aids, mined areas, and searched or swept channels.

Q-route--A system of preplanned shipping lanes in mined or potentially mined waters used to minimize the area the mine countermeasures commander has to keep clear of mines to provide safe passage for friendly shipping.

Q-ship--See decoy ship.

quadrant elevation(*)--The angle between the horizontal plane and the axis of the bore when the weapon is laid. (DOD) It is the algebraic sum of the elevation, angle of site, and complementary angle of site.

qualifying years creditable for retirement pay---The time a Guardsman or reservist must serve to be eligible for retired pay at age 60. Individuals must have at least 20 years of service in which they receive at least 50 retirement points a year, and the last eight years of service must be served in a reserve component.

quay--A structure of solid construction along a shore or bank which provides berthing and which generally provides cargo-handling facilities. A similar facility of open construction is called a wharf. See also pier; wharf.

quicktrans--Long-term contract airlift service within continental United States for the movement of cargo in support of the logistic system for the Military Services (primarily the Navy and Marine Corps) and Department of Defense agencies. See also logair.

R

radar--A radio detection device that provides information on range, azimuth and/or elevation of objects.

radar advisory--The term used to indicate that the provision of advice and information is based on radar observation.

radar altimetry area(*)--A large and comparatively level terrain area with a defined elevation which can be used in determining the altitude of airborne equipment by the use of radar.

radar altitude control mode--In an automatic flight control system, a control mode in which the height of an aircraft is maintained by reference to signals from a radar altimeter.

radar beacon--A receiver-transmitter combination which sends out a coded signal when triggered by the proper type of pulse, enabling determination of range and bearing information by the interrogating station or aircraft.

radar camouflage(*)--The use of radar absorbent or reflecting materials to change the radar echoing properties of a surface of an object.

radar clutter(*)--Unwanted signals, echoes, or images on the face of the display tube, which interfere with observation of desired signals.

radar countermeasures--See electronic warfare; chaff.

radar coverage(*)--The limits within which objects can be detected by one or more radar stations.

radar danning(*)--In naval mine warfare, a method of navigating by using radar to keep the required distance from a line of dan buoys.

radar deception--See electromagnetic deception.

radar exploitation report--A formatted statement of the results of a tactical radar imagery reconnaissance mission. The report includes the interpretation of the sensor imagery. Also called RADAREXREP.

RADAREXREP--See radar exploitation report.

radar fire(*)--Gunfire aimed at a target which is tracked by radar. See also fire.

radar guardship(*)--Any ship which has been assigned the task by the officer in tactical command of maintaining the radar watch.

radar horizon(*)--The locus of points at which the rays from a radar antenna become tangential to the Earth's surface. On the open sea this locus is horizontal but on land it varies according to the topographical features of the terrain.

radar imagery--Imagery produced by recording radar waves reflected from a given target surface.

radar intelligence--Intelligence derived from data collected by radar. Also called RADINT. See also intelligence.

radar netting(*)--The linking of several radars to a single center to provide integrated target information.

radar netting station(*)--A center which can receive data from radar tracking stations and exchange this data among other radar tracking stations, thus forming a radar netting system. See also radar netting unit; radar tracking station.

radar netting unit--Optional electronic equipment which converts the operations central of certain air defense fire distribution systems to a radar netting station. See also **radar netting station.**

radar picket(*)--Any ship, aircraft, or vehicle, stationed at a distance from the force protected, for the purpose of increasing the radar detection range.

radar picket CAP--Radar picket combat air patrol.

radar reconnaissance--Reconnaissance by means of radar to obtain information on enemy activity and to determine the nature of terrain.

radarscope overlay(*)--A transparent overlay for placing on the radarscope for comparison and identification of radar returns.

radarscope photography(*)--A film record of the returns shown by a radar screen.

radar signal film--The film on which is recorded all the reflected signals acquired by a coherent radar, and which must be viewed or processed through an optical correlator to permit interpretation.

radar silence(*)--An imposed discipline prohibiting the transmission by radar of electromagnetic signals on some or all frequencies.

radar spoking(*)--Periodic flashes of the rotating time base on a radial display. Sometimes caused by mutual interference.

radar tracking station--A radar facility which has the capability of tracking moving targets.

radiac(*)--An acronym derived from the words "radioactivity, detection, indication and computation" and used as an all-encompassing term to designate various types of radiological measuring instruments or equipment. (This word is normally used as an adjective.)

radiac dosimeter--An instrument used to measure the ionizing radiation absorbed by that instrument.

radial--A magnetic bearing extending from a very high frequency omni-range/tactical air navigation station.

radial displacement(*)--On vertical photographs, the apparent "leaning out," or the apparent displacement of the top of any object having height in relation to its base. The direction of displacement is radial from the principal point on a true vertical, or from the isocentre on a vertical photograph distorted by tip or tilt.

radiant exposure--See **thermal exposure.**

radiation dose(*)--The total amount of ionizing radiation absorbed by material or tissues, expressed in centigrays. (DOD) The term radiation dose is often used in the sense of the exposure dose expressed in roentgens, which is a measure of the total amount of ionization that the quantity of radiation could produce in air. This could be distinguished from the absorbed dose, also given in rads, which represents the energy absorbed from the radiation per gram of specified body tissue. Further, the biological dose, in rems, is a measure of the biological effectiveness of the radiation exposure. See also **absorbed dose; exposure dose.**

radiation dose rate(*)--The radiation dose (dosage) absorbed per unit of time. (DOD) A radiation dose rate can be set at some particular unit of time (e.g., H + 1 hour) and would be called H + 1 radiation dose rate.

radiation exposure state(*)--The condition of a unit, or exceptionally an individual, deduced from the cumulative whole body radiation dose(s) received. It is expressed as a symbol which indicates the potential for future operations and the degree of risk if exposed to additional nuclear radiation.

radiation intelligence--Intelligence derived from the collection and analysis of non-information-bearing elements extracted from the electromagnetic energy unintentionally emanated by foreign devices, equipments, and systems, excluding those generated by the detonation of atomic/nuclear weapons.

radiation intensity(*)--The radiation dose rate at a given time and place. It may be used, coupled with a figure, to denote the radiation intensity used at a given number of hours after a nuclear burst, e.g., RI-3 is a radiation intensity 3 hours after the time of burst.

radiation scattering(*)--The diversion of radiation (thermal, electromagnetic, or nuclear) from its original path as a result of interaction or collisions with atoms, molecules, or larger particles in the atmosphere or other media between the source of the radiation (e.g., a nuclear explosion) and a point at some distance away. As a result of scattering, radiation (especially gamma rays and neutrons) will be received at such a point from many directions instead of only from the direction of the source.

radiation sickness(*)--An illness resulting from excessive exposure to ionizing radiation. The earliest symptoms are nausea, vomiting, and diarrhea, which may be followed by loss of hair, hemorrhage, inflammation of the mouth and throat, and general loss of energy.

RADINT--See **radar intelligence.**

radioactive decay(*)--The decrease in the radiation intensity of any radioactive material with respect to time.

radioactive decay curve(*)--A graph line representing the decrease of radioactivity with the passage of time.

radioactive decay rate--The time rate of the disintegration of radioactive material generally accompanied by the emission of particles and/or gamma radiation.

radioactivity--The spontaneous emission of radiation, generally alpha or beta particles, often accompanied by gamma rays, from the nuclei of an unstable isotope.

radioactivity concentration guide(*)--The amount of any specified radioisotope that is acceptable in air and water for continuous consumption.

radio and wire integration--The combining of wire circuits with radio facilities.

radio approach aids(*)--Equipment making use of radio to determine the position of an aircraft with considerable accuracy from the time it is in the vicinity of an airfield or carrier until it reaches a position from which landing can be carried out.

radio beacon(*)--A radio transmitter which emits a distinctive, or characteristic, signal used for the determination of bearings, courses, or location. See also **beacon.**

radio countermeasures--See **electronic warfare.**

radio deception--The employment of radio to deceive the enemy. Radio deception includes sending false dispatches, using deceptive headings, employing enemy call signs, etc. See also **electronic warfare.**

radio detection(*)--The detection of the presence of an object by radio-location without precise determination of its position.

radio direction finding(*)--Radio-location in which only the direction of a station is determined by means of its emissions.

radio direction finding data base--The aggregate of information, acquired by both airborne and surface means, necessary to provide support to radio direction finding operations to produce fixes on target transmitters/emitters. The resultant bearings and fixes serve as a basis for tactical decisions concerning military operations, including exercises, planned or underway.

radio fix(*)--1. The locating of a radio transmitter by bearings taken from two or more direction finding stations, the site of the transmitter being at the point of intersection. 2. The location of a ship or aircraft by determining the direction of radio signals coming to the ship or aircraft from two or more sending stations, the locations of which are known.

radio guard--A ship, aircraft, or radio station designated to listen for and record transmissions, and to handle traffic on a designated frequency for a certain unit or units.

radiological defense(*)--Defensive measures taken against the radiation hazards resulting from the employment of nuclear and radiological weapons.

radiological environment(*)--Conditions found in an area resulting from the presence of a radiological hazard.

radiological monitoring--See **monitoring.**

radiological operation(*)--The employment of radioactive materials or radiation producing devices to cause casualties or restrict the use of terrain. It includes the intentional employment of fallout from nuclear weapons.

radiological survey(*)--The directed effort to determine the distribution and dose rates of radiation in an area.

radiological survey flight altitude--The altitude at which an aircraft is flown during an aerial radiological survey.

radio magnetic indicator(*)--An instrument which displays aircraft heading and bearing to selected radio navigation aids.

radio navigation(*)--Radio-location intended for the determination of position or direction or for obstruction warning in navigation.

radio range finding(*)--Radio-location in which the distance of an object is determined by means of its radio emissions, whether independent, reflected, or retransmitted on the same or other wave length.

radio range station(*)--A radio navigation land station in the aeronautical radio navigation service providing radio equi-signal zones. (In certain instances a radio range station may be placed on board a ship.)

radio recognition(*)--The determination by radio means of the friendly or enemy character, or the individuality, of another.

radio recognization and identification--See **Identification, Friend or Foe.**

radio silence(*)--A condition in which all or certain radio equipment capable of radiation is kept inoperative. (DOD) (Note: In combined or United States Joint or intra-Service communications the frequency bands and/or types of equipment affected will be specified.)

radio sonobuoy--See **sonobuoy.**

radio telegraphy(*)--The transmission of telegraphic codes by means of radio.

radio telephony(*)--The transmission of speech by means of modulated radio waves.

radius of action(*)--The maximum distance a ship, aircraft, or vehicle can travel away from its base along a given course with normal combat load and return without refueling, allowing for all safety and operating factors.

radius of damage--The distance from ground zero at which there is a 0.50 probability of achieving the desired damage.

radius of integration--The distance from ground zero which indicates the area within which the effects of both the nuclear detonation and conventional weapons are to be integrated.

radius of safety(*)--The horizontal distance from ground zero beyond which the weapon effects on friendly troops are acceptable.

raid(*)--An operation, usually small scale, involving a swift penetration of hostile territory to secure information, confuse the enemy, or to destroy installations. It ends with a planned withdrawal upon completion of the assigned mission.

raid report(*)--In air defense, one of a series of related reports that are made for the purpose of developing a plot to assist in the rapid evaluation of a tactical situation.

railhead(*)--A point on a railway where loads are transferred between trains and other means of transport. See also **navigation head.**

railway line capacity(*)--The maximum number of trains which can be moved in each direction over a specified section of track in a 24 hour period. See also **route capacity.**

railway loading ramp(*)--A sloping platform situated at the end or beside a track and rising to the level of the floor of the rail cars or wagons.

rainfall (nuclear)--The water that is precipitated from the base surge clouds after an underwater burst of a nuclear weapon. This rain is radioactive and presents an important secondary effect of such a burst.

rainout(*)--Radioactive material in the atmosphere brought down by precipitation.

ramjet(*)--A jet-propulsion engine containing neither compressor nor turbine which depends for its operation on the air compression accomplished by the forward motion of the engine. See also **pulsejet.**

random minelaying(*)--In land mine warfare, the laying of mines without regard to pattern.

range--1. The distance between any given point and an object or target. 2. Extent or distance limiting the operation or action of something, such as the range of an aircraft, ship, or gun. 3. The distance which can be covered over a hard surface by a ground vehicle, with its rated payload, using the fuel in its tank and its cans normally carried as part of the ground vehicle equipment. 4. Area equipped for practice in shooting at targets. In this meaning, also called **target range.**

range marker(*)--A single calibration blip fed onto the time base of a radial display. The rotation of the time base shows the single blips as a circle on the plan position indicator scope. It may be used to measure range.

range markers--Two upright markers which may be lighted at night, placed so that when aligned, the direction indicated assists in piloting. They may be used in amphibious operations to aid in beaching landing ships or craft.

Rangers--Rapidly deployable airborne light infantry organized and trained to conduct highly complex joint direct action operations in coordination with or in support of other special operations units of all Services. Rangers also can execute direct action operations in support of conventional nonspecial operations missions conducted by a combatant commander and can operate as conventional light infantry when properly augmented with other elements of combined arms.

range spread--The technique used to place the mean point of impact of two or more units 100 meters apart on the gun-target line.

ranging(*)--The process of establishing target distance. Types of ranging include echo, intermittent, manual, navigational, explosive echo, optical, radar, etc. See also **spot.**

rated load(*)--The designed safe operating load for the equipment under prescribed conditions.

rate of fire(*)--The number of rounds fired per weapon per minute.

rate of march(*)--The average number of miles or kilometers to be travelled in a given period of time, including all ordered halts. It is expressed in miles or kilometers in the hour. See also **pace.**

ratification--The declaration by which a nation formally accepts with or without reservation the content of a standardization agreement. See also **implementation; reservation; subscription.**

rationalization--Any action that increases the effectiveness of allied forces through more efficient or effective use of defense resources committed to the alliance. Rationalization includes consolidation, reassignment of national priorities to higher alliance needs, standardization, specialization, mutual support or improved interoperability, and greater cooperation. Rationalization applies to both weapons/materiel resources and non-weapons military matters.

ration dense--Foods which, through processing, have been reduced in volume and quantity to a small compact package without appreciable loss of food value, quality, or acceptance, with a high yield in relation to space occupied, such as dehydrates and concentrates.

ratio print--A print the scale of which has been changed from that of the negative by photographic enlargement or reduction.

ratline--An organized effort for moving personnel and/or material by clandestine means across a denied area or border.

R-day--See **times.**

RDD--See **required delivery date.**

reaction time--1. The elapsed time between the initiation of an action and the required response. 2. The time required between the receipt of an order directing an operation and the arrival of the initial element of the force concerned in the designated area.

readiness--See **military capability.**

readiness condition--See **operational readiness.**

readiness planning--Operational planning required for peacetime operations. Its objective is the maintenance of high states of readiness and the deterrence of potential enemies. It includes planning activities that influence day-to-day operations and the peacetime posture of forces. As such, its focus is on general capabilities and readiness rather than the specifics of a particular crisis, either actual or potential. The assignment of geographic responsibilities to combatant commanders, establishment of readiness standards and levels, development of peacetime deployment patterns, coordination of reconnaissance and surveillance assets and capabilities, and planning of joint exercises are examples of readiness planning. No formal joint planning system exists for readiness planning such as exists for contingency and execution planning.

ready(*)--The term used to indicate that a weapon(s) is loaded, aimed, and prepared to fire.

ready CAP--Fighter aircraft in condition of "standby."

ready position(*)--In helicopter operations, a designated place where a helicopter load of troops and/or equipment waits for pick-up.

ready reserve--The Selected Reserve and Individual Ready Reserve liable for active duty as prescribed by law. (10 U.S.C. 268, 672, and 673.)

Ready Reserve Force--A force composed of ships acquired by the Maritime Administration (MARAD) with Navy funding and newer ships acquired by the MARAD for the National Defense Reserve Fleet (NDRF). Although part of the NDRF, ships of the Ready Reserve Force are maintained in a higher state of readiness and can be made available without mobilization or congressionally declared state of emergency. Also called **RRF.** See also **National Defense Reserve Fleet.**

ready-to-load date--The day, relative to C-day, in a time-phased force and deployment data when the unit, nonunit equipment, and forces are prepared to depart their origin on organic transportation or are prepared to begin loading on US Transportation Command-provided transportation. Also called **RLD.**

reallocation authority(*)--The authority given to NATO commanders and normally negotiated in peacetime, to reallocate in an "emergency in war" national logistic resources controlled by

the combat forces under their command, and made available by nations, in order to influence the battle logistically. See also **reallocation of resources.**

real precession(*)--Precession resulting from an applied torque such as friction and dynamic imbalance.

real property--Lands, buildings, structures, utilities systems, improvements, and appurtenances thereto. Includes equipment attached to and made part of buildings and structures (such as heating systems) but not movable equipment (such as plant equipment).

real time(*)--Pertaining to the timeliness of data or information which has been delayed only by the time required for electronic communication. This implies that there are no noticeable delays. See also **near real time.**

real wander--See **real precession.**

rear area(*)--For any particular command, the area extending forward from its rear boundary to the rear of the area of responsibility of the next lower level of command. This area is provided primarily for the performance of combat service support functions. See also **Army service area.**

rear area operations center/rear tactical operations center--A command and control facility that serves as an area/subarea commander's planning, coordinating, monitoring, advising, and directing agency for area security operations.

rear echelon(*)--Elements of a force which are not required in the objective area.

rear guard--Security detachment that protects the rear of a column from hostile forces. During a withdrawal, it delays the enemy by armed resistance, destroying bridges, and blocking roads.

rearming--1. An operation that replenishes the prescribed stores of ammunition, bombs, and other armament items for an aircraft, naval ship, tank, or armored vehicle, including replacement of defective ordnance equipment, in order to make it ready for combat service. 2. Resetting the fuze on a bomb, or on an artillery, mortar, or rocket projectile, so that it will detonate at the desired time.

rebuild--The restoration of an item to a standard as nearly as possible to its original condition in appearance, performance, and life expectancy. See also **overhaul; repair.**

recce--See **reconnaissance.**

RECCEXREP--See **reconnaissance exploitation report.**

receipt(*)--A transmission made by a receiving station to indicate that a message has been satisfactorily received.

receipt into the supply system--That point in time when the first item or first quantity of the item of the contract has been received at or is en route to point of first delivery after inspection and acceptance. See also **procurement lead time.**

receiving ship(*)--The ship in a replenishment unit that receives the rig(s).

reception--1. All ground arrangements connected with the delivery and disposition of air or sea drops. Includes selection and preparation of site, signals for warning and approach, facilitation of secure departure of agents, speedy collection of delivered articles, and their prompt removal to storage places having maximum security. When a group is involved, it may be called a reception committee. 2.

Arrangements to welcome and provide secure quarters or transportation for defectors, escapees, evaders, or incoming agents.

receptivity(*)--The vulnerability of a target audience to particular psychological operations media.

reclama--A request to duly constituted authority to reconsider its decision or its proposed action.

recognition--1. The determination by any means of the individuality of persons, or of objects such as aircraft, ships, or tanks, or of phenomena such as communications-electronics patterns. 2. In ground combat operations, the determination that an object is similar within a category of something already known; e.g., tank, truck, man.

recognition signal--Any prearranged signal by which individuals or units may identify each other.

recoilless rifle (heavy)--A weapon capable of being fired from either a ground mount or from a vehicle, and capable of destroying tanks.

recompression chamber--See **hyperbaric chamber.**

reconnaissance(*)--A mission undertaken to obtain, by visual observation or other detection methods, information about the activities and resources of an enemy or potential enemy, or to secure data concerning the meteorological, hydrographic, or geographic characteristics of a particular area.

reconnaissance by fire(*)--A method of reconnaissance in which fire is placed on a suspected enemy position to cause the enemy to disclose a presence by movement or return of fire.

reconnaissance exploitation report(*)--A standard message format used to report the results of a tactical air reconnaissance mission. Whenever possible the report should include the interpretation of sensor imagery. Also called **RECCEXREP.**

reconnaissance in force(*)--An offensive operation designed to discover and/or test the enemy's strength or to obtain other information.

reconnaissance patrol--See **patrol.**

reconnaissance photography--Photography taken to obtain information on the results of bombing, or on enemy movements, concentrations, activities, and forces. The primary purposes do not include making maps, charts, or mosaics.

reconstitution site--A location selected by the surviving command authority as the site at which a damaged or destroyed headquarters can be reformed from survivors of the attack and/or personnel from other sources, predesignated as **replacements.**

record as target(*)--In artillery and naval gunfire support, the order used to denote that the target is to be recorded for future engagement or reference.

recorded(*)--In artillery and naval gunfire support, the response used to indicate that the action taken to "record as target" has been completed.

record information--All forms (e.g., narrative, graphic, data, computer memory) of information registered in either temporary or permanent form so that it can be retrieved, reproduced, or preserved.

recoverable item--An item which normally is not consumed in use and is subject to return for repair or disposal. See also **reparable item.**

recovery(*)--1. In air operations, that phase of a mission which involves the return of an aircraft to a base. 2. In naval mine warfare, salvage of a mine as nearly intact as possible to permit further investigation for intelligence and/or evaluation purposes. See also **salvage procedure.** 3. (DOD) In amphibious reconnaissance, the physical extraction of landed forces or their link-up with friendly forces.

recovery airfield--Any airfield, military or civil, at which aircraft might land post-H-hour. It is not expected that combat missions would be conducted from a recovery airfield. See also **airfield.**

recovery and reconstitution--Those actions taken by one nation prior to, during, and following an attack by an enemy nation to minimize the effects of the attack, rehabilitate the national economy, provide for the welfare of the populace, and maximize the combat potential of remaining forces and supporting activities.

recovery controller(*)--The air controller responsible for the correct execution of recovering aircraft to the appropriate terminal control agency.

recovery procedures--See **explosive ordnance disposal procedures.**

recovery site--In evasion and escape usage, an area from which an evader or an escaper can be evacuated.

recovery vehicle, medium--A full-tracked vehicle designed for crew rescue and recovery of tanks and other vehicles under battlefield conditions. Designated as **M88A1.**

recovery zone--A designated geographic area from which special operations forces can be extracted by air, boat, or other means. Also called **RZ.**

rectification(*)--In photogrammetry, the process of projecting a tilted or oblique photograph on to a horizontal reference plane.

rectified airspeed--See **calibrated airspeed.**

rectifier(*)--A device for converting alternating current into direct current. See also **inverter.**

recuperation--Not to be used. See **recovery and reconstitution.**

recurring demand--A request by an authorized requisitioner to satisfy a materiel requirement for consumption or stock replenishment that is anticipated to recur periodically. Demands for which the probability of future occurrence is unknown will be considered as recurring. Recurring demands will be considered by the supporting supply system in order to procure, store, and distribute materiel to meet similar demands in the future.

redeployment--The transfer of a unit, an individual, or supplies deployed in one area to another area, or to another location within the area, or to the zone of interior for the purpose of further employment.

redeployment airfield(*)--An airfield not occupied in its entirety in peacetime, but available immediately upon outbreak of war for use and occupation by units redeployed from their peacetime locations. It must have substantially the same standard of operational facilities as a main airfield. See also **airfield; alternative airfield; departure airfield; main airfield.**

redesignated site--A surviving facility that may be redesignated as **the command center to carry on the functions of an incapacitated alternate headquarters and/or facility.**

Redeye--A lightweight manportable, shoulder-fired air defense artillery weapon for low altitude air defense of forward combat area troops. Designated as **FIM-43.**

redistribution--The act of effecting transfer in control, utilization, or location of material between units or activities within or among the Military Services or between the Military Services and other Federal agencies.

reduced charge--1. The smaller of the two propelling charges available for naval guns. 2. Charge employing a reduced amount of propellant to fire a gun at short ranges as compared to a normal charge. See also **normal charge.**

reduced lighting(*)--The reduction in brightness of ground vehicle lights by either reducing power or by screening in such a way that any visible light is limited in output. See also **normal lighting.**

reduced operational status--Applies to the Military Sealift Command ships withdrawn from full operational status (FOS) because of decreased operational requirements. A ship in reduced operational status is crewed in accordance with shipboard maintenance and possible future operational requirements with crew size predetermined contractually. The condition of readiness in terms of calendar days required to attain full operational status is designated by the numeral following the acronym ROS (i.e., ROS-5). Also called **ROS.** See also **Military Sealift Command.**

reduction--The creation of lanes through a minefield or obstacle to allow passage of the attacking ground force.

reduction (photographic)--The production of a negative, diapositive, or print at a scale smaller than the original.

reefer--1. A refrigerator. 2. A motor vehicle, railroad freight car, ship, aircraft, or other conveyance, so constructed and insulated as to protect commodities from either heat or cold.

reentry phase--That portion of the trajectory of a ballistic missile or space vehicle where there is a significant interaction of the vehicle and the Earth's atmosphere. See also **boost phase; midcourse phase; terminal phase.**

reentry vehicle(*)--That part of a space vehicle designed to re-enter the Earth's atmosphere in the terminal portion of its trajectory. See also **maneuverable reentry vehicle; multiple reentry vehicle.**

reference box(*)--The identification box placed in the margin of a map or chart which contains the series designation, sheet number and edition number in a readily identified form. Also called **refer to box.** See also **information box.**

reference datum(*)--As used in the loading of aircraft, an imaginary vertical plane at or near the nose of the aircraft from which all horizontal distances are measured for balance purposes. Diagrams of each aircraft show this reference datum as "balance station zero."

reference diversion point(*)--One of a number of positions selected by the routing authority on both sides of the route of a convoy or independent to facilitate diversion at sea.

reference point(*)--A prominent, easily located point in the terrain.

refer-to box--See **reference box.**

reflected shock wave--When a shock wave traveling in a medium strikes the interface between this medium and a denser medium, part of the energy of the shock wave induces a shock wave in the denser medium and the remainder of the energy results in the formation of a reflected

shock wave that travels back through the less dense medium. See also **shock wave.**

reflection(*)--Energy diverted back from the interface of two media. The reflection may be specular (i.e. direct) or diffuse according to the nature of the contact surfaces.

reflex force--As applied to Air Force units, that part of the alert force maintained overseas or at zone of interior forward bases by scheduled rotations.

reflex sight(*)--An optical or computing sight that reflects a reticle image (or images) onto a combining glass for superimposition on the target.

refuge area(*)--A coastal area considered safe from enemy attack to which merchant ships may be ordered to proceed when the shipping movement policy is implemented. See also **safe anchorage.**

refugee--A civilian who, by reason of real or imagined danger, has left home to seek safety elsewhere. See also **displaced person; evacuee; expellee.**

regimental landing team--A task organization for landing comprised of an infantry regiment reinforced by those elements which are required for initiation of its combat function ashore.

register(*)--In cartography, the correct position of one component of a composite map image in relation to the other components, at each stage of production.

register marks(*)--In cartography, designated marks, such as small crosses, circles, or other patterns applied to original copy prior to reproduction to facilitate registration of plates and to indicate the relative positions of successive impressions.

registration--The adjustment of fire to determine firing data corrections.

registration fire(*)--Fire delivered to obtain accurate data for subsequent effective engagement of targets. See also **fire.**

registration point(*)--Terrain feature or other designated point on which fire is adjusted for the purpose of obtaining corrections to firing data.

regrade--To determine that certain classified information requires, in the interests of national defense, a higher or a lower degree of protection against unauthorized disclosure than currently provided, coupled with a changing of the classification designation to reflect such higher or lower degree.

regroup airfield--Any airfield, military or civil, at which post-H-hour reassembling of aircraft is planned for the express purpose of rearming, recocking, and resumption of armed alert, overseas deployment, or conducting further combat missions. See also **airfield.**

regular drill--See **unit training assembly.**

regulated item(*)--Any item whose issue to a user is subject to control by an appropriate authority for reasons that may include cost, scarcity, technical or hazardous nature, or operational significance. Also called **controlled item.** See also **critical supplies and materiel.**

regulating station--A command agency established to control all movements of personnel and supplies into or out of a given area.

rehabilitation(*)--1. The processing, usually in a relatively quiet area, of units or individuals recently withdrawn from combat or arduous duty, during which units recondition equipment and are rested, furnished special facilities, filled up with replacements, issued replacement supplies and equipment, given training, and generally made ready for employment in future operations. 2. The action performed in restoring an installation to authorized design standards.

reinforcement training unit--See **voluntary training unit.**

reinforcing(*)--In artillery usage, tactical mission in which one artillery unit augments the fire of another artillery unit.

reinforcing obstacles--Those obstacles specifically constructed, emplaced, or detonated through military effort and designed to strengthen existing terrain to disrupt, fix, turn, or block enemy movement. See also **obstacle.**

relateral tell(*)--The relay of information between facilities through the use of a third facility. This type of telling is appropriate between automated facilities in a degraded communications environment. See also **track telling.**

relative altitude--See **vertical separation.**

relative aperture--The ratio of the equivalent focal length to the diameter of the entrance pupil of photographic lens expressed f:4.5, etc. Also called **f-number; stop; aperture stop; or diaphragm stop.**

relative bearing(*)--The direction expressed as a horizontal angle normally measured clockwise from the forward point of the longitudinal axis of a vehicle, aircraft or ship to an object or body. See also **bearing; grid bearing.**

relative biological effectiveness--The ratio of the number of rads of gamma (or X) radiation of a certain energy which will produce a specified biological effect to the number of rads of another radiation required to produce the same effect is the relative biological effectiveness of the latter radiation.

release(*)--In air armament, the intentional separation of a free-fall aircraft store, from its suspension equipment, for purposes of employment of the store.

release altitude--Altitude of an aircraft above the ground at the time of release of bombs, rockets, missiles, tow targets, etc.

release point (road)--A well-defined point on a route at which the elements composing a column return under the authority of their respective commanders, each one of these elements continuing its movement towards its own appropriate destination.

releasing commander (nuclear weapons)--A commander who has been delegated authority to approve the use of nuclear weapons within prescribed limits. See also **commander(s); executing commander (nuclear weapons).**

releasing officer--A properly designated individual who may authorize the sending of a message for and in the name of the originator. See also **originator.**

reliability diagram(*)--In cartography, a diagram showing the dates and quality of the source material from which a map or chart has been compiled. See also **information box.**

reliability of source--See **evaluation.**

relief(*)--Inequalities of evaluation and the configuration of land features on the surface of the Earth which may be represented on maps or charts by contours, hypsometric tints, shading, or spot elevations.

relief in place(*)--An operation in which, by direction of higher authority, all or part of a unit is replaced in an area by the incoming unit. The responsibilities of the replaced elements for the mission and the assigned zone of operations are transferred to the incoming unit. The incoming unit continues the operation as ordered.

religious ministry support--The entire spectrum of professional duties to include providing for or facilitating essential religious needs and practices, pastoral care, family support programs, religious education, volunteer and community activities, and programs performed to enhance morale and moral, ethical, and personal well being. Enlisted religious support personnel assist the chaplain in providing religious ministry support. See also **command chaplain; command chaplain of the combatant command; lay leader or lay reader; religious ministry support plan; religious ministry support team; Service component command chaplain.**

religious ministry support plan--A plan that describes the way in which religious support personnel will provide religious support to all members of a joint force. When approved by the commander, it may be included as an annex to operation plans. See also **command chaplain; command chaplain of the combatant command; lay leader or lay reader; religious ministry support; religious ministry support team; Service component command chaplain.**

religious ministry support team--A team that is composed of a chaplain and an Army Chaplain Assistant or Navy Religious Program Specialist or Air Force Chaplain Service Support Personnel or Coast Guard yeoman. The team works together in designing, implementing, and executing the command religious program. See also **command chaplain; command chaplain of the combatant command; lay leader or lay reader; religious ministry support; religious ministry support plan; Service component command chaplain.**

remain-behind equipment--Unit equipment left by deploying forces at their bases when they deploy.

remaining forces--The total surviving United States forces at any given stage of combat operations.

remote delivery(*)--In mine warfare, the delivery of mines to a target area by any means other than direct emplacement. The exact position of mines so laid may not be known.

remotely piloted vehicle(*)--An unmanned vehicle capable of being controlled from a distant location through a communication link. It is normally designed to be recoverable. See also **drone.**

render safe procedures--See **explosive ordnance disposal procedures.**

rendezvous(*)--1. A pre-arranged meeting at a given time and location from which to begin an action or phase of an operation, or to which to return after an operation. See also **join-up.** 2. In land warfare, an easily found terrain location at which visitors to units, headquarters or facilities are met by personnel from the element to be visited. See also **contact point.**

rendezvous area--In an amphibious operation, the area in which the landing craft and amphibious vehicles rendezvous to form waves after being loaded, and prior to movement to the line of departure.

reorder cycle--The interval between successive reorder (procurement) actions.

reorder point--1. That point at which time a stock replenishment requisition would be submitted to maintain the predetermined or calculated stockage objective. 2. The sum of the safety level of supply plus the level for order

and shipping time equals the reorder point. See also **level of supply.**

repair--The restoration of an item to serviceable condition through correction of a specific failure or unserviceable condition. See also **overhaul; rebuild.**

repair cycle--The stages through which a reparable item passes from the time of its removal or replacement until it is reinstalled or placed in stock in a serviceable condition.

repair cycle aircraft--Aircraft in the active inventory that are in or awaiting depot maintenance, including those in transit to or from depot maintenance.

reparable item--An item that can be reconditioned or economically repaired for reuse when it becomes unserviceable. See also **recoverable item.**

repatriate--A person who returns to his or her country or citizenship, having left his or her native country, either against his or her will, or as one of a group who left for reason of politics, religion, or other pertinent reasons.

repeat(*)--In artillery and naval gunfire support, an order or request to fire again the same number of rounds with the same method of fire.

repeater-jammer(*)--A receiver transmitter device which amplifies, multiplies and retransmits the signals received, for purposes of deception or jamming.

replacement demand--A demand representing replacement of items consumed or worn out.

replacement factor(*)--The estimated percentage of equipment or repair parts in use that will require replacement during a given period due to wearing out beyond repair, enemy action, abandonment, pilferage, and other causes except catastrophes.

replacements--Personnel required to take the place of others who depart a unit.

replenishment at sea(*)--Those operations required to make a transfer of personnel and/or supplies when at sea.

reply(*)--An answer to a challenge. See also **challenge; countersign; password.**

reported unit--A unit designation that has been mentioned in an agent report, captured document, or interrogation report, but for which available information is insufficient to include the unit in accepted order of battle holdings.

reporting post(*)--An element of the control and reporting system used to extend the radar coverage of the control and reporting center. It does not undertake the control of aircraft.

reporting time interval--1. In surveillance, the time interval between the detection of an event and the receipt of a report by the user. 2. In communications, the time for transmission of data or a report from the originating terminal to the end receiver. See also **near real time.**

representative downwind direction(*)--During the forecast period, the mean surface downwind direction in the hazard area towards which the cloud travels.

representative downwind speed(*)--The mean surface downwind speed in the hazard area during the forecast period.

representative fraction--The scale of a map, chart, or photograph expressed as a fraction or ratio. See also **scale.**

request modify(*)--In artillery and naval gunfire support, a request by any person, other than

the person authorized to make modifications to a fire plan, for a modification.

required delivery date--A date, relative to C-day, when a unit must arrive at its destination and complete offloading to properly support the concept of operations. Also called **RDD.**

required supply rate (ammunition)--In Army usage, the amount of ammunition expressed in terms of rounds per weapon per day for ammunition items fired by weapons, and in terms of other units of measure per day for bulk allotment and other items, estimated to be required to sustain operations of any designated force without restriction for a specified period. Tactical commanders use this rate to state their requirements for ammunition to support planned tactical operations at specified intervals. The required supply rate is submitted through command channels. It is consolidated at each echelon and is considered by each commander in subsequently determining the controlled supply rate within the command.

requirements--See **military requirement.**

requirements capability--This capability provides a Joint Operation Planning and Execution System user the ability to identify, update, review, and delete data on forces and sustainment required to support an operation plan or course of action.

requisition(*)--1. An authoritative demand or request especially for personnel, supplies, or services authorized but not made available without specific request. (DOD) 2. To demand or require services from an invaded or conquered nation.

requisitioning objective--The maximum quantities of materiel to be maintained on hand and on order to sustain current operations. It will consist of the sum of stocks represented by the operating level, safety level, and the order and shipping time or procurement lead time, as appropriate. See also **level of supply.**

rescue combat air patrol--An aircraft patrol provided over a combat search and rescue objective area for the purpose of intercepting and destroying hostile aircraft. Its primary mission is to protect the search and rescue task forces during recovery operations. See also **combat air patrol.**

rescue coordination center--A primary search and rescue facility suitably staffed by supervisory personnel and equipped for coordinating and controlling search and rescue operations. The facility may be operated unilaterally by personnel of a single Service (**rescue coordination center**), jointly by personnel of two or more Services (**joint rescue coordination center**), or it may have a combined staff of personnel from two or more allied nations (**combined rescue coordination center**). Formerly called **Search and Rescue Coordination Center**. Also called **RCC or JRCC.**

rescue ship(*)--In shipping control, a ship of a convoy stationed at the rear of a convoy column to rescue survivors.

research--All effort directed toward increased knowledge of natural phenomena and environment and toward the solution of problems in all fields of science. This includes basic and applied research.

reseau(*)--A grid system of a standard size in the image plane of a photographic system used for mensuration purposes.

reservation--The stated qualification by a nation that describes the part of a standardization agreement that it will not implement or will implement only with limitations. See also **implementation; ratification; subscription.**

reserve--1. Portion of a body of troops which is kept to the rear, or withheld from action at the beginning of an engagement, available for a decisive movement. 2. Members of the Military Services who are not in active service but who are subject to call to active duty. 3. Portion of an appropriation or contract authorization held or set aside for future operations or contingencies and in respect to which administrative authorization to incur commitments or obligations has been withheld. See also **general reserve; operational reserve; reserve supplies.**

reserve aircraft--Those aircraft which have been accumulated in excess of immediate needs for active aircraft and are retained in the inventory against possible future needs. See also **aircraft.**

reserve component category--The category that identifies an individual's status in a reserve component. The three reserve component categories are Ready Reserve, Standby Reserve, and Retired Reserve. Each reservist is identified by a specific reserve component category designation.

Reserve Components--Reserve Components of the Armed Forces of the United States are: a. the Army National Guard of the United States; b. the Army Reserve; c. the Naval Reserve; d. the Marine Corps Reserve; e. the Air National Guard of the United States; f. the Air Force Reserve; and g. the Coast Guard Reserve.

reserved demolition target(*)--A target for demolition, the destruction of which must be controlled at a specific level of command because it plays a vital part in the tactical or strategic plan, or because of the importance of the structure itself, or because the demolition may be executed in the face of the enemy. See also **demolition target.**

reserved obstacles--Those demolition obstacles that are deemed critical to the plan for which the authority to detonate is reserved by the designating commander. See also **obstacle.**

reserved route(*)--In road traffic, a specific route allocated exclusively to an authority or formation. See also **route.**

reserve supplies--Supplies accumulated in excess of immediate needs for the purpose of ensuring continuity of an adequate supply. Also called **reserves.** See also **battle reserves; beach reserves; contingency retention stock; economic retention stock; individual reserves; initial reserves; unit reserves.**

residual contamination(*)--Contamination which remains after steps have been taken to remove it. These steps may consist of nothing more than allowing the contamination to decay normally.

residual forces--Unexpended portions of the remaining United States forces that have an immediate combat potential for continued military operations, and that have been deliberately withheld from utilization.

residual radiation(*)--Nuclear radiation caused by fallout, artificial dispersion of radioactive material, or irradiation which results from a nuclear explosion and persists longer than one minute after burst. See also **contamination; induced radiation; initial radiation.**

residual radioactivity--Nuclear radiation that results from radioactive sources and which persists for longer than one minute. Sources of residual radioactivity created by nuclear explosions include fission fragments and radioactive matter created primarily by neutron activation, but also by gamma and other radiation activation. Other possible sources of residual radioactivity include radioactive material created and dispersed by means other than nuclear explo-

sion. See also **contamination; induced radiation; initial radiation.**

resistance movement--An organized effort by some portion of the civil population of a country to resist the legally established government or an occupying power and to disrupt civil order and stability.

resolution(*)--A measurement of the smallest detail which can be distinguished by a sensor system under specific conditions.

resource and unit monitoring--Worldwide Military Command and Control System application systems that support approved requirements relating to resource and unit monitoring, readiness assessment, situation assessment, and operations by integrating data from functional areas such as operations, logistics, personnel, and medical.

resources--The forces, materiel, and other assets or capabilities apportioned or allocated to the commander of a unified or specified command.

response force--A mobile force with appropriate fire support designated, usually by the area commander, to deal with Level II threats in the rear area.

responsibility--1. The obligation to carry forward an assigned task to a successful conclusion. With responsibility goes authority to direct and take the necessary action to ensure success. 2. The obligation for the proper custody, care, and safekeeping of property or funds entrusted to the possession or supervision of an individual. See also **accountability.**

responsor(*)--An electronic device used to receive an electronic challenge and display a reply thereto.

rest(*)--In artillery, a command that indicates that the unit(s) or gun(s) to which it is addressed shall not follow up fire orders during the time that the order is in force.

rest and recuperation--The withdrawal of individuals from combat or duty in a combat area for short periods of rest and recuperation. Also called **R&R.** See also **rehabilitation.**

restart at . . .(*)--In artillery, a term used to restart a fire plan after "dwell at . . ." or "check firing" or "cease loading" has been ordered.

restitution(*)--The process of determining the true planimetric position of objects whose images appear on photographs.

restitution factor--See **correlation factor.**

restraint factor(*)--In air transport, a factor, normally expressed in multiples of the force of gravity, which determines the required strength of lashings and tie-downs to secure a particular load.

restraint of loads--The process of binding, lashing, and wedging items into one unit or into its transporter in a manner that will ensure immobility during transit.

restricted air cargo--See **cargo.**

restricted area--1. An area (land, sea, or air) in which there are special restrictive measures employed to prevent or minimize interference between friendly forces. 2. An area under military jurisdiction in which special security measures are employed to prevent unauthorized entry. See also **air surface zones; controlled firing area; restricted areas (air).**

restricted areas (air)--Designated areas established by appropriate authority over which flight of aircraft is restricted. They are shown on aeronautical charts and published in notices to airmen, and publications of aids to air navigation. See also **restricted area.**

restricted dangerous air cargo(*)--Cargo which does not belong to the highly dangerous category but which is hazardous and requires, for transport by cargo or passenger aircraft, extra precautions in packing and handling.

restricted data--All data (information) concerning: a. design, manufacture, or use of atomic weapons; b. the production of special nuclear material; or c. the use of special nuclear material in the production of energy, but shall not include data declassified or removed from the restricted data category pursuant to Section 142 of the Atomic Energy Act. (Section 11w, Atomic Energy Act of 1954, as amended.) See also **formerly restricted data.**

restricted operations area(*)--Airspace of defined dimensions, designated by the airspace control authority, in response to specific operational situations/requirements within which the operation of one or more airspace users is restricted.

restrictive fire plan(*)--A safety measure for friendly aircraft which establishes airspace that is reasonably safe from friendly surface delivered non-nuclear fires.

resume--In air intercept usage a code meaning, "Resume last patrol ordered."

resupply(*)--The act of replenishing stocks in order to maintain required levels of supply.

retain--When used in the context of deliberate planning, the directed command will keep the referenced operation plan, operation plan in concept format, or concept summary and any associated Joint Operation Planning System or Joint Operation Planning and Execution System automated data processing files in an inactive library or status. The plan and its associated files will not be maintained unless directed by follow-on guidance. See also **archive; maintain.**

retard--A request from a spotter to indicate that the illuminating projectile burst is desired later in relation to the subsequent bursts of high explosive projectiles.

reticle(*)--A mark such as a cross or a system of lines lying in the image plane of a viewing apparatus. It may be used singly as a reference mark on certain types of monocular instruments or as one of a pair to form a floating mark as in certain types of stereoscopes. See also **graticule.**

retirement(*)--An operation in which a force out of contact moves away from the enemy.

retirement route--The track or series of tracks along which helicopters move from a specific landing site or landing zone. See also **approach route; helicopter lane.**

retrofit action--Action taken to modify inservice equipment.

retrograde cargo--Cargo evacuated from a theater of operations.

retrograde movement--Any movement of a command to the rear, or away from the enemy. It may be forced by the enemy or may be made voluntarily. Such movements may be classified as withdrawal, retirement, or delaying action.

retrograde operation--See **retrograde movement.**

retrograde personnel--Personnel evacuated from a theater of operations who may include medical patients, noncombatants, and civilians.

returned to military control--The status of a person whose casualty status of duty status - whereabouts unknown or missing has been changed due to the person's return or recovery

by US military authority. Also called **RMC.** See also **casualty status; duty status - whereabouts unknown; missing.**

return load(*)--Personnel and/or cargo to be transported by a returning carrier.

return to base--Proceed to the point indicated by the displayed information. This point is being used to return the aircraft to a place at which the aircraft can land. Command heading, speed, and altitude may be used, if desired. Also called **RTB.**

revolutionary--An individual attempting to effect a social or political change through the use of extreme measures. See also **antiterrorism.**

revolving fund--A fund established to finance a cycle of operations to which reimbursements and collections are returned for reuse in a manner such as will maintain the principal of the fund, e.g., working capital funds, industrial funds, and loan funds.

RF-4--See **Phantom II.**

RGM-66D--See **Standard SSM (ARM).**

RGM-84--See **Harpoon.**

RH-53--See **Sea Stallion.**

right (left) bank--See **left (right) bank.**

right (or left)--See **left (or right).**

RIM-66--See **Standard Missile.**

RIM-67--See **Standard Missile.**

RINT--See **unintentional radiation intelligence.**

riot control agent--A chemical that produces temporary irritating or disabling effects when in contact with the eyes or when inhaled.

riot control operations--The employment of riot control agents and/or special tactics, formations and equipment in the control of violent disorders.

rising mine(*)--In naval mine warfare, a mine having positive buoyancy which is released from a sinker by a ship influence or by a timing device. The mine may fire by contact, hydrostatic pressure or other means.

risk--See **degree of risk (nuclear).**

riverine area--An inland or coastal area comprising both land and water, characterized by limited land lines of communication, with extensive water surface and/or inland waterways that provide natural routes for surface transportation and communications.

riverine operations--Operations conducted by forces organized to cope with and exploit the unique characteristics of a riverine area, to locate and destroy hostile forces, and/or to achieve, or maintain control of the riverine area. Joint riverine operations combine land, naval, and air operations, as appropriate, and are suited to the nature of the specific riverine area in which operations are to be conducted.

RLD--See **ready-to-load date.**

RMC--See **returned to military control.**

road block(*)--A barrier or obstacle (usually covered by fire) used to block, or limit the movement of, hostile vehicles along a route.

road capacity--The maximum traffic flow obtainable on a given roadway, using all available lanes, usually expressed in vehicles per hour or vehicles per day.

road clearance time(*)--The total time a column requires to travel over and clear a section of the road.

road hazard sign(*)--A sign used to indicate traffic hazards. Military hazard signs should be used in a communications zone area only in accordance with existing agreements with national authorities.

road net--The system of roads available within a particular locality or area.

road space(*)--The length of roadway allocated to, and/or actually occupied by, a column on a route, expressed in miles or kilometers.

rocket propulsion--Reaction propulsion wherein both the fuel and the oxidizer, generating the hot gases expended through a nozzle, are carried as part of the rocket engine. Specifically, rocket propulsion differs from jet propulsion in that jet propulsion utilizes atmospheric air as an oxidizer whereas rocket propulsion utilizes nitric acid or a similar compound as an oxidizer. See also **jet propulsion.**

ROE--See **rules of engagement.**

roentgen(*)--A unit of exposure dose of gamma (or X-) radiation. In field dosimetry, one roentgen is essentially equal to one rad.

roentgen equivalent mammal--One roentgen equivalent mammal is the quantity of ionizing radiation of any type which, when absorbed by man or other mammal, produces a physiological effect equivalent to that produced by the absorption of 1 roentgen of X-ray or gamma radiation. Also called **REM.**

Roland--See **US Roland.**

role number(*)--In the medical field, the classification of treatment facilities according to their different capabilities.

roll back--The process of progressive destruction and/or neutralization of the opposing defenses, starting at the periphery and working inward, to permit deeper penetration of succeeding defense positions.

roll-in-point--The point at which aircraft enter the final leg of the attack, e.g., dive, glide.

roll-up--The process for orderly dismantling of facilities no longer required in support of operations and available for transfer to other areas.

romper(*)--A ship which has moved more than 10 nautical miles ahead of its convoy, and is unable to rejoin it. See also **straggler.**

rope(*)--An element of chaff consisting of a long roll of metallic foil or wire which is designed for broad, low-frequency responses. See also **chaff.**

ROS--See **reduced operational status.**

rotor governing mode(*)--A control mode in which helicopter rotor speed is maintained automatically.

roundout--See **flare.**

rounds complete(*)--In artillery and naval gunfire support, the term used to report that the number of rounds specified in fire for effect have been fired. See also **shot.**

route(*)--The prescribed course to be traveled from a specific point of origin to a specific destination. See also **axial route; controlled route; dispatch route; lateral route; reserved route; signed route; supervised route.**

route capacity(*)--1. The maximum traffic flow of vehicles in one direction at the most restricted point on the route. 2. The maximum number of metric tons which can be moved in one direction over a particular route in one hour. It is the product of the maximum traffic flow and the average payload of the vehicles

using the route. See also **railway line capacity.**

route classification(*)--Classification assigned to a route using factors of minimum width, worst route type, least bridge, raft or culvert military load classification, and obstructions to traffic flow. See also **military load classification.**

route lanes(*)--A series of parallel tracks for the routing of independently sailed ships.

routine message--A category of precedence to be used for all types of messages that justify transmission by rapid means unless of sufficient urgency to require a higher precedence. See also **precedence.**

routing indicator--A group of letters assigned to indicate: a. the geographic location of a station; b. a fixed headquarters of a command, activity, or unit at a geographic location; and c. the general location of a tape relay or tributary station to facilitate the routing of traffic over the tape relay networks.

row marker(*)--In land mine warfare, a natural, artificial, or specially installed marker, located at the start and finish of a mine row where mines are laid by individual rows.

RRF--See **Ready Reserve Force.**

RTB--See **return to base.**

rules of engagement--Directives issued by competent military authority which delineate the circumstances and limitations under which United States forces will initiate and/or continue combat engagement with other forces encountered. Also called **ROE.** See also **law of war.**

run--1. That part of a flight of one photographic reconnaissance aircraft during which photographs are taken. 2. The transit of a sweeper-

-sweep combination or of a mine-hunter operating its equipment through a lap. This term may also be applied to a transit of any formation of sweepers.

running fix(*)--The intersection of two or more position lines, not obtained simultaneously, adjusted to a common time.

runway(*)--A defined rectangular area of an airfield, prepared for the landing and takeoff run of aircraft along its length.

runway visual range(*)--The maximum distance in the direction of takeoff or landing at which the runway, or specified lights or markers delineating it, can be seen from a position above a specified point on its center line at a height corresponding to the average eye level of pilots at touch-down.

rupture zone(*)--The region immediately adjacent to the crater boundary in which the stresses produced by the explosion have exceeded the ultimate strength of the medium. It is characterized by the appearance of numerous radial cracks of various sizes. See also **plastic zone.**

RUR-5A--See **antisubmarine rocket.**

RZ--See **recovery zone.**

S

S-3--See **Viking.**

SAAFR--See **standard use Army aircraft flight route.**

sabot(*)--Lightweight carrier in which a subcaliber projectile is centered to permit firing the projectile in the larger caliber weapon. The carrier fills the bore of the weapon from which the projectile is fired; it is normally discarded a short distance from the muzzle.

sabotage--An act or acts with intent to injure, interfere with, or obstruct the national defense of a country by willfully injuring or destroying, or attempting to injure or destroy, any national defense or war material, premises or utilities, to include human and natural resources.

sabotage alert team--See **security alert team.**

saboteur--One who commits sabotage. See also **antiterrorism; countersabotage; sabotage.**

safe anchorage(*)--An anchorage considered safe from enemy attack to which merchant ships may be ordered to proceed when the shipping movement policy is implemented. See also **refuge area.**

safe area--A designated area in hostile territory that offers the evader or escapee a reasonable chance of avoiding capture and of surviving until he can be evacuated.

safe burst height(*)--The height of burst at or above which the level of fallout, or damage to ground installations is at a predetermined level acceptable to the military commander. See also **types of burst.**

safe current(*)--In naval mine warfare, the maximum current that can be supplied to a sweep in a given waveform and pulse cycle which does not produce a danger area with respect to the mines being swept for.

safe depth(*)--In naval mine warfare, the shallowest depth of water in which a ship will not actuate a bottom mine of the type under consideration. Safe depth is usually quoted for conditions of ship upright, calm sea and a given speed.

safe distance(*)--In naval mine warfare, the horizontal range from the edge of the explosion damage area to the center of the sweeper.

Safeguard--A ballistic missile defense system.

safe haven--1. Designated area(s) to which noncombatants of the United States Government's responsibility, and commercial vehicles and materiel, may be evacuated during a domestic or other valid emergency. 2. Temporary storage provided Department of Energy classified shipment transporters at Department of Defense facilities in order to assure safety and security of nuclear material and/or nonnuclear classified material. Also includes parking for commercial vehicles containing Class A or Class B explosives.

safe house--An innocent-appearing house or premises established by an organization for the purpose of conducting clandestine or covert activity in relative security.

safe separation distance(*)--The minimum distance between the delivery system and the weapon beyond which the hazards associated with functioning (detonation) are acceptable.

safe speed(*)--In naval mine warfare, the speed at which a particular ship can proceed without actuating a given influence mine, at the depth under consideration, within the damage area.

safety and arming mechanism(*)--A dual function device which prevents the unintended activation of a main charge or propulsion unit prior to arming but allows activation thereafter upon receipt of the appropriate stimuli.

safety angle--See **angle of safety.**

safety device(*)--A device which prevents unintentional functioning.

safety distance(*)--In road transport, the distance between vehicles traveling in column specified by the command in light of safety requirements.

safety fuze(*)--A pyrotechnic contained in a flexible and weather-proof sheath burning at a timed and constant rate, used to transmit a flame to the detonator.

safety height--See **altitude; minimum safe altitude.**

safety lane(*)--Specified sea lane designated for use in transit by submarine and surface ships to prevent attack by friendly forces.

safety level of supply--The quantity of materiel, in addition to the operating level of supply, required to be on hand to permit continuous operations in the event of minor interruption of normal replenishment or unpredictable fluctuations in demand. See also **level of supply.**

safety line(*)--In land mine warfare, demarcation line for trip wire or wire-actuated mines in a minefield. It serves to protect the laying personnel. After the minefield is laid this line is neither marked on the ground nor plotted on the minefield record.

safety pin--See **arming wire.**

safety wire(*)--A cable, wire, or lanyard attached to the aircraft and routed to an expendable aircraft store to prevent arming initiation prior to store release. See also **arming wire.**

safety zone(*)--An area (land, sea, or air) reserved for noncombat operations of friendly aircraft, surface ships, submarines or ground forces. (Note: DOD does not use the word "submarines".)

safe working load(*)--In sea operations, the maximum load that can be safely applied to a fitting, and normally shown on a label plate adjacent to the fitting. See also **static test load.**

safing--As applied to weapons and ammunition, the changing from a state of readiness for initiation to a safe condition.

safing and arming mechanism(*)--A mechanism whose primary purpose is to prevent an unintended functioning of the main charge of the ammunition prior to completion of the arming delay and, in turn, allow the explosive train of the ammunition to function after arming.

Saint--A satellite inspector system designed to demonstrate the feasibility of intercepting, inspecting, and reporting on the characteristics of satellites in orbit.

salted weapon(*)--A nuclear weapon which has, in addition to its normal components, certain elements or isotopes which capture neutrons at the time of the explosion and produce radioactive products over and above the usual radioactive weapon debris. See also **minimum residual radioactivity weapon.**

salvage--1. Property that has some value in excess of its basic material content but which is in such condition that it has no reasonable prospect of use for any purpose as a unit and its repair or rehabilitation for use as a unit is clearly impractical. 2. The saving or rescuing of condemned, discarded, or abandoned property, and of materials contained therein for reuse, refabrication, or scrapping.

salvage group--In an amphibious operation, a naval task organization designated and equipped to rescue personnel and to salvage equipment and material.

salvage operation--1. The recovery, evacuation, and reclamation of damaged, discarded, condemned, or abandoned allied or enemy materiel, ships, craft, and floating equipment for reuse, repair, refabrication, or scrapping. 2. Naval salvage operations include harbor and channel clearance, diving, hazardous towing and rescue tug services and the recovery of materiel, ships, craft, and floating equipment sunk offshore or elsewhere stranded.

salvo--1. In naval gunfire support, a method of fire in which a number of weapons are fired at the same target simultaneously. 2. In close air support/air interdiction operations, a method of delivery in which the release mechanisms are operated to release or fire all ordnance of a specific type simultaneously.

Sam-D--An Army air defense artillery, surface-to-air missile system under development to replace Nike Hercules and the improved Hawk systems.

sanctuary--A nation or area near or contiguous to the combat area which by tacit agreement between the warring powers is exempt from attack and therefore serves as a refuge for staging, logistic, or other activities of the combatant powers.

sanitize--Revise a report or other document in such a fashion as to prevent identification of sources, or of the actual persons and places with which it is concerned, or of the means by which it was acquired. Usually involves dele-

tion or substitution of names and other key details.

SAP--See **special access program.**

SAR--See **search and rescue.**

satellite and missile surveillance--The systematic observation of aerospace for the purpose of detecting, tracking, and characterizing objects, events, and phenomena associated with satellites and inflight missiles, friendly and enemy. See also **surveillance.**

saunter--In air intercept, a term meaning, "Fly at best endurance."

S-bend distortion--See **S-curve distortion.**

SBR--See **special boat squadron.**

SBU--See **special boat unit.**

scale(*)--The ratio or fraction between the distance on a map, chart or photograph and the corresponding distance on the surface of the Earth. See also **conversion scale; graphic scale; photographic scale; principal scale.**

scale (photographic)--See **photographic scale.**

scaling law(*)--A mathematical relationship which permits the effects of a nuclear explosion of given energy yield to be determined as a function of distance from the explosion (or from ground zero) provided the corresponding effect is known as a function of distance for a reference explosion, e.g., of 1-kiloton energy yield.

scan--1. In air intercept, a term meaning: "Search sector indicated and report any contacts." 2. The path periodically followed by a radiation beam. 3. In electronics intelligence, the motion of an electronic beam through space looking for a target. Scanning is produced by the motion of the antenna or by lobe switching. See also **electronics intelligence.**

scan line(*)--The line produced on a recording medium frame by a single sweep of a scanner.

scan period--The period taken by a radar, sonar, etc., to complete a scan pattern and return to a starting point.

scan rate(*)--The rate at which individual scans are recorded.

scan type--The path made in space by a point on the radar beam; for example, circular, helical, conical, spiral, or sector.

scatterable mine(*)--In land mine warfare, a mine laid without regard to classical pattern and which is designed to be delivered by aircraft, artillery, missile, ground dispenser, or by hand. Once laid, it normally has a limited life. See also **mine.**

scene of action commander(*)--In antisubmarine warfare, the commander at the scene of contact. He is usually in a ship, or may be in a fixed wing aircraft, helicopter, or submarine.

scheduled arrival date--The projected arrival date of a specified movement requirement at a specified location.

scheduled fire(*)--A type of prearranged fire executed at a predetermined time.

scheduled maintenance--Periodic prescribed inspection and/or servicing of equipment accomplished on a calendar, mileage, or hours of operation basis. See also **organizational maintenance.**

scheduled service (air transport)--A routine air transport service operated in accordance with a timetable.

scheduled speed(*)--The planned sustained speed of a convoy through the water which determines the speed classification of that convoy. See also **convoy speed; critical speed; declared speed.**

scheduled target(*)--In artillery and naval gunfire support, a planned target on which fire is to be delivered at a specific time.

scheduled target (nuclear)--A planned target on which a nuclear weapon is to be delivered at a specific time during the operation of the supported force. The time is specified in terms of minutes before or after a designated time or in terms of the accomplishment of a predetermined movement or task. Coordination and warning of friendly troops and aircraft are mandatory.

scheduled wave--See **wave.**

schedule of fire--Groups of fires or series of fires fired in a definite sequence according to a definite program. The time of starting the schedule may be ON CALL. For identification purposes schedules may be referred to by a code name or other designation.

schedule of targets(*)--In artillery and naval gunfire support, individual targets, groups or series of targets to be fired on, in a definite sequence according to a definite program.

schedules--The carrier itinerary which may involve cargo and passengers.

scheduling and movement capability--The capability required by Joint Operation Planning and Execution System planners and operators to allow for review and update of scheduling and movement data before and during implementation of a deployment operation.

scheme of maneuver--The tactical plan to be executed by a force in order to seize assigned objectives.

SCI--See **sensitive compartmented information.**

scientific and technical intelligence--The product resulting from the collection, evaluation, analysis, and interpretation of foreign scientific and technical information which covers: a. foreign developments in basic and applied research and in applied engineering techniques; and b. scientific and technical characteristics, capabilities, and limitations of all foreign military systems, weapons, weapon systems, and materiel, the research and development related thereto, and the production methods employed for their manufacture.

scientific intelligence--See **scientific and technical intelligence.**

scram--In air intercept usage, a code meaning, "Am about to open fire. Friendly units keep clear or get clear of indicated contact, bogey or area." Direction of withdrawal may be indicated. Type of fire may be indicated (e.g., scram proximity: "Am about to open fire with proximity-fuzed ammunition" scram mushroom: "Am about to fire a special weapon.").

scramble(*)--An order directing takeoff of aircraft as quickly as possible, usually followed by mission instructions.

scram mushroom--See **scram.**

scram proximity--See **scram.**

screen(*)--1. An arrangement of ships, aircraft and/or submarines to protect a main body or convoy. 2. In cartography, a sheet of transparent film, glass or plastic carrying a "ruling" or other regularly repeated pattern which may be used in conjunction with a mask, either photographically or photomechanically, to produce areas of the pattern. See also **halftone screen.** 3. In surveillance, camouflage and concealment, any natural or artificial material, opaque to surveillance sensor(s), interposed between the sensor(s) and the object to be camouflaged or concealed. See also **conceal-ment.** 4. A security element whose primary task is to observe, identify and report information, and which only fights in self-protection. See also **flankguard; guard.**

screening group--In amphibious operations, a task organization of ships that furnishes protection to the task force en route to the objective area and during operations in the objective area.

scribing(*)--In cartography, a method of preparing a map or chart by cutting the lines into a prepared coating.

S-curve distortion(*)--The distortion in the image produced by a scanning sensor which results from the forward displacement of the sensor during the time of lateral scan.

S-Day--See **times.**

sea-air-land team--A naval force specially organized, trained, and equipped to conduct special operations in maritime, littoral, and riverine environments. Also called **SEAL team.**

Sea Cobra--A single-rotor, dual-crew, light attack helicopter armed with a variety of machine guns, rockets, grenade launchers, and anti-tank missiles. It is used for attack helicopter support. Designated as **AH-1J.**

sea control operations--The employment of naval forces, supported by land and air forces, as appropriate, to achieve military objectives in vital sea areas. Such operations include destruction of enemy naval forces, suppression of enemy sea commerce, protection of vital sea lanes, and establishment of local military superiority in areas of naval operations. See also **land control operations.**

SEAD--See **suppression of enemy air defenses.**

sea echelon(*)--A portion of the assault shipping which withdraws from, or remains out of, the transport area during an amphibious landing and operates in designated areas to seaward in an on-call or unscheduled status.

sea echelon area--In amphibious operations, an area to seaward of a transport area from which assault shipping is phased into the transport area, and to which assault shipping withdraws from the transport area.

sea echelon plan--In amphibious operations, the plan for reduction of concentration of amphibious shipping in the transport area, to minimize losses due to enemy attack by mass destruction weapons and to reduce the area to be swept of mines.

sea frontier--The naval command of a coastal frontier, including the coastal zone in addition to the land area of the coastal frontier and the adjacent sea areas.

Sea King--A single-rotor, medium-lift helicopter used for air/sea rescue and personnel/cargo transport in support of aircraft carrier operations. Some versions are equipped for antisubmarine operations. Designated as **H-3.**

Sea Knight--A twin-rotor, medium-lift helicopter used for personnel and cargo transport. Designated as **H-46.**

sea-launched ballistic missile--A ballistic missile launched from a submarine or surface ship.

sealed cabin(*)--The occupied space of an aircraft characterized by walls which do not allow any gaseous exchange between the ambient atmosphere and the inside atmosphere and contain-

ing its own ways of regenerating the inside atmosphere.

sealift readiness program--A formal agreement, pursuant to the Merchant Marine Act of 1936, as amended, between US-flag, dry-cargo carriers and the government for the acquisition of ships and related equipment under conditions of less that full mobilization.

SEAL team--See **sea-air-land team.**

sea projection operations--See **land, sea, or aerospace projection operations.**

search--1. An operation to locate an enemy force known or believed to be at sea. 2. A systematic reconnaissance of a defined area, so that all parts of the area have passed within visibility. 3. To distribute gunfire over an area in depth by successive changes in gun elevation.

search and attack priority--The lowest category of immediate mission request involving suspected targets related to the enemy tactical or logistical capabilities, e.g., those which are not inhibiting a unit's advance but by their fleeting nature and tactical importance should be located and destroyed. See also **immediate mission request; priority of immediate mission requests.**

search and rescue(*)--The use of aircraft, surface craft, submarines, specialized rescue teams and equipment to search for and rescue personnel in distress on land or at sea. See also **component search and rescue controller; joint rescue coordination center.** (DOD Note: Also called **SAR.**) See also **combat search and rescue.**

search and rescue alert notice--An alerting message used for United States domestic flights. It corresponds to the declaration of the alert phase. Also called **ALNOT.** See also **search and rescue incident classification, subpart b.**

search and rescue coordinator--The designated search and rescue representative of the area commander with overall responsibility and authority for operation of the joint rescue coordination center, and for joint search and rescue operations within the geographical area assigned.

search and rescue incident classification--Three emergency phases into which an incident may be classified or progress, according to the seriousness of the incident and its requirement for rescue service: **a. uncertainty phase**--Doubt exists as to the safety of a craft or person because of knowledge of possible difficulties or because of lack of information concerning progress or position. **b. alert phase**--Apprehension exists for the safety of a craft or person because of definite information that serious difficulties exist that do not amount to a distress or because of a continued lack of information concerning progress or position. **c. distress phase**--Immediate assistance is required by a craft or person because of being threatened by grave or imminent danger or because of continued lack of information concerning progress or position after procedures for the alert phase have been executed.

search and rescue mission coordinator--A search and rescue controller selected by the search and rescue coordinator to direct a specific mission.

search and rescue region--See **inland search and rescue region; maritime search and rescue region; overseas search and rescue region.**

search attack unit--The designation given to one or more ships separately organized or detached from a formation as a tactical unit to search for and destroy submarines.

searched channel(*)--In naval mine warfare, the whole or part of a route or a path which has been searched, swept or hunted, the width of the channel being specified.

searching fire(*)--Fire distributed in depth by successive changes in the elevation of a gun. See also **fire.**

search jammer--See **automatic search jammer.**

search mission(*)--In air operations, an air reconnaissance by one or more aircraft dispatched to locate an object or objects known or suspected to be in a specific area.

search radius--In search and rescue operations, a radius centered on a datum point having a length equal to the total probable error plus an additional safety length to ensure a greater than 50 percent probability that the target is in the search area.

search sweeping(*)--In naval mine warfare, the operation of sweeping a sample of route or area to determine whether poised mines are present.

Sea Sprite--A single rotor light lift helicopter used for air/sea rescue, personnel/cargo transport and antisubmarine operations from naval vessels. Designated as **H-2.**

Sea Stallion--A single-rotor heavy-lift helicopter used for personnel/cargo transport. Designated as **CH-53.** A mine countermeasures-equipped version is designated as **RH-53.**

sea superiority--That degree of dominance in the sea battle of one force over another that permits the conduct of operations by the former and its related land, sea, and air forces at a given time and place without prohibitive interference by the opposing force.

sea supremacy--That degree of sea superiority wherein the opposing force is incapable of effective interference.

sea surveillance(*)--The systematic observation of surface and subsurface sea areas by all available and practicable means primarily for the purpose of locating, identifying and determining the movements of ships, submarines, and other vehicles, friendly and enemy, proceeding on or under the surface of the world's seas and oceans. See also **surveillance.**

sea surveillance system(*)--A system for collecting, reporting, correlating and presenting information supporting and derived from the task of sea surveillance.

seavan--Commercial or Government owned (or leased) shipping containers which are moved via ocean transportation without bogey wheels attached, i.e., lifted on and off the ship.

seaward launch point--A designated point off the coast from which special operations forces will launch to proceed to the beach to conduct operations. Also called **SLP.** See also **seaward recovery point.**

seaward recovery point--A designated point off the coast to which special operations forces will proceed for recovery by submarine, or other means of recovery. Also called **SRP.** See also **seaward launch point.**

secondary armament--In ships with multiple-size guns installed, that battery consisting of guns next largest to those of the main battery.

secondary censorship--Armed forces censorship performed on the personal communications of officers, civilian employees, and accompanying civilians of the Armed Forces of the United States, and on those personal communications of enlisted personnel of the Armed Forces not subject to Armed Forces primary censorship or those requiring reexamination. See also **censorship.**

secondary imagery dissemination--See **electronic imagery dissemination.**

secondary imagery dissemination system--See **electronic imagery dissemination.**

secondary port(*)--A port one or more berths, normally at quays, which can accommodate ocean-going ships for discharge. See also **port.**

secondary rescue facilities--Local airbase-ready aircraft, crash boats, and other air, surface, subsurface, and ground elements suitable for rescue missions including government and privately operated units and facilities.

secondary road--A road supplementing a main road, usually wide enough and suitable for two-way all-weather traffic at moderate or slow speeds.

secondary targets--Alternative targets of lower publicity value that are attacked when the primary target is unattainable. See also **anti-terrorism; primary target.**

second strike--The first counterblow of a war. (Generally associated with nuclear operations.)

secret--See **security classification.**

Secretary of a Military Department--The Secretary of the Air Force, Army or Navy; or the Commandant of the Coast Guard when operating as a Department of Transportation Agency.

section--1. As applied to ships or naval aircraft, a tactical subdivision of a division. It is normally one-half of a division in the case of ships, and two aircraft in the case of aircraft. 2. A subdivision of an office, installation, territory, works, or organization; especially a major subdivision of a staff. 3. A tactical unit of the Army and Marine Corps. A section is smaller than a platoon and larger than a squad. In some organizations the section, rather than the squad, is the basic tactical unit. 4. An area in a warehouse extending from one wall to the next; usually the largest subdivision of one floor.

sector(*)--1. An area designated by boundaries within which a unit operates, and for which it is responsible. 2. One of the subdivisions of a coastal frontier. See also **area of influence; zone of action.**

sector of fire(*)--A defined area which is required to be covered by the fire of individual or crew served weapons or the weapons of a unit.

sector scan(*)--Scan in which the antenna oscillates through a selected angle.

secure(*)--In an operational context, to gain possession of a position or terrain feature, with or without force, and to make such disposition as will prevent, as far as possible, its destruction or loss by enemy action. See also **denial measure.**

security--1. Measures taken by a military unit, an activity or installation to protect itself against all acts designed to, or which may, impair its effectiveness. 2. A condition that results from the establishment and maintenance of protective measures that ensure a state of inviolability from hostile acts or influences. 3. With respect to classified matter, it is the condition that prevents unauthorized persons from having access to official information that is safeguarded in the interests of national security. See also **national security.**

security alert team--Two or more security force members who form the initial reinforcing element responding to security alarms, emergencies, or irregularities.

security assistance--Group of programs authorized by the Foreign Assistance Act of 1961, as amended, and the Arms Export Control Act of 1976, as amended, or other related statutes by which the United States provides defense articles, military training, and other defense--related services, by grant, loan, credit, or cash sales in furtherance of national policies and objectives.

security assistance organization--All Department of Defense elements located in a foreign country with assigned responsibilities for carrying out security assistance management functions. It includes military assistance advisory groups, military missions and groups, offices of defense and military cooperation, liaison groups, and defense attache personnel designated to perform security assistance functions. See also **security assistance.**

security certification(*)--A certification issued by competent national authority to indicate that a person has been investigated and is eligible for access to classified matter to the extent stated in the certification. (Note: The DOD definition does not use the word "national.")

security classification--A category to which national security information and material is assigned to denote the degree of damage that unauthorized disclosure would cause to national defense or foreign relations of the United States and to denote the degree of protection required. There are three such categories: **a. top secret**--National security information or material which requires the highest degree of protection and the unauthorized disclosure of which could reasonably be expected to cause exceptionally grave damage to the national security. Examples of "exceptionally grave damage" include armed hostilities against the United States or its allies; disruption of foreign relations vitally affecting the national security; the compromise of vital national defense plans or complex cryptologic and communications intelligence systems; the revelation of sensitive intelligence operations; and the disclosure of scientific or technological developments vital to national security. **b. secret**--National security information or material which requires a substantial degree of protection and the unauthorized disclosure of which could reasonably be expected to cause serious damage to the national security. Examples of "serious damage" include disruption of foreign relations significantly affecting the national security; significant impairment of a program or policy directly related to the national security; revelation of significant military plans or intelligence operations; and compromise of significant scientific or technological developments relating to national security. **c. confidential**--National security information or material which requires protection and the unauthorized disclosure of which could reasonably be expected to cause damage to the national security. See also **classification; security.**

security clearance(*)--An administrative determination by competent national authority that an individual is eligible, from a security stand--point, for access to classified information. (Note: The DOD definition does not use the word "national.")

security intelligence(*)--Intelligence on the identity, capabilities and intentions of hostile organizations or individuals who are or may be engaged in espionage, sabotage, subversion or terrorism. See also **counterintelligence; intelligence; security.**

security supporting assistance--Program by which economic assistance is provided on a loan or grant basis, to selected foreign governments having unique security problems. The funds are used to finance imports of commodities, capital, or technical assistance in accordance with terms of a bilateral agreement; counterpart funds thereby generated may be used as budgetary support. These funds enable a recipient to devote more of its own resources to defense and security purposes than it otherwise could do without serious economic or political consequences.

Selected Reserve--Those units and individuals within the Ready Reserve designated by their respective Services and approved by the Joint Chiefs of Staff as so essential to initial wartime missions that they have priority over all other Reserves. All Selected Reservists are in an active status. The Selected Reserve also includes persons performing initial active duty for training. See also **ready reserve.**

Selected Reserve strength--The total number of Guardsmen and reservists in the Selected Reserve who are subject to the 200K Presidential recall or mobilization under declaration of war or national emergency.

selective identification feature--A capability which, when added to the basic Identification Friend or Foe system, provides the means to transmit, receive, and display selected coded replies.

selective jamming--See **spot jamming.**

selective loading(*)--The arrangement and stowage of equipment and supplies aboard ship in a manner designed to facilitate issues to units. See also **loading.**

selective mobilization--See **mobilization.**

selective release process--The process involving requesting, analyzing, and obtaining approval for release of weapons to obtain specific, limited damage on selected targets.

selective unloading--In an amphibious operation, the controlled unloading from assault shipping, and movement ashore, of specific items of cargo at the request of the landing force commander. Normally, selective unloading parallels the landing of nonscheduled units during the initial unloading period of the ship-to-shore movement.

selenodesy--That branch of applied mathematics that determines, by observation and measurement, the exact positions of points and the figures and areas of large portions of the moon's surface, or the shape and size of the moon.

selenodetic--Of or pertaining to, or determined by selenodesy.

self-destroying fuze(*)--A fuze designed to burst a projectile before the end of its flight. See also **fuze.**

self-protection depth(*)--The depth of water where the aggregate danger width relative to mines affected by a minesweeping technique is zero. Safe depth is a particular self-protection depth.

semi-active homing guidance(*)--A system of homing guidance wherein the receiver in the missile utilizes radiations from the target which has been illuminated by an outside source.

semi-controlled mosaic(*)--A mosaic composed of corrected or uncorrected prints laid so that major ground features match their geographical coordinates. See also **mosaic.**

semi-fixed ammunition(*)--Ammunition in which the cartridge case is not permanently attached to the projectile. See also **ammunition.**

senior meteorological and oceanographic officer--Meteorological and oceanographic officer responsible for assisting the combatant commander and staff in developing and executing operational meteorological and oceanographic service concepts. Also called **SMO.**

senior officer present afloat--The senior line officer of the Navy, on active service, eligible for command at sea, who is present and in command of any unit of the operating forces afloat in the locality or within an area prescribed by competent authority. This officer is responsible for the administration of matters which collectively affect naval units of the operating forces afloat in the locality prescribed. Also called **SOPA.**

sensitive--Requiring special protection from disclosure which could cause embarrassment, compromise, or threat to the security of the sponsoring power. May be applied to an agency, installation, person, position, document, material, or activity.

sensitive compartmented information--All information and materials bearing special community controls indicating restricted handling within present and future community intelligence collection programs and their end products for which community systems of compartmentation have been or will be formally established. (These controls are over and above the provisions of DOD 5200.1-R, Information Security Program Regulation.) Also called **SCI.**

sensor(*)--An equipment which detects, and may indicate, and/or record objects and activities by means of energy or particles emitted, reflected, or modified by objects.

separate loading ammunition(*)--Ammunition in which the projectile and charge are loaded into a gun separately. See also **ammunition.**

separation zone(*)--An area between two adjacent horizontal or vertical areas into which units are not to proceed unless certain safety measures can be fulfilled.

sequence circuit(*)--In mine warfare, a circuit which requires actuation by a predetermined sequence of influences of predetermined magnitudes.

sequenced ejection system--See **ejection systems.**

Sergeant--A mobile, inertially guided, solid-propellant, surface-to-surface missile, with nuclear warhead capability, designed to attack targets up to a range of 75 nautical miles. Designated as **MGM-29A.**

serial(*)--An element or a group of elements within a series which is given a numerical or alphabetical designation for convenience in planning, scheduling, and control.

serial assignment table--A table that is used in amphibious operations and shows the serial number, the title of the unit, the approximate number of personnel; the material, vehicles, or equipment in the serial; the number and type of landing craft and/or amphibious vehicles required to boat the serial; and the ship on which the serial is embarked.

seriously ill or injured--The casualty status of a person whose illness or injury is classified by medical authority to be of such severity that there is cause for immediate concern, but there is not imminent danger to life. Also called **SII.** See also **casualty status.**

seriously wounded--A stretcher case. See also **wounded.**

service ammunition--Ammunition intended for combat, rather than for training purposes.

Service component command--A command consisting of the Service component commander and all those individuals, units, detachments, organizations and installations under the command that have been assigned to the unified command.

Service component command chaplain--The senior chaplain assigned to the staff of, or designated by, the Service component commander. The component command chaplain is responsible for supervising and coordinating religious ministries within the purview of the

component commander and may be supported by a staff of chaplains and enlisted religious support personnel. See also **command chaplain; command chaplain of the combatant command; lay leader or lay reader; religious ministry support; religious ministry support plan; religious ministry support team.**

service environment(*)--All external conditions, whether natural or induced, to which items of materiel are likely to be subjected throughout their life cycle.

service force--A naval task organization that performs missions for the logistic support of operations.

service group--A major naval administration and/or tactical organization, consisting of the commander and the staff, designed to exercise operational control and administrative command of assigned squadrons and units in executing their tasks of providing logistic support of fleet operations.

service mine(*)--A mine capable of a destructive explosion.

service squadron--An administrative and/or tactical subdivision of a naval service force or service group, consisting of the commander and the staff, organized to exercise operational control and administrative command of assigned units in providing logistic support of fleet units as directed.

service test--A test of an item, system of materiel, or technique conducted under simulated or actual operational conditions to determine whether the specified military requirements or characteristics are satisfied. See also **troop test.**

service troops--Those units designed to render supply, maintenance, transportation, evacuation, hospitalization, and other services required by air and ground combat units to carry out effectively their mission in combat. See also **combat service support elements; troops.**

servicing--See **common servicing; cross-servicing; joint servicing.** See also **inter-Service support.**

severe damage--See **nuclear damage (land warfare).**

SF--See **special forces.**

SFG--See **special forces group.**

SFOB--See **special forces operations base.**

shaded relief(*)--A cartographic technique that provides an apparent three-dimensional configuration of the terrain on maps and charts by the use of graded shadows that would be cast by high ground if light were shining from the northwest. Shaded relief is usually used in combination with contours. See also **hill shading.**

shadow--See **trailer aircraft.**

shallow fording--The ability of a self-propelled gun or ground vehicle equipped with built-in waterproofing, with its wheels or tracks in contact with the ground, to negotiate a water obstacle without the use of a special waterproofing kit. See also **deep fording; flotation.**

shaped charge(*)--A charge shaped so as to concentrate its explosive force in a particular direction.

sheaf--In artillery and naval gunfire support, planned planes (lines) of fire that produce a desired pattern of bursts with rounds fired by two or more weapons.

shear link assembly(*)--A device designed to break at a specified mechanical load.

sheet explosive(*)--Plastic explosive provided in a sheet form.

sheetlines--Those lines defining the geographic limits of the map or chart detail.

shelf life(*)--The length of time during which an item of supply, subject to deterioration or having a limited life which cannot be renewed, is considered serviceable while stored. See also **storage life.**

shelling report(*)--Any report of enemy shelling containing information on caliber, direction, time, density and area shelled.

shell (specify)(*)--A command or request indicating the type of projectile to be used.

shielding(*)--1. Material of suitable thickness and physical characteristics used to protect personnel from radiation during the manufacture, handling, and transportation of fissionable and radioactive materials. 2. Obstructions which tend to protect personnel or materials from the effects of a nuclear explosion.

shifting fire--Fire delivered at constant range at varying deflections; used to cover the width of a target that is too great to be covered by an open sheaf.

Shillelagh--A missile system mounted on the main battle tank and assault reconnaissance vehicle for employment against enemy armor, troops, and field fortifications. Designated as **MGM-51.**

ship combat readiness--See **combat ready.**

ship counter(*)--In naval mine warfare, a device in a mine which prevents the mine from detonating until a preset number of actuations has taken place.

ship haven--See **moving havens.**

ship influence(*)--In naval mine warfare, the magnetic, acoustic and pressure effects of a ship, or a minesweep simulating a ship, which is detectable by a mine or other sensing devices.

shipping control--See naval control of shipping.

shipping designator--A code word assigned to a particular overseas base, port, or area, for specific use as an address on shipments to the overseas location concerned. The code word is usually four letters and may be followed by a number to indicate a particular addressee.

shipping lane(*)--A term used to indicate the general flow of merchant shipping between two departure/terminal areas.

shipping time--The time elapsing between the shipment of materiel by the supplying activity and receipt of materiel by the requiring activity. See also order and shipping time.

ship-to-shore movement(*)--That portion of the assault phase of an amphibious operation which includes the deployment of the landing force from the assault shipping to designated landing areas.

ship will adjust--In naval gunfire support, a method of control in which the ship can see the target and, with the concurrence of the spotter, will adjust.

shock front(*)--The boundary between the pressure disturbance created by an explosion (in air, water, or earth) and the ambient atmosphere, water, or earth.

shock wave(*)--The continuously propagated pressure pulse formed by the blast from an explosion in air, under water or under ground. See also blast wave.

SHORADEZ--See short-range air defense engagement zone.

shoran--A precise short-range electronic navigation system which uses the time of travel of pulse-type transmission from two or more fixed stations to measure slant-range distance from the stations. Also, in conjunction with a suitable computer, used in precision bombing. (This term is derived from the words "short-range navigation.")

shore fire control party--A specially trained unit for control of naval gunfire in support of troops ashore. It consists of a spotting team to adjust fire and a naval gunfire liaison team to perform liaison functions for the supported battalion commander.

shoreline effect--See coastal refraction.

shore party(*)--A task organization of the landing force, formed for the purpose of facilitating the landing and movement off the beaches of troops, equipment, and supplies; for the evacuation from the beaches of casualties and enemy prisoners of war; and for facilitating the beaching, retraction, and salvaging of landing ships and craft. It comprises elements of both the naval and landing forces. Also called beach group. See also beachmaster unit; beach party; naval beach group.

shore-to-shore movement--The assault movement of personnel and materiel directly from a shore staging area to the objective, involving no further transfers between types of craft or ships incident to the assault movement.

short(*)--In artillery and naval gunfire support, a spotting, or an observation, used by an observer to indicate that a burst(s) occurred short of the target in relation to the spotting line.

shortfall--The lack of forces, equipment, personnel, materiel, or capability, reflected as the difference between the resources identified as a plan requirement and those apportioned to a combatant commander for planning, that would adversely affect the command's ability to accomplish its mission.

short-range air defense engagement zone--See **weapon engagement zone.**

short-range attack missile--An air-to-surface missile, armed with a nuclear warhead, launched from the B-52 and the FB-111 aircraft. The missile range, speed, and accuracy allow the carrier aircraft to "standoff" from its intended targets and launch missiles outside enemy defenses. Designated as **AGM-69.**

short-range ballistic missile--A ballistic missile with a range capability up to about 600 nautical miles. Also called **SRBM.**

short-range transport aircraft--See **transport aircraft.**

short round--1. The unintentional or inadvertent delivery of ordnance on friendly troops, installations, or civilians by a friendly weapon system. 2. A defective cartridge in which the projectile has been seated too deeply.

short scope buoy(*)--A buoy used as a navigational reference which remains nearly vertical over its sinker.

short supply--An item is in short supply when the total of stock on hand and anticipated receipts during a given period are less than the total estimated demand during that period.

short takeoff and landing(*)--The ability of an aircraft to clear a 50-foot (15 meters) obstacle within 1,500 feet (500 meters) of commencing takeoff or in landing, to stop within 1,500 feet (500 meters) after passing over a 50-foot (15 meters) obstacle.

short takeoff and vertical landing aircraft(*)--Fixed-wing aircraft capable of clearing a 15-meter (50-foot) obstacle within 450 meters (1500 feet) of commencing takeoff run, and capable of landing vertically. Also called **STOVL.** See also **short takeoff and landing; vertical/short takeoff and landing aircraft; vertical takeoff and landing.**

short title(*)--A short, identifying combination of letters, and/or numbers assigned to a document or device for purposes of brevity and/or security.

shot(*)--In artillery and naval gunfire support, a report that indicates a gun, or guns, have been fired. See also **rounds complete.**

Shrike--An air-launched antiradiation missile designed to home on and destroy radar emitters. Designated as **AGM-45.**

shuttered fuze(*)--A fuze in which inadvertent initiation of the detonator will not initiate either the booster or the burst charge. See also **fuze.**

shuttle bombing--Bombing of objectives using two bases. By this method, a bomber formation bombs its target, flies on to its second base, reloads, and returns to its home base, again bombing a target if required.

sick--In air intercept, a code meaning, "Equipment indicated is operating at reduced efficiency."

sidelay(*)--Device on the feed board of a printing machine for controlling the lateral alignment of the printing paper.

side looking airborne radar(*)--An airborne radar, viewing at right angles to the axis of the vehicle, which produces a presentation of terrain or moving targets. (DOD) Also called **SLAR.**

side oblique air photograph--An oblique photograph taken with the camera axis at right angles to the longitudinal axis of the aircraft.

side overlap--See **overlap.**

Sidewinder--A solid-propellant, air-to-air missile with nonnuclear warhead and infrared, heat--seeking homer. Designated as **AIM-9.** The ground-to-air version is designated as **Chaparral (MIM-72).**

sighting--Actual visual contact. Does not include other contacts, which must be reported by type, e.g., radar and sonar contacts. See also **contact report.**

SIGINT--See **signals intelligence.**

SIGINT direct service--A reporting procedure to provide signals intelligence (SIGINT) to a military commander or other authorized recipient in response to SIGINT requirements. The product may vary from recurring, serialized reports produced by the National Security Agency/Central Security Service to instantaneous aperiodic reports provided to the command or other recipient, usually from a fixed SIGINT activity engaged in collection and processing. See also **signals intelligence.**

SIGINT direct service activity--A signals intelligence (SIGINT) activity composed of collection and associated resources that normally performs in a direct service role under the SIGINT operational control of the Director, National Security Agency/Chief, Central Security Service. See also **signals intelligence.**

SIGINT direct support--The provision of signals intelligence (SIGINT) information to a military commander by a SIGINT direct support unit in response to SIGINT operational tasking levied by that commander. See also **signals intelligence.**

SIGINT direct support unit--A signals intelligence (SIGINT) unit, usually mobile, designed to perform a SIGINT direct support role for a military commander under delegated authority from the Director, National Security Agency/Chief, Central Security Service. See also **signals intelligence.**

SIGINT operational control--The authoritative direction of signals intelligence (SIGINT) activities, including tasking and allocation of effort, and the authoritative prescription of those uniform techniques and standards by which SIGINT information is collected, processed, and reported. See also **signals intelligence.**

SIGINT operational tasking--The authoritative operational direction of and direct levying of signals intelligence (SIGINT) information needs by a military commander on designated SIGINT resources. These requirements are directive, irrespective of other priorities, and are conditioned only by the capability of those resources to produce such information. Operational tasking includes authority to deploy all or part of the SIGINT resources for which SIGINT operational tasking authority has been delegated. See also **signals intelligence.**

SIGINT operational tasking authority--A military commander's authority to operationally direct and levy signals intelligence (SIGINT) requirements on designated SIGINT resources; includes authority to deploy and redeploy all or part of the SIGINT resources for which SIGINT operational tasking authority has been delegated. Also called **SOTA.** See also **signals intelligence.**

SIGINT resources--Personnel and equipment of any unit, activity, or organizational element engaged in signals intelligence (SIGINT) activities. See also **signals intelligence.**

SIGINT support plans--Plans prepared by the National Security Agency/Central Security Service, in coordination with concerned elements of the United States SIGINT system, which specify how the resources of the system will be aligned in crisis or war to support military operations covered by certain JCS and unified and specified command operation plans. See also **signals intelligence.**

signal(*)--1. As applied to electronics, any transmitted electrical impulse. 2. Operationally, a type of message, the text of which consists of one or more letters, words, characters, signal flags, visual displays, or special sounds with prearranged meaning, and which is conveyed or transmitted by visual, acoustical, or electrical means.

signal center--A combination of signal communication facilities operated by the Army in the field and consisting of a communications center, telephone switching central and appropriate means of signal communications. See also **communications center.**

signal letters--See **international call sign.**

signal operation instructions--A series of orders issued for technical control and coordination of the signal communication activities of a command. In Marine Corps usage, these instructions are designated communication operation instructions.

signal security--A generic term that includes both communications security and electronics security. See also **security.**

signals intelligence--1. A category of intelligence comprising either individually or in combination all communications intelligence, electronics intelligence, and foreign instrumentation signals intelligence, however transmitted. 2. Intelligence derived from communications, electronics, and foreign instrumentation signals. Also

called **SIGINT.** See also **communications intelligence; electronics intelligence; intelligence; foreign instrumentation signals intelligence.**

signal-to-noise ratio--The ratio of the amplitude of the desired signal to the amplitude of noise signals at a given point in time.

signature equipment(*)--Any item of equipment which reveals the type and nature of the unit or formation to which it belongs.

signed route--A route along which a unit has placed directional signs bearing its unit identification symbol. The signs are for the unit's use only and must comply with movement regulations.

significant track(*)--In air defense, the track of an aircraft or missile which behaves in an unusual manner which warrants attention and could pose a threat to a defended area.

SII--See **seriously ill or injured.**

simulative electromagnetic deception--See **electromagnetic deception.**

simultaneous engagement--The concurrent engagement of hostile targets by combination of interceptor aircraft and surface-to-air missiles.

single department purchase--A method of purchase whereby one Military Department buys commodities for another Military Department or Departments. See also **purchase.**

single flow route(*)--A route at least one-and-a--half lanes wide allowing the passage of a column of vehicles, and permitting isolated vehicles to pass or travel in the opposite direction at predetermined points. See also **limited access route; double flow route.**

single manager--A Military Department or Agency designated by the Secretary of Defense to be responsible for management of specified commodities or common service activities on a Department of Defense-wide basis.

sinker(*)--In naval mine warfare, a heavy weight to which a buoyant mine is moored. The sinker generally houses the mooring rope drum and depth-setting mechanism and for mines laid by ships, it also serves as a launching trolley.

SIT--See **special interest target.**

situation assessment--Assessment produced by combining military geography, weather, and threat data to provide a comprehensive projection of the situation for the decisionmaker. See also **assessment.**

situation map(*)--A map showing the tactical or the administrative situation at a particular time. See also **map.**

situation report(*)--A report giving the situation in the area of a reporting unit or formation.

skim sweeping(*)--In naval mine warfare, the technique of wire sweeping to a fixed depth over deep-laid moored mines to cut any shallow enough to endanger surface shipping.

skin paint--A radar indication caused by the reflected radar signal from an object.

skin tracking--The tracking of an object by means of a skin paint.

skip bombing--A method of aerial bombing in which a bomb is released from such a low altitude that it slides or glances along the surface of the water or ground and strikes the target at or above water level or ground level. See also **minimum-altitude bombing.**

skip it--In air intercept, a code meaning, "Do not attack"; "Cease attack"; "Cease interception."

Skyhawk--A single-engine, turbojet attack aircraft designed to operate from aircraft carriers, and capable of delivering nuclear and/or nonnuclear weapons, providing troop support, or conducting reconnaissance missions. It can act as a tanker, and can itself be air refueled. It possesses a limited all-weather attack capability, and can operate from short, unprepared fields. Designated as **A-4.**

slant range(*)--The line of sight distance between two points, not at the same level relative to a specific datum.

slated items--Bulk petroleum and packaged bulk petroleum items that are requisitioned for overseas use by means of a consolidated requirement document, prepared and submitted through joint petroleum office channels. Packaged petroleum items are requisitioned in accordance with normal requisitioning procedures.

slice--An average logistic planning factor used to obtain estimates of requirements for personnel and materiel. A personnel slice, e.g., generally consists of the total strength of the stated basic combatant elements, plus its proportionate share of all supporting and higher headquarters personnel.

slightly wounded--A casualty that is a sitting or a walking case. See also **wounded.**

slip indicator(*)--An instrument which displays a measure of the resultant of the inertial and gravity forces in the lateral and normal plane of aircraft.

SLP--See **seaward launch point.**

small arms--Man portable, individual, and crew--served weapon systems used mainly against personnel and lightly armored or unarmored equipment.

small arms ammunition--Ammunition for small arms, i.e., all ammunition up to and including 20 millimeters (.787 inches).

small-lot storage--Generally considered to be a quantity of less than one pallet stack, stacked to maximum storage height. Thus, the term refers to a lot consisting of from one container to two or more pallet loads, but is not of sufficient quantity to form a complete pallet column. See also **storage.**

small-scale map--A map having a scale smaller than 1:600,000. See also **map.**

SMO--See **senior meteorological and oceanographic officer.**

smoke screen(*)--Cloud of smoke used to mask either friendly or enemy installations or maneuvers.

SMU--See **special mission unit.**

snagline mine(*)--A contact mine with a buoyant line attached to one of the horns or switches which may be caught up and pulled by the hull or propellers of a ship.

snap report--Not to be used. See **Joint Tactical Air Reconnaissance/Surveillance Mission Report.**

snow--In air intercept, a term meaning sweep jamming.

SO--See **special operations.**

SOC--See **special operations command.**

SOCCT--See **special operations combat control team.**

SOF--See **special operations forces.**

sofar--The technique of fixing an explosion at sea by time difference of arrival of sound energy at several separate geographical locations. (The term is derived from the words "sound, fixing and ranging.")

soft missile base(*)--A launching base that is not protected against a nuclear explosion.

software--A set of computer programs, procedures, and associated documentation concerned with the operation of a data processing system, e.g., compilers, library routines, manuals, and circuit diagrams.

soil shear strength--The maximum resistance of a soil to shearing stresses.

solenoid sweep(*)--In naval mine warfare, a magnetic sweep consisting of a horizontal axis coil wound on a floating iron tube.

SOMPF--See **special operations mission planning folder.**

sonar--A sonic device used primarily for the detection and location of underwater objects. (This term is derived from the words "sound navigation and ranging.")

sonic--Of or pertaining to sound or the speed of sound. See also **speed of sound.**

SONMET--See **special operations naval mobile environment team.**

sonobuoy--A sonar device used to detect submerged submarines which when activated relays information by radio. It may be active directional or nondirectional, or it may be passive directional or nondirectional.

sortie(*)--In air operations, an operational flight by one aircraft.

sortie allotment message--The means by which the joint force commander allots excess sorties to meet requirements of his subordinate commanders which are expressed in their air employment/allocation plan.

sortie number(*)--A reference used to identify the images taken by all the sensors during one air reconnaissance sortie.

sortie plot--An overlay representing the area on a map covered by imagery taken during one sortie. **(NATO)**--See **master plot.**

sortie reference--See **sortie number.**

sorting--See **triage.**

source--1. A person, thing, or activity from which intelligence information is obtained. 2. In clandestine activities, a person (agent), normally a foreign national, in the employ of an intelligence activity for intelligence purposes. 3. In interrogation activities, any person who furnishes intelligence information, either with or without the knowledge that the information is being used for intelligence purposes. In this context, a controlled source is in the employment or under the control of the intelligence activity and knows that the information is to be used for intelligence purposes. An uncontrolled source is a voluntary contributor of information and may or may not know that the information is to be used for intelligence purposes. See also **agent; collection agency.**

SOW--See **special operations wing.**

SOWT/TE--See **special operations weather team/tactical element.**

space assignment--An assignment to the individual Departments/Services by the appropriate transportation operating agency of movement capability which completely or partially satisfies the stated requirements of the Depart-

ments/Services for the operating month and that has been accepted by them without the necessity for referral to the Joint Transportation Board for allocation.

space control operations--Operations that provide freedom of action in space for friendly forces while, when directed, denying it to an enemy, and include the broad aspects of protection of US and US allied space systems and negation of enemy space systems. Space control operations encompass all elements of the space defense mission.

space defense--All defensive measures designed to destroy attacking enemy vehicles (including missiles) while in space, or to nullify or reduce the effectiveness of such attack. See also **aerospace defense.**

space support operations--Operations required to ensure that space control and support of terrestrial forces are maintained. They include activities such as launching and deploying space vehicles, maintaining and sustaining space vehicles while on orbit, and recovering space vehicles if required.

space systems--All of the devices and organizations forming the space network. The network includes spacecraft, ground control stations, and associated terminals.

Spacetrack--A global system of radar, optical and radiometric sensors linked to a computation and analysis center in the North American Air Defense Command combat operations center complex. The Spacetrack mission is detection, tracking, and cataloging of all manmade objects in orbit of the Earth. It is the Air Force portion of the North American Air Defense Command Space Detection and Tracking system. See also **Spadats; Spasur.**

space weather--A term used to describe the environment and other natural phenomena occurring above 50 kilometers altitude.

Spadats--A space detection and tracking system capable of detecting and tracking space vehicles from the Earth, and reporting the orbital characteristics of these vehicles to a central control facility. See also **Spacetrack; Spasur.**

span of detonation (atomic demolition munition employment)--That total period of time, resulting from a timer error, between the earliest and the latest possible detonation time. **1. early time**--The earliest possible time that an atomic demolition munition can detonate; **2. fire time**--That time the atomic demolition munition will detonate should the timers function precisely without error; **3. late time**--The latest possible time that an atomic demolition munition can detonate.

Sparrow--An air-to-air solid-propellant missile with nonnuclear warhead and electronic-controlled homing. Designated as **AIM-7.** The ship-launched surface-to-air version is designated as **Sea Sparrow (RIM-7).**

Spartan--A nuclear surface-to-air guided missile formerly deployed as part of the Safeguard ballistic missile defense weapon system. It is designed to intercept strategic ballistic reentry vehicles in the exoatmosphere.

spasm war--Not to be used. See **general war.**

Spasur--An operational space surveillance system with the mission to detect and determine the orbital elements of all manmade objects in orbit of the Earth. The mission is accomplished by means of a continuous fan of continuous wave energy beamed vertically across the continental United States and an associated computational facility. It is the Navy portion of the North American Air Defense Command Space Detec-

tion and Tracking System. See also **Spacetrack; Spadats.**

special access program--A sensitive program, approved in writing by a head of agency with original top secret classification authority, which imposes need-to-know and access controls beyond those normally provided for access to confidential, secret, or top secret information. The level of controls is based on the criticality of the program and the assessed hostile intelligence threat. The program may be an acquisition program, an intelligence program, or an operations and support program. Also called **a SAP.**

special activities--Activities conducted in support of national foreign policy objectives which are planned and executed so that the role of the US Government is not apparent or acknowledged publicly. They are also functions in support of such activities but are not intended to influence United States political processes, public opinion, policies, or media and do not include diplomatic activities or the collection and production of intelligence or related support functions.

special agent--A person, either United States military or civilian, who is a specialist in military security or the collection of intelligence or counterintelligence information.

special air operation--An air operation conducted in support of special operations and other clandestine, covert, and psychological activities.

special ammunition supply point--A mobile supply point where special ammunition is stored and issued to delivery units.

special assignment airlift requirements--Airlift requirements, including JCS-directed/coordinated exercises, that require special consideration due to the number of passengers involved, weight or size of cargo, urgency of movement, sensitivity, or other valid factors that preclude the use of channel airlift.

special atomic demolition munition--A very low-yield, man-portable, atomic demolition munition that is detonated by a timer device.

special boat squadron--A permanent Navy echelon III major command to which two or more special boat units are assigned for some operational and all administrative purposes. The squadron is tasked with the training and deployment of these special boat units and may augment naval special warfare task groups and task units. Also called **SBR.**

special boat unit--Those US Navy forces organized, trained, and equipped to conduct or support naval special warfare, riverine warfare, coastal patrol and interdiction, and joint special operations with patrol boats or other combatant craft designed primarily for special operations support. Also called **SBU.**

special cargo--Cargo that requires special handling or protection, such as pyrotechnics, detonators, watches, and precision instruments. See also **cargo.**

special-equipment vehicle--A vehicle consisting of a general-purpose chassis with special-purpose body and/or mounted equipments designed to meet a specialized requirement. See also **vehicle.**

special flight(*)--An air transport flight, other than a scheduled service, set up to move a specific load.

special forces--US Army forces organized, trained, and equipped specifically to conduct special operations. Special forces have five primary missions: unconventional warfare, foreign internal defense, direct action, special reconnaissance, and counterterrorism. Coun-

terterrorism is a special mission for specially organized, trained, and equipped special forces units designated in theater contingency plans. Also called **SF.**

special forces group--A combat arms organization capable of planning, conducting, and supporting special operations activities in all operational environments in peace, conflict, and war. It consists of a group headquarters and headquarters company, a support company, and special forces battalions. The group can operate as a single unit, but normally the battalions plan and conduct operations from widely separated locations. The group provides general operational direction and synchronizes the activities of subordinate battalions. Although principally structured for unconventional warfare, special forces group units are capable of task-organizing to meet specific requirements. Also called **SFG.**

special forces operations base--A command, control, and support base established and operated by a special forces group or battalion from organic and attached resources. The base commander and his staff coordinate and synchronize the activities of subordinate and forward-deployed forces. A special forces operations base is normally established for an extended period of time to support a series of operations. Also called **SFOB.**

special hazard(*)--In aircraft crash rescue and fire-fighting activities: fuels, materials, components or situations that could increase the risks normally associated with military aircraft accidents and could require special procedures, equipment or extinguishing agents.

special interest target--In counterdrug operations, a contact that may be outside initial sorting criteria but still requires special handling, such as controlled deliveries or other unusual situations. Also called **SIT.** See also **suspect; track of interest.**

specialist intelligence report--A category of specialized, technical reports used in the dissemination of intelligence. See also **intelligence reporting**.

specialization--An arrangement within an alliance wherein a member or group of members most suited by virtue of technical skills, location, or other qualifications assume(s) greater responsibility for a specific task or significant portion thereof for one or more other members.

special mission unit--A generic term to represent a group of operations and support personnel from designated organizations that is task-organized to perform highly classified activities. Also called **SMU**.

special operations--Operations conducted by specially organized, trained, and equipped military and paramilitary forces to achieve military, political, economic, or psychological objectives by unconventional military means in hostile, denied, or politically sensitive areas. These operations are conducted during peacetime competition, conflict, and war, independently or in coordination with operations of conventional, nonspecial operations forces. Political-military considerations frequently shape special operations, requiring clandestine, covert, or low visibility techniques and oversight at the national level. Special operations differ from conventional operations in degree of physical and political risk, operational techniques, mode of employment, independence from friendly support, and dependence on detailed operational intelligence and indigenous assets. Also called **SO**.

special operations combat control team--A team of Air Force personnel organized, trained, and equipped to conduct and support special operations. Under clandestine, covert, or low-visibility conditions, these teams establish and control air assault zones; assist aircraft by verbal control, positioning, and operating navigation aids; conduct limited offensive direct action and special reconnaissance operations; and assist in the insertion and extraction of special operations forces. Also called **SOCCT**. See also **combat control team**.

special operations command--A subordinate unified or other joint command established by a joint force commander to plan, coordinate, conduct, and support joint special operations within the joint force commander's assigned area of operations. Also called **SOC**.

special operations forces--Those active and reserve component forces of the military Services designated by the Secretary of Defense and specifically organized, trained, and equipped to conduct and support special operations. Also called **SOF**. See also **Air Force special operations forces; Army special operations forces; naval special warfare forces**.

special operations mission planning folder--The package that contains the materials required to execute a given special operations mission. It will include the mission tasking letter, mission tasking package, original feasibility assessment (as desired), initial assessment (as desired), target intelligence package, plan of execution, infiltration and exfiltration plan of execution, and other documentation as required or desired. Also called **SOMPF**.

special operations naval mobile environment team--A team of Navy personnel organized, trained, and equipped to support naval special warfare forces by providing weather, oceanographic, mapping, charting, and geodesy support. Also called **SONMET**.

special operations peculiar--Equipment, materials, supplies, and services required for special operations mission support for which there is no broad conventional force requirement. It often includes nondevelopmental or special category items incorporating evolving technology but may include stocks of obsolete weapons and equipment designed to support indigenous personnel who do not possess sophisticated operational capabilities.

special operations weather team/tactical element--A task-organized team of Air Force personnel organized, trained, and equipped to collect critical weather observations from data-sparse areas. These teams are trained to operate independently in permissive or semi-permissive environments, or as augmentation to other special operations elements in nonpermissive environments, in direct support of special operations. Also called **SOWT/TE.**

special operations wing--An Air Force special operations wing. Also called **SOW.**

special (or project) equipment--Equipment not authorized in standard equipment publications but determined as essential in connection with a contemplated operation, function, or mission. See also **equipment.**

special-purpose vehicle--A vehicle incorporating a special chassis and designed to meet a specialized requirement. See also **vehicle.**

special reconnaissance--Reconnaissance and surveillance actions conducted by special operations forces to obtain or verify, by visual observation or other collection methods, information concerning the capabilities, intentions, and activities of an actual or potential enemy or to secure data concerning the meteorological, hydrographic, or geographic characteristics of a particular area. It includes target acquisition, area assessment, and post-strike reconnaissance. Also called **SR.**

special sheaf--In artillery and naval gunfire support, any sheaf other than parallel, converged, or open.

special staff--All staff officers having duties at a headquarters and not included in the general (coordinating) staff group or in the personal staff group. The special staff includes certain technical specialists and heads of services, e.g., quartermaster officer, antiaircraft officer, transportation officer, etc. See also **staff.**

special tactics team--An Air Force team composed primarily of special operations combat control and pararescue personnel. The team supports joint special operations by selecting, surveying, and establishing assault zones; providing assault zone terminal guidance and air traffic control; conducting direct action missions; providing medical care and evacuation; and, coordinating, planning, and conducting air, ground, and naval fire support operations.

special unloading berth--Berths established in the vicinity of the approach lanes into which transports may move for unloading, thus reducing the running time for landing craft and assisting in the dispersion of transports.

special weapons--A term sometimes used to indicate weapons grouped for special procedures, for security, or other reasons. Specific terminology, e.g., nuclear weapons, guided missiles, is preferable.

specific intelligence collection requirement--An identified gap in intelligence holdings that may be satisfied only by collection action, and that has been validated by the appropriate requirements control authority. Also called **SICR.**

specific search--Reconnaissance of a limited number of points for specific information.

specified command--A command that has broad continuing missions and that is established by the President through the Secretary of Defense with the advice and assistance of the Chairman of the Joint Chiefs of Staff. It normally is composed of forces from a single Military Department. Also called **specified combatant command.**

spectrozonal photography(*)--A photographic technique whereby the natural spectral emissions of all objects are selectively filtered in order to image only those objects within a particular spectral band or zone and eliminate the unwanted background.

spectrum management--Planning, coordinating, and managing joint use of the electromagnetic spectrum through operational, engineering, and administrative procedures, with the objective of enabling electronic systems to perform their functions in the intended environment without causing or suffering unacceptable interference. See also **electromagnetic spectrum; electronic warfare.**

spectrum of war--A term which encompasses the full range of conflict; cold, limited, and general war.

speed--See **airspeed; convoy speed; critical speed; declared speed; endurance speed; maximum sustained speed (transport vehicle); scheduled speed; speed of advance; speed of sound.**

speed of advance(*)--In naval usage, the speed expected to be made good over the ground. See also **pace; rate of march.**

speed of sound(*)--The speed at which sound travels in a given medium under specified conditions. The speed of sound at sea level in the International Standard Atmosphere is 1108 ft/second, 658 knots, 1215 km/hour. See also **hypersonic; sonic; subsonic; supersonic; transonic.**

spin stabilization--Directional stability of a projectile obtained by the action of gyroscopic forces that result from spinning of the body about its axis of symmetry.

spitting--In air antisubmarine warfare operations, a code meaning, "I am about to lay, or am laying, sonobuoys. I may be out of radio contact for a few minutes." If transmitted from the submarine it indicates that the submarine has launched a sonobuoy.

splash(*)--1. In artillery and naval gunfire support, word transmitted to an observer or spotter five seconds before the estimated time of the impact of a salvo or round. 2. In air interception, target destruction verified by visual or radar means.

splashed--In air intercept, a code meaning, "Enemy aircraft shot down," (followed by number and type).

split cameras(*)--An assembly of two cameras disposed at a fixed overlapping angle relative to each other.

split pair--See **split vertical photography.**

split-up--See **break-up.**

split vertical photography(*)--Photographs taken simultaneously by two cameras mounted at an angle from the vertical, one tilted to the left and one to the right, to obtain a small side overlap.

spoiling attack--A tactical maneuver employed to seriously impair a hostile attack while the enemy is in the process of forming or assembling for an attack. Usually employed by armored units in defense by an attack on enemy assembly positions in front of a main line of resistance or battle position.

sponsor--Military member or civilian employee with dependents.

spoofer--In air intercept, a code meaning, "A contact employing electronic or tactical deception measures."

spot(*)--1. To determine by observation, deviations of ordnance from the target for the purpose of supplying necessary information for the adjustment of fire. 2. To place in a proper location. See also **adjustment of fire.**

spot elevation(*)--A point on a map or chart whose elevation is noted.

spot jamming(*)--The jamming of a specific channel or frequency. See also **barrage jamming; electronic warfare; jamming.**

spot net--Radio communication net used by a spotter in calling fire.

spot report--A concise narrative report of essential information covering events or conditions that may have an immediate and significant effect on current planning and operations that is afforded the most expeditious means of transmission consistent with requisite security. (Note: In reconnaissance and surveillance usage, spot report is not to be used. See **Joint Tactical Air Reconnaissance/Surveillance Mission Report.**)

spot size(*)--The size of the electron spot on the face of the cathode ray tube.

spotter--An observer stationed for the purpose of observing and reporting results of naval gunfire to the firing agency and who also may be employed in designating targets. See also **field artillery observer; naval gunfire spotting team.**

spotting(*)--A process of determining by visual or electronic observation, deviations of artillery or naval gunfire from the target in relation to a spotting line for the purpose of supplying

necessary information for the adjustment or analysis of fire.

spotting line(*)--Any straight line to which the fall of shot of projectiles is related by an observer or a spotter. See also **gun-target line; observer-target line.**

spray dome(*)--The mound of water spray thrown up into the air when the shock wave from an underwater detonation of a nuclear weapon reaches the surface.

spreading fire--A notification by the spotter or the naval gunfire ship, depending on who is controlling the fire, to indicate that fire is about to be distributed over an area.

Sprint--A high acceleration, nuclear surface-to-air guided missile formerly deployed as part of the Safeguard ballistic missile defense weapon system. It is designed to intercept strategic ballistic reentry vehicles in the endoatmosphere.

sprocket(*)--In naval mine warfare, an anti-sweep device included in a mine mooring to allow a sweep wire to pass through the mooring without parting the mine from its sinker.

squadron--1. An organization consisting of two or more divisions of ships, or two or more divisions (Navy) or flights of aircraft. It is normally, but not necessarily, composed of ships or aircraft of the same type. 2. The basic administrative aviation unit of the Army, Navy, Marine Corps, and Air Force.

squawk--A code meaning, "Switch Identification Friend or Foe master control to 'normal' (Mode and Code as directed) position."

squawk flash--A code meaning, "Actuate Identification Friend or Foe I/P switch."

squawking--A code meaning, "Showing Identification Friend or Foe in Mode (and Code) indicated."

squawk low--A code meaning, "Switch Identification Friend or Foe master control to 'low' position."

squawk may day--A code meaning, "Switch Identification Friend or Foe master control to 'emergency' position."

squawk mike--A code meaning, "Actuate Identification Friend or Foe MIC switch and key transmitter as directed."

squawk standby--A code meaning, "Switch Identification Friend or Foe master control to 'standby' position."

squib--A small pyrotechnic device that may be used to fire the igniter in a rocket or for some similar purpose. Not to be confused with a detonator that explodes.

squirt(*)--In air-to-air refuelling, a means of providing visual detection of a nearby aircraft. In practice this is achieved by the donor aircraft dumping fuel and/or the receiver aircraft selecting afterburners, if so equipped.

SR--See **special reconnaissance.**

SRP--See **seaward recovery point.**

SS--See **submarine.**

SSBN--See **fleet ballistic missile submarine.**

SSG--See **guided missile submarine.**

SSGN--See **guided missile submarine.**

SSN--See **submarine.**

staballoy--Designates metal alloys made from high-density depleted uranium with other metals for use in kinetic energy penetrators for armor-piercing munitions. Several different metals such as titanium or molybdenum can be used for the purpose. The various staballoy metals have low radioactivity that is not considered to be a significant health hazard.

stable base film(*)--A particular type of film having a high stability in regard to shrinkage and stretching.

staff--See **combined staff; general staff; integrated staff; joint staff; parallel staff; special staff.**

staff estimates--Assessments of courses of action by the various staff elements of a command that serve as the foundation of the commander's estimate.

staff supervision--The process of advising other staff officers and individuals subordinate to the commander of the commander's plans and policies, interpreting those plans and policies, assisting such subordinates in carrying them out, determining the extent to which they are being followed, and advising the commander thereof.

stage(*)--1. An element of the missile or propulsion system that generally separates from the missile at burnout or cut-off. Stages are numbered chronologically in order of burning. 2. To process, in a specified area, troops which are in transit from one locality to another. See also **marshalling; staging area.**

staged crews--Aircrews specifically positioned at intermediate airfields to take over aircraft operating on air routes, thus relieving complementary crews of flying fatigue and speeding up the flow rate of the aircraft concerned.

staging area(*)--1. Amphibious or airborne--A general locality between the mounting area and the objective of an amphibious or airborne expedition, through which the expedition or parts thereof pass after mounting, for refueling, regrouping of ships, and/or exercise, inspection, and redistribution of troops. 2. Other movements--A general locality established for the concentration of troop units and transient personnel between movements over the lines of communications. See also **marshalling; stage.**

staging base--1. An advanced naval base for the anchoring, fueling, and refitting of transports and cargo ships, and for replenishing mobile service squadrons. 2. A landing and takeoff area with minimum servicing, supply, and shelter provided for the temporary occupancy of military aircraft during the course of movement from one location to another.

standard(*)--An exact value, a physical entity, or an abstract concept, established and defined by authority, custom, or common consent to serve as a reference, model, or rule in measuring quantities or qualities, establishing practices or procedures, or evaluating results. A fixed quantity or quality.

standard advanced base units--Personnel and materiel organized to function as advanced base units, including the functional components which are employed in the establishment of naval advanced bases. Such advanced base units may establish repair bases, supply bases, supply depots, airfields, air bases, or other naval shore establishments at overseas locations; e.g., Acorns, Cubs, Gropacs, and Lions.

Standard Arm--An air-launched antiradiation missile designed to home on and destroy radar emitters. Designated as **AGM-78.**

standardization--The process by which the Department of Defense achieves the closest practicable cooperation among the Services and

Defense agencies for the most efficient use of research, development, and production resources, and agrees to adopt on the broadest possible basis the use of: a. common or compatible operational, administrative, and logistic procedures; b. common or compatible technical procedures and criteria; c. common, compatible, or interchangeable supplies, components, weapons, or equipment; and d. common or compatible tactical doctrine with corresponding organizational compatibility.

Standard Missile--A shipboard, surface-to-surface/air missile with solid propellant rocket engine. It is equipped with nonnuclear warhead and semi-active or passive homing. Designated RIM-66 Medium Range (Tartar replacement) and RIM-67 Extended Range (Terrier replacement).

standard operating procedure--See **standing operating procedure.**

standard parallel(*)--A parallel on a map or chart along which the scale is as stated for that map or chart.

standard pattern(*)--In landmine warfare, the agreed pattern to which mines are normally laid.

standard route(*)--In naval control of shipping, a pre-planned single track, assigned a code name, connecting positions within the main shipping lanes.

Standard SSM (ARM)--A surface-to-surface anti-radiation missile equipped with a conventional warhead. It is planned for anti-ship missions and is carried by the FFG-1 class, 8 DDG-2 class units and the PG 98 and 100. Designated as **RGM-66D.**

standard use Army aircraft flight route--Routes established below the coordinating altitude to facilitate the movement of Army aviation assets. Routes are normally located in the corps through brigade rear areas of operation and do not require approval by the airspace control authority. Also called **SAAFR**.

standby reserve--Those units and members of the Reserve Components (other than those in the Ready Reserve or Retired Reserve) who are liable for active duty only as provided in 10 U.S.C. 273, 672 and 674. **1. active status, standby reserve**--Reservists who (a) are completing their statutory military service obligation, or (b) are being retained in an active status under 10 U.S.C. 1006, or (c) were screened from the Ready Reserve as being key personnel and request assignment to the Active Status List, or (d) may be temporarily assigned to the Standby Reserve for hardship or other cogent reason determined by the secretary concerned, with the expectation of their being returned to the Ready Reserve. **2. inactive status, standby reserve**--Individuals who are not required by law or regulation to remain members of an active status program but who (a) desire to retain their Reserve affiliation in a non-participating status, and (b) have skills that may be of possible future use to the Military Department concerned.

stand fast(*)--In artillery, the order at which all action on the position ceases immediately.

standing operating procedure(*)--A set of instructions covering those features of operations which lend themselves to a definite or standardized procedure without loss of effectiveness. The procedure is applicable unless ordered otherwise. Also called **SOP**.

standing order(*)--A promulgated order which remains in force until amended or cancelled.

Starlifter--A large cargo transport powered by four turbofan engines, capable of intercontinental range with heavy payloads and airdrops. Designated as **C-141**.

state and regional defense airlift--The program for use during an emergency of civil aircraft other than air carrier aircraft.

state chicken--In air intercept, a code meaning, "I am at a fuel state requiring recovery, tanker service, or diversion to an airfield."

state lamb--In air intercept, a code meaning, "I do not have enough fuel for an intercept plus reserve required for carrier recovery."

state of readiness--See **defense readiness conditions; weapons readiness state**.

state of readiness--state 1--safe--The state of a demolition target upon or within which the demolition charge has been placed and secured. The firing or initiating circuits have been installed, but not connected to the demolition charge. Detonators or initiators have not been connected nor installed. See also **state of readiness--state 2--armed**.

state of readiness--state 2--armed(*)--The state of a demolition target in which the demolition charges are in place, the firing and priming circuits are installed and complete, ready for immediate firing. See also **state of readiness--state 1--safe**.

state tiger--In air intercept, a code meaning, "I have sufficient fuel to complete my mission as assigned."

static air temperature(*)--The temperature at a point at rest relative to the ambient air.

static line (air transport)--A line attached to a parachute pack and to a strop or anchor cable in an aircraft so that when the load is dropped the parachute is deployed automatically.

static line cable--See **anchor cable**.

static marking(*)--Marks on photographic negatives and other imagery caused by unwanted discharges of static electricity.

static test load(*)--In sea operations, twice the safe working load. See also **safe working load.**

station--1. A general term meaning any military or naval activity at a fixed land location. 2. A particular kind of activity to which other activities or individuals may come for a specific service, often of a technical nature, e.g., aid station. 3. An assigned or prescribed position in a naval formation or cruising disposition; or an assigned area in an approach, contact, or battle disposition. 4. Any place of duty or post or position in the field to which an individual, or group of individuals, or a unit may be assigned. 5. One or more transmitters or receivers or a combination of transmitters and receivers, including the accessory equipment necessary at one location, for carrying on radio communication service. Each station will be classified by the service in which it operates permanently or temporarily.

station authentication--A security measure designed to establish the authenticity of a transmitting or receiving station.

station time(*)--In air transport operations, the time at which crews, passengers, and cargo are to be on board and ready for the flight.

status-of-forces agreement--An agreement which defines the legal position of a visiting military force deployed in the territory of a friendly state. Agreements delineating the status of visiting military forces may be bilateral or multilateral. Provisions pertaining to the status of visiting forces may be set forth in a separate agreement, or they may form a part of a more comprehensive agreement. These provisions describe how the authorities of a visiting force may control members of that force and the amenability of the force or its members to the local law or to the authority of local officials. To the extent that agreements delineate matters affecting the relations between a military force and civilian authorities and population, they may be considered as civil affairs agreements. Also called **SOFA.** See also **civil affairs agreement.**

stay behind--Agent or agent organization established in a given country to be activated in the event of hostile overrun or other circumstances under which normal access would be denied.

stay behind force(*)--A force which is left in position to conduct a specified mission when the remainder of the force withdraws or retires from the area.

steady--In air intercept, a code meaning, "Am on prescribed heading," or, "Straighten out immediately on present heading or heading indicated."

steer--In air intercept, close air support and air interdiction, a code meaning, "Set magnetic heading indicated to reach me (or _____)."

stellar guidance--A system wherein a guided missile may follow a predetermined course with reference primarily to the relative position of the missile and certain preselected celestial bodies. See also **guidance.**

stepped-up separation(*)--The vertical separation in a formation of aircraft measured from an aircraft ahead upward to the next aircraft behind or in echelon.

stereographic coverage--Photographic coverage with overlapping air photographs to provide a three-dimensional presentation of the picture; 60 percent overlap is considered normal and 53 percent is generally regarded as the minimum.

sterilize(*)--1. In naval mine warfare, to permanently render a mine incapable of firing by means of a device (e.g., sterilizer) within the mine. (DOD) 2. To remove from material to be used in covert and clandestine operations, marks or devices which can identify it as emanating from the sponsoring nation or organization.

sterilizer(*)--In mine warfare, a device included in mines to render the mine permanently inoperative on expiration of a pre-determined time after laying.

stern attack--In air intercept, an attack by an interceptor aircraft that terminates with a heading crossing angle of 45 degrees or less. See also **heading crossing angle.**

stick (air transport)--A number of paratroopers who jump from one aperture or door of an aircraft during one run over a drop zone.

stick commander (air transport)--A designated individual who controls parachutists from the time they enter the aircraft until their exit. See also **jumpmaster.**

Stinger--A lightweight, man-portable, shoulder-fired, air defense artillery missile weapon for low altitude air defense of forward area combat troops. Designated as **FIM-92A.**

stockage objective--The maximum quantities of materiel to be maintained on hand to sustain current operations. It will consist of the sum of stocks represented by the operating level and the safety level. See also **level of supply.**

stock control(*)--Process of maintaining inventory data on the quantity, location, and condition of supplies and equipment due-in, on-hand, and due-out, to determine quantities of material and equipment available and/or required for issue and to facilitate distribution and management of materiel. See also **inventory control.**

stock coordination--A supply management function exercised usually at department level that controls the assignment of material cognizance for items or categories of material to inventory managers.

stock fund--A revolving fund established to finance costs of inventories of supplies. It is authorized by specific provision of law to finance a continuing cycle of operations. Reimbursements and collections derived from such operations are available for use by the fund without further action by the Congress.

stock level--See **level of supply.**

Stock Number--See **National Stock Number.**

stockpile to target sequence--1. The order of events involved in removing a nuclear weapon from storage, and assembling, testing, transporting, and delivering it on the target. 2. A document that defines the logistical and employment concepts and related physical environments involved in the delivery of a nuclear weapon from the stockpile to the target. It may also define the logistical flow involved in moving nuclear weapons to and from the stockpile for quality assurance testing, modification and retrofit, and the recycling of limited life components.

stock record account--A basic record showing by item the receipt and issuance of property, the balances on hand and such other identifying or stock control data as may be required by proper authority.

stop squawk--A code meaning, "Turn identification friend or foe master control to 'off.' "

stopway(*)--A defined rectangular area on the ground at the end of a runway in the direction of takeoff designated and prepared by the competent authority as a suitable area in which an aircraft can be stopped in the case of an interrupted takeoff. It must be capable of supporting aircraft of approximately 23,000 kilograms (50,000 lbs.).

storage--1. The retention of data in any form, usually for the purpose of orderly retrieval and documentation. 2. A device consisting of electronic, electrostatic, electrical, hardware or other elements into which data may be entered, and from which data may be obtained as desired. See also **ammunition and toxic material open space; bin storage; bulk storage; igloo space; large-lot storage; medium-lot storage; open improved storage space; open unimproved wet space; small-lot storage.**

storage life(*)--The length of time for which an item of supply including explosives, given specific storage conditions, may be expected to remain serviceable and, if relevant, safe. See also **shelf life.**

stores--See **naval stores; supplies.**

stowage--The method of placing cargo into a single hold or compartment of a ship to prevent damage, shifting, etc.

stowage diagram(*)--A scaled drawing included in the loading plan of a vessel for each deck or platform showing the exact location of all cargo. See also **stowage plan.**

stowage factor--The number which expresses the space, in cubic feet, occupied by a long ton of any commodity as prepared for shipment, including all crating or packaging.

stowage plan--A completed stowage diagram showing what materiel has been loaded and its stowage location in each hold, between-deck compartment, or other space in a ship, including deck space. Each port of discharge is indicated by colors or other appropriate means. Deck and between-deck cargo normally is shown in perspective, while cargo stowed in the lower hold is shown in profile, except that vehicles usually are shown in perspective regardless of stowage. See also **stowage diagram.**

strafing--The delivery of automatic weapons fire by aircraft on ground targets.

straggler(*)--Any personnel, vehicles, ships or aircraft which, without apparent purpose or assigned mission, become separated from their unit, column or formation. See also **romper.**

stranger (bearing, distance, altitude)--In air intercept, a code meaning, "An unidentified aircraft, bearing, distance, and altitude as indicated relative to you."

strangle--A code meaning, "Switch off equipment indicated."

strangle parrot--A code meaning, "Switch off Identification Friend or Foe equipment."

strapping--1. An operation by which supply containers, such as cartons or boxes, are reinforced by bands, metal straps, or wire, placed at specified intervals around them, drawn taut, and then sealed or clamped by a machine. 2. Measurement of storage tanks and calculation of volume to provide tables for conversion of depth of product in linear units of measurement to volume of contents.

strategic advantage--The overall relative power relationship of opponents that enables one nation or group of nations effectively to control the course of a military/political situation.

strategic air transport--The movement of personnel and materiel by air in accordance with a strategic plan.

strategic air transport operations(*)--The carriage of passengers and cargo between theaters by means of: a. scheduled service; b. special flight; c. air logistic support; d. aeromedical evacuation.

strategic air warfare--Air combat and supporting operations designed to effect, through the systematic application of force to a selected series of vital targets, the progressive destruction and disintegration of the enemy's war-making capacity to a point where the enemy no longer retains the ability or the will to wage war. Vital targets may include key manufacturing systems, sources of raw material, critical material, stockpiles, power systems, transportation systems, communication facilities, concentration of uncommitted elements of enemy armed forces, key agricultural areas, and other such target systems.

Strategic Army Forces--See **United States Strategic Army Forces.**

strategic concentration(*)--The assembly of designated forces in areas from which it is intended that operations of the assembled force shall begin so that they are best disposed to initiate the plan of campaign.

strategic concept(*)--The course of action accepted as the result of the estimate of the strategic situation. It is a statement of what is to be done in broad terms sufficiently flexible to permit its use in framing the military, diplomatic, economic, psychological and other measures which stem from it. See also **basic undertakings.**

strategic estimate--The estimate of the broad strategic factors that influence the determination of missions, objectives, and courses of action. The estimate is continuous and includes the strategic direction received from the National Command Authorities or the authoritative body of an alliance or coalition. See also **commander's estimate of the situation; estimate; logistic estimate of the situation; national intelligence estimate.**

strategic intelligence--Intelligence that is required for the formulation of strategy, policy, and military plans and operations at national and theater levels. See also **intelligence; operational intelligence; tactical intelligence.**

strategic level of war--The level of war at which a nation, often as a member of a group of nations, determines national or multinational (alliance or coalition) security objectives and guidance, and develops and uses national resources to accomplish these objectives. Activities at this level establish national and multinational military objectives; sequence initiatives; define limits and assess risks for the use of military and other instruments of national power; develop global plans or theater war plans to achieve these objectives; and provide military forces and other capabilities in accordance with strategic plans. See also **operational level of war; tactical level of war.**

strategic map--A map of medium scale, or smaller, used for planning of operations, including the movement, concentration, and supply of troops. See also **map.**

strategic material (critical)--A material required for essential uses in a war emergency, the procurement of which in adequate quantity, quality, or time, is sufficiently uncertain, for any reason, to require prior provision of the supply thereof.

strategic mining--A long-term mining operation designed to deny the enemy the use of specific sea routes or sea areas.

strategic mission--A mission directed against one or more of a selected series of enemy targets with the purpose of progressive destruction and disintegration of the enemy's warmaking capacity and his will to make war. Targets include key manufacturing systems, sources of raw material, critical material, stockpiles, power systems, transportation systems, communication facilities, and other such target systems. As opposed to tactical operations, strategic operations are designed to have a long-range, rather than immediate, effect on the enemy and its military forces.

strategic mobility--The capability to deploy and sustain military forces worldwide in support of national strategy. See also **mobility.**

strategic plan--A plan for the overall conduct of a war.

strategic psychological activities(*)--Planned psychological activities in peace and war which normally pursue objectives to gain the support and cooperation of friendly and neutral countries and to reduce the will and the capacity of hostile or potentially hostile countries to wage war.

strategic sealift--The afloat prepositioning and ocean movement of military material in support of US and allied forces. Sealift forces include organic and commercially acquired shipping and shipping services, including chartered foreign-flag vessels.

strategic transport aircraft(*)--Aircraft designed primarily for the carriage of personnel and/or cargo over long distances.

strategic vulnerability--The susceptibility of vital elements of national power to being seriously decreased or adversely changed by the application of actions within the capability of another nation to impose. Strategic vulnerability may pertain to political, geographic, economic, scientific, sociological, or military factors.

strategic warning--A warning prior to the initiation of a threatening act. See also **strategic warning lead time; strategic warning post-decision time; strategic warning predecision time; tactical warning; warning; warning of war.**

strategic warning lead time--That time between the receipt of strategic warning and the beginning of hostilities. This time may include two action periods: strategic warning pre-decision time and strategic warning post-decision time. See also **commander's estimate of the situation; strategic concept; strategic warning.**

strategic warning post-decision time--That time which begins after the decision, made at the highest levels of government(s) in response to strategic warning, is ordered executed and ends with the start of hostilities or termination of the threat. It is that part of strategic warning lead time available for executing pre-hostility actions to strengthen the national strategic posture; however, some preparatory actions may be initiated in the predecision period. See also **strategic warning; strategic warning lead time.**

strategic warning pre-decision time--That time which begins upon receipt of strategic warning and ends when a decision is ordered executed. It is that part of strategic warning lead time available to the highest levels of government(s) to determine that strategic course of action to be executed. See also **strategic warning; strategic warning lead time.**

strategy--The art and science of developing and using political, economic, psychological, and military forces as necessary during peace and war, to afford the maximum support to policies, in order to increase the probabilities and favorable consequences of victory and to lessen the chances of defeat. See also **military strategy; national strategy.**

strategy determination--The Joint Operation Planning and Execution System function in which analysis of changing events in the international environment and the development of national strategy to respond to those events is conducted. In joint operation planning, the responsibility for recommending military strategy to the National Command Authorities lies with the Chairman of the Joint Chiefs of Staff, in consultation with the other members of the Joint Chiefs of Staff and in concert with supported commanders. In the deliberate planning process, the Joint Strategic Capabilities Plan is produced as a result of this process. In the Crisis Assessment Phase of the crisis action planning process, Crisis Action Planning procedures are used to formulate decisions for direct development of possible military courses of action.

Stratofortress--An all-weather, intercontinental, strategic heavy bomber powered by eight turbojet engines. It is capable of delivering nuclear and nonnuclear bombs, air-to-surface missiles, and decoys. Its range is extended by in-flight refueling. Designated as **B-52.**

stratosphere--The layer of the atmosphere above the troposphere in which the change of temperature with height is relatively small. See also **atmosphere.**

Stratotanker--A multipurpose aerial tanker-transport powered by four turbojet engines. It is equipped for high-speed, high-altitude refueling of bombers and fighters. Designated as **KC-135.**

stream--Dispensing of chaff (solid/random interval/bursts).

stream takeoff(*)--Aircraft taking off in trail/column formation.

strength--See **economic potential; unit strength.**

strength group--A surface action group (unit) (element) composed of the heaviest combatant ships available with their aircraft and assigned screen.

stretcher--See **litter.**

stretch out--A reduction in the delivery rate specified for a program without a reduction in the total quantity to be delivered.

strike(*)--An attack which is intended to inflict damage on, seize, or destroy an objective.

strike cruiser--A warship designed to operate offensively with carrier strike forces or surface action groups against surface, air and subsurface threats. Planned armaments include the Aegis missile system, a major caliber gun, surface-to-surface missiles and advanced anti-submarine warfare weapons and sensors. Capability to operate helicopters or vertical takeoff and landing aircraft is planned.

strike force--A force composed of appropriate units necessary to conduct strikes, attack or assault operations. See also **task force.**

strike photography(*)--Air photographs taken during an air strike.

strip marker(*)--In land mine warfare, a marker, natural, artificial, or specially installed, located at the start and finish of a mine strip. See also **marker.**

strip plot(*)--A portion of a map or overlay on which a number of photographs taken along a flight line is delineated without defining the outlines of individual prints.

strong point(*)--A key point in a defensive position, usually strongly fortified and heavily armed with automatic weapons, around which other positions are grouped for its protection.

structured message text(*)--A message text composed of paragraphs ordered in a specified sequence, each paragraph characterized by an identifier and containing information in free form. It is designed to facilitate manual handling and processing. See also **formatted message text; free form message text.**

subassembly(*)--In logistics, a portion of an assembly, consisting of two or more parts, that can be provisioned and replaced as an entity. See also **assembly; component; part.**

subgravity(*)--A condition in which the resultant ambient acceleration is between 0 and 1 G.

subkiloton weapon(*)--A nuclear weapon producing a yield below one kiloton. See also **kiloton weapon; megaton weapon; nominal weapon.**

sublimited war--Not to be used. No substitute recommended.

submarine--A warship designed for under-the-surface operations with primary mission of locating and destroying ships, including other submarines. It is capable of various other naval missions. SSNs are nuclear powered. Designated as **SS and SSN.** See also **fleet ballistic missile submarine.**

submarine launched missile--See **sea-launched ballistic missile.**

submarine locator acoustic beacon(*)--An electronic device, used by submarines in distress, for emitting a repetitive sonic pulse underwater.

submarine patrol area(*)--A restricted area established to allow submarine operations: a. unimpeded by the operation of, or possible attack from, friendly forces in wartime; b. without submerged mutual interference in peacetime.

submarine rocket--Submerged, submarine-launched, surface-to-surface rocket with nuclear depth charge or homing torpedo payload, primarily antisubmarine. Also called **SUBROC.** Designated as **WM-44A.**

submarine safety lanes--See **safety lanes.**

submarine sanctuaries--Restricted areas that are established for the conduct of noncombat submarine or antisubmarine exercises. They may be either stationary or moving and are normally designated only in rear areas. See also **moving havens.**

submarine striking forces--Submarines having guided or ballistic missile launching and/or guidance capabilities formed to launch offensive nuclear strikes.

submunition(*)--Any munition that, to perform its task, separates from a parent munition.

subordinate command--A command consisting of the commander and all those individuals, units, detachments, organizations, or installations that have been placed under the command by the authority establishing the subordinate command.

SUBROC--See **submarine rocket.**

subscription--An agreement by a nation's Military Services to agree to accept and abide by, with or without reservation, the details of a standardization agreement. See also **implementation; ratification; reservation.**

subsidiary landing(*)--In an amphibious operation, a landing usually made outside the designated landing area, the purpose of which is to support the main landing.

subsonic--Of or pertaining to speeds less than the speed of sound. See also **speed of sound.**

substitute transport-type vehicle--A wheeled vehicle designed to perform, within certain limitations, the same military function as military transport vehicles, but not requiring all the special characteristics thereof. They are developed from civilian designs by addition of certain features, or from military designs by deletion of certain features. See also **vehicle.**

subversion--Action designed to undermine the military, economic, psychological, or political strength or morale of a regime. See also **unconventional warfare.**

subversion of DOD personnel--Actions designed to undermine the loyalty, morale or discipline of Department of Defense military and civilian personnel.

subversive activity--Anyone lending aid, comfort, and moral support to individuals, groups or organizations that advocate the overthrow of incumbent governments by force and violence is subversive and is engaged in subversive activity. All willful acts that are intended to be detrimental to the best interests of the government and that do not fall into the categories of treason, sedition, sabotage, or espionage will be placed in the category of subversive activity.

subversive political action--A planned series of activities designed to accomplish political objectives by influencing, dominating, or displacing individuals or groups who are so placed as to affect the decisions and actions of another government.

suitability--Operation plan review criterion. The determination that the course of action will reasonably accomplish the identified objectives, mission, or task if carried out successfully. See also **acceptability; adequacy; completeness; feasibility.**

summit--The highest altitude above mean sea level that a projectile reaches in its flight from the gun to the target; the algebraic sum of the maximum ordinate and the altitude of the gun.

supersonic--Of or pertaining to speed in excess of the speed of sound. See also **speed of sound.**

supervised route(*)--In road traffic, a roadway over which limited control is exercised by means of traffic control posts, traffic patrols or both. Movement authorization is required for its use by a column of vehicles or a vehicle of exceptional size or weight. See also **route.**

supplementary facilities(*)--Facilities required at a particular location to provide a specified minimum of support for reinforcing forces, which exceed the facilities required to support in-place forces.

supplies--In logistics, all materiel and items used in the equipment, support, and maintenance of military forces. See also **assembly; component; equipment; part; subassembly.**

supply--The procurement, distribution, maintenance while in storage, and salvage of supplies, including the determination of kind and quantity of supplies. **a. producer phase**--That phase of military supply which extends from determination of procurement schedules to acceptance of finished supplies by the Military Services. **b. consumer phase**--That phase of military supply which extends from receipt of finished supplies by the military Services through issue for use or consumption.

supply by air--See **airdrop; air movement.**

supply control--The process by which an item of supply is controlled within the supply system, including requisitioning, receipt, storage, stock control, shipment, disposition, identification, and accounting.

supplying ship(*)--The ship in a replenishment unit that provides the personnel and/or supplies to be transferred.

supply management--See **inventory control.**

supply point(*)--Any point where supplies are issued in detail.

supply transaction reporting--Reporting on individual transactions affecting the stock status of materiel to the appropriate supply accounting activity as they occur.

support--1. The action of a force which aids, protects, complements, or sustains another force in accordance with a directive requiring such action. 2. A unit which helps another unit in battle. Aviation, artillery, or naval gunfire may be used as a support for infantry. 3. A part of any unit held back at the beginning of an attack as a reserve. 4. An element of a command which assists, protects, or supplies other forces in combat. See also **close support; direct support; general support; interdepartmental/agency support; international logistic support; inter-Service support; mutual support.**

supported commander--The commander having primary responsibility for all aspects of a task assigned by the Joint Strategic Capabilities Plan or other joint operation planning authority. In the context of joint operation planning, this term refers to the commander who prepares operation plans or operation orders in response to requirements of the Chairman of the Joint Chiefs of Staff.

support helicopter--See **assault aircraft; utility helicopter (maneuver); assault aircraft.**

supporting aircraft--All active aircraft other than unit aircraft. See also **aircraft.**

supporting arms--Air, sea, and land weapons of all types employed to support ground units.

supporting arms coordination center--A single location on board an amphibious command ship in which all communication facilities incident to the coordination of fire support of the artillery, air, and naval gunfire are centralized. This is the naval counterpart to the fire support coordination center utilized by the landing force. See also **fire support coordination center.**

supporting artillery--Artillery which executes fire missions in support of a specific unit, usually infantry, but remains under the command of the next higher artillery commander.

supporting attack(*)--An offensive operation carried out in conjunction with a main attack and designed to achieve one or more of the following: a. deceive the enemy; b. destroy or pin down enemy forces which could interfere with the main attack; c. control ground whose occupation by the enemy will hinder the main attack; or d. force the enemy to commit reserves prematurely or in an indecisive area.

supporting commander--A commander who provides augmentation forces or other support to a supported commander or who develops a supporting plan. Includes the designated combatant commands and Defense agencies as appropriate.

supporting fire(*)--Fire delivered by supporting units to assist or protect a unit in combat. See also **close supporting fire; deep supporting fire; direct supporting fire.**

supporting forces--Forces stationed in, or to be deployed to, an area of operations to provide support for the execution of an operation order. Combatant command (command authority) of supporting forces is not passed to the supported commander.

supporting operations(*)--In amphibious operations, those operations conducted by forces other than those assigned to the amphibious task force. They are ordered by higher authority at the request of the amphibious task force commander and normally are conducted outside the area for which the amphibious task force commander is responsible at the time of their execution.

supporting plan--An operation plan prepared by a supporting commander or a subordinate commander to satisfy the requests or requirements of the supported commander's plan.

support items--Items subordinate to, or associated with an end item (i.e., spares, repair parts, tools, test equipment and sundry materiel) and required to operate, service, repair or overhaul an end item.

support site--In the Air Force, a facility operated by an active, reserve, or Guard unit that provides general support to the Air Force mission and does not satisfy the criteria for a major or minor installation. Examples of support sites are missile tracking sites; radar bomb scoring sites; Air Force-owned, contractor-operated plants; radio relay sites, etc. See also **installation complex; major installation; minor installation; other activity.**

suppression--Temporary or transient degradation by an opposing force of the performance of a weapons system below the level needed to fulfill its mission objectives.

suppression mission--A mission to suppress an actual or suspected weapons system for the purpose of degrading its performance below the level needed to fulfill its mission objectives at a specific time for a specified duration.

suppression of enemy air defenses--That activity which neutralizes, destroys, or temporarily degrades surface-based enemy air defenses by destructive and/or disruptive means. Also called **SEAD.** See also **electromagnetic spectrum; electronic warfare.**

suppressive fire--Fires on or about a weapons system to degrade its performance below the level needed to fulfill its mission objectives, during the conduct of the fire mission. See also **fire.**

surface burst--See **nuclear surface burst.**

surface code--See **panel code.**

surface combatant--A ship constructed and armed for combat use with the capability to conduct operations in multiple maritime roles against air, surface and subsurface threats, and land targets.

surface striking forces (naval)--Forces that are organized primarily to do battle with enemy forces or to conduct shore bombardment. Units comprising such a force are generally incorporated in and operate as part of another force, but with provisions for their formation into a surface striking force should such action appear likely and/or desirable.

surface-to-air guided missile(*)--A surface-launched guided missile for use against air targets.

surface-to-air missile envelope--That air space within the kill capabilities of a specific surface-to-air missile system.

surface-to-air missile installation--A surface-to-air missile site with the surface-to-air missile system hardware installed.

surface-to-air missile site--A plot of ground prepared in such a manner that it will readily accept the hardware used in surface-to-air missile system.

surface-to-surface guided missile(*)--A surface-launched guided missile for use against surface targets.

surface zero--See **ground zero.**

surplus property--Any excess property not required for the needs and for the discharge of the responsibilities of all federal agencies, including the Department of Defense, as determined by the General Services Administration.

surprise dosage attack(*)--A chemical operation which establishes on target a dosage sufficient to produce the desired casualties before the troops can mask or otherwise protect themselves.

surveillance(*)--The systematic observation of aerospace, surface or subsurface areas, places, persons, or things, by visual, aural, electronic, photographic, or other means. See also **air surveillance; satellite and missile surveillance; sea surveillance.**

surveillance approach--An instrument approach conducted in accordance with directions issued by a controller referring to the surveillance radar display.

survey control point--A survey station used to coordinate survey control.

survey information center--A place where survey data are collected, correlated, and made available to subordinate units.

survey photography--See **air cartographic photography.**

susceptibility(*)--The vulnerability of a target audience to particular forms of psychological operations approach.

suspect--In counterdrug operations, a track of interest where correlating information actually ties the track of interest to alleged narcotics operations. See also **special interest target; track of interest.**

suspension equipment(*)--All aircraft devices such as racks, adapters, missile launchers and pylons used for carriage, employment and jettison of aircraft stores.

suspension strop(*)--A length of webbing or wire rope between the helicopter and cargo sling.

sustainability--See **military capability.**

sustained attrition minefield(*)--In naval mine warfare, a minefield which is replenished to maintain its danger to the enemy in the face of countermeasures.

sustained rate of fire(*)--Actual rate of fire that a weapon can continue to deliver for an indefinite length of time without seriously overheating.

sustaining stocks(*)--Stocks to support the execution of approved operational plans beyond the initial predetermined period covered by basic stocks until resupply is available for support of continued operations.

sustainment--The provision of personnel, logistic, and other support required to maintain and prolong operations or combat until successful accomplishment or revision of the mission or of the national objective.

sweep--To employ technical means to uncover planted microphones or other surveillance devices. See also **technical survey.**

sweeper track--See **hunter track.**

sweep jamming(*)--A narrow band of jamming that is swept back and forth over a relatively wide operating band of frequencies.

swept path(*)--In naval mine warfare, the width of the lane swept by the mechanical sweep at all depths less than the sweep depth.

switch horn(*)--In naval mine warfare, a switch in a mine operated by a projecting spike. See also **horn.**

sympathetic detonation(*)--Detonation of a charge by exploding another charge adjacent to it.

synchronization--1. The arrangement of military actions in time, space, and purpose to produce maximum relative combat power at a decisive place and time. 2. In the intelligence context, application of intelligence sources and methods in concert with the operational plan.

synthesis--In intelligence usage, the examining and combining of processed information with other information and intelligence for final interpretation.

synthetic exercise(*)--An exercise in which enemy and/or friendly forces are generated, displayed and moved by electronic or other means on simulators, radar scopes or other training devices.

system--Any organized assembly of resources and procedures united and regulated by interaction or interdependence to accomplish a set of specific functions.

system manager--A general term of reference to those organizations directed by individual managers, exercising authority over the planning, direction, and control of tasks and associated functions essential for support of designated weapons or equipment systems. The authority vested in this organization may include such functions as research, development, procurement, production, materiel distribution, and logistic support, when so assigned. When intended to relate to a specific system manager, this term will be preceded by the appropriate designation (e.g., Chinook System Manager, Sonar System Manager, F-4 System Manager). This term will normally be used in lieu of system support manager, weapon system manager, program manager, and project manager when such organizations perform these functions.

systems design(*)--The preparation of an assembly of methods, procedures, or techniques united by regulated interaction to form an organized whole.

system support manager--See **system manager.**

T

table of allowance--An equipment allowance document which prescribes basic allowances of organizational equipment, and provides the control to develop, revise, or change equipment authorization inventory data.

table of organization--See **establishment.**

table of organization and equipment--See **establishment.**

tacan(*)--An ultra-high frequency electronic air navigation system, able to provide continuous bearing and slant range to a selected station. The term is derived from tactical air navigation.

tacit arms control agreement--An arms control course of action in which two or more nations participate without any formal agreement having been made.

tac-log group--Representatives designated by troop commanders to assist Navy control officers aboard control ships in the ship-to-shore movement of troops, equipment, and supplies.

TACON--See **tactical control.**

tactical aeromedical evacuation(*)--That phase of evacuation which provides airlift for patients from the combat zone to points outside the combat zone, and between points within the communications zone.

tactical air command center--The principal United States Marine Corps air operation installation from which aircraft and air warning functions of tactical air operations are directed. It is the senior agency of the Marine Corps air command and control system from which the Marine Corps tactical air commander can direct and control tactical air operations and coordinate such air operations with other Services.

tactical air commander (ashore)--The officer (aviator) responsible to the landing force commander for control and coordination of air operations within the landing force commander's area of responsibility when control of these operations is passed ashore.

tactical air control center(*)--The principal air operations installation (land- or ship-based) from which all aircraft and air warning functions of tactical air operations are controlled.

tactical air control group--**1. land-based**--A flexible administrative and tactical component of a tactical air organization which provides aircraft control and warning functions ashore for offensive and defensive missions within the tactical air zone of responsibility. **2. ship-based**--An administrative and tactical component of an amphibious force which provides aircraft control and warning facilities afloat for offensive and defensive missions within the tactical air command area of responsibility.

tactical air controller--**(NATO)** The officer in charge of all operations of the tactical air control center. He is responsible to the tactical air commander for the control of all aircraft and air warning facilities within his area of responsibility. See also **air controller.**

tactical air control operations team--A team of ground environment personnel assigned to certain allied tactical air control units/elements.

tactical air control party(*)--A subordinate operational component of a tactical air control system designed to provide air liaison to land forces and for the control of aircraft.

tactical air control party support team--An Army team organized to provide armored combat and/or special purpose vehicles and crews to certain tactical air control parties.

tactical air control squadron--1. **land-based**--A flexible administrative component of a tactical air control group, known as TACRON, which provides the control mechanism for a land-based tactical air control center, a tactical air direction center, or tactical air control parties. 2. **ship-based**--An administrative and tactical component of the tactical air control group, known as TACRON, which provides the control mechanism for the ship-based tactical air direction center or the ship-based tactical air control center.

tactical air control system(*)--The organization and equipment necessary to plan, direct, and control tactical air operations and to coordinate air operations with other Services. It is composed of control agencies and communications-electronics facilities which provide the means for centralized control and decentralized execution of missions.

tactical air coordinator (airborne)--An officer who coordinates, from an aircraft, the action of combat aircraft engaged in close support of ground or sea forces. See also **forward observer.**

tactical air direction center--An air operations installation under the overall control of the tactical air control center (afloat)/tactical air command center, from which aircraft and air warning service functions of tactical air operations in an area of responsibility are directed. See also **tactical air director.**

tactical air director--The officer in charge of all operations of the tactical air direction center. This officer is responsible to the tactical air controller for the direction of all aircraft and air warning facilities assigned to the area of responsibility. When operating independently of a tactical air control center (afloat), the tactical air director assumes the functions of the tactical air controller. See also **tactical air direction center.**

tactical air doctrine(*)--Fundamental principles designed to provide guidance for the employment of air power in tactical air operations to attain established objectives.

tactical air force(*)--An air force charged with carrying out tactical air operations in coordination with ground or naval forces.

tactical air groups (shore-based)--Task organizations of tactical air units assigned to the amphibious task force that are to be land-based within, or sufficiently close to, the objective area to provide tactical air support to the amphibious task force.

tactical air observer--An officer trained as an air observer whose function is to observe from airborne aircraft and report on movement and disposition of friendly and enemy forces, on terrain, weather, and hydrography and to execute other missions as directed.

tactical air officer (afloat)--The officer (aviator) under the amphibious task force commander who coordinates planning of all phases of air participation of the amphibious operation and air operations of supporting forces en route to and in the objective area. Until control is passed ashore, this officer exercises control over all operations of the tactical air control center (afloat) and is charged with: a. control of all aircraft in the objective area assigned for tactical air operations, including offensive and defensive air; b. control of all other aircraft entering or passing through the objective area; and c. control of all air warning facilities in the objective area.

tactical air operation--An air operation involving the employment of air power in coordination with ground or naval forces to: a. gain and maintain air superiority; b. prevent movement of enemy forces into and within the objective area and to seek out and destroy these forces and their supporting installations; c. join with ground or naval forces in operations within the objective area, in order to assist directly in attainment of their immediate objective.

tactical air operations center--A subordinate operational component of the Marine air command and control system designed for direction and control of all en route air traffic and air defense operations, to include manned interceptors and surface-to-air weapons, in an assigned sector. It is under the operational control of the tactical air command center.

tactical air reconnaissance--The use of air vehicles to obtain information concerning terrain, weather, and the disposition, composition, movement, installations, lines of communications, electronic and communication emissions of enemy forces. Also included are artillery and naval gunfire adjustment, and systematic and random observation of ground battle areas, targets, and/or sectors of airspace.

tactical air support(*)--Air operations carried out in coordination with surface forces and which directly assist land or maritime operations. See also **air support.**

tactical air support element--An element of a United States Army division, corps, or field army tactical operations center consisting of G-2 and G-3 air personnel who coordinate and integrate tactical air support with current tactical ground operations.

tactical air transport operations(*)--The carriage of passengers and cargo within a theater by means of: a. Airborne operations: (1) Parachute assault, (2) Helicopter borne assault, (3)

Air landing; b. Air logistic support; c. Special missions; d. Aeromedical evacuation missions.

tactical area of responsibility--A defined area of land for which responsibility is specifically assigned to the commander of the area as a measure for control of assigned forces and coordination of support. Also called **TAOR.**

tactical call sign(*)--A call sign which identifies a tactical command or tactical communication facility. See also **call sign.**

tactical combat force--A combat unit, with appropriate combat support and combat service support assets, that is assigned the mission of defeating Level III threats.

tactical concept(*)--A statement, in broad outline, which provides a common basis for future development of tactical doctrine. See also **tactical sub-concept.**

tactical control(*)--The detailed and, usually, local direction and control of movements or maneuvers necessary to accomplish missions or tasks assigned. (DOD) Also called **TACON.** See also **combatant command (command authority); operational control.**

tactical deception group--A task organization that conducts deception operations against the enemy, including electronic, communication, visual, and other methods designed to misinform and confuse the enemy.

tactical digital information link--A Joint Staff approved, standardized communication link suitable for transmission of digital information. Current practice is to characterize a tactical digital information link (TADIL) by its standardized message formats and transmission characteristics. TADILs interface two or more command and control or weapon systems via a single or multiple network architecture and multiple communication media for exchange of tactical information. a. TADIL-A--A secure, half-duplex, netted digital data link utilizing parallel transmission frame characteristics and standard message formats at either 1364 or 2250 bits per second. It is normally operated in a roll-call mode under control of a net control station to exchange digital information among airborne, land-based, and shipboard systems. NATO's equivalent is Link 11. b. TADIL-B--A secure, full-duplex, point-to-point digital data link utilizing serial transmission frame characteristics and standard message formats at either 2400, 1200, or 600 bits per second. It interconnects tactical air defense and air control units. NATO's equivalent is Link 11B. c. TADIL-C--An unsecure, time-division digital data link utilizing serial transmission characteristics and standard message formats at 5000 bits per second from a controlling unit to controlled aircraft. Information exchange can be one-way (controlling unit to controlled aircraft) or two-way. NATO's equivalent is Link 4. d. TADIL-J--A secure, high capacity, jam-resistant, nodeless data link which uses the Joint Tactical Information Distribution System (JTIDS) transmission characteristics and the protocols, conventions, and fixed-length message formats defined by the JTIDS Technical Interface Design Plan (TIDP). NATO's equivalent is Link 16. e. Army Tactical Data Link 1 (ATDL-1)--A secure, full-duplex, point-to-point digital data link utilizing serial transmission frame characteristics and standard message formats at a basic speed of 1200 bits per second. It interconnects tactical air control systems and Army or Marine tactical air defense oriented systems. f. Interim JTIDS Message Specification (IJMS)--A secure, high capacity, jam-resistant, nodeless interim message specification that uses the Joint Tactical Information Distribution System (JTIDS) transmission characteristics and the protocols, conventions, and fixed-length message formats defined by the IJMS. See also **airborne tactical data system; data link.**

tactical diversion--See **diversion.**

tactical information processing and interpretation system--A tactical, mobile, land-based, automated information-handling system designed to store and retrieve intelligence information and to process and interpret imagery or nonimagery data. Also called **TIPI.**

tactical intelligence--Intelligence that is required for planning and conducting tactical operations. See also **intelligence; operational intelligence; strategic intelligence.**

tactical intelligence and related activities--Those activities outside the National Foreign Intelligence Program that: a. respond to operational commanders' tasking for time-sensitive information on foreign entities; b. respond to national intelligence community tasking of systems whose primary mission is support to operating forces; c. train personnel for intelligence duties; d. provide an intelligence reserve; or e. are devoted to research and development of intelligence or related capabilities. Specifically excluded are programs which are so closely integrated with a weapon system that their primary function is to provide immediate-- use targeting data. Also called **TIARA.**

tactical level of war--The level of war at which battles and engagements are planned and executed to accomplish military objectives assigned to tactical units or task forces. Activities at this level focus on the ordered arrangement and maneuver of combat elements in relation to each other and to the enemy to achieve combat objectives. See also **operational level of war; strategic level of war.**

tactical loading--See **combat loading; unit loading.**

tactical locality(*)--An area of terrain which, because of its location or features, possesses a tactical significance in the particular circumstances existing at a particular time.

tactical map--A large-scale map used for tactical and administrative purposes. See also **map.**

tactical minefield(*)--A minefield which is part of a formation obstacle plan and is laid to delay, channel or break up an enemy advance.

tactical mining(*)--In naval mine warfare, mining designed to influence a specific operation or to counter a known or presumed tactical aim of the enemy. Implicit in tactical mining is a limited period of effectiveness of the minefield.

tactical nuclear weapon employment--The use of nuclear weapons by land, sea, or air forces against opposing forces, supporting installations or facilities, in support of operations which contribute to the accomplishment of a military mission of limited scope, or in support of the military commander's scheme of maneuver, usually limited to the area of military operations.

tactical obstacles--Those obstacles employed to disrupt enemy formations, to turn them into a desired area, to fix them in position under direct and indirect fires, and to block enemy penetrations.

tactical operations area--That area between the fire support coordination line and the rear operations area where maximum flexibility in the use of airspace is needed to assure mission accomplishment. The rear boundary of the tactical operations area should normally be at or near the rear boundary of the frontline divisions.

tactical operations center--A physical groupment of those elements of an Army general and special staff concerned with the current tactical operations and the tactical support thereof. Also called **TOC.**

tactical range(*)--A range in which realistic targets are in use and a certain freedom of maneuver is allowed.

tactical reserve(*)--A part of a force, held under the control of the commander as a maneuvering force to influence future action.

tactical security(*)--In operations, the measures necessary to deny information to the enemy and to ensure that a force retains its freedom of action and is warned or protected against an unexpected encounter with the enemy or an attack. See also **physical security; protective security; security.**

tactical sub-concept(*)--A statement, in broad outline, for a specific field of military capability within a tactical concept which provides a common basis both for equipment and weapon system development and for future development of tactical doctrine. See also **tactical concept.**

tactical transport aircraft(*)--Aircraft designed primarily for the carriage of personnel and/or cargo over short or medium distances.

tactical troops--Combat troops, together with any service troops required for their direct support, who are organized under one commander to operate as a unit and engage the enemy in combat. See also **troops.**

tactical unit--An organization of troops, aircraft, or ships which is intended to serve as a single unit in combat. It may include service units required for its direct support.

tactical vehicle--See **military designed vehicle.**

tactical warning--1. A warning after initiation of a threatening or hostile act based on an evaluation of information from all available sources. 2. In satellite and missile surveillance, a notification to operational command centers that a specific threat event is occurring. The

374

component elements that describe threat events are: **Country of origin**--country or countries initiating hostilities. **Event type and size**--identification of the type of event and determination of the size or number of weapons. **Country under attack**--determined by observing trajectory of an object and predicting its impact point. **Event time**--time the hostile event occurred. Also called **integrated tactical warning**. See also **attack assessment; strategic warning.**

tactical warning and assessment--A composite term. See separate definitions for tactical warning and for attack assessment.

tactical warning and attack assessment--A composite term. See separate definitions for tactical warning and for attack assessment.

tactical warning/attack assessment--A composite term. See separate definitions for tactical warning and for attack assessment.

tactics--1. The employment of units in combat. 2. The ordered arrangement and maneuver of units in relation to each other and/or to the enemy in order to use their full potentialities.

TADIL--See **tactical digital information link.**

TAI--See **International Atomic Time.**

tail hook--See **aircraft arresting hook.**

tally ho--A code meaning, "Target visually sighted" (presumably the target I have been ordered to intercept). This should be followed by initial contact report as soon as possible. The sighting should be amplified if possible (e.g., "tally ho pounce," or "tally ho heads up").

tank, combat, full-tracked, 105-mm gun--A heavy, fully armored combat vehicle providing mobile fire power and crew protection for offensive combat, armed with one 105-mm

gun, one 7.62-mm gun and one 50-caliber machine gun. Designated as **M-60.**

tank, combat, full-tracked, 152-mm gun--A heavily armored vehicle providing mobile firepower and crew protection for offensive combat armed with one 152-mm gun/launcher, capable of firing Shillelagh missiles or conventional combustible ammunition, one 50-caliber machine gun and one 7.62-mm machine gun.

tank, combat, full-tracked, 90-mm gun--A fully armored combat vehicle providing mobile fire power and crew protection for offensive combat, armed with one 90-mm gun, one 50-caliber machine gun, and one 7.62-mm machine gun. Designated as **M48A3.**

tank landing ship--A naval ship designed to transport and land amphibious vehicles, tanks, combat vehicles, and equipment in amphibious assault. Designated as **LST.**

tank, main battle--A tracked vehicle providing mobile firepower and crew protection for offensive combat.

TAOR--See **tactical area of responsibility.**

target--1. A geographical area, complex, or installation planned for capture or destruction by military forces. 2. In intelligence usage, a country, area, installation, agency, or person against which intelligence operations are directed. 3. An area designated and numbered for future firing. 4. In gunfire support usage, an impact burst which hits the target. See also **objective area.**

target acquisition(*)--The detection, identification, and location of a target in sufficient detail to permit the effective employment of weapons. See also **target analysis.**

target analysis(*)--An examination of potential targets to determine military importance, priority of attack, and weapons required to obtain a desired level of damage or casualties. See also **target acquisition.**

target approach point(*)--In air transport operations, a navigational check point over which the final turn into the drop zone/landing zone is made. See also **initial point.**

target area survey base(*)--A base line used for the locating of targets or other points by the intersection of observations from two stations located at opposite ends on the line.

target array--A graphic representation of enemy forces, personnel, and facilities in a specific situation, accompanied by a target analysis.

target audience(*)--An individual or group selected for influence or attack by means of psychological operations.

target base line--A line connecting prime targets along the periphery of a geographic area.

target bearing--1. **true**--The true compass bearing of a target from a firing ship. 2. **relative**--The bearing of a target measured in the horizontal from the bow of one's own ship clockwise from 0 degrees to 360 degrees, or from the nose of one's own aircraft in hours of the clock.

target CAP--See **target combat air patrol.**

target classification--A grouping of targets in accordance with their threat to the amphibious task force and its component elements: targets not to be fired upon prior to D-day and targets not to be destroyed except on direct orders.

target combat air patrol--A patrol of fighters maintained over an enemy target area to destroy enemy aircraft and to cover friendly shipping in the vicinity of the target area in

amphibious operations. See also **combat air patrol.**

target complex(*)--A geographically integrated series of target concentrations. See also **target.**

target component--A major element of a target complex or target. It is any machinery, structure, personnel, or other productive asset that contributes to the operation or output of the target complex or target. See also **target; target critical damage point.**

target concentration(*)--A grouping of geographically proximate targets. See also **target; target complex.**

target critical damage point--The part of a target component that is most vital. Also called **critical node.** See also **target; target component.**

target data inventory--A basic targeting program which provides a standardized target data in support of the requirements of the Joint Chiefs of Staff, Military Departments, and unified and specified commands for target planning coordination and weapons application.

target date(*)--The date on which it is desired that an action be accomplished or initiated.

target description--See **description of target.**

target director post--A special control element of the tactical air control system. It performs no air warning service but is used to position friendly aircraft over predetermined target coordinates, or other geographical locations, under all weather conditions.

target discrimination(*)--The ability of a surveillance or guidance system to identify or engage any one target when multiple targets are present.

target dossier(*)--A file of assembled target intelligence about a specific geographic area.

target folder(*)--A folder containing target intelligence and related materials prepared for planning and executing action against a specific target.

targeting--1. The process of selecting targets and matching the appropriate response to them, taking account of operational requirements and capabilities. 2. The analysis of enemy situations relative to the commander's mission, objectives, and capabilities at the commander's disposal, to identify and nominate specific vulnerabilities that, if exploited, will accomplish the commander's purpose through delaying, disrupting, disabling, or destroying enemy forces or resources critical to the enemy. See also **joint targeting coordination board.**

target intelligence(*)--Intelligence which portrays and locates the components of a target or target complex and indicates its vulnerability and relative importance.

target list--The listing of targets maintained and promulgated by the senior echelon of command; it contains those targets that are to be engaged by supporting arms, as distinguished from a "list of targets" that may be maintained by any echelon as confirmed, suspected, or possible targets for informational and planning purposes. See also **joint target list; list of targets.**

target materials--Graphic, textual, tabular, digital, video, or other presentations of target intelligence, primarily designed to support operations against designated targets by one or more weapon(s) systems. Target materials are suitable for training, planning, executing, and evaluating military operations. See also **air target materials program.**

target of opportunity--1. A target visible to a surface or air sensor or observer, which is within range of available weapons and against which fire has not been scheduled or requested. **2. nuclear**--A nuclear target observed or detected after an operation begins that has not been previously considered, analyzed or planned for a nuclear strike. Generally fleeting in nature, it should be attacked as soon as possible within the time limitations imposed for coordination and warning of friendly troops and aircraft.

target overlay(*)--A transparent sheet which, when superimposed on a particular chart, map, drawing, tracing or other representation, depicts target locations and designations. The target overlay may also show boundaries between maneuver elements, objectives and friendly forward dispositions.

target pattern--The flight path of aircraft during the attack phase. Also called **attack pattern.**

target priority--A grouping of targets with the indicated sequence of attack.

target range--See **range.**

target response (nuclear)(*)--The effect on men, material, and equipment of blast, heat, light, and nuclear radiation resulting from the explosion of a nuclear weapon.

target signature(*)--1. The characteristic pattern of a target displayed by detection and identification equipment. 2. In naval mine warfare, the variation in the influence field produced by the passage of a ship or sweep.

target stress point--The weakest point (most vulnerable to damage) on the critical damage point. Also called **vulnerable node.** See also **target critical damage point.**

target system(*)--1. All the targets situated in a particular geographic area and functionally related. (DOD) 2. A group of targets which are so related that their destruction will produce some particular effect desired by the attacker. See also **target complex.**

target system component--A set of targets belonging to one or more groups of industries and basic utilities required to produce component parts of an end product such as periscopes, or one type of a series of interrelated commodities, such as aviation gasoline.

task component--A subdivision of a fleet, task force, task group, or task unit, organized by the respective commander or by higher authority for the accomplishment of specific tasks.

task element--A component of a naval task unit organized by the commander of a task unit or higher authority.

task fleet--A mobile command consisting of ships and aircraft necessary for the accomplishment of a specific major task or tasks which may be of a continuing nature.

task force(*)--1. A temporary grouping of units, under one commander, formed for the purpose of carrying out a specific operation or mission. 2. Semi-permanent organization of units, under one commander, formed for the purpose of carrying out a continuing specific task. 3. A component of a fleet organized by the commander of a task fleet or higher authority for the accomplishment of a specific task or tasks. See also **force(s).**

task group--A component of a naval task force organized by the commander of a task force or higher authority.

task organization--1. In the Navy, an organization which assigns to responsible commanders the means with which to accomplish their assigned tasks in any planned action. 2. An organization table pertaining to a specific naval directive.

task-organizing--The act of designing an operating force, support staff, or logistics package of specific size and composition to meet a unique task or mission. Characteristics to examine when task-organizing the force include, but are not limited to: training, experience, equipage, sustainability, operating environment, enemy threat, and mobility.

task unit--A component of a naval task group organized by the commander of a task group or higher authority.

taxiway(*)--A specially prepared or designated path on an airfield for the use of taxiing aircraft.

TCC--See **transportation component command.**

T-day--See **times.**

tear line--A physical line on an intelligence message or document which separates categories of information that have been approved for foreign disclosure and release. Normally, the intelligence below the tear line is that which has been previously cleared for disclosure or release.

TECDOC--See **technical documentation.**

technical analysis(*)--In imagery interpretation, the precise description of details appearing on imagery.

technical assistance--The providing of advice, assistance, and training pertaining to the installation, operation, and maintenance of equipment.

technical characteristics--Those characteristics of equipment which pertain primarily to the engineering principles involved in producing equipment possessing desired military characteristics, e.g., for electronic equipment, technical characteristics include such items as circuitry, and types and arrangement of components.

technical documentation--Visual information documentation (with or without sound as an integral documentation component) of an actual event made for purposes of evaluation. Typically, technical documentation contributes to the study of human or mechanical factors, procedures, and processes in the fields of medicine, science, logistics, research, development, test and evaluation, intelligence, investigations, and armament delivery. Also called **TECDOC.** See also **visual information documentation.**

technical escort--An individual technically qualified and properly equipped to accompany designated material requiring a high degree of safety or security during shipment.

technical evaluation--The study and investigations by a developing agency to determine the technical suitability of material, equipment, or a system, for use in the Military Services. See also **operational evaluation.**

technical information--Information, including scientific information, which relates to research, development, engineering, test, evaluation, production, operation, use, and maintenance of munitions and other military supplies and equipment.

technical intelligence--Intelligence derived from exploitation of foreign materiel, produced for strategic, operational, and tactical level commanders. Technical intelligence begins when an individual service member finds something new on the battlefield and takes the proper steps to report it. The item is then exploited at succeedingly higher levels until a countermeasure is produced to neutralize the adversary's

technological advantage. Also called **TECHINT.** See also **intelligence; scientific and technical intelligence.**

technical operational intelligence--A Defense Intelligence Agency initiative to provide enhanced scientific and technical intelligence to the commanders of unified commands and their subordinates through a closed loop system involving all Service and Defense Intelligence Agency scientific and technical intelligence centers. Through a system manager in the National Military Joint Intelligence Center, the technical operational intelligence program provides timely collection, analysis, and dissemination of area of responsibility specific scientific and technical intelligence to combatant commanders and their subordinates for planning, training, and executing joint operations. Also called **TOPINT.**

technical review authority--The organization tasked to provide specialized technical or administrative expertise to the primary review authority or coordinating review authority for joint publications. See also **coordinating review authority; joint publication; primary review authority.**

technical specification(*)--A detailed description of technical requirements stated in terms suitable to form the basis for the actual design development and production processes of an item having the qualities specified in the operational characteristics. See also **operational characteristics.**

technical supply operations--Operations performed by supply units or technical supply elements of supply and maintenance units in acquiring, accounting for, storing, and issuing Class II and IV items needed by supported units and maintenance activities.

technical survey--A complete electronic and physical inspection to ascertain that offices, conference rooms, war rooms, and other similar locations where classified information is discussed are free of monitoring systems. See also **sweep**.

telecommunication(*)--Any transmission, emission, or reception of signs, signals, writings, images, sounds, or information of any nature by wire, radio, visual, or other electromagnetic systems.

telecommunications center--A facility, normally serving more than one organization or terminal, responsible for transmission, receipt, acceptance, processing and distribution of incoming and outgoing messages.

teleconference(*)--A conference between persons remote from one another but linked by a telecommunications system.

telemetry intelligence--Technical intelligence derived from the intercept, processing, and analysis of foreign telemetry. Telemetry intelligence is a category of foreign instrumentation signals intelligence. Also called **TELINT**. See also **electronics intelligence; intelligence; foreign instrumentation signals intelligence**.

teleprocessing--The combining of telecommunications and computer operations interacting in the automatic processing, reception, and transmission of data and/or information.

teleran system--A navigational system which: a. employs ground-based search radar equipment along an airway to locate aircraft flying near that airway; b. transmits, by television means, information pertaining to these aircraft and other information to the pilots of properly equipped aircraft; and c. provides information to the pilots appropriate for use in the landing approach.

television imagery--Imagery acquired by a television camera and recorded or transmitted electronically.

TELINT--See **telemetry intelligence**.

telling--See **track telling**.

temperature gradient--At sea, a temperature gradient is the change of temperature with depth; a positive gradient is a temperature increase with an increase in depth, and a negative gradient is a temperature decrease with an increase in depth.

temporary cemetery(*)--A cemetery for the purpose of: a. The initial burial of the remains if the circumstances permit or b. The reburial of remains exhumed from an emergency burial.

terminal clearance capacity--The amount of cargo or personnel that can be moved through and out of a terminal on a daily basis.

terminal control area--A control area or portion thereof normally situated at the confluence of air traffic service routes in the vicinity of one or more major airfields. See also **airway; control area; controlled airspace; control zone**.

terminal guidance--1. The guidance applied to a guided missile between midcourse guidance and arrival in the vicinity of the target. 2. Electronic, mechanical, visual, or other assistance given an aircraft pilot to facilitate arrival at, operation within or over, landing upon, or departure from an air landing or airdrop facility. See also **guidance**.

terminal operations--The reception, processing, and staging of passengers, the receipt, transit storage and marshalling of cargo, the loading and unloading of ships or aircraft, and the manifesting and forwarding of cargo and passengers to destination.

terminal phase--That portion of the trajectory of a ballistic missile between reentry into the atmosphere or the end of the mid-course phase and impact or arrival in the vicinity of the target. See also **boost phase; midcourse phase; reentry phase.**

terminal velocity(*)--1. Hypothetical maximum speed a body could attain along a specified flight path under given conditions of weight and thrust if diving through an unlimited distance in air of specified uniform density. 2. Remaining speed of a projectile at the point in its downward path where it is level with the muzzle of the weapon.

terrain analysis(*)--The collection, analysis, evaluation, and interpretation of geographic information on the natural and manmade features of the terrain, combined with other relevant factors, to predict the effect of the terrain on military operations.

terrain avoidance system(*)--A system which provides the pilot or navigator of an aircraft with a situation display of the ground or obstacles which project above either a horizontal plane through the aircraft or a plane parallel to it, so that the pilot can maneuver the aircraft to avoid the obstruction.

terrain clearance system(*)--A system which provides the pilot, or autopilot, of an aircraft with climb or dive signals such that the aircraft will maintain a selected height over flat ground and clear the peaks of undulating ground within the selected height in a vertical plane through the flight vector. This system differs from terrain following in that the aircraft need not descend into a valley to follow the ground contour.

terrain exercise--An exercise in which a stated military situation is solved on the ground, the troops being imaginary and the solution usually being in writing.

terrain flight(*)--Flight close to the Earth's surface during which airspeed, height and/or altitude are adapted to the contours and cover of the ground in order to avoid enemy detection and fire.

terrain following system(*)--A system which provides the pilot or autopilot of an aircraft with climb or dive signals such that the aircraft will maintain as closely as possible, a selected height above a ground contour in a vertical plane through the flight vector.

terrain intelligence--Processed information on the military significance of natural and manmade characteristics of an area.

terrain study--An analysis and interpretation of natural and manmade features of an area, their effects on military operations, and the effect of weather and climate on these features.

terrestrial environment--The Earth's land area, including its manmade and natural surface and sub-surface features, and its interfaces and interactions with the atmosphere and the oceans.

terrestrial reference guidance--The technique of providing intelligence to a missile from certain characteristics of the surface over which the missile is flown, thereby achieving flight along a predetermined path. See also **guidance.**

terrorism--The calculated use of violence or threat of violence to inculcate fear; intended to coerce or to intimidate governments or societies in the pursuit of goals that are generally political, religious, or ideological. See also **antiterrorism; combatting terrorism; counterterrorism; terrorist; terrorist groups; terrorist threat conditions.**

terrorist--An individual who uses violence, terror, and intimidation to achieve a result. See also **terrorism.**

terrorist groups--Any element regardless of size or espoused cause, which repeatedly commits acts of violence or threatens violence in pursuit of its political, religious, or ideological objectives. See also **terrorism**.

terrorist threat conditions--A Chairman of the Joint Chiefs of Staff-approved program standardizing the Military Services' identification of and recommended responses to terrorist threats against US personnel and facilities. This program facilitates inter-Service coordination and support for antiterrorism activities. Also called **THREATCONS**. There are four THREATCONS above normal: **a. THREATCON ALPHA**--This condition applies when there is a general threat of possible terrorist activity against personnel and facilities, the nature and extent of which are unpredictable, and circumstances do not justify full implementation of THREATCON BRAVO measures. However, it may be necessary to implement certain measures from higher THREATCONS resulting from intelligence received or as a deterrent. The measures in this THREATCON must be capable of being maintained indefinitely. **b. THREATCON BRAVO**--This condition applies when an increased and more predictable threat of terrorist activity exists. The measures in this THREATCON must be capable of being maintained for weeks without causing undue hardship, affecting operational capability, and aggravating relations with local authorities. **c. THREATCON CHARLIE**--This condition applies when an incident occurs or intelligence is received indicating some form of terrorist action against personnel and facilities is imminent. Implementation of measures in this THREATCON for more than a short period probably will create hardship and affect the peacetime activities of the unit and its personnel. **d. THREATCON DELTA**--This condition applies in the immediate area where a terrorist attack has occurred or when intelligence has been received that terrorist action against a specific location or person is likely. Normally, this THREATCON is declared as a localized condition. See also **antiterrorism**.

test depth(*)--The depth to which the submarine is tested by actual or simulated submergence. See also **maximum operating depth**.

tests--See **service test; troop test**.

theater--The geographical area outside the continental United States for which a commander of a combatant command has been assigned responsibility.

theater of focus--A theater in which operations are most critical to national interests and are assigned the highest priority for allocation of resources. See also **economy of force theater**.

theater of operations--See **area of operations**.

theater of war--See **area of war**.

theater strategy--The art and science of developing integrated strategic concepts and courses of action directed toward securing the objectives of national and alliance or coalition security policy and strategy by the use of force, threatened use of force, or operations not involving the use of force within a theater. See also **military strategy; national military strategy; national security strategy; strategy**.

thermal crossover--The natural phenomenon which normally occurs twice daily when temperature conditions are such that there is a loss of contrast between two adjacent objects on infrared imagery.

thermal energy--The energy emitted from the fireball as thermal radiation. The total amount of thermal energy received per unit area at a specified distance from a nuclear explosion is generally expressed in terms of calories per square centimeter.

thermal exposure(*)--The total normal component of thermal radiation striking a given surface throughout the course of a detonation; expressed in calories per square centimeter and/or megajoules per square meter.

thermal imagery(*)--Imagery produced by sensing and recording the thermal energy emitted or reflected from the objects which are imaged.

thermal pulse--The radiant power versus time pulse from a nuclear weapon detonation.

thermal radiation(*)--1. The heat and light produced by a nuclear explosion. (DOD) 2. Electromagnetic radiations emitted from a heat or light source as a consequence of its temperature; it consists essentially of ultraviolet, visible, and infrared radiations.

thermal shadow(*)--The tone contrast difference of infrared linescan imagery which is caused by a thermal gradient which persists as a result of a shadow of an object which has been moved.

thermal X-rays(*)--The electromagnetic radiation, mainly in the soft (low-energy) X-ray region, emitted by the debris of a nuclear weapon by virtue of its extremely high temperature.

thermonuclear(*)--An adjective referring to the process (or processes) in which very high temperatures are used to bring about the fusion of light nuclei, with the accompanying liberation of energy.

thermonuclear weapon(*)--A weapon in which very high temperatures are used to bring about the fusion of light nuclei such as those of hydrogen isotopes (e.g., deuterium and tritium) with the accompanying release of energy. The high temperatures required are obtained by means of fission.

third area conflict--Not to be used. See **cold war; general war; guerrilla warfare; limited war; low intensity conflict.**

thorough decontamination(*)--Decontamination carried out by a unit, with or without external support, to reduce contamination on personnel, equipment, materiel and/or working areas to the lowest possible levels, to permit the partial or total removal of individual protective equipment and to maintain operations with minimum degradation. This may include terrain decontamination beyond the scope of operational decontamination. See also **decontamination; immediate decontamination; operational decontamination.**

threat analysis--In antiterrorism, threat analysis is a continual process of compiling and examining all available information concerning potential terrorist activities by terrorist groups which could target a facility. A threat analysis will review the factors of a terrorist group's existence, capability, intentions, history, and targeting, as well as the security environment within which friendly forces operate. Threat analysis is an essential step in identifying probability of terrorist attack and results in a threat assessment. See also **antiterrorism.**

threat and vulnerability assessment--In antiterrorism, the pairing of a facility's threat analysis and vulnerability analysis. See also **antiterrorism.**

THREATCON ALPHA--See **terrorist threat conditions.**

THREATCON BRAVO--See **terrorist threat conditions.**

THREATCON CHARLIE--See **terrorist threat conditions.**

THREATCON DELTA--See **terrorist threat conditions.**

THREATCONS--See **terrorist threat conditions.**

threat identification and assessment--The Joint Operation Planning and Execution System function that provides timely warning of potential threats to US interests; intelligence collection requirements; the effects of environmental, physical, and health hazards, and cultural factors on friendly and enemy operations; and determines the enemy military posture and possible intentions.

threat-oriented munitions(*)--In stockpile planning, munitions intended to neutralize a finite assessed threat and for which the total requirement is determined by an agreed mathematical model. See also **level-of-effort munitions.**

threshold(*)--The beginning of that portion of the runway usable for landing.

throughput--The average quantity of cargo and passengers that can pass through a port on a daily basis from arrival at the port to loading onto a ship or plane, or from the discharge from a ship or plane to the exit (clearance) from the port complex. Throughput is usually expressed in measurement tons, short tons, or passengers. Reception and storage limitation may affect final throughput.

Thunderbolt II--A twin-engine, subsonic, turbofan, tactical fighter/bomber. It is capable of employing a variety of air-to-surface-launched weapons in the close air support role. Short--field, unimproved surfaces are considered normal takeoff/landing operating areas. This aircraft is also capable of long endurance in the target area and is supplemented by air refueling. An internally mounted 30-mm cannon is capable of destroying a wide variety of armor--protected vehicles. Designated as **A-10.**

TIARA--See **tactical intelligence and related activities.**

tied on--In air intercept, a code meaning, "The aircraft indicated is in formation with me."

tie down(*)--The fastening or securing of a load to its carrier by use of ropes, cables or other means to prevent shifting during transport. Also used (as a noun) to describe the material employed to secure a load.

tie down diagram(*)--A drawing indicating the prescribed method of securing a particular item of cargo within a specific type of vehicle.

tie down point(*)--An attachment point provided on or within a vehicle for securing cargo.

tie down point pattern(*)--The pattern of tie down points within a vehicle.

tilt angle(*)--The angle between the optical axis of an air camera and the vertical at the time of exposure. See also **angle of depression.**

time--An epoch, i.e., the designation of an instant on a selected time scale, astronomical or atomic. It is used in the sense of time of day.

time and frequency standard--A reference value of time and time interval. Standards of time and frequency are determined by astronomical observations and by the operation of atomic clocks and other advanced timekeeping instruments. They are disseminated by transport of clocks, radio transmissions, satellite relay, and other means.

time and material contract--A contract providing for the procurement of supplies or services on the basis of: a. direct labor hours at specified fixed hourly rates (which rates include direct and indirect labor, overhead, and profit); and b. material at cost.

time fuze(*)--A fuze which contains a graduated time element to regulate the time interval after which the fuze will function. See also **fuze.**

time interval--Duration of a segment of time without reference to when the time interval begins or ends. Time intervals may be given in seconds of time or fractions thereof.

time of attack--The hour at which the attack is to be launched. If a line of departure is prescribed, it is the hour at which the line is to be crossed by the leading elements of the attack.

time of delivery--The time at which the addressee or responsible relay agency receipts for a message.

time of flight(*)--In artillery and naval gunfire support, the time in seconds from the instant a weapon is fired, launched, or released from the delivery vehicle or weapons system to the instant it strikes or detonates.

time of origin--The time at which a message is released for transmission.

time of receipt--The time at which a receiving station completes reception of a message.

time on target--1. Time at which aircraft are scheduled to attack/photograph the target. 2. The actual time at which aircraft attack/photograph the target. 3. The time at which a nuclear detonation is planned at a specified desired ground zero.

time on target (air)--See **time on target--Parts 1 and 2.**

time over target conflict--A situation wherein two or more delivery vehicles are scheduled such that their proximity violates the established separation criteria for yield, time, distance, or all three.

time over target (nuclear)--See **time on target---Part 3.**

time-phased force and deployment data--The computer-supported data base portion of an operation plan; it contains time-phased force data, non-unit-related cargo and personnel data, and movement data for the operation plan, including: a. In-place units. b. Units to be deployed to support the operation plan with a priority indicating the desired sequence for their arrival at the port of debarkation. c. Routing of forces to be deployed. d. Movement data associated with deploying forces. e. Estimates of non-unit-related cargo and personnel movements to be conducted concurrently with the deployment of forces. f. Estimate of transportation requirements that must be fulfilled by common-user lift resources as well as those requirements that can be fulfilled by assigned or attached transportation resources. Also called **TPFDD.**

time-phased force and deployment data maintenance--The deliberate planning process that requires a supported commander to incorporate changes to a time-phased force and deployment data (TPFDD) that occur after the TPFDD becomes effective for execution. TPFDD maintenance is conducted by the supported combatant commander in coordination with the supporting combatant commanders, Service components, US Transportation Command, and other agencies as required. At designated intervals, changes to data in the TPFDD, including force structure, standard reference files, and Services' type unit characteristics file, are updated in Joint Operation Planning and Execution System (JOPES) to ensure currency of deployment data. TPFDD maintenance may also be used to update the TPFDD for Chairman of the Joint Chiefs of Staff or Joint Strategic Capabilities Plan submission in lieu of refinement during the JOPES plan development phase. Also called **TPFDD maintenance.** See also **time-phased force and deployment data; time-phased force and deployment data refinement.**

time-phased force and deployment data refinement--For both global and regional operation plan development, the process consists of several discrete phases time-phased force and deployment data (TPFDD) that may be conducted sequentially or concurrently, in whole or in part. These phases are Concept, Plan Development, and Review. The Plan Development Phase consists of several subphases: Forces, Logistics, and Transportation, with shortfall identification associated with each phase. The Plan Development phases are collectively referred to as TPFDD refinement. The normal TPFDD refinement process consists of sequentially refining forces, logistics (non-unit-related personnel and sustainment), and transportation data to develop a TPFDD file that supports a feasible and adequate overlapping of several refinement phases. The decision is made by the supported commander, unless otherwise directed by the Chairman of the Joint Chiefs of Staff. For global planning, refinement conferences are conducted by the Joint Staff in conjunction with US Transportation Command. TPFDD refinement is conducted in coordination with supported and supporting commanders, Services, the Joint Staff, and other supporting agencies. Commander in Chief, US Transportation Command, will normally host refinement conferences at the request of the Joint Staff or the supported commander. Also called **TPFDD refinement.** See also **time-phased force and deployment data; time-phased force and deployment data maintenance.**

time-phased force and deployment list--Appendix 1 to Annex A of the operation plan. It identifies types and/or actual units required to support the operation plan and indicates origin and ports of debarkation or ocean area. It may also be generated as a computer listing from the time-phased force and deployment data. Also called **TPFDL.**

times--(C-, D-, M-days end at 2400 hours Universal Time (zulu time) and are assumed to be 24 hours long for planning.) The Chairman of the Joint Chiefs of Staff normally coordinates the proposed date with the commanders of the appropriate unified and specified commands, as well as any recommended changes to C-day. L-hour will be established per plan, crisis, or theater of operations and will apply to both air and surface movements. Normally, L-hour will be established to allow C-day to be a 24-hour day. **a. C-day.** The unnamed day on which a deployment operation commences or is to commence. The deployment may be movement of troops, cargo, weapon systems, or a combination of these elements using any or all types of transport. The letter "C" will be the only one used to denote the above. The highest command or headquarters responsible for coordinating the planning will specify the exact meaning of C-day within the aforementioned definition. The command or headquarters directly responsible for the execution of the operation, if other than the one coordinating the planning, will do so in light of the meaning specified by the highest command or headquarters coordinating the planning. **b. D-day.** The unnamed day on which a particular operation commences or is to commence. **c. F-hour.** The effective time of announcement by the Secretary of Defense to the Military Departments of a decision to mobilize Reserve units. **d. H-hour.** The specific hour on D-day at which a particular operation commences. **e. L-hour.** The specific hour on C-day at which a deployment operation commences or is to commence. **f. M-day.** The term used to designate the unnamed day on which full mobilization commences or is due to commence. **g. N-day.** The unnamed day an active duty unit is notified for deployment or redeployment. **h. R-day.** Redeployment day. The day on which redeployment of major combat, combat support, and combat service support forces begins in an operation. **i. S-day.** The day the President authorizes

Selected Reserve callup (not more than 200,000). **j. T-day.** The effective day coincident with Presidential declaration of National Emergency and authorization of partial mobilization (not more than 1,000,000 personnel exclusive of the 200,000 callup). **k. W-day.** Declared by the National Command Authorities, W-day is associated with an adversary decision to prepare for war (unambiguous strategic warning).

time-sensitive special operations planning--The planning for the deployment and employment of assigned, attached, and allocated forces and resources that occurs in response to an actual situation. Time-sensitive planners base their plan on the actual circumstances that exist at the time planning occurs. See also **deliberate planning.**

time-sensitive targets--Those targets requiring immediate response because they pose (or will soon pose) a clear and present danger to friendly forces or are highly lucrative, fleeting targets of opportunity.

time slot(*)--Period of time during which certain activities are governed by specific regulations.

time-to-go--During an air intercept, the time to fly to the offset point from any given interceptor position; after passing the offset point, the time to fly to the intercept point.

tip--See **pitch.**

tips--External fuel tanks.

title block--See **information box.**

TNT equivalent(*)--A measure of the energy released from the detonation of a nuclear weapon, or from the explosion of a given quantity of fissionable material, in terms of the amount of TNT (Trinitrotoluene) which could

release the same amount of energy when exploded.

TOI--See **track of interest.**

tolerance dose--The amount of radiation which may be received by an individual within a specified period with negligible results.

Tomahawk--An air-, land-, ship-, or submarine-launched cruise missile with three variants: land attack with conventional or nuclear capability, and tactical anti-ship with conventional warhead.

Tomcat--A twin turbofan, dual-crew, supersonic, all-weather, long-range interceptor designed to operate from aircraft carriers. It carries a wide assortment of air-to-air and air-to-ground missiles and conventional ordnance. Primary mission is long-range fleet air defense with secondary close air support capability. Designated as **F-14.**

tone down--See **attenuation.**

TOPINT--See **technical operational intelligence.**

topographic base--See **chart base.**

topographic map--A map which presents the vertical position of features in measurable form as well as their horizontal positions. See also **map.**

top secret--See **defense classification.**

torpedo defense net(*)--A net employed to close an inner harbor to torpedoes fired from seaward or to protect an individual ship at anchor or underway.

toss bombing--A method of bombing where an aircraft flies on a line towards the target, pulls up in a vertical plane, releasing the bomb at an angle that will compensate for the effect of gravity drop on the bomb. Similar to loft bombing; unrestricted as to altitude. See also **loft bombing; over-the-shoulder bombing.**

total active aircraft authorization--The sum of the primary and backup aircraft authorizations.

total active aircraft inventory--The sum of the primary and backup aircraft assigned to meet the total active aircraft authorization.

total dosage attack(*)--A chemical operation which does not involve a time limit within which to produce the required toxic level.

total materiel assets--The total quantity of an item available in the military system worldwide and all funded procurement of the item with adjustments to provide for transfers out of or into the inventory through the appropriation and procurement lead-time periods. It includes peacetime force materiel assets and war reserve stock.

total materiel requirement--The sum of the peacetime force material requirement and the war reserve material requirement.

total mobilization--See **mobilization.**

total nuclear war--Not to be used. See **general war.**

total overall aircraft inventory--The sum of the total active aircraft inventory and the inactive aircraft inventory.

total pressure(*)--The sum of dynamic and static pressures.

total war--Not to be used. See **general war.**

touchdown(*)--The contact, or moment of contact, of an aircraft or spacecraft with the landing surface.

touchdown zone(*)--1. For fixed wing aircraft--- The first 3,000 feet or 1,000 meters of runway beginning at the threshold. 2. For rotary wings and vectored thrust aircraft--That portion of the helicopter landing area or runway used for landing.

TOW (missile)--A component of a tube-launched, optically tracked, wire-command link guided missile weapon system which is crew-portable.

toxic chemical, biological, or radiological attack--An attack directed at personnel, animals, or crops, using injurious agents of radiological, biological, or chemical origin.

toxin agent--A poison formed as a specific secretion product in the metabolism of a vegetable or animal organism as distinguished from inorganic poisons. Such poisons can also be manufactured by synthetic processes.

TPFDD--See **time-phased force and deployment data.**

TPFDD maintenance--See **time-phased force and deployment data maintenance.**

TPFDD refinement--See **time-phased force and deployment data refinement.**

track--1. A series of related contacts displayed on a plotting board. 2. To display or record the successive positions of a moving object. 3. To lock onto a point of radiation and obtain guidance therefrom. 4. To keep a gun properly aimed, or to point continuously a target-locating instrument at a moving target. 5. The actual path of an aircraft above, or a ship on, the surface of the Earth. The course is the path that is planned; the track is the path that is actually taken. 6. One of the two endless belts on which a full-track or half-track vehicle runs. 7. A metal part forming a path for a moving object, e.g., the track around the inside of a vehicle for moving a mounted machine gun.

track correlation--Correlating track information for identification purposes using all available data.

track crossing angle--In air intercept, the angular difference between interceptor track and target track at the time of intercept.

tracking(*)--1. Precise and continuous position--finding of targets by radar, optical, or other means. (DOD) 2. In air intercept, a code meaning, "By my evaluation, target is steering true course indicated."

track mode--In a flight control system, a control mode in which the ground track of an aircraft is maintained automatically.

track of interest--In counterdrug operations, contacts that meet the initial sorting criteria applicable in the area where the contacts are detected. Also called **TOI**. See also **special interest target; suspect.**

track production area(*)--An area in which tracks are produced by one radar station.

track symbology(*)--Symbols used to display tracks on a data display console or other display device.

track telling--The process of communicating air surveillance and tactical data information between command and control systems or between facilities within the systems. Telling may be classified into the following types: back tell; cross tell; forward tell; lateral tell; overlap tell and relateral tell.

tractor group--A group of landing ships in an amphibious operation which carries the amphibious vehicles of the landing force.

trafficability(*)--Capability of terrain to bear traffic. It refers to the extent to which the terrain will permit continued movement of any and/or all types of traffic.

traffic circulation map--A map showing traffic routes and the measures for traffic regulation. It indicates the roads for use of certain classes of traffic, the location of traffic control stations, and the directions in which traffic may move. Also called **a circulation map.** See also **map.**

traffic control police(*)--Any persons ordered by a military commander and/or by national authorities to facilitate the movement of traffic and to prevent and/or report any breach of road traffic regulations.

traffic density(*)--The average number of vehicles that occupy one mile or one kilometer of road space, expressed in vehicles per mile or per kilometer.

traffic flow(*)--The total number of vehicles passing a given point in a given time. Traffic flow is expressed as vehicles per hour.

traffic flow security--The protection resulting from features, inherent in some cryptoequipment, which conceal the presence of valid messages on a communications circuit, normally achieved by causing the circuit to appear busy at all times.

traffic information (radar)--Information issued to alert an aircraft to any radar targets observed on the radar display which may be in such proximity to its position or intended route of flight to warrant its attention.

traffic management--The direction, control, and supervision of all functions incident to the procurement and use of freight and passenger transportation services.

traffic pattern--The traffic flow that is prescribed for aircraft landing at, taxiing on, and taking off from an airport. The usual components of a traffic pattern are upwind leg, crosswind leg, downwind leg, base leg, and final approach.

trail(*)--1. A term applied to the manner in which a bomb trails behind the aircraft from which it has been released, assuming the aircraft does not change its velocity after the release of the bomb. (DOD) 2. Track (or shadow). (The words "landward" or "seaward" may be used to indicate from which side of enemy unit to shadow.)

trailer aircraft(*)--Aircraft which are following and keeping under surveillance a designated airborne contact.

train--1. A service force or group of service elements which provides logistic support, e.g., an organization of naval auxiliary ships or merchant ships or merchant ships attached to a fleet for this purpose; similarly, the vehicles and operating personnel which furnish supply, evacuation, and maintenance services to a land unit. 2. Bombs dropped in short intervals or sequence.

trained strength in units--Those reservists assigned to units who have completed initial active duty for training of 12 weeks or its equivalent and are eligible for deployment overseas on land when mobilized under proper authority. Excludes personnel in non-deployable accounts or a training pipeline.

train headway--The interval of time between two trains boarded by the same unit at the same point.

training aids--Any item which is developed and/or procured with the primary intent that it shall assist in training and the process of learning.

training and retirement category--The category identifying (by specific training and retirement

category designator) a reservist's training or retirement status in a reserve component category and reserve component.

training-pay category--A designation identifying the number of days of training and pay required for members of the Reserve Components.

training period--An authorized and scheduled regular inactive duty training period. A training period must be at least two hours for retirement point credit and four hours for pay. Previously used interchangeably with other common terms such as drills, drill period, assemblies, periods of instruction, etc.

training pipeline--A reserve component category designation that identifies untrained officer and enlisted personnel who have not completed initial active duty for training of 12 weeks or its equivalent. See also **nondeployable account.**

training unit--A unit established to provide military training to individual reservists or to reserve component units.

train path(*)--In railway terminology, the timing of a possible movement of a train along a given route. All the train paths on a given route constitute a timetable.

trajectory--See **ballistic trajectory.**

transattack period--1. In nuclear warfare, the period from the initiation of the attack to its termination. 2. As applied to the Single Integrated Operational Plan, the period which extends from execution (or enemy attack, whichever is sooner) to termination of the Single Integrated Operational Plan. See also **postattack period.**

transfer area--In an amphibious operation, the water area in which the transfer of troops and supplies from landing craft to amphibious vehicles is effected.

transfer loader(*)--A wheeled or tracked vehicle with a platform capable of vertical and horizontal adjustment used in the loading and unloading of aircraft, ships, or other vehicles.

transient--1. Personnel, ships, or craft stopping temporarily at a post, station or port to which they are not assigned or attached, and having destination elsewhere. 2. An independent merchant ship calling at a port and sailing within 12 hours, and for which routing instructions to a further port have been promulgated. (DOD, NATO) 3. An individual awaiting orders, transport, etc., at a post or station to which he is not attached or assigned.

transient forces--Forces which pass or stage through, or base temporarily within, the area of responsibility of another command but are not under its operational control.

transit area--See **staging area.**

transit bearing(*)--A bearing determined by noting the time at which two features on the Earth's surface have the same relative bearing.

transition altitude--The altitude at or below which the vertical position of an aircraft is controlled by reference to true altitude.

transition layer(*)--The airspace between the transition altitude and the transition level.

transition level(*)--The lowest flight level available for use above the transition altitude. See also **altitude; transition altitude.**

transit route(*)--A sea route which crosses open waters normally joining two coastal routes.

transmission factor (nuclear)--The ratio of the dose inside the shielding material to the outside

(ambient) dose. Transmission factor is used to calculate the dose received through the shielding material. See also **half thickness; shielding.**

transmission security--See **communications security.**

transonic(*)--Of or pertaining to the speed of a body in a surrounding fluid when the relative speed of the fluid is subsonic in some places and supersonic in others. This is encountered when passing from subsonic to supersonic speeds and vice versa. See also **speed of sound.**

transparency(*)--An image fixed on a clear base by means of a photographic, printing, chemical or other process, especially adaptable for viewing by transmitted light. See also **diapositive.**

transponder(*)--A receiver-transmitter which will generate a reply signa, upon proper interrogation. See also **responsor.**

transponder india--International Civil Aviation Organization/secondary surveillance radar.

transponder sierra--Identification Friend or Foe mark X (selective identification feature).

transponder tango--Identification Friend or Foe mark X (basic).

transportability--The capability of material to be moved by towing, self-propulsion, or carrier via any means, such as railways, highways, waterways, pipelines, oceans, and airways.

transport aircraft(*)--Aircraft designed primarily for the carriage of personnel and/or cargo. Transport aircraft may be classed according to range, as follows: **a.** **Short-range**--Not to exceed 1200 nautical miles at normal cruising conditions (2222 Km). **b.** **Medium-range**--

Between 1200 and 3500 nautical miles at normal cruising conditions (2222 and 6482 Km). **c. Long-range**--Exceeds 3500 nautical miles at normal cruising conditions (6482 Km). See also **strategic transport aircraft; tactical transport aircraft.**

transport area--In amphibious operations, an area assigned to a transport organization for the purpose of debarking troops and equipment. See also **inner transport area; outer transport area.**

transportation closure--The actual arrival date of a specified movement requirement at port of debarkation.

transportation component command--The three component commands of USTRANSCOM: Air Force Air Mobility Command, Navy Military Sealift Command, and Army Military Traffic Management Command. Each transportation component command remains a major command of its parent Service and continues to organize, train, and equip its forces as specified by law. Each transportation component command also continues to perform Service-unique missions. Also called **TCC.**

transportation emergency--A situation created by a shortage of normal transportation capability and of a magnitude sufficient to frustrate military movement requirements, and which requires extraordinary action by the President or other designated authority to ensure continued movement of essential Department of Defense Traffic.

transportation operating agencies--Those Federal agencies having responsibilities under national emergency conditions for the operational direction of one or more forms of transportation. Also called **Federal Modal Agencies; Federal Transport Agencies.**

transportation priorities--Indicators assigned to eligible traffic which establish its movement precedence. Appropriate priority systems apply to the movement of traffic by sea and air. In times of emergency, priorities may be applicable to continental United States movements by land, water, or air.

transportation system--All the land, water, and air routes and transportation assets engaged in the movement of US forces and their supplies during peacetime training, conflict, or war, involving both mature and contingency theaters and at the strategic, operational, and tactical levels of war.

transport control center (air transport)--The operations center through which the air transport force commander exercises control over the air transport system.

transport group--An element that directly deploys and supports the landing of the landing force (LF) is functionally designated as **a transport group in the amphibious task force organization.** A transport group provides for the embarkation, movement to the objective, landing, and logistic support of the LF. Transport groups comprise all sealift and airlift in which the LF is embarked. They are categorized as follows: a. airlifted groups. b. Navy amphibious ship transort groups. c. strategic sealift shipping groups.

transport stream(*)--Transport aircraft flying in single file, either in formation or singly, at defined intervals. See also **column formation; trail formation.**

transport vehicle--A motor vehicle designed and used without modification to the chassis, to provide general transport service in the movement of personnel and cargo. See also **vehicle.**

transshipment point(*)--A location where material is transferred between vehicles.

traverse(*)--1. To turn a weapon to the right or left on its mount. 2. A method of surveying in which lengths and directions of lines between points on the earth are obtained by or from field measurements, and used in determining positions of the points.

traverse level(*)--That vertical displacement above low-level air defense systems, expressed both as a height and altitude, at which aircraft can cross the area.

treason--Violation of the allegiance owed to one's sovereign or state; betrayal of one's country.

trench burial--A method of burial resorted to when casualties are heavy whereby a trench is prepared and the individual remains are laid in it side by side, thus obviating the necessity of digging and filling in individual graves. See also **burial.**

trend--The straying of the fall of shot, such as might be caused by incorrect speed settings of the fire support ship.

triage(*)--The evaluation and classification of casualties for purposes of treatment and evacuation. It consists of the immediate sorting of patients according to type and seriousness of injury, and likelihood of survival, and the establishment of priority for treatment and evacuation to assure medical care of the greatest benefit to the largest number.

triangulation station(*)--A point on the Earth, the position of which is determined by triangulation. Also called **trig point.**

tri-camera photography(*)--Photography obtained by simultaneous exposure of three cameras systematically disposed in the air vehicle at fixed overlapping angles relative to each other in order to cover a wide field. See also **fan camera photography.**

Trident--A general descriptive term for the sea-based strategic weapon system consisting of the highly survivable nuclear-powered Trident submarine, long-range Trident ballistic missiles and the integrated refit facilities required to support the submarine and missile subsystems as well as associated personnel.

Trident I--A three-stage, solid propellant ballistic missile capable of being launched from a Trident submarine either surfaced or submerged. It is sized to permit backfit into Poseidon submarines and is equipped with advanced guidance, nuclear warheads and a maneuverable bus which can deploy these warheads to separate targets. It is capable of carrying a full payload to 4000 nautical miles with greater ranges achievable in reduced payload configurations. Designated as **UGM-96A.**

Trident II--A solid propellant ballistic missile capable of being launched from a Trident submarine. It is larger than the Trident I missile and will replace these missiles in Ohio-class submarines. It will provide the option to deploy a higher throw weight, more accurate, submarine-launched ballistic missile.

trig list--A list published by certain Army units which includes essential information of accurately located survey points.

trim for takeoff feature--A flight control system feature in which the control surfaces of an aircraft are automatically trimmed to a predetermined takeoff position.

trim size(*)--The size of a map or chart sheet when the excess paper outside the margin has been trimmed off after printing.

triple point--The intersection of the incident, reflected, and fused (or Mach) shock fronts accompanying an air burst. The height of the triple point above the surface, i.e., the height of the Mach stem, increases with increasing distance from a given explosion.

troop basis--An approved list of those military units and individuals (including civilians) required for the performance of a particular mission by numbers, organization and equipment, and, in the case of larger commands, by deployment.

troops--A collective term for uniformed military personnel (usually not applicable to naval personnel afloat). See also **airborne troops; combat service support elements; combat support troops; combat troops; service troops; tactical troops.**

troop safety (nuclear)--An element which defines a distance from the proposed burst location beyond which personnel meeting the criteria described under degree of risk will be safe to the degree prescribed.

troop space cargo--Cargo such as sea or barracks bags, bedding rolls or hammocks, locker trunks, and office equipment, which is normally stowed in an accessible place. This cargo will also include normal hand-carried combat equipment and weapons to be carried ashore by the assault troops. See also **cargo.**

troop test--A test conducted in the field for the purpose of evaluating operational or organizational concepts, doctrine, tactics, and techniques, or to gain further information on material. See also **service test.**

tropical storm--A tropical cyclone in which the surface wind speed is at least 34, but not more than 63 knots.

tropopause(*)--The transition zone between the stratosphere and the troposphere. The tropopause normally occurs at an altitude of about 25,000 to 45,000 feet (8 to 15 kilometers) in polar and temperate zones, and at 55,000 feet

(20 kilometers) in the tropics. See also **atmosphere.**

troposphere(*)--The lower layers of atmosphere, in which the change of temperature with height is relatively large. It is the region where clouds form, convection is active, and mixing is continuous and more or less complete. See also **atmosphere.**

tropospheric scatter(*)--The propagation of electromagnetic waves by scattering as a result of irregularities or discontinuities in the physical properties of the troposphere.

true airspeed indicator(*)--An instrument which displays the speed of the aircraft relative to the ambient air.

true altitude--The height of an aircraft as measured from mean sea level.

true bearing(*)--The direction to an object from a point, expressed as a horizontal angle measured clockwise from true north.

true convergence--The angle at which one meridian is inclined to another on the surface of the Earth. See also **convergence.**

true horizon(*)--1. The boundary of a horizontal plane passing through a point of vision. 2. In photogrammetry, the boundary of a horizontal plane passing through the perspective center of a lens system.

true north(*)--The direction from an observer's position to the geographic North Pole. The north direction of any geographic meridian.

turbojet--A jet engine whose air is supplied by a turbine-driven compressor, the turbine being activated by exhaust gases.

turn and slip indicator(*)--An instrument which combines the functions of a turn indicator and a slip indicator.

turnaround(*)--The length of time between arriving at a point and being ready to depart from that point. It is used in this sense for the loading, unloading, re-fueling and re-arming, where appropriate, of vehicles, aircraft and ships. See also **turnaround cycle.**

turnaround cycle(*)--A term used in conjunction with vehicles, ships and aircraft, and comprising the following: loading time at departure point; time to and from destination; unloading and loading time at destination; unloading time at returning point; planned maintenance time, and where applicable, time awaiting facilities. See also **turnaround.**

turn indicator(*)--An instrument which displays the aircraft's rate and direction of turn.

turning movement(*)--A variation of the envelopment in which the attacking force passes around or over the enemy's principal defensive positions to secure objectives deep in the enemy's rear to force the enemy to abandon his position or divert major forces to meet the threat.

turning point(*)--In land mine warfare, a point of the centerline of a mine strip or row where strips or rows change direction.

turn-in point(*)--The point at which an aircraft starts to turn from the approach direction to the line of attack. See also **contact point; pull-up point.**

turn-off guidance(*)--Information which enables the pilot of a landing aircraft to select and follow the correct taxiway from the time the aircraft leaves the runway until it may safely be brought to a halt clear of the active runway.

two-person rule--A system designed to prohibit access by an individual to nuclear weapons and certain designated components by requiring the presence at all times of at least two authorized persons, each capable of detecting incorrect or unauthorized procedures with respect to the task to be performed.

type command--An administrative subdivision of a fleet or force into ships or units of the same type, as differentiated from a tactical subdivision. Any type command may have a flagship, tender, and aircraft assigned to it.

types of burst--See **airburst; fallout safe height of burst; height of burst; high airburst; high altitude burst; low airburst; nuclear airburst; nuclear exoatmospheric burst; nuclear surface burst; nuclear underground burst; nuclear underwater burst; optimum height of burst; safe burst height.**

type unit--A type of organizational or functional entity established within the Armed Forces and uniquely identified by a five-character, alpha-numeric code called a unit type code.

type unit data file--A file that provides standard planning data and movement characteristics for personnel, cargo, and accompanying supplies associated with type units.

U

UAV--See **unmanned aerial vehicle.**

UGM-27--See **Polaris.**

UGM-73A--See **Poseidon.**

UGM-84A--See **Harpoon.**

UGM-96A--See **Trident I.**

UH-1--See **Iroquois.**

ULN--See **unit line number.**

ultraviolet imagery--That imagery produced as a result of sensing ultraviolet radiations reflected from a given target surface.

UNAAF--See **Unified Action Armed Forces.**

unaccounted for--An inclusive term (not a casualty status) applicable to personnel whose person or remains are not recovered or otherwise accounted for following hostile action. Commonly used when referring to personnel who are killed in action and whose bodies are not recovered. See also **casualty; casualty category; casualty status; casualty type.**

uncertain environment--See **operational environment.**

uncharged demolition target(*)--A demolition target for which charges have been calculated, prepared, and stored in a safe place, and for which execution procedures have been established. See also **demolition; demolition target.**

unclassified matter(*)--Official matter which does not require the application of security safeguards, but the disclosure of which may be subject to control for other reasons. See also **classified matter.**

uncontrolled mosaic(*)--A mosaic composed of uncorrected photographs, the details of which have been matched from print to print, without ground control or other orientation. Accurate measurement and direction cannot be accomplished. See also **controlled mosaic.**

unconventional warfare--A broad spectrum of military and paramilitary operations, normally of long duration, predominantly conducted by indigenous or surrogate forces who are organized, trained, equipped, supported, and directed in varying degrees by an external source. It includes guerrilla warfare and other direct offensive, low visibility, covert, or clandestine operations, as well as the indirect activities of subversion, sabotage, intelligence activities, and evasion and escape. Also called **UW.**

unconventional warfare forces--United States forces having an existing unconventional warfare capability consisting of Army Special Forces and such Navy, Air Force, and Marine units as are assigned for these operations.

understowed cargo--See **flatted cargo.**

underwater demolition(*)--The destruction or neutralization of underwater obstacles; this is normally accomplished by underwater demolition teams.

underwater demolition team--A group of officers and men specially trained and equipped for making hydrographic reconnaissance of approaches to prospective landing beaches; for effecting demolition of obstacles and clearing mines in certain areas; locating, improving, and marking of useable channels; channel and harbor clearance; acquisition of pertinent data during pre-assault operations, including military information; visual observation of the hinterland to gain information useful to the landing force; and for performing miscellaneous underwater and surface tasks within their capabilities.

underway replenishment--See **replenishment at sea.**

underway replenishment force(*)--A task force of fleet auxiliaries (consisting of oilers, ammunition ships, stores issue ships, etc.) adequately protected by escorts furnished by the responsible operational commander. The function of this force is to provide underway logistic support for naval forces. See also **force(s).**

underway replenishment group--A task group configured to provide logistic replenishment of ships underway by transfer-at-sea methods.

unexploded explosive ordnance(*)--Explosive ordnance which has been primed, fused, armed or otherwise prepared for action, and which has been fired, dropped, launched, projected or placed in such a manner as to constitute a hazard to operations, installations, personnel or material and remains unexploded either by malfunction or design or for any other cause.

Unified Action Armed Forces--A publication setting forth the principles, doctrines, and functions governing the activities and performance of the Armed Forces of the United States when two or more Services or elements thereof are acting together. Also called **UNAAF.**

unified command--A command with broad continuing missions under a single commander and composed of forces from two or more Military Departments, and which is established by the President, through the Secretary of Defense with the advice and assistance of the Chairman

of the Joint Chiefs of Staff. Also called **unified combatant command.**

unified operation--A broad generic term that describes the wide scope of actions taking place within unified commands under the overall direction of the commanders of those commands.

uniformed services--The Army, Navy, Air Force, Marine Corps, Coast Guard, National Oceanic and Atmospheric Administration, and Public Health Service. See also **Military Department; Military Service.**

unilateral arms control measure--An arms control course of action taken by a nation without any compensating concession being required of other nations.

unintentional radiation exploitation--Exploitation for operational purposes of noninformation-bearing elements of electromagnetic energy unintentionally emanated by targets of interest.

unintentional radiation intelligence--Intelligence derived from the collection and analysis of noninformation-bearing elements extracted from the electromagnetic energy unintentionally emanated by foreign devices, equipment, and systems, excluding those generated by the detonation of nuclear weapons. Also called **RINT.** See also **intelligence.**

uni-Service command--A command comprised of forces of a single Service.

unit(*)--1. Any military element whose structure is prescribed by competent authority, such as a table of organization and equipment; specifically, part of an organization. 2. An organization title of a subdivision of a group in a task force. 3. A standard or basic quantity into which an item of supply is divided, issued, or used. In this meaning, also called **unit of issue.** 4. With regard to reserve components of the Armed Forces, denotes a Selected Reserve unit organized, equipped and trained for mobilization to serve on active duty as a unit or to augment or be augmented by another unit. Headquarters and support functions without wartime missions are not considered units.

unit aircraft--Those aircraft provided an aircraft unit for the performance of a flying mission. See also **aircraft.**

unit combat readiness--See **combat readiness.**

unit commitment status(*)--The degree of commitment of any unit designated and categorized as a force allocated to NATO.

unit designation list--A list of actual units by unit identification code designated to fulfill requirements of a force list.

United States Armed Forces--Used to denote collectively only the regular components of the Army, Navy, Air Force, Marine Corps, and Coast Guard. See also **Armed Forces of the United States.**

United States Civil Authorities--Those elected and appointed public officials and employees who constitute the governments of the 50 States, District of Columbia, Commonwealth of Puerto Rico, United States possessions and territories, and political subdivisions thereof.

United States Civilian Internee Information Center--The national center of information in the United States for enemy and United States civilian internees.

United States controlled shipping--That shipping under United States flag plus those selected ships under foreign flag which are considered to be under "effective United States control," i.e., which can reasonably be expected to be made available to the United States in time of national emergency.

United States country team--The senior, in-country, US coordinating and supervising body, headed by the chief of the US diplomatic mission, usually an ambassador, and composed of the senior member of each represented US department or agency, as desired by the chief of the US diplomatic mission.

United States Military Service Funded Foreign Training--Training which is provided to foreign nationals in United States Military Service schools and installations under authority other than the Foreign Assistance Act of 1961.

United States Naval Ship--A public vessel of the United States in the custody of the Navy and is: a. Operated by the Military Sealift Command and manned by a civil service crew. b. Operated by a commercial company under contract to the Military Sealift Command and manned by a merchant marine crew. Also called **USNS.** See also **Military Sealift Command.**

United States Prisoner of War Information Center--The national center of information in the United States for enemy and United States prisoners of war.

United States Strategic Army Forces--That part of the Army, normally located in the Continental United States, which is trained, equipped, and maintained for employment at national level in accordance with current plans.

unit identification code--A six-character, alphanumeric code that uniquely identifies each Active, Reserve, and National Guard unit of the Armed Forces. Also called **UIC.**

unitized load--A single item, or a number of items packaged, packed or arranged in a specified manner and capable of being handled as a unit. Unitization may be accomplished by placing the item or items in a container or by banding them securely together. See also **palletized unit load.**

unit line number--A seven-character, alphanumeric field that uniquely describes a unit entry (line) in a Joint Operation Planning and Execution System time-phased force and deployment data. Also called **ULN.**

unit loading(*)--The loading of troop units with their equipment and supplies in the same vessels, aircraft, or land vehicles. See also **loading.**

unit of issue--In its special storage meaning, refers to the quantity of an item; as each number, dozen, gallon, pair, pound, ream, set, yard. Usually termed unit of issue to distinguish from "unit price." See also **unit.**

unit personnel and tonnage table--A table included in the loading plan of a combat-loaded ship as a recapitulation of totals of personnel and cargo by type, listing cubic measurements and weight.

unit price--The cost or price of an item of supply based on the unit of issue.

unit reserves--Prescribed quantities of supplies carried by a unit as a reserve to cover emergencies. See also **reserve supplies.**

unit training assembly--An authorized and scheduled period of unit inactive duty training of a prescribed length of time.

unit type code--A five-character, alphanumeric code that uniquely identifies each type unit of the Armed Forces. Also called **UTC.**

universal polar stereographic grid--A military grid prescribed for joint use in operations in limited areas and used for operations requiring precise position reporting. It covers areas between the 80 degree parallels and the poles.

Universal Time--A measure of time that conforms, within a close approximation, to the mean diurnal rotation of the Earth and serves as the basis of civil timekeeping. Universal Time (UT1) is determined from observations of the stars, radio sources, and also from ranging observations of the Moon and artificial Earth satellites. The scale determined directly from such observations is designated Universal Time Observed (UTO); it is slightly dependent on the place of observation. When UTO is corrected for the shift in longitude of the observing station caused by polar motion, the time scale UT1 is obtained. When an accuracy better than one second is not required, Universal Time can be used to mean Coordinated Universal Time (UTC). Also called **ZULU time.** Formerly called Greenwich Mean Time.

universal transverse mercator grid(*)--A grid coordinate system based on the transverse mercator projection, applied to maps of the Earth's surface extending to 84 degrees N and 80 degrees S latitudes. Also called **UTM Grid.**

unknown--1. A code meaning information not available. 2. An unidentified target.

unlimited war--Not to be used. See **general war.**

unmanned aerial vehicle--A powered, aerial vehicle that does not carry a human operator, uses aerodynamic forces to provide vehicle lift, can fly autonomously or be piloted remotely, can be expendable or recoverable, and can carry a lethal or nonlethal payload. Ballistic or semiballistic vehicles, cruise missiles, and artillery projectiles are not considered unmanned aerial vehicles. Also called **UAV.**

unobserved fire(*)--Fire for which the points of impact or burst are not observed. See also **fire.**

unpremeditated expansion of a war--Not to be used. See **escalation.**

unpremeditated war--Not to be used. See **accidental attack.**

unscheduled convoy phase(*)--The period in the early days of war when convoys are instituted on an ad hoc basis before the introduction of convoy schedules in the regular convoy phase.

unwanted cargo(*)--A cargo loaded in peacetime which is not required by the consignee country in wartime. See also **cargo.**

unwarned exposed(*)--The vulnerability of friendly forces to nuclear weapon effects. In this condition, personnel are assumed to be standing in the open at burst time, but have dropped to a prone position by the time the blast wave arrives. They are expected to have areas of bare skin exposed to direct thermal radiation, and some personnel may suffer dazzle. See also **warned exposed; warned protected.**

up(*)--In artillery and naval gunfire support: 1. A term used in a call for fire to indicate that the target is higher in altitude than the point which has been used as a reference point for the target location. 2. A correction used by an observer or a spotter in time fire to indicate that an increase in height of burst is desired.

urgent mining(*)--In naval mine warfare, the laying of mines with correct spacing but not in the ordered or planned positions. The mines may be laid either inside or outside the allowed area in such positions that they will hamper the movements of the enemy more than those of our own forces.

urgent priority--A category of immediate mission request which is lower than emergency priority but takes precedence over ordinary priority, e.g., enemy artillery or mortar fire which is falling on friendly troops and causing casualties or enemy troops or mechanized units moving up in such force as to threaten a break-through. See also **immediate mission request; priority of immediate mission request.**

USNS--See **United States Naval Ship.**

US Roland--A short range, low-altitude, all-weather, Army air defense artillery surface-to-air missile system which is based upon the Franco-German Roland III missile system.

US Transportation Command coordinating instructions--Instructions of the US Transportation Command that establish suspense dates for selected members of the joint planning and execution community to complete updates to the operation plan data base. Instructions will ensure the target date movement requirements will be validated and available for scheduling.

UTC--See **Universal Time.**

utility helicopter(*)--Multi-purpose helicopter capable of lifting troops but may be used in a command and control, logistics, casualty evacuation or armed helicopter role.

UTM-grid--See also **universal transverse mercator grid.**

UUM-44A--See **submarine rocket.**

UW--See **unconventional warfare.**

V

validate--Execution procedure used by combatant command components, supporting combatant commanders, and providing organizations to confirm to the supported commander and US Transportation Command that all the information records in a time-phased force and deployment data not only are error-free for automation purposes, but also accurately reflect the current status, attributes, and availability of units and requirements. Unit readiness, movement dates, passengers, and cargo details should be confirmed with the unit before validation occurs.

validation--1. A process normally associated with the collection of intelligence that provides official status to an identified requirement and confirms that the requirement is appropriate for a given collector and has not been previously satisfied. 2. In computer modeling and simulation, the process of determining the degree to which a model or simulation is an accurate representation of the real world from the perspective of the intended uses of the model or simulation. See also **accreditation; configuration management; independent review; verification.**

valuable cargo(*)--Cargo which may be of value during a later stage of the war. See also **cargo.**

value engineering--An organized effort directed at analyzing the function of Department of Defense systems, equipment, facilities, procedures and supplies for the purpose of achieving the required function at the lowest total cost of effective ownership, consistent with requirements for performance, reliability, quality, and maintainability.

variability(*)--The manner in which the probability of damage to a specific target decreases with the distance from ground zero; or, in damage assessment, a mathematical factor introduced to average the effects of orientation, minor shielding and uncertainty of target response to the effects considered.

variable safety level--See **safety level of supply.**

variant--1. One of two or more cipher or code symbols which have the same plain text equivalent. 2. One of several plain text meanings that are represented by a single code group. Also called **alternative.**

vector--In air intercept, close air support and air interdiction usage, a code meaning, "Alter heading to magnetic heading indicated." Heading ordered must be in three digits; e.g., "vector" zero six zero (for homing, use "steer").

vectored attack(*)--Attack in which a weapon carrier (air, surface, or subsurface) not holding contact on the target, is vectored to the weapon delivery point by a unit (air, surface or subsurface) which holds contact on the target.

vehicle(*)--A self-propelled, boosted, or towed conveyance for transporting a burden on land or sea or through air or space. See also **air cushion vehicle; amphibious vehicle; combat vehicle; commercial vehicle; ground effect machine; remotely piloted vehicle; special-- equipment vehicle; special-purpose vehicle; substitute transport-type vehicle; transport vehicle.**

vehicle cargo--Wheeled or tracked equipment, including weapons, which require certain deck space, head room, and other definite clearance. See also **cargo.**

vehicle distance(*)--The clearance between vehicles in a column which is measured from the rear of one vehicle to the front of the following vehicle.

vehicle summary and priority table--A table listing all vehicles by priority of debarkation from a combat-loaded ship. It includes the nomenclature, dimensions, square feet, cubic feet, weight, and stowage location of each vehicle, the cargo loaded in each vehicle, and the name of the unit to which the vehicle belongs.

verification--1. In arms control, any action, including inspection, detection, and identification, taken to ascertain compliance with agreed measures. 2. In computer modeling and simulation, the process of determining that a model or simulation implementation accurately represents the developer's conceptual description and specifications. See also **accreditation; configuration management; independent review; validation.**

verify(*)--1. To ensure that the meaning and phraseology of the transmitted message conveys the exact intention of the originator. (DOD) 2. A request from an observer, a spotter, or a fire-control agency to reexamine firing data and report the results of the reexamination.

vertex(*)--In artillery and naval gunfire support, the highest point in the trajectory of a projectile.

vertex height--See **maximum ordinate.**

vertical air photograph(*)--An air photograph taken with the optical axis of the camera perpendicular to the surface of the Earth.

vertical and/or short takeoff and landing--Vertical and/or short takeoff and landing capability for aircraft.

vertical envelopment--A tactical maneuver in which troops, either air-dropped or air-landed, attack the rear and flanks of a force, in effect cutting off or encircling the force.

vertical interval--Difference in altitude between two specified points or locations, e.g., the battery or firing ship and the target; observer location and the target; location of previously fired target and new target; observer and a height of burst; battery or firing ship and a height of burst, etc.

vertical loading(*)--A type of loading whereby items of like character are vertically tiered throughout the holds of a ship, so that selected items are available at any stage of the unloading. See also **loading.**

vertical probable error--The product of the range probable error and the slope of fall.

vertical replenishment(*)--The use of a helicopter for the transfer of materiel to or from a ship.

vertical separation(*)--Separation between aircraft expressed in units of vertical distance.

vertical/short takeoff and landing aircraft(*)--An aircraft capable of executing a vertical takeoff and landing, a short takeoff and landing or any combination of these modes of operation. Also called **V/STOL.** See also **short takeoff and landing; short takeoff and vertical landing aircraft; vertical takeoff and landing.**

vertical situation display(*)--An electronically generated display on which information on aircraft attitude and heading, flight director commands, weapon aiming and terrain following can be presented, choice of presentation being under the control of the pilot.

vertical strip--A single flightline of overlapping photos. Photography of this type is normally taken of long, narrow targets such as beaches or roads.

vertical takeoff and landing(*)--The capability of an aircraft to take off and land vertically and to transfer to or from forward motion at heights required to clear surrounding obstacles.

very high--A height above fifty thousand feet.

very low--A height below five hundred feet.

very seriously ill or injured--The casualty status of a person whose illness or injury is classified by medical authority to be of such severity that life is imminently endangered. Also called **VSII**. See also **casualty status**.

vesicant agent--See **blister agent**.

VIDOC--See **visual information documentation**.

vignetting(*)--A method of producing a band of color or tone on a map or chart, the density of which is reduced uniformly from edge to edge.

Viking--A twin turbofan engine, multicrew anti-submarine aircraft capable of operating off aircraft carriers. It is designed to detect, locate, and destroy submarines using an integrated, computer-controlled attack system and a variety of conventional and/or nuclear ordnance. Designated as **S-3**.

visibility--In air intercept usage, "Visibility (in miles) is _____."

visibility range--The horizontal distance (in kilometers or miles) at which a large dark object can just be seen against the horizon sky in daylight.

visual call sign(*)--A call sign provided primarily for visual signaling. See also **call sign**.

visual identification(*)--In a flight control system, a control mode in which the aircraft follows a radar target and is automatically positioned to allow visual identification.

visual information--Use of one or more of the various visual media with or without sound. Generally, visual information includes still photography, motion picture photography, video or audio recording, graphic arts, visual aids, models, display, visual presentation services, and the support processes.

visual information documentation--Motion media, still photography, and audio recording of technical and nontechnical events while they occur, usually not controlled by the recording crew. Visual information documentation encompasses Combat Camera, operational documentation, and technical documentation. Also called **VIDOC**. See also **Combat Camera; operational documentation; technical documentation**.

visual interceptor(*)--An interceptor which has no special equipment to enable it to intercept its target in dark or daylight conditions by other than visual means.

visual meteorological conditions--Weather conditions in which visual flight rules apply; expressed in terms of visibility, ceiling height, and aircraft clearance from clouds along the path of flight. When these criteria do not exist, instrument meteorological conditions prevail and instrument flight rules must be complied with. Also called **VMC**. See also **instrument meteorological conditions**.

visual mine firing indicator(*)--A device used with exercise mines to indicate that the mine would have detonated had it been poised.

visual report--Not to be used. See **inflight report**.

vital area(*)--A designated area or installation to be defended by air defense units. See also **area.**

vital ground(*)--Ground of such importance that it must be retained or controlled for the success of the mission. See also **key terrain.**

VMC--See **visual meteorological conditions.**

voice call sign(*)--A call sign provided primarily for voice communication. See also **call sign.**

voluntary training--Training in a non-pay status for Individual Ready Reservists and active status Standby Reservists. Participation in voluntary training is for retirement points only and may be achieved by training with Selected Reserve or voluntary training units; by active duty for training; by completion of authorized military correspondence courses; by attendance at designated courses of instruction; by performing equivalent duty; by participation in special military and professional events designated by the Military Departments; or by participation in authorized Civil Defense activities. Retirees may voluntarily train with organizations to which they are properly preassigned by orders for recall to active duty in a national emergency or declaration of war. Such training shall be limited to that training made available within the resources authorized by the Secretary concerned.

voluntary training unit--A unit formed by volunteers to provide reserve component training in a non-pay status for Individual Ready Reservists and active status Standby Reservists attached under competent orders and participating in such units for retirement points. Also called **reinforcement training unit.**

VOR(*)--An air navigational radio aid which uses phase comparison of a ground transmitted signal to determine bearing. This term is derived from the words "very high frequency omnidirectional radio range."

VSII--See **very seriously ill or injured.**

V/STOL--See **vertical/short takeoff and landing aircraft.**

Vulcan--An Army air defense artillery gun which provides low-altitude air defense and has a direct fire capability against surface targets. The gun is a 6-barreled, air-cooled, 20-mm rotary-fired weapon.

vulnerability--1. The susceptibility of a nation or military force to any action by any means through which its war potential or combat effectiveness may be reduced or its will to fight diminished. 2. The characteristics of a system which cause it to suffer a definite degradation (incapability to perform the designated mission) as a result of having been subjected to a certain level of effects in an unnatural (manmade) hostile environment.

vulnerability program--A program to determine the degree of, and to remedy insofar as possible, any existing susceptibility of nuclear weapon systems to enemy countermeasures, accidental fire, and accidental shock.

vulnerability study--An analysis of the capabilities and limitations of a force in a specific situation to determine vulnerabilities capable of exploitation by an opposing force.

vulnerable area--See **vital area.**

vulnerable node--See **target stress point.**

vulnerable point--See **vital area.**

W

wading crossing--See **deep fording; deep fording capability; shallow fording; shallow fording capability.**

walking patient(*)--A patient not requiring a litter while in transit.

Walleye--A guided air-to-surface glide bomb for the stand-off destruction of large semi-hard targets. It incorporates a contrast-tracking television system for guidance.

wanted cargo(*)--In naval control of shipping, a cargo which is not immediately required by the consignee country but will be needed later. See also **cargo.**

war air service program--The program designed to provide for the maintenance of essential civil air routes and services, and to provide for the distribution and re-distribution of air-carrier aircraft among civil air transportation carriers after withdrawal of aircraft allocated to the Civil Reserve Air Fleet.

warble(*)--In naval mine warfare, the process of varying the frequency of sound produced by a narrow band noisemaker to ensure that the frequency to which the mine will respond is covered.

warehouse chart--See **planograph.**

war game(*)--A simulation, by whatever means, of a military operation involving two or more opposing forces, using rules, data, and procedures designed to depict an actual or assumed real life situation.

warhead(*)--That part of a missile, projectile, torpedo, rocket, or other munition which contains either the nuclear or thermonuclear system, high explosive system, chemical or biological agents or inert materials intended to inflict damage.

warhead mating--The act of attaching a warhead section to a rocket or missile body, torpedo, airframe, motor or guidance section.

warhead section(*)--A completely assembled warhead including appropriate skin sections and related components.

WARM--See **wartime reserve modes.**

WARMAPS--See **wartime manpower planning system.**

war materiel procurement capability--The quantity of an item which can be acquired by orders placed on or after the day an operation commences (D-day) from industry or from any other available source during the period prescribed for war materiel procurement planning purposes.

war materiel requirement--The quantity of an item required to equip and support the approved forces specified in the current Secretary of Defense guidance through the period prescribed for war materiel planning purposes.

warned exposed(*)--The vulnerability of friendly forces to nuclear weapon effects. In this condition, personnel are assumed to be prone with all skin covered and with thermal protection at least that provided by a two-layer summer uniform. See also **unwarned exposed; warned protected.**

warned protected(*)--The vulnerability of friendly forces to nuclear weapon effects. In this condition, personnel are assumed to have some protection against heat, blast, and radiation such as that afforded in closed armored vehicles or crouched in fox holes with improvised overhead shielding. See also **unwarned exposed; warned exposed.**

warning--A communication and acknowledgment of dangers implicit in a wide spectrum of activities by potential opponents ranging from routine defense measures to substantial increases in readiness and force preparedness and to acts of terrorism or political, economic, or military provocation. See also **strategic warning; strategic warning lead time; strategic warning post-decision time; strategic warning pre-decision time; tactical warning; warning of attack; warning of war.**

warning area--See **danger area.**

warning net--A communication system established for the purpose of disseminating warning information of enemy movement or action to all interested commands.

warning of attack--A warning to national policymakers that an adversary is not only preparing its armed forces for war, but intends to launch an attack in the near future. See also **tactical warning; warning; warning of war.**

warning of war--A warning to national policymakers that a state or alliance intends war, or is on a course that substantially increases the risks of war and is taking steps to prepare for war. See also **strategic warning; warning; warning of attack.**

warning order(*)--A preliminary notice of an order or action which is to follow.

WARNING ORDER (Chairman of the Joint Chiefs of Staff)--A crisis action planning directive issued by the Chairman of the Joint Chiefs of Staff that initiates the development and evaluation of courses of action by a supported commander and requests that a commander's estimate be submitted. See also **warning order.**

warning red--See **air defense warning conditions.**

warning white--See **air defense warning conditions.**

warning yellow--See **air defense warning conditions.**

war reserve materiel requirement--That portion of the war materiel requirement required to be on hand on D-day. This level consists of the war materiel requirement less the sum of the peacetime assets assumed to be available on D-day and the war materiel procurement capability.

war reserve materiel requirement, balance--That portion of the war reserve materiel requirement which has not been acquired or funded. This level consists of the war reserve materiel requirement less the war reserve materiel requirement, protectable.

war reserve materiel requirement, protectable----That portion of the war reserve materiel requirement that is either on hand and/or previously funded which shall be protected; if issued for peacetime use, it shall be promptly reconstituted. This level consists of the pre-positioned war reserve materiel requirement, protectable, and the other war reserve materiel requirement, protectable.

war reserve (nuclear)--Nuclear weapons materiel stockpiled in the custody of the Department of Energy or transferred to the custody of the Department of Defense and intended for employment in the event of war.

war reserves(*)--Stocks of materiel amassed in peacetime to meet the increase in military requirements consequent upon an outbreak of war. War reserves are intended to provide the interim support essential to sustain operations until resupply can be effected.

war reserve stock(s)--That portion of total materiel assets which is designated to satisfy the war reserve materiel requirement.

war reserve stocks for allies--A Department of Defense program to have the Services procure or retain in their inventories those minimum stockpiles of materiel such as munitions, equipment, and combat essential consumables to ensure support for selected allied forces in time of war, until future in-country production and external resupply can meet the estimated combat consumption.

wartime load(*)--The maximum quantity of supplies of all kinds which a ship can carry. The composition of the load is prescribed by proper authority. See also **combat load.**

wartime manpower planning system--A standardized DOD-wide procedure, structure, and data base for computing, compiling, projecting, and portraying the time-phased wartime manpower requirements, demand, and supply of the DOD components. Also called **WARMAPS.** See also **S-day.** (NATO: See also **designation of days and hours.**)

wartime reserve modes--Characteristics and operating procedures of sensor, communications, navigation aids, threat recognition, weapons, and countermeasures systems that will contribute to military effectiveness if unknown to or misunderstood by opposing commanders before they are used, but could be exploited or neutralized if known in advance. Wartime reserve modes are deliberately held in reserve for wartime or emergency use and seldom, if ever, applied or intercepted prior to such use. Also called **WARM.**

watching mine(*)--In naval mine warfare, a mine secured to its mooring but showing on the surface, possibly only in certain tidal conditions. See also **floating mine; mine.**

waterspace management(*)--The allocation of surface and underwater spaces into areas and the implementation of agreed procedures to permit the coordination of assets, with the aim of preventing mutual interference between submarines or between submarines and other assets, while enabling optimum use to be made of all antisubmarine warfare assets involved.

water suit--A G-suit in which water is used in the interlining thereby automatically approximating the required hydrostatic pressure-gradient under G forces. See also **pressure suit.**

water terminal--A facility for berthing ships simultaneously at piers, quays, and/or working anchorages, normally located within sheltered coastal waters adjacent to rail, highway, air, and/or inland water transportation networks.

wave(*)--A formation of forces, landing ships, craft, amphibious vehicles or aircraft, required to beach or land about the same time. Can be classified as to type, function or order as shown: a. Assault wave; b. Boat wave; c. Helicopter wave; d. Numbered wave; e. On-call wave; f. Scheduled wave.

way point--In air operations, a point or a series of points in space to which an aircraft may be vectored.

W-day--See **times.**

weapon and payload identification--1. The determination of the type of weapon being used in an attack. 2. The discrimination of a re-entry vehicle from penetration aids being utilized with the re-entry vehicle. See also **attack assessment.**

weapon debris (nuclear)(*)--The residue of a nuclear weapon after it has exploded; that is, materials used for the casing and other components of the weapon, plus unexpended plutonium or uranium, together with fission products.

weaponeering--The process of determining the quantity of a specific type of lethal or nonlethal weapons required to achieve a specific level of damage to a given target, considering target vulnerability, weapon effect, munitions delivery accuracy, damage criteria, probability of kill, and weapon reliability.

weapon engagement zone--In air defense, airspace of defined dimensions within which the responsibility for engagement of air threats normally rests with a particular weapon system. Also called **WEZ.** a. fighter engagement zone. In air defense, that airspace of defined dimensions within which the responsibility for engagement of air threats normally rests with fighter aircraft. Also called **FEZ.** b. high-altitude missile engagement zone. In air defense, that airspace of defined dimensions within which the responsibility for engagement of air threats normally rests with high-altitude surface-to-air missiles. Also called **HIMEZ.** c. low-altitude missile engagement zone. In air defense, that airspace of defined dimensions within which the responsibility for engagement of air threats normally rests with low- to medium-altitude surface-to-air missiles. Also called **LOMEZ.** d. short-range air defense engagement zone. In air defense, that airspace of defined dimensions within which the responsibility for engagement of air threats normally rests with short-range air defense weapons. It may be established within a low- or high-altitude missile engagement zone. Also called **SHORADEZ.** e. joint engagement zone. In air defense, that airspace of defined dimensions within which multiple air defense systems (surface-to-air missiles and aircraft) are simultaneously employed to engage air threats. Also called **JEZ.**

weapons assignment(*)--In air defense, the process by which weapons are assigned to individual air weapons controllers for use in accomplishing an assigned mission.

weapons free(*)--In air defense, a weapon control order imposing a status whereby weapons systems may be fired at any target not positively recognized as friendly. See also **weapons hold; weapons tight.**

weapons free zone--An air defense zone established for the protection of key assets or facilities, other than air bases, where weapon systems may be fired at any target not positively recognized as friendly. See also **weapons free.**

weapons hold(*)--In air defense, a weapon control order imposing a status whereby weapons systems may only be fired in self-defense or in response to a formal order. See also **weapons free; weapons tight.**

weapons of mass destruction--In arms control usage, weapons that are capable of a high order of destruction and/or of being used in such a manner as to destroy large numbers of people. Can be nuclear, chemical, biological, and radiological weapons, but excludes the means of transporting or propelling the weapon where such means is a separable and divisible part of the weapon.

weapons readiness state--The degree of readiness of air defense weapons which can become airborne or be launched to carry out an assigned task. Weapons readiness states are expressed in numbers of weapons and numbers of minutes. Weapon readiness states are defined as follows: a. **2 minutes**--Weapons can be launched within two minutes. b. **5 minutes**--Weapons can be launched within five minutes. c. **15 minutes**--Weapons can be launched within fifteen minutes. d. **30 minutes**--Weapons can be launched within thirty minutes. e. **1 hour**--Weapons can be launched within one hour. **f. 3 hours**--Weapons can be launched within three hours. **g. released**--Weapons are released from defense commitment for a specified period of time.

weapons recommendation sheet(*)--A sheet or chart which defines the intention of the attack, and recommends the nature of weapons, and resulting damage expected, tonnage, fuzing, spacing, desired mean points of impact, and intervals of reattack.

weapons state of readiness--See **weapons readiness state.**

weapon(s) system(*)--A combination of one or more weapons with all related equipment, materials, services, personnel and means of delivery and deployment (if applicable) required for self-sufficiency.

weapons tight(*)--In air defense, a weapon control order imposing a status whereby weapons systems may be fired only at targets recognized as hostile. See also **weapons free; weapons hold.**

weapon system employment concept(*)--A description in broad terms, based on established outline characteristics, of the application of a particular equipment or weapon system within the framework of tactical concept and future doctrines.

weapon system manager--See **system manager.**

weapon-target line--An imaginary straight line from a weapon to a target.

weather central--An organization which collects, collates, evaluates, and disseminates meteorological information in such manner that it becomes a principal source of such information for a given area.

weather forecast--A prediction of weather conditions at a point, along a route, or within an area, for a specified period of time.

weather map--A map showing the weather conditions prevailing, or predicted to prevail, over a considerable area. Usually, the map is based upon weather observations taken at the same time at a number of stations. See also **map.**

weather minimum--The worst weather conditions under which aviation operations may be conducted under either visual or instrument flight rules. Usually prescribed by directives and standing operating procedures in terms of minimum ceiling, visibility, or specific hazards to flight.

weather (VAT B)--Short form weather report, giving:
a. V--Visibility in miles.
b. A--Amount of clouds, in eights.
c. T--Height of cloud top, in thousands of feet.
d. B--Height of cloud base, in thousands of feet.
(The reply is a series of four numbers preceded by the word "weather." An unknown item is reported as "unknown.")

weight and balance sheet(*)--A sheet which records the distribution of weight in an aircraft and shows the center of gravity of an aircraft at takeoff and landing.

well--As used in air intercept, a code meaning, "Equipment indicated is operating efficiently."

WEZ--See **weapon engagement zone.**

wharf--A structure built of open rather than solid construction along a shore or a bank which provides cargo-handling facilities. A similar facility of solid construction is called a quay. See also **pier; quay.**

what luck--As used in air intercept, a code meaning, "What are/were the results of assigned mission?"

what state--As used in air intercept, a code meaning, "Report amount of fuel, ammunition, and oxygen remaining."

what's up--As used in air intercept, a code meaning, "Is anything the matter?"

wheelbase(*)--The distance between the centers of two consecutive wheels. In the case of vehicles with more than two axles or equivalent systems, the successive wheelbases are all given in the order front to rear of the vehicle.

wheel load capacity--The capacity of airfield runways, taxiways, parking areas, or roadways to bear the pressures exerted by aircraft or vehicles in a gross weight static configuration.

which transponder--A code meaning report type of transponder fitted--Identification Friend or Foe, Air Traffic Control Radar Beacon System, or Secondary Surveillance Radar.

whiteout(*)--Loss of orientation with respect to the horizon caused by sun reflecting on snow and overcast sky.

white propaganda--Propaganda disseminated and acknowledged by the sponsor or by an accredited agency thereof. See also **propaganda.**

WIA--See **wounded in action.**

width of sheaf--Lateral interval between center of flank bursts or impacts. The comparable naval gunfire term is deflection pattern.

wild weasel(*)--An aircraft specially modified to identify, locate, and physically suppress or destroy ground based enemy air defense systems that employ sensors radiating electromagnetic energy.

will not fire--A term sent to the spotter or other requesting agency to indicate that the target will not be engaged by the fire support ship.

Wilson cloud--See **condensation cloud.**

window--See **chaff.**

wind shear--A change of wind direction and magnitude.

wind velocity(*)--The horizontal direction and speed of air motion.

wing--1. An Air Force unit composed normally of one primary mission group and the necessary supporting organizations, i.e., organizations designed to render supply, maintenance, hospitalization, and other services required by the primary mission groups. Primary mission groups may be functional, such as combat, training, transport, or service. 2. A fleet air wing is the basic organizational and administrative unit for naval-, land-, and tender-based aviation. Such wings are mobile units to which are assigned aircraft squadrons and tenders for administrative organization control. 3. A balanced Marine Corps task organization of aircraft groups/squadrons together with appropriate command, air control, administrative, service, and maintenance units. A standard Marine Corps aircraft wing contains the aviation elements normally required for the air support of a Marine division. 4. A flank unit; that part of a military force to the right or left of the main body.

wingman--An aviator subordinate to and in support of the designated section leader; also, the aircraft flown in this role.

withdrawal operation(*)--A planned operation in which a force in contact disengages from an enemy force.

withhold (nuclear)--The limiting of authority to employ nuclear weapons by denying their use within specified geographical areas or certain countries.

wooden bomb--A concept which pictures a weapon as being completely reliable and having an infinite shelf life while at the same time requiring no special handling, storage or surveillance.

working anchorage(*)--An anchorage where ships lie to discharge cargoes over-side to coasters or lighters. See also **emergency anchorage.**

working capital fund--A revolving fund established to finance inventories of supplies and other stores, or to provide working capital for industrial-type activities.

work order--A specific or blanket authorization to perform certain work--usually broader in scope than a job order. It is sometimes used synonymously with job order.

world geographic reference system--See **georef.**

wounded--See **seriously wounded; slightly wounded.** See also **battle casualty.**

wounded in action--A casualty category applicable to a hostile casualty, other than the victim of a terrorist activity, who has incurred an injury due to an external agent or cause. The term encompasses all kinds of wounds and other injuries incurred in action, whether there is a piercing of the body, as in a penetration or perforated wound, or none, as in the contused wound. These include fractures, burns, blast concussions, all effects of biological and chemical warfare agents, and the effects of an exposure to ionizing radiation or any other destructive weapon or agent. The hostile casualty's status may be very seriously ill or injured, seriously ill or injured, incapacitating illness or injury, or not seriously injured. Also called **WIA.** See also **casualty category.**

wreckage locator chart--A chart indicating the geographic location of all known aircraft wreckage sites, and all known vessel wrecks which show above low water or which can be seen from the air. It consists of a visual plot of each wreckage, numbered in chronological order, and cross referenced with a wreckage locator file containing all pertinent data concerning the wreckage.

wrong--A proword meaning, "Your last transmission was incorrect, the correct version is _____."

411

X

x axis--A horizontal axis in a system of rectangular coordinates; that line on which distances to the right or left (east or west) of the reference line are marked, especially on a map, chart, or graph.

X-scale(*)--On an oblique photograph, the scale along a line parallel to the true horizon.

Y

yaw(*)--1. The rotation of an aircraft, ship or missile about its vertical axis so as to cause the longitudinal axis of the aircraft, ship or missile to deviate from the flight line or heading in its horizontal plane. 2. The rotation of a camera or a photograph coordinate system about either the photograph z-axis or the exterior z-axis. 3. Angle between the longitudinal axis of a projectile at any moment and the tangent to the trajectory in the corresponding point of flight of the projectile.

y-axis--A vertical axis in a system of rectangular coordinates; that line on which distances above and below (north or south) the reference line are marked, especially on a map, chart, or graph.

yield--See **nuclear yields.**

Y-scale(*)--On an oblique photograph, the scale along the line of the principal vertical, or any other line inherent or plotted, which, on the ground, is parallel to the principal vertical.

Z

zero-length launching(*)--A technique in which the first motion of the missile or aircraft removes it from the launcher.

zero point--The location of the center of a burst of a nuclear weapon at the instant of detonation. The zero point may be in the air, or on or beneath the surface of land or water, dependent upon the type of burst, and it is thus to be distinguished from ground zero.

zippers--Target dawn and dusk combat air patrol.

Z marker beacon(*)--A type of radio beacon, the emissions of which radiate in a vertical cone shaped pattern.

zone--See **air defense identification zone; air surface zone; combat zone; communications zone; control zone; dead zone; demilitarized zone; drop zone; landing zone; rupture zone; safety zone.** See also **area.**

zone fire--Artillery or mortar fires that are delivered in a constant direction at several quadrant elevations. See also **fire.**

zone III (nuclear)--A circular area (less zones I and II), determined by using minimum safe distance III as the radius and the desired ground zero as the center, in which all personnel require minimum protection. Minimum protection denotes that armed forces personnel are prone on open ground with all skin areas covered and with an overall thermal protection at least equal to that provided by a two-layer uniform.

zone II (nuclear)--A circular area (less zone I), determined by using minimum safe distance II as the radius and the desired ground zero as the center, in which all personnel require maximum protection. Maximum protection denotes

that armed forces personnel are in "buttoned up" tanks or crouched in foxholes with improvised overhead shielding.

zone I (nuclear)--A circular area, determined by using minimum safe distance I as the radius and the desired ground zero as the center, from which all armed forces are evacuated. If evacuation is not possible or if a commander elects a higher degree of risk, maximum protective measures will be required.

zone of action(*)--A tactical subdivision of a larger area, the responsibility for which is assigned to a tactical unit; generally applied to offensive action. See also **sector.**

zone of fire--An area into which a designated ground unit or fire support ship delivers, or is prepared to deliver, fire support. Fire may or may not be observed. See also **contingent zone of fire.**

Z-scale(*)--On an oblique photograph, the scale used in calculating the height of an object. Also the name given to this method of height determination.

ZULU time--See **Universal Time.**

Listing of NATO Only Terms

A

ABAC scale
acceleration error
acceptable product
acceptance trial
access procedures
active
active public information policy
add
ad hoc movement
administration
administrative plan
advanced fleet anchorage
advanced guard
advisory control
aerodrome
aerodrome damage repair
aeromedical evacuation
aeromedical evacuation coordinating officer
aeromedical evacuation operations officer
aeromedical evacuation system
aeromedical staging unit
aeronautical chart
aeronautical topographic chart
aeropause
after-flight inspection
after-flight servicing
agent
airborne force liaison officer
airborne operation
airborne radio relay
aircraft climb corridor
aircraft control unit
aircraft dispersal area
aircraft flat pallet
aircraft guide
aircraft handover
aircraft inspection
aircraft marshaller

aircraft marshalling area
aircraft picketing
aircraft servicing connector
air defense
air defense area
air defense command
air defense commander
air defense identification zone
air defense operations area
air defense ship
airdrop
air freighting
air interception
air liaison officer
air movement officer
air movement section
air movement traffic section
air policing
air reconnaissance
airspace control
air surveillance plotting board
air terminal
air transported force
air trooping
alighting area
alignment
allied commander
allied press information center
allocation
allotment
all-source intelligence
alternate escort operating base
alternate water terminal
amphibious assault
amphibious group
amphibious operation
amphibious transport group
analysis staff
angular velocity sight
antiaircraft operations center

antisubmarine carrier group
antisubmarine minefield
apportionment
appreciation of the situation
approach route
approach time
apron
area of operational interest
area search
armed reconnaissance
arming
army
army corps
army group
artificial daylight
artificial moonlight
artillery preparation
aspect change
assault
assault aircraft
associated product
astro altitude
astro compass
astronomical twilight
astro-tracker
atomic demolition munition
attach
attack position
augmentation force
authentication
authenticator
authentic document
autonomous operation
auxiliary contours
available supply rate
average heading
axis

B

barometric vertical speed indicator
barrage jamming
barrier

base ejection shell
base fuze
base map symbol
base symbol
basic intelligence
battery control center
battery left (or right)
battle casualty
beachhead '
beam rider
before-flight servicing
bi-margin format
biological agent
black forces
blast wave
block time
blue forces
blue key
body of a map or chart
bombing errors
bomb sighting systems
bottom sweep
boundary
boundary disclaimer
branch
bridgehead

C

calibrated airspeed
calibrated altitude
camera magazine
camera window
cartesian coordinates
cascade image intensifier
cassette
casualty
casualty staging unit
caution area
cell
center of gravity limits
central analysis team
central planning team

NATO Only Terms

chaff
charging point
chemical mine
circular error probable
civil affair
civil defense
civilian preparedness for war
civil-military cooperation
civil-military relations
civil twilight
clandestine operation
close air support
close control
cocooning
collation
collimating mark
combat available aircraft
combat control team
combat day of supply
combat information
combat intelligence
combat load
combat patrol
combat readiness
combat ready
combat ready aircraft
combat service support
combat zone
command
command, control and information system
commonality
communications and information system
compartment marking
compilation
compilation diagram
complete round
component life
concentrated fire
concept of operations
confidential
constant of the cone
consultation
contamination
contamination control

contamination control line
contamination control point
contingency plan
continuously computed release point
continuously set vector
continuous processor
continuous strip photography
control
control and reporting center
control and reporting system
controlled interception
converge
convoy assembly port
coordinated attack
coordinated illumination fire
coordinating authority
corps
corrective maintenance
counter air operation
counter command, control and communications
counter-espionage
counter-insurgency
counterintelligence
counter-sabotage
counter-subversion
countersurveillance
crested
critical item
cross-servicing
cryptanalysis
cryptomaterial
currency
current intelligence

D

date-time group
days
debarkation
defector
defense readiness condition
defense shipping authority
defensive fire

defensive mine countermeasures
deferred maintenance
degree of nuclear risk
departure aerodrome
deployment
deployment operating base
derived information
designation of days and hours
detail
detection
detonating cord amplifier
diaphragm
died of wounds received in action
direct damage assessment
direct fire
directional radar prediction
direct support
diversion
doctrine
door bundle
doppler radar
downgrade
drainage system
drawing key
drone
dry gap bridge

E

early resupply
earmarked for assignment
easting
E-day
edition
edition designation
electromagnetic compatibility
electromagnetic environment
electromagnetic interference
electromagnetic radiation hazard
electromagnetic vulnerability
electronic deception
electronic imitative deception
electronic manipulative deception

electronic simulative deception
electronic warfare
electronic warfare support measures
elevation of security
emergency complement
emergency destruction of nuclear weapons
emergency establishment
emergency fleet operating base
emergency in war
emergency nuclear risk
emission control
emission control policy
endurance speed
endurance time
equal area projection
essential cargo
essential supply
estimate of the situation
evacuees
evaluation
examination
executing commander
exercise commander
exercise planning directive
exercise program
expendable supplies and materials
explosive
explosive ordnance reconnaissance

F

face of a map or chart
fair drawing
false color film
false parallax
fiducial mark
fighter
fighting patrol
filler point
filtering
flight readiness firing
floating lines
floating mark or dot

NATO Only Terms

fluxgate
fluxvalve
follow-on echelon
follow-up
forces allocated to NATO
forming up place
forward air controller
forward line of own troops
forward observer
found shipment
fragmentary order
free fall
full command
fully planned movement
functional command
fusion

G

G-day
go around
great circle route
grid bearing
grid convergence
gripper edge
ground effect machine
ground liaison officer
ground liaison section
ground observer organization
ground position indicator
guardship
guided missile
guide signs
gun direction

H

heading
height
H-hour
high density airspace control zone
holiday
hook operation

horse collar
host nation post
host nation support
hovercraft
hunter-killer group

I

identification
illumination fire
image degradation
image displacement
immediate destination (merchant shipping)
immediate operational readiness
impact area
implosion weapon
incapacitating agent
index contour line
index to adjoining sheets
indirect air support
indirect damage assessment
indirect fire
individual nuclear, biological and chemical
    protection
infill
infiltration
inflight report
influence mine
infrared film
initial point
initiation
in-place force
inset
instantaneous vertical speed indicator
instrument recording photography
integrated logistic support
integration
intelligence
intelligence cycle
interceptor controller
inter-command exercise
interdiction fire
intermediate area illumination

intermediate contour line
international actual strength
international civilian personnel with NATO
   status
international job description
international manpower ceiling
international map of the world
international military personnel
international military post
international personnel
international post
interpretation
interrupted line
isocentre
isogriv

J

jettison
joint amphibious task force
joint staff

K

K-day
key
key symbol
killed in action

L

landing group
large ship
launching site
laying-up position
lead aircraft
legend
lens coating
lens distortion
line astern
line gauge
line of arrival

line of impact
lines of communications
line weight
live exercise
local wage rate NATO civilian employee
location diagram
logistic assistance

M

magnetic minehunting
mainguard
maintainability
maintenance
Major NATO Commanders
Major Subordinate Commanders
major water terminal
maneuvering area
manpower scaling guide
maritime area
maritime operation
marking team
mass casualties
M-day
mean sea level
measuring magnifier
message
midcourse guidance
military geographic documentation
military geographic information
military strategy
minefield marking
minehunting
mine warfare
minimum quality surveillance
mission
mobile air movements team
mobile support group
mobilizable reinforcing force
mobilization
moderate nuclear risk
movement control
movement control officer

NATO Only Terms

movement priority
multi-agent munition
mutual aid

N

nadir
nadir point
national command
national commander
national component
national force commander
national forces for the defense of the NATO area
nationality undetermined post
national military authority
national shipping authority
national territorial commander
NATO airspace
NATO assigned forces
NATO code number
NATO commander
NATO command forces
NATO earmarked forces
NATO forces
NATO intelligence subject code
NATO international civilian post
NATO military authority
NATO preparation time
NATO standardization agreement
NATO warning time
NATO-wide exercise
nautical twilight
naval augmentation group
naval control of shipping liaison officer
negligible nuclear risk
net weight
neutralization fire
nickname
non-battle casualty
non-expendable supplies and material
non-quota post
northing
nuclear, biological, chemical area of observation

nuclear, biological, chemical collection center
nuclear, biological, chemical control center
nuclear, biological, chemical zone of observation
nuclear incident
nuclear logistic movement
nuclear weapon(s) accident
nuclear yield

O

observer-target distance
obstruction
occupation of position
offensive mine countermeasures
officer conducting the exercise
officer conducting the serial
officer scheduling the exercise
offset distance
offset post
on call
O-O line
operational aircraft cross-servicing requirement
operational characteristics
operational command
operational control
operational interchangeability
operational level of war
operational stocks
operation order
operation plan
operations security
orange forces
organic
other forces for NATO
overlap
overrun control
overshoot

P

packaged petroleum product
parallax
part

partially planned movement
passive public information policy
pathfinder aircraft
pathfinder team
peacetime complement
peacetime establishment
pecked line
photographic filter
piece part
plan for landing
planned resupply
planning staff
plan range
plastic explosive
plastic spray packaging
point target
positive control
post-flight inspection
poststrike damage estimation
preassault operation
precedence
preplanned mission request (reconnaissance)
pre-set vector
press information center
preventive maintenance
priming charge
Principal Subordinate Commanders
prisoner of war camp
prisoner of war collecting point
prohibited area
projectile
projection print
propaganda
protective security
provisional unit
psychological media
psychological operations
psychological operations approach
psychological situation
psychological theme
public information
purple commander
purple forces

Q

quick search procedure
quota post

R

radar echo
radar return
radiation situation map
radioactive decay rate
range
range resolution
reallocation of resources
rear guard
reciprocal jurisdiction
recognition
reconnaissance patrol
refugees
regional reinforcing force
regional reserve
registered matter
registered publication
register glass
regulatory sign
reimbursable NATO military personnel
reinforcing force
reinforcing nation
relative biological effectiveness
released
release point
releasing commander
reliability
report line
representative fraction
reproduction material
required military force
required supply rate
rescue strop
restricted area
resupply of Europe
reverse slope
riot control agent

ripe
roamer

S

rocket
roll
roller conveyor
rotational post
route reconnaissance
rules of engagement
run
run-up area
salvage
salvage procedure
scale of an exercise
scan
screen coordinator
sea skimmer
secondary water terminal
second strike capability
section
sector commander
sector controller
security
security classification
selective identification feature
selective unloading
series of targets
seriously ill
servicing
severely threatened coastline
shadower
shadow factor
shallow fording capability
shipping movement policy
shore bombardment line
short distance navigational aid
sighting angle
signal area
signals support
small ship
snake mode

sonobuoy
source
spare
special air operation
special job cover map
spigot
spoiling attack
sprag
stability augmentation feature
stage
staged crew
STANAG
standard day of supply
standardization
standardized product
standard load
standard NATO data message
standing patrol
start point
state of readiness--state 1--safe
step-up
stereogram
stereoscopic cover
stereoscopic model
stereoscopic pair
stick
stockpile to target sequence
stocks
strategic air warfare
strategic intelligence
strategic mining
strategic reserve
strategic warning
strip search
sub-collection center
submarine base
submarine exercise area coordinator
submarine havens
submarine movement advisory authority
submarine notice
submarine operating authority
Subordinate Area Commanders
subversion
superimposed

supernumerary NATO civilian personnel
supplement
supplemental programmed interpretation report
supplies
support
suppression of enemy air defenses
sustainability

T

tachometric or synchronous sights
tactical air controller
tactical air operation
tactical command
tactical intelligence
tactical warning
tan alt
target
target allocation
target grid
target illustration print
target information sheet
targeting
target list
target number
target of opportunity
target status board
tasking
technical intelligence
telebrief
temporarily filled military post
temporary civilian personnel
terminal control area
terminal guidance
terrorism
theater operational stocks
tilt
time on target
titling strip
tone
track
track handover
track production

track telling
trail formation
transition altitude
transport capacity
true convergence
trunk air route
twilight
two-up
type load

U

unconventional warfare
underslung load
unit emplaning officer
unit equipment
unit of issue
unit strength
unsurveyed area

V

vector sights
vertical interval
vertical scale instrument systems
vertical speed indicator
very deep draught ship
very seriously ill

W

waiting position
water terminal
wave-off
weapon engagement zone
white forces
wingman
wounded in action

X-Y-Z

ZULU time

Acronyms and Abbreviations

A

A	analog
A2C2	Army airspace command and control
A/C	aircraft
A/D	analog-to-digital
A/DACG	arrival/departure airfield control group
A/G	air to ground
A/M	approach and moor
A/NM	administrative/network management
AAA	antiaircraft artillery; assign alternate area
AABFS	amphibious assault bulk fuel system
AABWS	amphibious assault bulk water system
AADC	area air defense commander
AAFES	Army and Air Force Exchange System
AAFSF	amphibious assault fuel supply facility
AAP	Allied Administrative Publication; assign alternate parent
AAR	after-action report
AAT	automatic analog test
AAU	analog applique unit
AAV	assault amphibious vehicle
AAW	antiair warfare
AAWC	antiair warfare commander
AB	airbase
ABCA	American, British, Canadian, Australian Armies Standardization Program
ABCCC	airborne battlefield command and control center
ABFC	advanced base functional components
ABFS	amphibious bulk fuel system
ABN	airborne
ABO	air base operability
AC	Active component; alternating current
ACA	airspace control authority
ACAA	automatic chemical agent alarm
ACAPS	area communications electronics capabilities
ACC	Air Combat Command; air component commander; area coordination center
ACCHAN	Allied Command Channel
ACCS	air command and control system

ACCSA	Allied Communications and Computer Security Agency
ACDO	Assistant Command Duty Officer
ACE	aviation combat element (MAGTF); airborne command element (USAF); air combat element (NATO); Allied Command Europe
ACF	air contingency force
ACI	assign call inhibit
ACINT	acoustic intelligence
ACK	acknowledgement
ACL	allowable cabin load; access control list
ACLANT	Allied Command Atlantic
ACN	assign commercial network
ACO	airspace control order
ACOC	area communications operating center
ACP	Allied Communications Publication; assign common pool; airspace control plan
ACR	armored cavalry regiment (Army); assign channel reassignment
ACS	airspace control system
ACSA	Allied Communications Security Agency; acquisition cross-Service agreement
ACT	activity
ACU	assault craft unit
ACV	air cushion vehicle; armored combat vehicle
AD	advanced deployability; priority add-on
ADA	air defense artillery
A/DACG	arrival/departure control group
ADC	area damage control
ADCOM	Air (Aerospace) Defense Command
ADCON	administrative control
ADD	assign on-line diagnostic
ADDO (MS)	Associate Deputy Director for Operations/Military Support
ADE	assign digit editing
ADF	automatic direction finding
ADIZ	air defense identification zone
ADKC/RCU	Automatic Key Distribution Center/Rekeying Control Unit
ADL	assign XX (SL) routing
ADMIN	administration
ADN	Allied Command Europe (ACE) desired ground zero (DGZ) number
ADOC	air defense operations center
ADP	automated data process(ing)
ADPE	automated data processing equipment
ADPS	automatic data processing system

Acronyms and Abbreviations

ADR	aircraft damage repair; armament delivery recording
ADSIA	Allied Data Systems Interoperability Agency
ADT	automatic digital tester; assign digital transmission group
ADVCAP	advanced capability
ADVON	advanced echelon
AE	assault echelon; aeromedical evacuation; attenuation equalizer
AECA	Arms Export Control Act
AECC	aeromedical evacuation control center
AELT	aeromedical evacuation liaison team
AEPS	aircrew escape propulsion system
AES	aeromedical evacuation system
AEU	assign essential user bypass
AEW	airborne early warning
AF/DP	Deputy Chief of Staff for Personnel, United States Air Force
AF/IN	Deputy Chief of Staff for Intelligence, United States Air Force
AF/LG	Deputy Chief of Staff for Logistics, United States Air Force
AF/SC	Deputy Chief of Staff for Command, Control, Communications, and Computers, United States Air Force
AF/XO	Deputy Chief of Staff for Plans and Operations, United States Air Force
AF/XOO	Director of Operations, United States Air Force
AFB	Air Force Base
AFC	automatic frequency control
AFCC	Ai Force Component Commander
AFCS	automatic flight control system
AFD	assign fixed directory
AFDC	Air Force Doctrine Center
AFDIGS	Air Force digital graphics system
AFEES	Armed Forces Examining and Entrance Station
AFFOR	Air Force forces
AFID	anti-fratricide identification device
AFLC	Air Force Logistics Command
AFM	Air Force Manual
AFMLO	Air Force Medical Logistics Office
AFMPC	United States Air Force Military Personnel Center
AFOE	assault follow-on echelon
AFOSI	Air Force Office of Special Investigations
AFP	Armed Forces Publication; Air Force Pamphlet
AFR	Air Force regulation; assign frequency for network reporting
AFRC	Armed Forces Recreation Center
AFRCC	Air Force rescue coordination center
AFRRI	Armed Forces Radiological Research Center
AFRTS	Armed Forces Radio and Television Service

AFS	aeronautical fixed service
AFSATCOM	Air Force satellite communications
AFSC	Armed Forces Staff College; United States Air Force specialty code
AFSOB	Air Force special operations base
AFSOC	Air Force special operations component
AFSOCC	Air Force special operations control center
AFSOD	Air Force special operations detachment
AFSOE	Air Force special operations element
AFSOF	Air Force special operations forces
AFSPOC	Air Force Space Operations Center
AFSOUTH	Allied Forces, South (NATO)
AFTAC	Air Force Technical Applications Center
AFTN	Aeronautical Feed Telecommunications Network
AG	Adjutant General (Army)
AGARD	Advisory Group for Aerospace Research and Development
AGIL	airborne general illumination lightself
AGL	above ground level
AHA	American Hospital Association
AHIP	Army Helicopter Improvement Program
AI	air interdiction
AIASA	annual integrated assessment for security assistance
AIC	assign individual compressed dial
AID	Agency for International Development
AIF	automated installation intelligence file
AIG	addressee indicator group
AIM	Airman's Information Manual
AIMD	aviation intermediate maintenance department
AIQC	Antiterrorism Instructor Qualification Course
AIRREQRECON	air request reconnaissance
AIRREQSUP	air request support
AIREVACCONFIRM	air evacuation confirmation
AIREVACREQ	air evacuation request
AIREVACRESP	air evacuation response
AIS	automated information systems
AIU	AUTODIN Interface Unit
AJ	anti-jam
AJ/CM	anti-jam control modem
AJBPO	Area Joint Blood Program Office
AK	commercial cargo ship
ALCC	airlift control center
ALCE	airlift control element
ALCG	analog line conditioning group

Acronyms and Abbreviations

ALCM	air launched cruise missile
ALCOM	United States Alaskan Command
ALCON	all concerned
ALD	airborne laser designator; available-to-load date at POE; accounting line designator
ALERFA	alert phase (ICAO)
ALLA	Allied Long Lines Agency
ALNOT	search and rescue alert notice; alert notice
ALO	air liaison officer
ALOC	air lines of communications
ALSS	naval advanced logistic support site
AM	amplitude modulation
AMA	American Medical Association
AMB	air mobility branch
AMC	Air Mobility Command; Army Materiel Command: midpoint compromise search area
AME	air mobility element; antenna mounted electronics
AMEMB	American Embassy
AMH	automated message handler
AMMO	ammunition
AMOPS	Army Mobilization Operations System; Army Mobilization and Operations Planning System
AMP	amplifier
AMPN	amplification
AMPSSO	Automated Message Processing System Security Office (or Officer)
AMRAAM	advanced medium-range air-to-air missile
AMS	Army management structure
AMVER	Automated Mutual-Assistance Vessel Rescue System
AMW	amphibious warfare
AN	alphanumeric; analog nonsecure
ANCA	Allied Naval Communications Agency
ANDVT	advanced narrowband digital voice terminal
ANG	Air National Guard
ANGLICO	air/naval gunfire liaison company
ANMCC	Alternate National Military Command Center
ANN	assign NNX routing
ANX	assign NNXX routing
ANY	assign NYX routing
ANZUS	Australia-New Zealand-United States Treaty
AO	area of operations; aviation ordnance person; air officer
AO&M	administration, operation, and maintenance
AOA	amphibious objective area

Acronyms and Abbreviations

AOB	advanced operations base
AOC	air operations center (USAF)
AOCC	Air Operations Control Center
AOCU	analog orderwire control unit
AOD	on-line diagnostic
AOI	area of interest
AOPA	Aircraft Owner's and Pilot's Association
AOR	area of responsibility
AOSS	aviation ordnance safety supervisor
AP	average power
APC	armored personnel carrier; aerial port commander; assign preprogrammed conference list
APF	afloat pre-positioning force
APO	afloat pre-positioning operations; Army Post Office
APOD	aerial port of debarkation
APOE	aerial port of embarkation
APORTS	aerial ports
APORTSREP	air operations bases report
APR	assign PR routing
APS	afloat prepositioned ship
APU	auxiliary power unit
AR	Army Regulation
ARB	assign receive bypass lists
ARBS	angle rate bombing system
ARC	American (National) Red Cross; air reserve components
ARDF	automatic radio direction finding
AREC	air resource element coordinator
ARF	Air Reserve forces
ARFA	Allied Radio Frequency Agency
ARFOR	Army forces
ARG	amphibious ready group
ARINC	Aeronautical Radio Incorporated
ARM	antiradiation missiles
ARNG	Army National Guard
ARPERCEN	United States Army Reserve Personnel Center
ARQ	automatic request-repeat
ARRDATE	arrival date
ARSOA	Army special operations aviation
ARSOC	Army special operations component
ARSOF	Army special operations forces
ARSOTF	Army special operations task force
ARSPOC	Army space operations center
ARTCC	Air Route Traffic Control Center

Acronyms and Abbreviations

ARTS III	Automated Radar Tracking System
AS	analog secure
ASAS	All Source Analysis System
ASBPO	Armed Services Blood Program Office
ASC	Air Systems Command; assign switch classmark; AUTODIN Switching Center
ASCC	Air Standardization Coordinating Committee
ASCII	American Standard Code for Information Interchange
ASD(A&L)	Assistant Secretary of Defense (Acquisition and Logistics)
ASD(C)	Assistant Secretary of Defense (Comptroller)
ASD(C3I)	Assistant Secretary of Defense (Command, Control, Communications, and Intelligence)
ASD(FM&P)	Assistant Secretary of Defense (Force Management and Personnel)
ASD(HA)	Assistant Secretary of Defense (Health Affairs)
ASD(ISA)	Assistant Secretary of Defense (International Security Affairs)
ASD(ISP)	Assistant Secretary of Defense (International Security Policy)
ASD(LA)	Assistant Secretary of Defense (Legislative Affairs)
ASD(P&L)	Assistant Secretary of Defense (Production and Logistics)
ASD(PA)	Assistant Secretary of Defense (Public Affairs)
ASD(PA&E)	Assistant Secretary of Defense for program Analysis and Evaluation
ASD(RA)	Assistant Secretary of Defense (Reserve Affairs)
ASD(RSA)	Assistant Secretary of Defense (Regional Security Affairs)
ASD(SO/LIC)	Assistant Secretary of Defense (Special Operations and Low Intensity Conflict)
ASDI	analog simple data interface
ASE	automated stabilization equipment; aircraft survivability equipment
ASF	aeromedical staging facility
ASG	area support group
ASI	assign and display switch initialization
ASIF	Airlift Support Industrial Fund
ASL	assign switch locator (SL) routing; allowable supply list; authorized stockage list (Army)
ASM	automated scheduling message; armored scout mission
ASMD	antiship missile defense
ASMRO	Armed Services Medical Regulating Office
ASOC	air support operations center
ASPPO	Armed Service Production Planning Office
ASROC	antisubmarine rocket
ASRT	air support radar team
ASSETREP	Transportation Assets Report

AST	assign secondary traffic channels
ASTOR	antisubmarine torpedo
ASUW	antisurface warfare
ASUWC	antisurface warfare commander
ASW	antisubmarine warfare; average surface wind
ASWBPL	Armed Services Whole Blood Processing Laboratories
ASWC	antisubmarine warfare commander
AT	antiterrorism
At	total attainable search area
ATA	airport traffic area
ATACMS	Army Tactical Missile System
ATACO	air tactical actions control officer
ATACS	Army Tactical Communications System
ATAF	Allied Tactical Air force (NATO)
ATBM	antitactical ballistical missile
ATC	air traffic control; Air Training Command; air transportable clinic (USAF)
ATCA	Allied Tactical Communications Agency
ATCAA	air traffic control assigned airspace
ATCALS	air traffic control and landing systems
ATCC	air traffic control center; Antiterrorism Coordinating Committee
ATDL1	Army tactical data link 1
ATDM	adaptive time division multiplexer
ATDS	airborne tactical data system
ATF	amphibious task force
ATG	assign trunk group cluster
ATGM	anti-tank guided missile; anti-tank guided munition
ATH	assign thresholds; air transportable hospital
ATHS	Airborne Target Handover System
ATM	assign traffic metering
ATN	assign thresholds
ATO	air tasking order
ATP	Allied Tactical Pub
ATS	Air Traffic Service; assign terminal service
ATSD(AE)	Assistant to the Secretary of Defense (Atomic Energy)
ATT	assign terminal type
AUTODIN	Automatic Digital Network
AUTOSEVOCOM	Automatic Secure Voice Communications Network
AUX	auxiliary
AV	air vehicle
AV/VI	audiovisual/visual information
AVDTG	analog via digital trunk group
AVIM	aviation intermediate maintenance

Acronyms and Abbreviations

AVL	assign variable location
AVOU	analog voice orderwire unit
AVOW	analog voice orderwire
AVS	Audiovisual Squadron
AVUM	aviation unit maintenance
AWACS	Airborne Warning and Control System
AWADS	adverse weather aerial delivery system
AWCAP	airborne weapons corrective action program
AWN	Automated Weather Network
AWOL	absent without leave
AWS	Air Weather Service
AWSE	armament weapons support equipment
AWSR	Air Weather Service Regulation
AXX	assign XXX routing
AZR	assign zone restriction lists

B

B	cross-over barrier pattern
B&A	boat and aircraft
B/B	break bulk; baseband
BAF	backup alert force
BAG	baggage
BAI	battlefield air interdiction
BB	breakbulk
BC	bottom current
BCE	battlefield coordination element
BCI	bit count integrity
BCN	beacon
BCOC	base cluster operations center
BDA	bomb or battle damage assessment
BDC	blood donor centers
BDE	brigade
BDL	beach discharge lighter
BDOC	base defense operations center
BDZ	base defense zone
BE	basic encyclopedia
BE Number	basic encyclopedia number
BER	bit error ratio
BES	budget estimate submission
BGC	boat group commander

BI	battle injury
BIAS	Battlefield Illumination Assistance System
BIDDS	Base Information Digital Distribution System
BIDE	basic identity data element
BIFC	Boise Interagency Fire Center
BIT	built-in test
BITE	built in test equipment
BIU	beach interface unit
BLCP	beach lighterage control point
BLDREP	blood report
BLDSHIPREP	blood shipment report
BLS	beach landing site
BLT	battalion landing team
BMU	beachmaster unit
BN	battalion
BOC	base operations center
BOCCA	Bureau of Coordination of Civil Aircraft
BOH	bottom of hill
BP	block parity
BPD	Blood Products Depots
BPO	Blood Program Office
BPS	Basic PSYOP Study; bits per second
BPSK	biphase shift keying
BPT	beach party team
BRC	base recovery course
BS	battle staff; broadcast source
BSA	beach support area
BSC	black station clock
BSC ro	black station clock receive out
BSSG	brigade service support group
BSU	blood supply unit
BTC	blood transshipment center
BTU	beach termination unit
BULK	bulk cargo
BVR	beyond visual range
BW	bandwidth; biological warfare

C

C	coverage factor; creeping line pattern; clock
C di	conditioned diphase

Acronyms and Abbreviations

C2	command and control
C2S	command and control system
C2W	command and control warfare
C3	command, control, and communications
C3AG	Command, Control, and Communications Advisory Group
C3CM	command, control, and communications countermeasures
C3I	command, control, communications, and intelligence
C3IC	coalition coordination, communications, and integration center
C3SMP	Command, Control, and Communications Systems Master Plan
C4	command, control, communications, and computers
C4CM	command, control, communications, and computer countermeasures
C4I	command, control, communications, computers, and intelligence
C4IFTW	command, control, communications, computers, and intelligence for the Warrior
C4S	command, control, communications, and computer systems
C&LAT	cargo and loading analysis table
C-B	chemical-biological
C-day	unnamed day on which a deployment operation begins
C-E	communications-electronics
C-level	category level
C/C	cabin cruiser; cast off and clear
CA	civil affairs; combat assessment
CAA	civil air augmentation; Command Arrangement Agreements
CAB	combat aviation brigade
CADRS	Concern and Deficiency Reporting System
CADS	cartridge actuated devices
CAF	Canadian Air Force
CAFMS	Computer-Assisted Force Management System
CAG	Civil Affairs Group
CAIMS	Conventional Ammunition Integrated Management System
CAL	caliber
CAM	crisis action module; chemical agent monitor
CANA	convalescent antidote for nerve agent
CANUS	Canada-United States
CAO	counter air operation
CAO SOP	Standing Operating Procedures for Coordination of Atomic Operations
CAP	combat air patrol; Civil Air Patrol; crisis action planning; configuration and alarm panel
CAR	Chief of the Army Reserve
CARP	contingency alternate route plan
CARS	Combat Arms Regimental System

CARVER	criticality, accessibility, recuperability, vulnerability, effect, recognizability
CAS	close air support
CASP	computer-aided search planning
CASPER	Contact Area Summary Position Report
CAT	crisis action team; category
CATCC	carrier air traffic control center
CATF	commander, amphibious task force
CAU	crypto ancillary unit; cryptographic auxiliary unit
CAVU	ceiling and visibility unlimited
CB	chemical-biological
CBBLS	hundreds of barrels
CBFS	cesium beam frequency standard
CBLTU	common battery LTU
CBPO	Consolidated Base Personnel Office
CBPS	chemical biological protective shelter
CBR	chemical, biological, and radiological
CBS	common battery signaling
CBT	common battery terminal
CBU	cluster bomb unit; conference bridge unit
CCA	contingency capabilities assessment; carrier-controlled approach; circuit card assembly
CCB	Community Counterterrorism Board; Configuration Control Board
CCC	critical control circuit
CCD	camouflage, concealment, and deception
CCEB	Combined Communications-Electronics Board
CCF	Collection Coordination Facility
CCG	Crisis Coordination Group
CCIR	International Radio Consultative Committee
CCIS	Common Channel Interswitch Signaling
CCITT	International Telegraph and Telephone Consultative Committee
CCIU	CEF control interface unit
CCL	communications/computer link
CCO	combat cargo officer; central control officer
CCP	casualty collection point
CCRD	CINCs required delivery date
CCS	central control ship
CCSA	containership cargo stowage adapter
CCSD	Command Communications Service Designator; control communications service designator
CCT	combat control team
CCTV	closed circuit television

Acronyms and Abbreviations

CD	counterdrug; channel designator
CDB	chemical, biological defense
CDF	combined distribution frame
CDI	cargo disposition instructions; conditioned diphase
CDIP	Combined Defense Improvement Project
CDM	cable driver modem
CDMGM	cable driver modem group buffer
CDN	compressed dial number
CDO	command duty officer
CDP	landing craft air cushion (LCAC) departure point
CDR	continuous data recording; commander
CDRESC	Commander, Electronic Security Command
CDREUDAC	Commander, European Command Defense Analysis Center (ELINT) or European Data Analysis Center
CDRMTMC	Commander, Military Traffic Management Command
CD-ROM	compact disc read only memory
CDRFORSCOM	Commander, Forces Command
CDRUSAINSCOM	Commander, United States Army Intelligence and Security Command
CDRUSELEMNORAD	Commander, United States Element, North American Air Defense Command
CE	communications-electronics; command element (MAGTF)
CEDREP	Communications-Electronics Deployment Report
CEE	captured enemy equipment
CEF	Civil Engineering File; common equipment facility
CEG	common equipment group
CEI	critical employment indicator
CEM	combined effects munition
CEMC	Communications-Electronics Management Center
CEOI	communications-electronics operating instructions
CEP	circular error probable; cable entrance panel
CEPOD	Communications-Electronics Post-Deployment Report
CES	Coast Earth Station
CESE	civil engineering support equipment; communications equipment support element
CESG	Communications Equipment Support Group
CESP	Civil Engineering Support Plan
CESPG	Civil Engineering Support Planning Generator
CF	drift error confidence factor; causeway ferry
CFC	Combined Forces Command, Korea
CFL	Contingency Planning Facilities List
CFM	cubic feet per minute
CFR	Code of Federal Regulations

CG	Chairman's Guidance; center of gravity; guided missile cruiser; Coast Guard; Comptroller General
CG CAP	Coast Guard Capabilities Plan
CG FMFLANT	Commanding General, Fleet Marine Forces, Atlantic
CG FMFPAC	Commanding General, Fleet Marine Forces, Pacific
CG LSMP	Coast Guard Logistic Support and Mobilization Plan
CGAS	Coast Guard Air Station
CGAUX	Coast Guard Auxiliary
CH	channel
CHAMPUS	Civilian Health and Medical Program for the Uniformed Services
CHB	Navy cargo handling battalion
CHCSS	Chief, Central Security Service
CHE	cargo handing equipment
CHOP	change of operational control
CHRIS	Chemical Hazard Response Information System
CHSTR	Characteristics of Transportation Resources
CHSTREP	Characteristics of Transportation Resources Report
CI	counterintelligence; civilian internees
CIA	Central Intelligence Agency
CIAP	command intelligence architecture plan
CIC	combat information center; content indicator code; communications interface controller
CID	Criminal Investigation Division
CIDC	Criminal Investigation Division Command
CIF	CINC Initiative Fund
CIG	Communications Interface Group
CIL	critical items list
CIN	cargo increment number
CINC	commander of a combatant command; commander in chief
CINCAFLANT	Commander in Chief, Air Forces Atlantic
CINCARLANT	Commander in Chief, Army Forces Atlantic
CINCCFC	Commander in Chief, Combined Forces Command
CINCHAN	Allied Commander-in-Chief Channel
CINCLANTFLT	Commander in Chief, Atlantic Fleet
CINCNET	CINCs' network
CINCNORAD	CINC, North American Aerospace Defense Command
CINCPACAF	Commander in Chief, Pacific Air Forces
CINCPACFLT	Commander in Chief, Pacific Fleet
CINCUNC	Commander in Chief, United Nations Command
CINCUSACOM	Commander in Chief, United States Atlantic Command
CINCUSAFE	Commander in Chief, United States Air Force in Europe
CINCUSAREUR	Commander in Chief, United States Army, Europe

Acronyms and Abbreviations

CINCUSNAVEUR	Commander in Chief, Naval Forces Europe
CIO	Central Imagery Office
CIP	communications interface processor
CIPSU	communications interface processor pseudo line
CIR	continuing intelligence requirement
CIRM	International Radio-Medical Center
CIRV	common interswitch rekeying variable
CIRVIS	communications instructions for reporting vital intelligence sightings
CIS	common item support; communications interface shelter; Commonwealth of Independent States
CISO	counter intelligence support officer
CJATF	Commander, Joint Amphibious Task Force
CJCS	Chairman of the Joint Chiefs of Staff
CJCSAN	Chairman of the Joint Chiefs of Staff Alerting Network
CJCSI	Chairman of the Joint Chiefs of Staff Instruction
CJDA	critical joint duty assignment
CJMAO	Chief, Joint Mortuary Affairs Office
CJTF	commander, joint task force
CKT	circuit
CLA	LCAC launch area
CLF	commander, landing force; combat logistics force
CLGP	cannon-launched guided projectile
CLIPS	Communications Link Interface Planning System
CLPSB	CINC Logistic Procurement Support Board
CLS	contractor logistic support
CLZ	cushion landing zone
CM	Chairman's Memorandum; control modem
Cm	mean coverage factor
CMC	Commandant of the Marine Corps
Cmc	midpoint compromise coverage factor
CMD	command
CML	CIP/message processor line
CMO	civil-military operations
CMOS	complimentary metal-oxide semiconductor
CMP	communications message processor
CMS	community management staff; crisis management system; cockpit management system
CMTS	comments
CMTU	cartridge magnetic tape unit
CN	counternarcotic
CNASP	Chairman's Net Assessment for Strategic Planning
CNCE	communications nodal control element

CNGB	Chief, National Guard Bureau
CNO	Chief of Naval Operations
CNRF	Commander, Naval Reserve Forces
CNTY	country
CNWDI	critical nuclear weapons design information
CO	commanding officer
COA	course of action
COBOL	common business-oriented language
COC	combat operations center
COCOM	combatant command (command authority)
COE	Army Corps of Engineers
COGARD	Coast Guard
COLDS	cargo offload and discharge system
COLT	combat observation and lasing team
COM	collection operations management; commander
COMAFFOR	Commander, Air Force Forces
COMAFSOC	Commander, Air Force Special Operations Command
COMALF	Commander Airlift Forces
COMARFOR	Commander, Army Forces
COMCAM	combat camera
COMCARGRU	Commander, Carrier Group
COMCRUDESGRU	Commander, Cruiser Destroyer Group
COMDESRON	Commander Destroyer Squadron
COMCEN	communications center
COMDCAEUR	Commander, Defense Communications Agency Europe
COMDT COGARD	Commandant, United States Coast Guard
COMDTINST	Commandant, United States Coast Guard Instruction
COMIDEASTFOR	Commander, Middle East Forces
COMINEWARCOM	Commander, Mine Warfare Command
COMINT	communications intelligence
COMJCSE	Commander, Joint Communications Support Element
COMJIC	Commander, Joint Intelligence Center
COMJSOTF	Commander, Joint Special Operations Task Force
COMJTF	Commander, joint task force
COMLANDFOR	Commander, land forces
COMLANTAREACOGARD	Commander, Coast Guard Atlantic Area
COMLOGFOR	Combat Logistics Force
COMM	Communications
COMMARFOR	Commander, Marine Forces
COMMZ	communications zone
COMNAV	Committee for European Airspace Coordination Working Group on Communications and Navigation Aids
COMNAVAIRLANT	Commander, Naval Air Force, Atlantic

Acronyms and Abbreviations

COMNAVAIRPAC	Commander, Naval Air Force, Pacific
COMNAVCOMTELCOM	Commander, Naval Computer and Telecommunications Command
COMNAVFOR	Commander, Naval Forces
COMNAVSEASYSCOM	Commander, Naval Sea Systems Command
COMNAVSECGRP	Commander, United States Navy Security Group
COMNAVSURFLANT	Commander, Naval Surface Force, Atlantic
COMNAVSURFPAC	Commander, Naval Surface Force, Pacific
COMP	component
COMPACAREACOGARD	Commander, Coast Guard Pacific Area
COMPLAN	communications plan
COMPUSEC	computer security
COMSAT	communications satellite
COMSC	Commander, Military Sealift Command
COMSEC	communications security
COMSOCCENT	Commander, Special Operations Command, United States Central Command
COMSOCEUR	Commander, Special Operations Command, United States European Command
COMSOCLANT	Commander Special Operations Command, United States Atlantic Command
COMSOCSOUTH	Commander Special Operations Command, United States Southern Command
COMSOCPAC	Commander Special Operations Command, United States Pacific Command
COMSOF	Commander, Special Operations Forces
COMSTAT	communications status
COMSUBLANT	Commander Submarine Force, United States Atlantic Fleet
COMSUBPAC	Commander Submarine Force, United States Pacific Fleet
COMSUPNAVFOR	Commander, Supporting Naval Forces
COMTAC	tactical communications
COMUSARCENT	Commander, United States Army Forces, Central Command
COMUSCENTAF	Commander, United States Air Force, Central Command
COMUSFORAZ	Commander United States Forces Azores
COMICEDEFOR	Commander United States Forces Iceland
COMUSJ	Commander United States Forces Japan
COMUSK	Commander United States Forces Korea
COMUSMARCENT	Commander, United States Marine Forces, Central Command
COMUSNAVCENT	Commander, United States Navy, Central Command
CONEXPLAN	contingency and exercise plan
CONOPS	concept of operations
CONPLAN	operation plan in concept format
CONUS	continental United States

CONUSA	Continental United States Army
COOP	continuity of operations plan
COPS	Communications Operational Planning System
COS	critical occupational specialty
COSCOM	corps support command
COSMIC	North Atlantic Treaty Organization (NATO) security category
COSPAS	cosmicheskaya sistyema poiska avariynch sudov - space system for search of distressed vessels (Russian satellite system)
COT	commanding officer of troops
COTP	captain of the port
COTS	cargo offload and transfer system; commercial-off-the-shelf
COU	cable orderwire unit
counter C3	counter command, control, and communications countermeasures
CP	command post
CP&I	coastal patrol and interdiction
CPA	Chairman's Program Assessment; closest point of approach
CPE	customer premise equipment
CPFL	contingency planning facilities list
CPG	Contingency Planning Guidance; central processor group
CPI	crash position indicator
CPM	Civilian Personnel Manual
CPO	chief petty officer; complete provisions only
CPR	cardiopulmonary resuscitation
CPS	characters per second
CPU	central processing unit
CPX	command post exercise
CR-UAV	close-range unmanned aerial vehicle
CRA	coordinating review authority; command relationships agreement
CRAF	Civil Reserve Air Fleet
CRAM	control random access memory
CRB	configuration review board
CRC	control and reporting center, circuit routing chart; CONUS replacement center
CRCC	combat rubber raiding craft
CRD	Commander in Chief's (CINC's) required delivery date
CRF	channel reassignment function
CRI	collective RI
CRIF	Cargo Routing Information File
CRITIC	critical information; critical intelligence communication; critical message (intelligence)
CRITICOMM	Critical Intelligence Communications System
CRM	collection operations management

Acronyms and Abbreviations

CRS	coastal radio station
CRT	cathode ray tube
CRTS	casualty receiving and treatment ship
CRYPTO	cryptographic
CS	combat support; call sign; coastal station; creeping line single-unit; controlled space; circuit switch
CSA	Chief of Staff, United States Army
CSAAS	Combat Support Agency Assessment System
CSADR	Combat Support Agency Director's Report
CSAF	Chief of Staff, United States Air Force
CSAM	computer security for acquisition managers
CSAR	combat search and rescue
CSAR3	Combat Support Agency Responsiveness and Readiness Report
CSC	creeping line single-unit coordinated
CSCC	coastal sea control commander
CSEL	circuit switch select line
CSG	cryptologic support group (NSA)
CSH	combat support hospital
CSI	critical safety item; critical sustainability item
CSIF	communications service industrial fund
CSIPG	Circuit Switch Interface Planning Guide
CSNP	causeway section, nonpowered
CSOA	combined special operations area
CSNP(BE)	causeway section nonpowered (beach end)
CSNP(I)	causeway section, nonpowered (intermediate)
CSNP(SE)	causeway section, nonpowered (sea end)
CSO	communications support organization
CSOB	command systems operations branch
CSOD	command systems operation division
CSP	commence search point; causeway section, powered; crisis staffing procedures (JCS); cryptologic support package; call service position
CSPAR	Commander in Chief's (CINC's) Preparedness Assessment Report
CSR	Commander in Chief's (CINC's) Summary Report
CSS	coordinator surface search; combat service support; central security service; communications subsystem
CSSA	combat service support area
CSSE	combat service support element (MAGTF)
CSW	compartment stowage worksheet
CT	counterterrorism; communications terminal; control telemetry
CTF	combined task force
CTG	commander, task group

442

CTID	communications transmission identifier
CTOC	corp tactical operations center
CU	cubic capacity; common unit
CV	curriculum vitae; carrier
CVBG	carrier battle group
CVISC	combat visual information support center
CVN	aircraft carrier, nuclear
CVSD	continuous variable slope delta
CW	chemical warfare; continuous wave; carrier wave
CWC	composite warfare commander
CWPD	Conventional War Plans Division, Joint Staff (J-7)
CWR	calm water ramp
CY	calendar year

D

D	total drift, data
d	surface drift
D-day	unnamed day on which operations commence or are scheduled to commence
D/A	digital-to-analog
D/C	downconverter
DA	Department of Army; data administrator; direct action; data adapter
Da	aerospace drift
DA&M	Director of Administration and Management
DAA	designated approving authority; display alternate area routing lists
DACB	data adapter control block
DACM	data adapter control mode
DADCAP	dawn and dusk combat air patrol
DAMA	demand assigned multiple access
DAN	Diver's Alert Network
DAO	department/agency/organization
DAR	distortion adaptive receiver
DARPA	Defense Advanced Research Projects Agency
DART	disaster assistance response team
DAS	direct access subscriber; direct air support
DAS3	decentralized automated service support system
DASA	Department of the Army (DA) staff agencies
DASC	direct air support center

Acronyms and Abbreviations

DASC-A	direct air support center airborne
DASPS-E	Department of the Army Standard Port System--Enhanced
DASSS	decentralized automated service support system
DAT	deployment action item
DATT	Defense Attache
DATU	data adapter termination unit
dB	decibel
DBA	data base administrator
dBA	noise measurement unit
DBG	data base generation
DBI	defense budget issue
dBm	decibels referred to 1 milliwatt; decibel millivolts
dBmCo	decibels referred to 1 milliwatt, C-message
dBmO	noise power in dbm at a point of zero relative transmission level
DBMS	data base management system
DBOF-T	defense business operations fund-transportation
dBr	decibels level reference
DC	direct current; Depuuties Committee
DC/S for RA	Deputy Chief of Staff for Reserve Affairs
DCA	defensive counterair
DCAA	Defense Contract Audit Agency
DCC	damage control center
DCCC	defense collection coordination center
DCCEP	Developing Country Combined Exercise Program
DCCLR	direct current closure adapter
DCI	Director of Central Intelligence; dual channel interchange
DCID	Director of Central Intelligence
DCM	data channel multiplexer
DCMC	Office of Deputy Chairman, Military Committee
DCNO	Deputy Chief of Naval Operations
DCO	dial central office
DCPA	Defense Civil Preparedness Agency
DCPG	digital clock pulse generator
DCS	Defense Communications System; Defense Courier Service; digital computer system
DCSCU	dual capability servo control unit
DCSLOG	Deputy Chief of Staff for Logistics, US Army
DCSOPS	Deputy Chief of Staff for Operations and Plans, United States Army
DCSPER	Deputy Chief of Staff for Personnel, United States Army
DCTN	Defense Commercial Telecommunications Network
DD	navy destroyer

DDA	Designated Development Activity
DDG	guided missile destroyer
DDI	Deputy Director of Intelligence (CIA)
DDM	digital data modem
DDN	Defense Data Network
DDO	Director of Operations (CIA)
DDR&E	Director of Defense Research and Engineering
DDS	dry deck shelter; Defense Dissemination System
DE	directed energy; delay equalizer
De	total drift error
de	individual drift error
de max	maximum drift error
de min	minimum drift error
de minimax	minimax drift error
DEA	Drug Enforcement Administration
dea	aerospace drift error
DEARAS	Department of Defense (DOD) Emergency Authorities Retrieval and Analysis System
DeCA	Defense Commissary Agency
DECL	declassify
DEFCON	defense readiness condition
DEFSMAC	Defense Special Missile and Astronautics Center
DEMARC	demarcation
DEMUX	demultiplex
DEP	deployed
DEPCJTF	Deputy Commander, Joint Task Force
DEPMEDS	deployable medical systems
DepOpsDeps	Service Deputy Operations Deputies
DESCOM	Depot System Command (Army)
DEST	destination (GELOC)
DET	detachment; detainee
DETRESFA	distress phase (lCAO)
DEW	directed-energy warfare
DF	Direction Finding; disposition form; dispersion factor
DFE	division force equivalent
DFM	deterrent force module
DFR/E	Defense Fuel Region, Europe
DFR/ME	Defense Fuel Region, Middle East
DFSC	Defense Fuel Supply Center
DFSP	Defense Fuel Support Point
DFT	deployment for training
DG	defense guidance
DGM	digital group multiplex

Acronyms and Abbreviations

DGZ	desired ground zero
DHHS	Department of Health and Human Services
DIA	Defense Intelligence Agency
DIAC	Defense Intelligence Analysis Center
DIAM	Defense Intelligence Agency Memorandum; Defense Intelligence Agency Manual
DIAR	Defense Intelligence Agency (DIA) Regulation
DIBITS	digital in-band interswitch trunk signaling
DIBTS	digital in-band trunk signaling
DICO	Data Information Coordination Office
DIDHS	Deployable Intelligence Data Handling System
DIG	digital
DILPA	diphase loop modem-A
DIN	defense intelligence notice
DINET	Defense Industrial Net
DIPC	Defense Industrial Plant Equipment Center
DIPGM	diphase supergroup modem
DIRLAUTH	direct liaison authorized
DIRM	Directorate for Information and Resource Management
DIRMOBFOR	Director of Mobilization Forces
DIRNSA	Director, National Security Agency
DIS	Defense Investigative Service
DISA	Defense Information Systems Agency
DISANMOC	Defense Information Systems Agency Network Management and Operations Center
DISCOM	division support command (Army)
DISGM	diphase supergroup
DISN	Defense Information Systems Network
DISUM	daily intelligence summary
DJS	Director, Joint Staff
DJSM	Director, Joint Staff, memorandum
DLA	Defense Logistics Agency
DLAR	Defense Logistics Agency Regulation
DLED	dedicated loop encryption device
DLPMA	diphase loop modem A
DLQ	deck landing qualification
DLR	depot-level repairable
DLSA	Defense Legal Services Agency
DLTM	digital line termination module
DLTU	digital line termination unit
DMA	Defense Mapping Agency
DMAHT	Defense Mapping Agency Hydrographic Topographic Center
DMAINST	Defense Mapping Agency Instruction

dmax	maximum drift distance
DMB	datum marker buoy
DMC	data mode control
DMD	digital message device
DME	distance measuring equipment
dmin	minimum drift distance
DML	data manipulation language
DMO	directory maintenance official
DMPI	designated mean point of impact
DMOS	duty military occupational specialty
DMRD	defense management resource decision
DMRIS	Defense Medical Regulating Information System
DMS	Defense Meteorological System; Defense Message System
DMSOs	directors of major staff offices
DMSP	Defense Meteorological Satellite Program
DMSSC	Defense Medical System Support Center
DMU	disk memory unit
DN	digital nonsecure
DNA	Defense Nuclear Agency
DNBI	disease and nonbattle injury
DNVT	digital nonsecure voice terminal
DNY	display area code (NYX) routing
DO	Air Force component operations officer (staff)
DOA	dead on arrival
DOC	Department of Commerce
DOCDIV	Documents Division
DOCEX	document exploitation
DOD	Department of Defense
DODAAC	Department of Defense activity address code
DODD	Department of Defense Directive
DODDS	Department of Defense Dependent Schools
DODEX	Department of Defense Intelligence System Information System Extension
DODI	Department of Defense Instruction
DODIC	Department of Defense identification code
DODID	Department of Defense Intelligence Digest
DODIIS	Department of Defense Intelligence Information System
DOD-JIC	Department of Defense Joint Intelligence Center
DODM	data orderwire diphase modem
DOE	Department of Energy
DOI	Department of Interior; Defense Special Security Communications System (DSSCS) Operating Instructions
DOJ	Department of Justice

Acronyms and Abbreviations

DOL	Department of Labor
DOM	day-of-month
DOMS	Director of Military Support
DON	Department of the Navy
DOS	Department of State; disk operating system; day of supply
DOT	Department of Transportation
DOW	died of wounds; Data Orderwire
DOY	day-of-year
DP	Air Force component plans officer (staff)
dp	parachute drift
DPG	Defense Planning Guidance
DPLSM	dipulse group modem
DPP	Distributed Production Program; data patch panel
DPR	display NN routing
DPRB	Defense Planning and Resources Board
DPS	data processing system
DPSC	Defense Personnel Support Center
DPSK	differential phase shift keying
DR	dead reckoning; digital receiver
DRB	Defense Resources Board
DRe	dead reckoning error
DRMO	Defense Reutilization Marketing Office
DRT	dead reckoning tracer
DRTC	designated reporting technical control
DS	doctrine sponsor
DSA	defense special assessment (DIA)
DSAA	Defense Security Assistance Agency
DSAR	Defense Supply Agency Regulation
DSB	digital in-band trunk signaling (DIBTS) signaling buffer
DSC	digital selective calling
DSCS	Defense Satellite Communications System
DSCSOC	Defense Satellite Communications System operations center
DSDI	digital simple data interface
DSG	digital signal generator
DSL	display switch locator (SL) routing
DSMAC	digital scene-matching area correlation
DSN	Defense Switched Network
DSNET	Defense Secure Network
DSNET-2	Defense Secure Network-2
DSNET-3	Defense Secure Network-3
DSP	Defense Support Program
DSPL	Display System Programming Language
DSSCS	Defense Special Security Communications System

DSSO	Data System Support Organization; defense systems support organization
DSTP	Director of Strategic Target Planning
DSTR	destroy
DSVT	digital subscriber voice terminal
DTE	data terminal equipment
DTED	digital terrain elevation data
DTG	date-time group; digital trunk group (digital transmission group)
DTMF	dual tone multi-frequency
DTMR	Defense Traffic Management Regulation
DTO	division transportation office
DTS	Defense Transportation System; Diplomatic Telecommunications Service
DUSDP	Deputy Under Secretary of Defense for Policy
DVITS	Digital Video Imagery Transmission System
DVOW	digital voice orderwire
DWT	deadweight tonnage
DX	direct exchange
DZ	drop zone
DZST	drop zone support team

E

E	total probable error
E3	electromagnetic environmental effects
E&DCP	Evaluation and Data Collection Plan
E&E	evasion and escape
E&M	ear and mouth; special signaling leads
E-ARTS	En Route Automated Radar Tracking System
E-UAV	endurance unmanned aerial vehicle
EA	electronic attack, executive assistant
ea	each
EAC	emergency action console; echelons above corps
EAD	extended active duty; Evaluation and Analysis Division; earliest arrival date at port of debarkation
EALT	earliest anticipated launch time
EAM	emergency action message
EAP	emergency action procedures
EAP-CJCS	Emergency Action Procedures of the Chairman of the Joint Chiefs of Staff

Acronyms and Abbreviations

EBCDIC	Extended Binary Coded Decimal Interchange Code
EC	electronic combat; error control; European Community
ECAC	Electromagnetic Compatibility Analysis Center
ECCM	electronic counter-countermeasures
ECM	electronic countermeasures
ECN	Minimum Essential Emergency Communications Network
ECP	engineering change proposal; emergency command precedence
ECU	environmental control unit
ED	evaluation directive; envelope delay
EDC	estimated date of completion of loading at POE
EDD	earliest delivery date
EE	emergency establishment
EEBD	emergency escape breathing device
EED	electro-explosive device
EEFI	essential elements of friendly information
EEI	essential elements of information
EEPROM	electronic erasable programmable read-only memory
EFTO	encrypt for transmission only
EHF	extremely high frequency
EIA	Electronic Industries Association
ELBA	emergency location beacon
ELCAS	elevated causeway system
ELECTRO-OPTINT	electro-optical intelligence
ELINT	electronics intelligence
ELPP	equal level patch panel
ELR	extra-long-range aircraft
ELSEC	electronic security
ELT	emergency locator transmitter
ELVA	emergency low visibility approach
EMC	electromagnetic compatibility
EMCON	emission control
EME	electromagnetic environment
EMI	electromagnetic interface
EMP	electromagnetic pulse
EMR Hazards	electromagnetic radiation hazards
EMS	emergency medical services
EMSEC	emanations security
EMT	emergency medical technician
EMV	electromagnetic vulnerability
ENDEX	exercise termination
EO	end office; electro-optical; eyes only
EO-IR	electro-optical-infrared
EOB	enemy order of battle; electronic order of battle

450

EOD	explosive ordnance disposal
EOL	end of link
EOM	end of message
EOP	emergency operating procedures
EOW	engineering orderwire
EP	electronic protection; execution planning
EPBX	electronic private branch exchange
EPDS	Electronic Processing and Dissemination System
EPIRB	emergency position-indicating radio beacon
EPROM	erasable programmable read only memory
EPW	enemy prisoner of war
ES	electronic warfare support
ESD	Electronics Systems Division
ESF	Economic Support Fund
ESM	electronic warfare support measures; electronic surveillance measures
ESO	embarkation staff officer
EST	emergency service team
ETA	estimated time of arrival
ETAC	emergency tactical air control
ETD	estimated time of departure
ETD	electronic transfer device
ETI	estimated time of intercept
ETIC	estimated time for completion
ETPL	endorsed TEMPEST products list
ETS	European Telephone System
ETSS	extended training service specialist
ETX	end of text
EUB	essential user bypass
EURV	essential user rekeying variable
EUSC	effective United States controlled shipping
EW	electronic warfare; early warning
EW/GCI	early warning/ground-controlled intercept
EWC	electronic warfare coordinator
EWCS	electronic warfare control ship
EWIR	electronic warfare integrated reprogramming
EWO	electronic warfare officer
EXCOM	extended communications search
EXORD	execute order
EXPLAN	exercise plan
EZ	extraction zone

Acronyms and Abbreviations

F

F	flare patterns; flash
F-hour	effective time of announcement by the Secretary of Defense to the Military Departments of a decision to mobilize Reserve units
F/V	fishing vessel
FA	field artillery; feasibility assessment
FAA	Federal Aviation Administration; Foreign Assistance Act
FAC	forward air controller
FAC(A)	forward air controller (airborne)
FACSFAC	Fleet Area Control and Surveillance Facility
FAD	force activity designator; feasible arrival date
FAE	fuel air explosive
FAPES	Force Augmentation Planning and Execution System
FAR	Federal Aviation Regulation
FARP	forward arming and refueling point
FASCAM	family of scatterable mines
FAX	facsimile
FBI	Federal Bureau of Investigation
FC	floating craft; field circular
FCC	Federal Communications Commission
FCT	firepower control team
FDA	Food and Drug Administration
FDBM	functional data base manager
FDC	fire direction center
FDESC	force description
FDL	fast deployment logistics
FDLP	flight deck landing practice
FDM	frequency division multiplexing
FDO	flight deck officer; flexible deterrent option
FDR/FA	flight data recorder/fault analyzer
FDS	fault detection system
FDSL	fixed directory subscriber list
FDSS	fault detection subsystem
FDSSS	flight deck status and signalling system
FDUL	fixed directory unit list
FDX	full duplex
FEBA	forward edge of the battle area
FEC	forward error correction
FED-STD	federal standard
FEK	frequency exchange keying

FEMA	Federal Emergency Management Agency
FEP	fleet satellite (FLTSAT) extremely high frequency (EHF) package
FEU	forty-foot equivalent unit
FEWS	Follow-on Early Warning System
FEZ	fighter engagement zone
FF	navy fast frigate
Ff	fatigue correction factor
FFE	flame field expedients
FFG	guided missile frigate
FFH	fast frequency hopping
FFH-net	fast-frequency-hopping net
FFHT-net	fast-frequency-hopping training net
FFTU	forward freight terminal unit
FGMDSS	Future Global Maritime Distress and Safety System
FIC	force indicator code
FID	foreign internal defense
FIDAF	foreign internal defense augmentation force
FIE	fly-in echelon
FIFO	first-in-first-out
FIR	flight information region; first-impressions report
FIRCAP	foreign intelligence requirements capabilities and priorities
FIS	Flight Information Service
FIST	Fleet imagery support terminal; fire support team
FISINT	foreign instrumentation signals intelligence
FISS	Foreign Intelligence Security Service
FIST	fire support team
FIXe	navigational fix error
FLAR	forward-looking airborne radar
FLIP	Flight Information Publication; flight instruction procedures
FLIR	forward-looking infrared radar
FLO/FLO	float-on/float-off
FLOLS	fresnel lens optical landing system
FLOT	forward line of own troops
FLS	naval forward logistic site
FLTCINC	fleet commander in chief
FLTSAT	fleet satellite
FLTSATCOM	fleet satellite communications
FM	flare multiunit; frequency modulation; functional manager; force module(s); field manual
FMA-net	frequency management A-net
FMAS	foreign media analysis subsystem
FMCC	force movement control center

Acronyms and Abbreviations

FMCR	Fleet Marine Corps Reserve
FMF	Fleet Marine Force
FMFM	Fleet Marine Force Manual
FMFP	foreign military financing program
FMFRP	Fleet Marine Force Reference Publication
FMID	force module identifier
FMO	Frequency Management Office
FMP	force module package
FMS	foreign military sales; force module subsystem
FMT-net	frequency management training net
FNOC	Fleet Numerical Oceanographic Command
FNS	foreign nation support
FO	forward observer; flash override; fiber optic
FOB	forward operations base
FOC	full operational capability
FOD	foreign object damage
FOI	fault detection isolation
FOIU	fiber optic interface unit
FOL	forward operating location
FORSCOM	United States Army Forces Command
FORSTAT	Force Status and Identity Report
FOUO	for official use only
FOV	field of view
FPC	field press censorship
FPM	Federal Personnel Manual
FPO	Fleet Post Office
FR	final report; frequency response
FRAG	fragmentation code
FREQ	frequency
FROG	free rocket over ground
FRN	force requirement number
FS	file server; flare single-unit; fighter squadron; file separator
fs	search radius safety factor
FSA	fire support area
FSB	forward staging report
FSC	fire support coordinator
FSCC	fire support coordination center
FSCL	fire support coordination line
FSE	fire support element
FSK	frequency shift key
FSN	foreign service national
FSO	fire support officer
FSS	flight service station; fast sealift ships

FSSG	force service support group (Marine air-ground task force)
FSW	feet of seawater
ft	foot; feet
FTC	Federal Trade Commission
FTRG	Fleet Tactical Readiness Group
FTS	Federal Telecommunications System; Federal Telephone Service; file transfer service
FTX	field training exercise
FUAC	functional area code
Fv	aircraft speed correction factor
FW	weather correction factor; fighter wing
FWD	forward
FY	fiscal year
FYDP	Five Year Defense Program; future years defense program

G

G-1	Army or Marine Corps component manpower or personnel staff officer (Army division or higher staff, Marine Corps brigade or higher staff)
G-2	Army or Marine Corps component intelligence staff officer (Army division or higher staff, Marine Corps brigade or higher staff)
G-3	Army or Marine Corps component operations staff officer (Army division or higher staff, Marine Corps brigade or higher staff)
G-4	Army or Marine Corps component logistics staff officer (Army division or higher staff, Marine Corps brigade or higher staff)
G/A	ground to air
G/VLLD	ground/vehicle laser locator designator
GA	Tabun, a nerve agent
GAA	general agency agreement
GB	group buffer; Sarin, a nerve agent
GC	Geneva Convention
GC3A	global command, control, and communications assessment
GC4A	global command, control, communications, and computer assessment
GCE	ground combat element (MAGTF)
GCI	ground control intercept
GCRI	general collective routing indicator (RI)

Acronyms and Abbreviations

GCS	ground control station
GD	Soman, a nerve agent
GDIP	General Defense Intelligence Program
GDIPP	General Defense Intelligence Proposed Program
GDP	General Defense Plan (SACEUR)
GDSS	Global Decision Support System
GENSER	general service (message)
GENTEXT	general text
GEOCODE	geographic code
GEOCODES	geographic codes
GEOFILE	geolocation code file; standard specified geographic location file
GEOLOC	geographic location code
GEOREF	geographic reference
GF	a nerve agent
GFE	government furnished equipment
GFOAR	Global Family of OPLANs Assessment Report
GFU	group framing unit
GHz	gigahertz
GIC	gabarit international de chargement (international loading gauge)
GM	group modem
GMD	group mux/demux
GMDSS	Global Maritime Distress and Safety System
GMF	ground mobile forces
GMI	general military intelligence
GMR	graduated mobilization response
GMT	Greenwich Mean Time
GOCO	government owned contractor operated
GOES	geostationary operational environmental satellite
GOS	grade of service
GOSG	general officer steering group
GP	group
GPD	gallons per day
GPEE	general purpose encryption equipment
GPMDM	group modem
GPS	global positioning system
GS	ground speed; general support; group separator
GSA	General Services Administration
GSE	ground support equipment
GSI	glide slope indicator
GSM	ground station module
GSR	ground surveillance radar
GRCA	ground reference coverage area

gt	gross tons
GTL	gun-target line
GTN	Global Transportation Network
GSM	ground station module
GW	guerrilla warfare
GWC	global weather central

H

H&I	harassing and interdicting
H-hour	specific time an operation or exercise begins; seaborne assault landing hour
HA	humanitarian assistance
HAC	helicopter aircraft commander
HAHO	high-altitude high-opening parachute technique
HALO	high-altitude low-opening parachute technique
HARM	high-speed anti-radiation missile
HAZ	hazardous cargo
HB	heavy boat
HCA	humanitarian and civic assistance
HCO	helicopter control officer
HCP	hardcopy printer
HCS	helicopter coordination section; helicopter control station
HD	harmonic distortion; a mustard agent
HDC	helicopter direction center; harbor defense commander
HDPLX	half duplex
HE	high explosives
HEAT	high explosive antitank
HEC	helicopter employment coordinator
HEFOE	hydraulic electrical fuel oxygen engine
HEL-H	heavy helicopter
HEL-L	light helicopter
HEL-M	medium helicopter
HELO	helicopter
HEMP	high-altitude electromagnetic pulse
HEMTT	heavy expanded mobile tactical truck
HERF	hazards of electromagnetic radiation to fuel
HERO	electromagnetic radiation hazards; hazards of electromagnetic radiation to ordnance
HERP	hazards of electromagnetic radiation to personnel
HET	heavy equipment transporter

Acronyms and Abbreviations

HF	high frequency
HFDF	high frequency direction-finding
HH	homing pattern
HICAP	high-capacity firefighting foam station
HIDACZ	high-density airspace control zone
HIFR	helicopter in-flight refueling
HIMEZ	high-altitude missile engagement zone
HIRSS	hover infrared suppressor subsystem
HJ	crypto key change
HLPS	heavy-lift preposition ship
HLZ	helicopter landing zone
HMMWV	high mobility multipurpose wheeled vehicle
HMW	health, morale, and welfare
HN	host nation
HNS	host nation support
HOB	height of burst
HOGE	hover out of ground effect
HOIS	hostile intelligence service
HOSTAC	helicopter operations from ships other than aircraft carriers (USN publication)
HOTPHOTOREP	hot photo interpretation report
HPA	high power amplifier
HPMSK	high priority mission support kit
HQ	headquarters; HAVE QUICK
HQCOMDT	headquarters commandant
HQDA	Headquarters, Department of the Army
HQFM-net	HAVE QUICK Frequency Modulation net
HQFMT-net	HAVE QUICK Frequency Modulation training net
HQMC	Headquarters Marine Corps
HRS	horizon reference system
HRT	hostage rescue team
HS	homing single-unit
HSB	high speed boat
HSC	Health Services Command
HSCDM	high speed cable driver modem
HSEP	Hospital Surgical Expansion Package (USAF)
HSPR	high speed pulse restorer
HSS	health service support
HSSDB	high speed serial data buffer
HST	helicopter support team
HT	hatch team
HU	hospital unit
HUD	heads-up display

HUMINT	human intelligence; human resources intelligence
HUS	hardened unique storage
HVAA	high value airborne assets
HWM	high water mark
Hz	hertz

I

I	individual; immediate
I&W	indication and warning
I/B	inboard
I/O	input/output
IA	initial assessment
IADB	Inter-American Defense Board
IADS	Integrated Air Defense System
IAP	international airport
IATACS	Improved Army Tactical Communications System
IAW	in accordance with
IBM	International Business Machines
IC	incident commander; intercept
ICAD	individual concern and deficiency
ICAO	International Civil Aviation Organization
ICBM	intercontinental ballistic missile
ICECON	control of ice information
ICEDEFOR	United States Forces Iceland
ISMMP	Integrated CONUS Medical Mobilization Plan
ICN	idle channel noise
ICOD	intelligence cutoff data
ICON	imagery communications and operations node; intermediate coordination node
ICP	incident control point; inventory control point; Intertheater Communications Security (COMSEC) Package
ICRC	International Committee of the Red Cross
ICRI	interswitch collective routing indicator (RI)
ICS	internal communications system; incident command system; inter-Service chaplain support
ICSAR	interagency committee on search and rescue
ICU	interface control unit
ID	identification
IDAD	internal defense and development
IDB	integrated data base

Acronyms and Abbreviations

IDCA	International Development Cooperation Agency
IDDF	intermediate data distribution facility
IDF	intermediate distribution frame
IDHS	Intelligence Data Handling System
IDS	interface design standards; intrusion detection system
IDSS	Interoperability Decision Support System
IED	improved explosives devices
IEL	illustrative evaluation scenario
IEMATS	Improved Emergency Message Automatic Transmission System
IES	Imagery Exploitation System
IF	intermediate frequency
IFF	identification, friend or foe
IFF/SIF	identification, friend or foe/selective identification feature
IFFN	identification, friend, foe, or neutral
IFR	instrument flight rules
IG	inspector general
IGPS	global positioning system
IGSM	interim ground station module (JSTARS)
IHADSS	Integrated Helmet and display Sight System (Army)
IIP	Interoperability Improvement Program
IIR	Intelligence Information Report; imaging infrared
IJC3S	Initial Joint Command, Control Communications System
IJC3S	Integrated Joint Command and Control Communications System
IMA	individual mobilization augmentee
IMC	instrument meteorological conditions
IMET	international military education and training
IMINT	imagery intelligence
IMLTU	intermatrix line termination unit
IMO	International Maritime Organization
IMOSAR	IMO Search and Rescue Manual
IMP	implementation
IMS	international military staff
IMU	intermatrix unit
IN	Air Force component intelligence officer (staff); instructor; impulse noise
INCERFA	uncertainty phase (ICAO)
INCNR	increment number
IND	improvised nuclear device
INF	infantry
INFOSEC	information security
ING	Inactive National Guard
INID	intercept network in dialing
INMARSAT	international maritime satellite

INR	Bureau of Intelligence and Research, Department of State
INREQ	information request
INS	inertial navigation system; insert code
INSCOM	United States Army Intelligence and Security Command
INTAC	Individual Terrorism Awareness Course
INTACS	Integrated Tactical Communications System
INTELSITSUM	daily intelligence summary
INTERCO	International Code of Signals
INTREP	intelligence report
INTSUM	intelligence summary
INV	invalid
IO	information objectives
IOC	initial operational capability; investigations operations center
IOM	installation, operation, and maintenance
IOU	input/output unit
IP	instructor pilot; initial point; initial position
IPA	intelligence production agency
IPB	intelligence preparation of the battlespace
IPDS	Inland Petroleum Distribution System; Imagery Processing and Dissemination System
IPE	industrial plant equipment
IPL	integrated priority list
IPP	impact point prediction; industrial preparedness program
IPS	illustrative planning scenario; interoperability planning system
IPSP	intelligence priorities for strategic planning
IR	infrared; information requirement; incident report; information rate
IRC	International Red Cross
IRDS	infrared detection set
IRINT	infrared intelligence
IRR	integrated readiness report; individual ready reserve
IS	interswitch
ISA	inter-Service agreement
ISB	intermediate staging base
ISDB	integrated satellite communications (SATCOM) database
ISE	intelligence support element
ISO	International Standards Organization; isolation
ISOO	Information Security Oversight Office
ISSA	inter-Service support agreement
IST	integrated system test; interswitch trunk
ITA	international telegraphic alphabet
ITAAD	installation/the army authorization document
ITALD	improved tactical air-launched decoy

Acronyms and Abbreviations

ITAR	international traffic in arms regulation (coassembly)
ITF	intelligence task force (DIA)
ITU	International Telecommunications Union
ITV	in-transit visibility
ITW/AA	integrated tactical warning/attack assessment
IUWG	inshore undersea warfare group
IVSN	Initial Voice Switched Network
IWG	Interagency Working Group

J

J-1	Manpower and Personnel Directorate of a joint staff
J-2	Intelligence Directorate of a joint staff; Intelligence Directorate, Joint Staff, Defense Intelligence Agency
J-3	Operations Directorate of a joint staff
J-4	Logistics Directorate of a joint staff
J-5	Plans Directorate of a joint staff
J-6	Command, Control, Communications, and Computer Systems Directorate of a joint staff
J-7	Operational Plans and Interoperability Directorate, Joint Staff
J-8	Director for Force Structure, Resource, and Assessment, Joint Staff
J-CAT	joint crisis action team
J-SEAD	joint suppression of enemy air defenses
J-STARS	Joint Surveillance Target Attack Radar System
JAARS	Joint After-Action Reporting System
JAAT	joint air attack team
JACC/CP	joint airborne communications center/command post
JADO	joint air defense operations
JADREP	Joint Resource Assessment Database Report
JAG	Judge Advocate General
JAI	joint administrative instruction
JAIC	joint air intelligence center
JAIEG	Joint Atomic Information Exchange Group
JAMPS	JINTACCS Automated Message Preparation System
JANAP	Joint Army, Navy, Air Force Publication
JAO	joint area of operations
JAOC	joint air operations center
JAPO	joint area petroleum office
JATF	joint amphibious task force
JBP	Joint Blood Program

JBPO	Joint Blood Program Office
JCAT	joint crisis action team
JCCC	joint communications control center; joint combat camera center
JCCP	joint casualty collection point
JCEOI	joint communications-electronics operating instructions
JCEWS	joint force commander's electronic warfare staff
JCGRO	joint central graves registration office
JCLL	joint center for lessons learned
JCMC	joint crisis management capability
JCMEB	Joint Civil-Military Engineering Board
JCMEC	joint captured materiel exploitation center
JCN	Joint Communications Network
JCS	Joint Chiefs of Staff
JCSC	joint communications satellite center
JCSE	Joint Communications Support Element
JCSM	Joint Chiefs of Staff Memorandum
JCSS	Joint Communications Support Squadron
JDA	joint duty assignment
JDAL	joint duty assignment list
JDC	Joint Doctrine Center; joint deployment community
JDD	Joint Doctrine Division
JDEC	joint document exploitation center
JDISS	Joint Deployable Intelligence Support System
JDS	Joint Deployment System
JDSS	Joint Decision Support System
JDSSC	Joint Data Systems Support Center
JDWP	Joint Doctrine Working Party
JECG	joint exercise control group
JEL	Joint Electronic Library
JEM	joint exercise manual
JEPES	Joint Engineer Planning and Execution System
JETD	Joint Exercise and Training Division, Joint Staff (J-7)
JEWC	Joint Electronic Warfare Center
JEZ	joint engagement zone
JFACC	joint force air component commander
JFAST	Joint Flow and Analysis System
JFC	joint force commander
JFIP	Japanese Facilities Improvement Project
JFLCC	joint force land component commander
JFMCC	joint force maritime component commander
JFSOCC	joint force special operations component commander
JFTR	Joint Federal Travel Regulations

Acronyms and Abbreviations

JFUB	Joint Facilities Utilization Board
JIB	Joint Information Bureau
JIC	Joint Intelligence Center
JIEO	Joint Interoperability Engineering Organization
JIEP	Joint Intelligence Estimate for Planning
JIF	Joint Interrogation Facility
JILE	Joint Intelligence Liaison Element (CIA)
JIMPP	Joint Industrial Mobilization Planning Process
JINTACCS	Joint Interoperability of Tactical Command and Control Systems
JIOP	joint interface operational procedures
JIOP-MTF	joint interface operating procedures-message text formats
JIPC	joint imagery production complex
JLOTS	joint logistics over-the-shore
JLRSA	Joint Long-Range Strategic Appraisal
JMAO	Joint Mortuary Affairs Office or Officer
JMAS	Joint Manpower Automation System
JMC	joint movement center; joint military command
JMED	Joint military Education Division, Joint Staff (J-7)
JMEM-SO	Joint Munitions Effectiveness Manual-Special Operations
JMET	joint mission essential task
JMETL	Joint Mission Essential Task List
JMNA	Joint Military Net Assessment
JMO	joint maritime operations
JMO(AIR)	joint maritime operations (air)
JMP	Joint Manpower Program; Joint Manpower System
JMPA	Joint Military Satellite Communications (MILSATCOM) Panel Administrator
JMPAB	Joint Materiel Priorities and Allocation Board
JMRC	Joint Mobile Relay Center
JMRO	Joint Medical Regulating Office
JMSWG	Joint Multi-Tactical Digital Information Link (Multi-TADIL) Standards Working Group
JMTG	Joint Military Terminology Group
JNOCC	Joint Operation Planning and Execution System (JOPES) Network Operation Control Center
JOA	joint operations area
JOC	Joint Operations Center
JOGS	Joint Operation Graphics System
JOPES	Joint Operation Planning and Execution System
JOPESIR	Joint Operation Planning and Execution System Incident Reporting System
JOPESREP	Joint Operation Planning and Execution System Reporting System

JOTS	Joint Operational Tactical System
JP	joint pub
JPB	Joint Blood Program
JPD	joint planning document
JPEC	Joint Planning and Execution Community
JPERSTAT	joint personnel status and casualty report
JPO	Joint Petroleum Office
JPOC	Joint Planning Orientation Course
JPOTF	joint psychological operations task force
JPOTG	joint psychological operations task group
JRA	joint rear area
JRAC	joint rear area coordinator
JRACO	joint rear area communications officer
JRADS	Joint Resource Assessment Data System
JRC	joint reconnaissance center
JRCC	joint reception coordination center; joint rescue coordination center
JRFL	joint restricted frequency list
JROC	Joint Requirements Oversight Council
JRS	Joint Reporting Structure
JRSC	joint rescue sub-center
JRTC	Joint Readiness Training Center
JRTOC	joint rear tactical operations center
JS	Joint Staff
JSAM	Joint Security Assistance Memorandum
JSCAT	Joint Staff Crisis Action Team
JSCC	Joint Services Coordination Committee
JSCP	Joint Strategic Capabilities Plan
JSO	joint specialty officer/joint specialist
JSOA	joint special operations area
JSOACC	joint special operations air component commander
JSOC	Joint Special Operations Command
JSOTF	joint special operations task force
JSPD	Joint Strategic Planning Document
JSPDSA	Joint Strategic Planning Document Supporting Analyses
JSPS	Joint Strategic Planning System
JSR	Joint Strategy Review
JSTARS	joint surveillance, target attack radar system
JTA	joint table of allowances
JTAO	joint tactical air operations
JTAO OPDAT	joint tactical air operations data
JTB	Joint Transportation Board
JTCB	Joint Targeting Coordination Board

Acronyms and Abbreviations

JTCG-ME	Joint Technical Coordinating Group for Munitions Effectiveness
JTD	joint table of distribution
JTF	joint task force
JTF HQ	joint task force headquarters
JTFP	Joint Tactical Fusion Program
JTIC	joint transportation intelligence center
JTIDS	Joint Tactical Information Distribution System
JTL	joint target list
JTMD	joint table of mobilization distribution; joint theater missile defense
JTR	Joint Travel Regulations
JTRB	Joint Telecommunication Resources Board
JTTP	joint tactics, techniques, and procedures
JTU	Japan telephone upgrade
JUH-MTF	Joint User Handbook-Message Text Formats
JUIC	joint unit identification code
JULL	Joint Universal Lessons Learned (Report)
JULLS	Joint Universal Lessons Learned System
JUSMAG	Joint United States Military Advisory Group
JVIDS	Joint Visual Integrated Display System
JWC	Joint Warfare Center
JWFC	Joint Warfighting Center
JWG	joint working group
JWICS	Joint Worldwide Intelligence Communications System

K

KAL	key assets list
KAPP	Key Assets Protection Program
kb	kilobits
kbps	kilobits per second
KEK	key encryption key
KG	key generator
kHz	kilohertz
KIA	killed in action
km	kilometer
KP	key pulse
kt	knot (nautical miles per hour); kiloton(s)
KTU	Korea telephone upgrade
KVG	key variable generator

kw	kilowatt
KWOC	keyword-out-of-context

L

L	length
l	search subarea length
L-hour	specific hour on C-day at which a deployment operation commences or is to commence
LA	lead agent; loop key generator (LKG) adapter; line amplifier
LACH	lightweight amphibious container handler
LACV	lighter, air cushioned vehicle
LAD	latest arrival date at port of debarkation
LAMPS	Light Airborne Multipurpose System; helicopter)
LAN	local area network
LANTIRN	low-altitude navigation and targeting infrared for night
LAO	limited attack option
LARC	lighter, amphibious resupply cargo
LARC-60	lighter, amphibious resupply cargo
LASH	lighter aboard ship
LASINT	laser intelligence
LAT	latitude
LAV	light armored vehicle
lb	pound
LC	lake current
LCAC	landing craft air cushion
LCB	line of constant bearing
LCC	amphibious command ship; lighterage control center; link communications circuit; land component commander
LCE	logistics capability estimator
LCES	line conditioning equipment scanner
LCM	landing craft, mechanized
LCO	lighterage control officer
LCP	lighterage control point
LCPL	landing craft personnel (large)
LCU	landing craft, utility
LCVP	landing craft, vehicle, personnel
LDF	lightweight digital facsimile
LDI	line driver interface
LDO	laser designator operator
LDR	low data rate

Acronyms and Abbreviations

LDRS	leaders
LEASAT	leased satellite
LED	light emitting diode
LEDET	Law Enforcement Detachment (USCG)
LERTCON	alert condition
LES	Lincoln Laboratories Experimental Satellite
LET	light equipment transport
LF	landing force; low frequency
LFM	Landing Force Manual
LFORM	landing force operational reserve material
LFSP	landing force support party
LG	deputy chief of staff for logistics
LGB	laser-guided bomb
LGM	laser-guided missile; loop group multiplexer
LGW	laser-guided weapon
LHA	general purpose amphibious assault ship
LHD	general purpose amphibious assault ship (with internal dock)
LHT	line-haul tractor
LIC	low intensity conflict; logistics indicator code
LIMDIS	limited distribution
LKA	attack cargo ship
LKG	loop key generator
LKP	last known position
LLLGB	low-level laser-guided bomb
LLTR	low-level transit route
LM	loop modem
LMAV	laser MAVERICK
LMF	language media format
LNA	low voice amplifier
LNO	liaison officer
LO/LO	lift-on/lift-off
LO/RO	lift-on/roll-off
LOA	letter of offer and acceptance; logistics over-the-shore (LOTS) operation area; Lead Operational Authority
LOAC	law of armed conflict
LOAL	lock-on after launch
LOBL	lock-on before launch
LOC	lines of communications
LOC ACC	location accuracy
LOD	line of departure
LOGAIS	logistics automated information system
LOGCAP	logistics civilian augmentation program
LOGEX	logistics exercise

LOGMARS	logistics applications of automated marking and reading symbols
LOGSAFE	logistic sustainment analysis and feasibility estimator
LOI	loss-of-input
LOMEZ	low-altitude missile engagement zone
LONG	longitude
LOP	line of position
LORAN	long-range aid to navigation
LOROP	long range oblique photography
LOS	line of sight
LOTS	logistics over-the-shore
LP	listening post
LPCLK	loop clock (card)
LPD	amphibious transport dock
LPH	amphibious assault ship, landing platform helicopter
LPI/D	low probability of intercept/detection
LPI/LPD	low probability of intercept/low probability of detection
LPU	line printer unit
LPV	laser-protective visor
LRC	logistics readiness center
LRD	laser range finder-detector
LRG	long-range aircraft
LRM	low rate multiplexer
LRRP	long range reconnaissance patrol
LRS	launch and recovery site
LSA	logistic sustainability analysis
LSB	landing support battalion
LSCDM	low speed cable driver modem
LSD	landing ship, dock
LSE	landing signal enlisted
LSO	landing safety officer; landing signal officer
LSPR	low speed pulse restorer
LST	landing ship, tank; laser spot tracker
LSV	logistics support vessel
LT	long ton; local terminal
LTD	laser target designator
LTD/R	laser target designator/ranger
LTG	local timing generator
LTU	line termination unit
LUA	launch under attack
LUT	local user terminal
LVS	Logistics Vehicle System (USMC)
LW	leeway

LZ	landing zone
LZSA	landing zone support area

M

M&R	acquisition and maintenance repair
M-DARC	military direct access radar channel
M-day	mobilization day; unnamed day on which mobilization of forces begins
M/T	measurement tons
MA	master
ma	milliampere(s)
MAAG	military assistance advisory group
MAC	multiple access conference call (AN/TTC-30)
MACCS	Marine Corps Air Command and Control System
MACG	Marine Air Control Group
MACOM	major Army command
MACSAT	multiple access commercial satellite
MAD	military air distress
MAG	Marine aircraft group
MAGTF	Marine air-ground task force
MAJCOM	major command (USAF)
MAP	Military Assistance Program; missed approach procedure
MARAD	Maritime Administration (National Defense Reserve Fleet)
MARDIV	Marine division
MARFOR	Marine Corps forces
MAROP	marine operators
MARS	Military Affiliate Radio System
MARSA	military assumes responsibility for separation of aircraft
MART	mobile Automatic Digital Network (AUTODIN) remote terminal
MAS	Military Agency for Standardization
MASF	mobile aeromedical staging facility
MASH	mobile Army surgical hospital
MASINT	measurement and signature intelligence
MAST	military assistance to safety and traffic
MAW	military airlift wing (USAF); Marine air wing
MAX	maximum
MB	medium boat; megabyte
MBBLs	thousands of barrels
MBI	major budget issue
Mbs	megabits per second

MC	military community
MC&G	mapping, charting, and geodesy
MCA	maximum calling area; mission concept approval; movement control agency; military civic action
MCAP	maximum calling area precedence
MCAS	Marine Corps Air Station
MCC	master control center; mission control center; military coordinating committee; movement control center; Military Cooperation Committee; Marine component commander
MCCDC	Marine Corps Combat Development Command
MCCISWG	Military Command, Control, and Information Systems Working Group
MCEB	Military Communications-Electronics Board
MCEWG	Military Communications-Electronics Working Group
MCM	mine countermeasures
MCMG	meteorological group
MCMU	mass core memory unit
MCO	Marine Corps Order
MCP	Marine Corps Capabilities Plan
MCSF	mobile cryptologic support facility
MCU	maintenance communications unit
MCW	modulated carrier wave
MDF	main distribution frame
MDR	medium data rate
MDSU	mobile diving and salvage unit
MEA	munitions effect assessment
MEB	Marine expeditionary brigade
MED	manipulative electronic deception
MEDEVAC	medical evacuation
MEDICO	international word meaning a radio medical situation
MEDINT	medical intelligence
MEDNEO	medical evacuation of noncombatants
MEDREGREP	medical regulating report
MEDS	meteorological data system
MEDSOM	medical supply, optical, and maintenance unit
MEDSTAT	medical status
MEF	Marine expeditionary force
MEP	mobile electric power
MEPES	Medical Planning and Execution System
MEQPT	major equipment
MERINT	merchant intelligence
MERSAR	Merchant Vessel Search and Rescue Manual
MESAR	minimum-essential security assistance requirements

Acronyms and Abbreviations

MET	medium equipment transporter
METCON	control of meteorological information
METL	mission-essential task list
METSAT	meteorological satellite
METT-T	mission, enemy, terrain and weather, troops and support available, time available
MEU	Marine expeditionary unit
MEU(SOC)	Marine expeditionary unit (special operations capable)
MEWSG	Multi-Service Electronic Warfare Support Group (NATO)
MEWSS	Mobile Electronic Warfare Support System
MEZ	missile engagement zone
MF	medium frequency; multi-frequency
MFL	master force list
MFO	multinational force and observers
MFP	major force program
MFPF	minefield planning folder
MFS	multifunction switch
MGM	master group multiplexer
MGRS	Military Grid Reference System
MHC	management headquarters ceiling
MHE	material handling equipment
MHW	mean high water
MHz	megahertz
MI	movement instructions; military intelligence
MIB	Military Intelligence Board
MIC	multiple inlet conference call (AN/TTC-30)
MICON	mission concept
MIF	maritime interception force
MIIDS	Military Intelligence Integrated Data System
MIIDS/IDB	military integrated intelligence database system/integrated database
MIJI	meaconing, interference, jamming, and intrusion
MILCON	military construction
MILDEP	military department
MILGP	military group (assigned to American Embassy in host nation)
MILOC	oceanography group
MILSATCOM	military satellite communications
MILSTAMP	military standard transportation and movement procedures
Milstar	military strategic and tactical relay system
MILSTRIP	military standard requisitioning and issue procedure
MIM	maintenance instruction manual
MINEOPS	joint minelaying operations
MIO	maritime intercept operations

MIPE	mobile intelligence processing element
MISCAP	mission capability
MISREP	joint tactical air reconnaissance/surveillance mission report
MITASK	mission tasking
MITT	mobile integrated tactical terminal
MIUW	mobile inshore undersea warfare
MIUWU	mobile inshore undersea warfare unit
MIW	mine warfare
MJCOM	major command (joint)
MJCS	Joint Chiefs of Staff Memorandum
MLA	mission load allowance
MLO	military liaison office
MLP	message load plan
MLPP	multilevel precedence and preemption
MLRS	Multiple Launch Rocket System
MLS	multilevel security
MLW	mean low water
MMC	materiel management center
MMI	man/machine interface
MNC	major North Atlantic Treaty Organization (NATO) commander
MNS	mission needs statements; Mine Neutralization System (USN)
MO	month
MOA	military operating area; memorandum of agreement
MOB	main operations base
MOBREP	military manpower mobilization and accession status report
MOD	Minister (Ministry) of Defense
MOD 8	Modulo 8 (TRI-TAC rate family)
MOD 9	Modulo 9 (ATACS rate family)
MODE	transportation mode
MODEM	modulator/demodulator
MODLOC	miscellaneous operational details, local operations
MOG	maximum (aircraft) on the ground
MOGAS	motor gasoline
MOLE	multichannel operational line evaluator
MOMAT	mobility matting
MOMSS	mode and message selection system
MOP	memorandum of policy
MOPP	mission-oriented protective posture
MOR	memorandum of record
MOS	military occupational specialty
MOU	memorandum of understanding
MOVREP	movement report
MOW	maintenance orderwire

Acronyms and Abbreviations

MP	military police
MPA	mission planning agent; maritime patrol aircraft
MPAT	military patient administration team
MPC	Military Personnel Center
MPF	maritime pre-positioning force
MPLAN	Marine Corps Mobilization Management Plan
MPM	medical planning module
MPS	maritime prepositioning ships; message processor shelter
MPSA	Military Postal Service Agency
MR	milliradian; mobile reserve
MR-UAV	medium-range unmanned aerial vehicle
MRA	Mountain Rescue Association
MRC	major regional contingency
MRCI	maximum rescue coverage intercept
MRE	meal, ready to eat
MRG	movement requirements generator
MRL	minimum-risk level
MRO	medical regulating office
MROC	multicommand required operational capability
MRR	minimum-risk route
MRS	movement report system
MRSA	Materiel Readiness Support Agency
MRU	mountain rescue unit
MS	message switch
ms	milliseconds
MS-DOS	Microsoft-disk operating system
MSC	Military Sealift Command; mission support confirmation; major subordinate command
MSCD	military support to civil defense
MSCO	Military Sealift Command Office
MSD	marginal support date
MSE	mobile subscriber equipment
MSECR	HIS 6000 security module
MSEL	master scenario events list
MSF	multiplex signal format
MSI	multi-spectral imagery
MSIS	Marine Safety Information System
MSL	mean sea level; master station log
MSNAP	Merchant Ship Naval Augmentation Program
MSO	military satellite communications (MILSATCOM) systems organization; marine safety office(r)
MSP	mission support plan
MSPF	maritime special purpose force

MSR	mission support request; main supply route
MSRV	message switch rekeying variable
MSSG	Marine expeditionary unit service support group (MAGTF)
MT	measurement ton
MTBF	mean time between failures
MT Bn	motor transport battalion
MTF	message text formats; medical treatment facility
MTG	master timing generator
MTI	moving target indicator
MTL	mission tasking letter
MTMC	Military Traffic Management Command
MTMCMEA	Military Traffic Management Command Transportation Engineering Agency
MTO	message to observer; mission type order
MTOE	modified table of organization and equipment (TOE)
MTON	measurement ton(s)
MTP	mission tasking package
MTT	mobile training team; magnetic tape transport
MTX	message text format
MU	marry up
MUL	Department of Defense master urgency list
MULE	modular universal laser equipment
MUREP	munitions report
MUSARC	major United States Army reserve commands
MUST	medical unit, self contained, transportable
MUX	multiplex
MV	merchant vessel; motor vessel
mV	millivolt
MWD	military working dog
MWF	medical working file
MWOD	multiple word-of-day
MWR	morale, welfare, and recreation
MYX	area code

N

N	number of search and rescue units (SRUs); number of required track spacings
N-1	Navy component manpower or personnel staff officer
N-2	Navy component intelligence staff officer
N-3	Navy component operations staff officer

Acronyms and Abbreviations

N-4	Navy component logistics staff officer
N-5	Navy component plans staff officer
N-6	Navy component communications staff officer
N-day	day an active duty unit is notified for deployment or redeployment
NAAG	North Atlantic Treaty Organization (NATO) Army Armaments Group
NACISA	North Atlantic Treaty Organization (NATO) Communications and Information Systems Agency
NACISC	North Atlantic Treaty Organization (NATO) Communications and Information Systems Committee
NACSEM	National Communications Security/Emanations Security (COMSEC/EMSEC) Information Memorandum
NACSI	National Communications Security (COMSEC) Instruction
NACSIM	National Communications Security (COMSEC) Information Memorandum
NADEFCOL	North Atlantic Treaty Organization (NATO) Defense College
NADEP	naval aircraft depot
NAF	nonappropriated fund; naval air facility
NAFAG	North Atlantic Treaty Organization (NATO) Air Force Armaments Group
NAK	negative acknowledgement
NALC	naval ammunition logistics code
NALSS	naval advanced logistic support sites
NAMP	North Atlantic Treaty Organization (NATO) Annual Manpower Plan
NAMTO	Navy material transportation office
NAPCAP	North Atlantic Treaty Organization (NATO) Allied Pre-Committed Civil Aircraft Program
NAPMA	North Atlantic Treaty Organization (NATO) Airborne Early Warning and Control Program Management Agency
NAR	notice of ammunition reclassification
NARC	non-automatic relay center
NAS	naval air station
NAS computer	national airspace system computer
NASA	National Aeronautics and Space Administration
NASAR	National Association for Search and Rescue
NAT	non-air-transportable (cargo)
NATO	North Atlantic Treaty Organization
NATOPS	naval air training and operating procedures standardization
NAVAIDS	navigational aids
NAVAIRSYSCOM	Naval Air Systems Command (Also called NAVAIR)
NAVATAC	Navy Antiterrorism Analysis Center

NAVCHAPDET	naval cargo handling and port group detachment
NAVCHAPGRU	Navy Cargo Handling and Port Group (Navy Cargo Handling Battalion)
NAVCOMSTA	Naval Communications Station
NAVFACENGCOM	Naval Facilities Engineering Command
NAVFOR	Navy forces
NAVMAG	naval magazine
NAVMEDCOMINST	Navy Medical Command Instruction
NAVMEDLOGCOM	Navy Medical Logistical Command
NAVMEDP	Navy Medical Pamphlet
NAVMTO	naval military transportation office
NAVORD	naval ordnance
NAVORDSTA	naval ordnance station
NAVSAFECEN	naval safety center
NAVSAT	navigation satellite
NAVSEALOGCEN	naval sea logistics center
NAVSEASYSCOM	Naval Sea Systems Command (Also called NAVSEA)
NAVSO	United States Navy Forces, Southern Command
NAVSOC	naval special warfare special operations component
NAVSOF	naval special warfare forces
NAVSPACECOM	Naval space Command
NAVSPECWARCOM	Naval Special Warfare Command
NAVSPOC	Naval Space Operations Center
NAVSUP	naval supply; Naval Supply Systems Command
NAVSUPINST	Navy Support Instruction
NAVSUPSYSCOM	Naval Supply Systems Command
NAWCAD	Naval Air Warfare Center, Aircraft Division
NB	narrowband
NBC	nuclear, biological, and chemical
NBDP	narrow band direct printing
NBG	naval beach group
NBI	nonbattle injury
NBS	National Bureau of Standards
NBST	narrowband secure terminal
NC3A	nuclear C3 assessment
NCA	National Command Authorities
NCAA	NATO Civil Airlift Agency
NCC	naval component commander; Navy component command
NCD	net control device
NCHB	Navy cargo handling battalion
NCHF	Navy cargo handling force
NCIC	National Crime Information Center
NCMD	nine channel MUX/DEMUX

Acronyms and Abbreviations

NCMP	Navy Capabilities and Mobilization Plan
NCO	noncommissioned officer
NCOIC	noncommissioned officer-in-charge
NCR	National Security Agency/Central Security Service Representative
NCS	National Communications System; net control station; naval control of shipping
NCS/DCAOC	National Communications System and/or Defense Communication Agency Operations Center
NCSC	National Computer Security Center
NCSORG	Naval Control of Shipping Organization
NCT	network control terminal
NCTAM	Naval Computer and Telecommunications Area Master Station
NCW	naval coastal warfare
NCWC	naval coastal warfare commander
NDA	national defense area
NDB	nondirectional beacon
NDC	Naval Doctrine Command
NDCS	national drug control strategy
NDHQ	National Defence Headquarters, Canada
NDMC	NATO Defense Manpower Committee
NDMS	National Disaster Medical System
NDP	national disclosure policy
NDRF	National Defense Reserve Fleet
NDU	National Defense University
NEA	Northeast Asia
NEACP	National Emergency Airborne Command Post
NEAT	naval embarked advisory team
NEO	noncombatant evacuation operation
NEREP	Nuclear Execution and Reporting Plan
NETS	Nationwide Emergency Telecommunications System
NETT	new equipment training team
NEW	net explosive weight
NEWAC	North Atlantic Treaty Organization (NATO) Electronic Warfare Advisory Committee
NFA	no-fire area
NFD	nodal fault diagnostics
NFO	naval flight officer
NGFS	naval gunfire support
NGO	nongovernmental organization
NI	national identification (number)
NIC	National Intelligence Council

NIDMS	National Military Command System (NMCS) Information for Decision Makers System
NIDS	National Military Command Center Information Display System
NIE	national intelligence estimates
NIEX	no-notice interoperability exercise
NIEXPG	No-notice Interoperability Exercise Planning Group
NISCOM	Naval Investigative Service Command
NISP	Nuclear Weapons Intelligence Support Plan
NIST	national intelligence support team
NITF	national imagery transmission format
NIU	North Atlantic Treaty Organization (NATO) interface unit
.NL.	not less than
NL	Navy lighterage
NLT	not later than
NM	nautical mile
NMBs	North Atlantic Treaty Organization (NATO) military bodies
NMCC	National Military Command Center
NMCS	National Military Command System
NMET	naval mobile environmental team
NMIC	National Military Intelligence Center
NMICC	National Military Intelligence Collection Center
NMIPC	National Military Intelligence Production Center
NMISC	National Military Intelligence Support Center
NMIST	National Military Intelligence Support Team (DIA)
NMJIC	National Military Joint Intelligence Center
NMOC	network management operations center
NMR	news media representative
NMS	National Military Strategy
NMSA	North Atlantic Treaty Organization (NATO) Mutual Support Act
NMSD	National Military Strategy Document
NNAG	North Atlantic Treaty Organization (NATO) Naval Armaments Group
NNT	non-nodal terminal
NNX	local exchange
NOAA	National Oceanic and Atmospheric Administration
NOC	network operations center
NOFORN	not releasable to foreign nationals
NOIC	Naval Operational Intelligence Center
NOK	next of kin
NOP	nuclear operations
NOPLAN	no operation plan available or prepared
NOPO	Nuclear Operations Planning Office
NORAD	North American Aerospace Defense Command

Acronyms and Abbreviations

NORM	normal; not operationally ready, maintenance
NORS	not operationally ready, supply
NOTAM	notice to airmen
NOTMAR	notice to mariners
NPES	Nuclear Planning and Execution System
NPIC	National Photographic Interpretation Center
NPWIC	National Prisoner of War Information Center
NPG	nonunit personnel generator
NPS	National Park Service; Nuclear Planning System
NQ	nonquota
NR	number
NR CHF	naval reserve cargo handling force
NRC	non-unit-related cargo
NRCHB	naval reserve cargo handling battalion (navy cargo handling battalion)
NRCHTB	naval reserve cargo handling training battalion (navy cargo handling battalion)
NRDM	Nuclear Weapons (NUWEP) Reconnaissance Data Manual
NRFI	not ready for issue
NRL	nuclear weapons (NUWEP) reconnaissance list
NRO	National Reconnaissance Office
NRP	non-unit-related personnel
NRPC	Naval Reserve Personnel Center
NRPM	Nuclear Weapons (NUWEP) Reconnaissance Planning Manual
NRT	near-real-time
NRZ	non-return-to-zero
NS/EP	national security and emergency preparedness
NSA	National Security Agency
NSA/CSS	National Security Agency/Central Security Service
NSC	National Security Council
NSCID	National Security Council Intelligence Directive
NSCS	National Security Council System
NSD	National Security Directive
NSDD	National Security Decision Directive
NSE	Navy support element
NSEP	national security emergency preparedness
NSF	National Science Foundation
NSFS	naval surface fire support
NSM	national search and rescue (SAR) manual
NSN	national stock number
NSNF	nonstrategic nuclear forces
NSO	non-SIOP option

NSOC	National Signals Intelligence Operations Center; Navy Satellite Operations Center
NSP	National Search and Rescue Plan
NSRL	national signals intelligence (SIGINT) requirements list
NSS	non-self-sustaining
NSSCS	non-self-sustaining containership
NSTISSC	National Security Telecommunications and Information Systems Security Committee
NSTL	national strategic targets list
NSW	naval special warfare
NSWG	naval special warfare group
NSWTG	naval special warfare task group
NSWTG/TU	naval special warfare task group/unit
NSWTU	naval special warfare task unit
NSWU	naval special warfare unit
NT	nodal terminal
NTACS	Navy tactical air control system
NTAP	National Track Air-is Program
NTB	national target base
NTC	National Training Center
NTDS	Naval Tactical Data System
NTIC	Navy Tactical Intelligence Center
NTISS	National Telecommunications and Information Security System
NTISSI	National Telecommunications and Information Security System (NTISS) Instruction
NTISSP	National Telecommunications and Information Security System (NTISS) Policy
NTPS	near-term prepositioned ships
NTSB	National Transportation Safety Board
NUC	non-unit-related cargo
NUCINT	nuclear intelligence
NUDET	nuclear detonation
NUP	non-unit-related personnel
NURC	non-unit related cargo
NUWEP	policy guidance for the employment of nuclear weapons
NVD	night vision device
NVG	night vision goggles
NW	not waiverable
NWB	normal wideband
NWBLTU	normal wideband line termination unit
NWP	naval warfare publication
NWREP	nuclear weapons report
NWS	National Weather Service

Acronyms and Abbreviations

NWT	normal wideband terminal
NYX	area code

O

O	contour pattern
O&M	operation and maintenance
O/B	outboard
OA	objective area; operating assembly
OADR	originating agency's determination required
OAE	operational area evaluation
OAR	Chairman of the Joint Chiefs of Staff Operation Plans Assessment Report
OASD(MI&L)	Office of the Assistant Secretary of Defense (Manpower, Installations, and Logistics)
OASD(PA)	Office of the Assistant Secretary of Defense (Public Affairs)
OASD(SO/LIC)	Office of the Assistant Secretary of Defense (Special operations/ Low Intensity Conflict)
OB	order of battle
OBA	oxygen breathing apparatus
OBFS	offshore bulk fuel system
OBST	obstacle
OCA	offensive counterair
OCAKA	observation and fields of fire, cover and concealment, obstacles, key terrain, and avenues of approach
OCC	Coast Guard Operations Computer Center
OCCA	Ocean Cargo Clearance Authority
OCD	orderwire clock distributor
OCEANCON	control of oceanographic information
OCJCS	Office of the Chairman of the Joint Chiefs of Staff
OCMI	Officer in Charge, Marine Inspection
OCO	offload control officer
OCONUS	outside the continental United States
OCOP	Outline Contingency Operation Plan
OCP	operational configuration processing
OCU	orderwire control unit (Types I, II, and III)
OCU-1	orderwire control unit-1
OD	operational detachment
ODA	operational detachment-Alpha
ODATE	organization date
ODB	operational detachment-Bravo

ODC	Office of Defense Cooperation
ODCSLOG	Office of the Deputy Chief of Staff for Logistics (Army)
ODCSOPS	Office of the Deputy Chief of Staff for Operations and Plans (Army)
ODCSPER	Office of the Deputy Chief of Staff for Personnel (Army)
ODIN	Operational Digital Network
ODR	Office of Defense Representative
OER	operational electronic intelligence (ELINT) requirements
OET	Office of Emergency Transportation (DOT)
OF	officer(s) (NATO)
OFDA	Office of Foreign Disaster Assistance
OGA	other government agencies
OH	overhead
OIC	officer in charge
OICC	Operational Intelligence Coordination Center
OID	operation order (OPORD) identification
OIR	other intelligence requirements; operational intelligence requirements
OLD	on-line tests and diagnostics
OLS	optical landing system
OM	contour multiunit
OMA	Office of Military Affairs (CIA)
OMB	Office of Management and Budget
OMC	Office of Military Cooperation
OMT	orthogonal mode transducer
ONDCP	Office of National Drug Control Policy
OOB	order of battle
OOD	officer of the deck
OOS	out-of-service
OP3	Overt Peacetime Psychological Operations Program
OP	observation post; operational publication (USN); ordnance pamphlet
OPCEN	Coast Guard Operations Center
OPCOM	operational command (NATO)
OPCON	operational control
OPDEC	operational deception
OPDOC	operational documentation
OPDS	offshore petroleum discharge system
OPELINT	operational electronic intelligence
OPLAN	operation plan
OPLAW	operational law
OPLIFT	opportune lift
OPM	operations per minute

Acronyms and Abbreviations

OPNAVINST	Chief of Naval Operations Instruction
OPORD	operation order
OPP	orderwire patch panel
OPR	office of primary responsibility
OPREP	operational report
OPS	operations
OPSDEPS	Service Operations Deputies
OPSEC	operations security
OPSTK	operational stock
OPSUM	operation summary
OPTAR	operating target
OPTEMPO	operating tempo
OPTINT	optical intelligence
OPZONE	operation zone
OR	operational readiness; other rank(s) (NATO)
ORD	operational requirements document
ORG	origin (GELOC)
OS	operating system; contour single-unit
OSAT	out-of-service analog test
OSC	on-scene commander
OSD	Office of the Secretary of Defense
OSE	on scene endurance
OSI	operational subsystem interface
OSINT	open-source intelligence
OSRI	originating station routing indicator
OSS	Office of Strategic Services
OSTP	Office of Science and Technology Policy
OSV	ocean station vessel
OTC	officer in tactical command; over-the-counter
OTH	over the horizon
OTH-T	over the horizon targeting
OTS	Oahu Telephone System
OUT	outsize cargo
OVE	on-vehicle equipment
OVER	oversize cargo
OW	orderwire

P

P	parallel pattern; priority
p-p	peak-to-peak

P-STATIC	precipitation static
P/C	pleasure craft
P/O	part of
P/P	patch panel
PA	public affairs; probability of arrival; parent relay
PABX	private automatic branch exchange
PAG	public affairs guidance
PAL	personnel allowance list
PAM	pulse amplitude modulation
PANS	procedures for air navigation services
PAO	Public Affairs Office; public affairs officer
PAR	performance assessment report; CJCS Preparedness Assessment Report; precision approach radar; population at risk
PARKHILL	high frequency cryptological device
PARPRO	peacetime application of reconnaissance programs
PAS	personnel accounting symbol
PAT	public affairs team
PAX	passengers
PB	particle beam; patrol boat
PBA	production base analysis
PBCR	portable bar code recorder
PBD	program budget decisions
PBX	private branch exchange
Pc	cumulative probability of detection
PC	personal computer; Principals Committee
PCA	printed card assembly
PCB	printed circuit board
PCC	primary control center
PCF	personnel control facilities
PC-LITE	processor, laptop imagery transmission equipment
PCM	pulse code modulation
PCO	primary control officer
PCS	permanent change of station; primary control ship; processor controlled strapping
PCZ	physical control zone
Pd	drift compensated parallelogram pattern
PD	Presidential Directive; procedures description; probability of damage; probability of detection
PDF	Panamanian Defense Forces
PDM	Program Decision Memorandum
PDS	protected distribution system
PE	peacetime establishment
PEAD	Presidential Emergency Action Document

Acronyms and Abbreviations

PEAS	psychological operations (PSYOP) effects analysis subsystem
PEC	program element code
PEDB	planning and execution data base
PEP	personnel exchange program
PERINTSUM	periodic intelligence summary
PERSCOM	Personnel Command (Army)
PERSINS	personnel information system
PES	preparedness evaluation system
PFD	personal flotation device
PGM	precision-guided munitions
PHIBCB	amphibious construction battalion
PHIBGRU	amphibious group
PHIBOP	amphibious operation
PHIBRON	amphibious squadron
PHOTINT	photographic intelligence
PHSD	post security and harbor defense
PIC	person identification code; parent indicator code
PID	plan identification number
PIDD	planned inactivation or discontinued date
PIN	personnel increment number
PIPS	Plans Integration Partitioning System
PIR	priority intelligence requirements
PIRAZ	positive identification radar advisory zone
PIW	person in water
PJ	individual pararescue specialist; project code
PKG-POL	packaged petroleum, oils, and lubricants
PKO	peacekeeping operations
PKP	purple k powder
PL	Public Law
PLA	plain language address
PLAD	plain language address directory
PLAT	pilot's landing aid television
PLB	personal locator beacon
PLC	power line conditioner
PLGR	precise lightweight Global Positioning System receiver
PLL	phase locked loop
PLL/ASL	prescribed load list/authorized stock level
PLRS	Position Location Reporting System
PLS	pillars of logistics support; Palletized Load System
PLT	platoon; program library tape
PM	parallel track multiunit
PMC	parallel multiunit circle
PMIS	psychological operations management information subsystem

PMN	parallel track multiunit non-return
PMO	program management office
PMOS	primary military occupational specialty
PMR	parallel track multiunit return
PN	pseudonoise
PNID	precedence network in dialing
PNVS	Pilot Night Vision System
PO	petty officer
POADS	Psychological Operations Automated Data System
POAS	Psychological Operations Automated System
POAT	psychological operations assessment team
POB	persons on board; psychological operations battalion
POC	point of contact
POD	plan of the day; probability of detection; port of debarkation
POE	port of embarkation
POG	psychological operations group; port operations group
POL	petroleum, oils, and lubricants
POLAD	political advisor
POM	Program Objective Memorandum
POMCUS	pre-positioning of materiel configured to unit sets
POPS	port operational performance simulator
PORTS	Portable Remote Telecommunications System
POS	primary operating stocks; probability of success; position
POTF	psychological operations task force
POTG	psychological operations task group
POV	privately owned vehicle
POW	prisoner of war
PPA	personnel information system (PERSINS) personnel activity
PPAG	proposed public affairs guidance
PPBS	Planning, Programming, and Budgeting System
PPDB	point positioning data base
ppm	parts per million
PPP	priority placement program; primary patch panel
PR	primary zone
PRA	primary review authority
PRBS	pseudo random binary sequence
PRC	populace and resources control
PRECOM	preliminary communications search
PREPO	prepositioned force, equipment, or supplies
PRF	pulse repetition frequency
PRG	program review group
PRI	movement priority for forces having the same LAD; priority; progressive routing indicator

Acronyms and Abbreviations

PRIFLY	primary flight control
PRM	Presidential Review Memorandum
PRMFL	perm file
PROC	processor
PROFIS	Professional Officer Filler Information System
PROM	programmable read-only memory
PROPIN	caution--proprietary information involved
PROVORG	providing organization
proword	procedure word
PRSL	primary zone/switch location
PRT	pararescue team
PRU	pararescue unit
PS	parallel track single-unit; processing subsystem
PSA	port support activity
PSB	poststrike base
PSC	principal subordinate command
PSD	planning systems division
PSE	peculiar support equipment
PSHD	port security and harbor defense
PSHDGRU	port security and harbor defense group
PSK	phase shift keying
PSL	parallel track single-unit LORAN
PSN	packet switching node
PSP	perforated steel planking
PSPS	psychological operations (PSYOP) studies program subsystem
PSS	parallel single-unit spiral
PSU	port security unit
PSV	psuedo-synthetic video
PSYOP	psychological operations
PSYWAR	psychological warfare
PTC	primary traffic channel
PTT	public telephone and telegraph; postal telephone and telegraph; push-to-talk
pub	publication
PUK	packup kit
PUL	parent unit level
PW	prisoner of war
pW	picowatt
PWB	printed wiring board (assembly)
PWIC	Prisoner of War Information Center
PWIS	Prisoner of War Information System
PwP	pW, psophometrically weighted
pWpo	pWp, referenced to O TLP

PWR	pre-positioned wartime reserves
PWRMR	pre-positioned war materiel requirement
PWRMS	pre-positioned war reserve materiel stock
PWRS	pre-positioned war reserve stocks
PZ	pickup zone

Q

Q-ship	decoy ship
QA	quality assurance
QAM	quadrature amplitude modulation
QAT	quality assurance team
QDR	quality deficiency report
QEEM	quick erect expandable mast
QM	quartermaster
QPSK	quadrature phase shift keying
QRA	quick reaction antenna
QRE	quick reaction element
QRSA	quick reaction satellite antenna
QSTAG	quadripartite standing agreement
QTY	quantity

R

R	search radius; routine
R&D	research and development
R&R	rest and recuperation
R-AFF	regimental affiliation
R-day	redeployment day
R/O	receive only
R/T	receive/transmit
RA	risk analysis
RABFAC	radar beacon forward air controller
RADAREXREP	radar exploitation report
RADAY	radio day
RADF	radarfind
RADHAZ	electromagnetic radiation hazards; hazards from electromagnetic radiation
RADINT	radar intelligence

Acronyms and Abbreviations

RAE	right of assistance entry
RAF	Royal Air Force (UK)
RAM	random access memory
RAOC	rear area operations center
RAP	Remedial Action Projects Program (JCS); rear area protection
RAST	Recovery Assistance, Securing, and Traversing Systems
RASU	random access storage unit
RATT	radio teletype
RB	radar beacon; short-range coastal or river boat
RB std	rubidium standard
RBE	remain-behind equipment
RBI	RED/BLACK Isolator
RC	river current; Reserve component; receive clock
RCA	riot control agents
RCC	rescue coordination center
RCP	resynchronization control panel
RCSP	remote call service position
RCU	rate changes unit; remote control unit
RCVR	receiver
RD	ringdown; receive data
RDD	required delivery date (at destination)
RDF	radio direction finder; rapid deployment force
RDT&E	research, development, test and evaluation
RECCE	reconnaissance
RECCEXREP	reconnaissance exploitation report
RECON	reconnaissance
REF	reference(s)
REL	relative
RELCAN	releasable to Canada
rem	roentgen equivalent mammal
REPOL	petroleum damage and deficiency report; reporting emergency petroleum, oils, and lubricants
REQCONF	request confirmation
RF	radio frequency; response force
RF/EMPINT	radio frequency/electromagnetic pulse intelligence
RFA	restricted fire area
RFC	response force commander
RFI	request for information; ready for issue
RFP	request for proposal
RFS	request for service
RFW	request for waiver
RGR	ranger
RH	reentry home

RHIB	rigid hull inflatable boat
RI	routing indicator
RIB	rubberized inflatable boat
RIC	routing indicator code
RINT	unintentional radiation intelligence
RIP	register of intelligence publications
RISOP	Red Integrated Strategic Offensive Plan
RIT	remote imagery transceiver
RLD	ready to load date (at origin)
RLGM	remote loop group multiplexer
RLGM/CD	remote loop group multiplexer/cable driver
RLP	remote line printer
RMC	Resource Management Committee (CSIF); remote multiplexer combiner
rms	root mean square
RMU	receiver matrix unit
RNAV	area navigation
RNP	remote network processor
Ro	search radius rounded to next highest whole number
ROA	restricted operations area
ROC	required operational capability
ROCU	remote orderwire control unit
ROE	rules of engagement
ROEX	rules of engagement exercise
ROK	Republic of Korea
ROM	read-only memory
RON	remain overnight
RO/RO	roll-on/roll-off
ROS	reduced operational status
ROTC	Reserve Officer Training Corps
ROWPU	reverse osmosis water purification unit
ROZ	restricted operations zone
RP	release point
RPTOR	reporting organization
RPV	remotely piloted vehicle
RQMT	requirement
RRDF	roll-on/roll-off (RO/RO) discharge facility
RRF	Ready Reserve Force; ready reserve fleet
RS	requirement submission; rate synthesizer
RSC	rescue sub-center; red station clock
RSG	reference signal generator
RSI	rationalization, standardization, and interoperability
RSL	received signal level

Acronyms and Abbreviations

RSP	Department of Defense Red Switch Project
RSS	radio subsystem
RSSC	regional space support center; regional satellite support cell; Regional Signals Intelligence Support Center (NSA)
RSTA	reconnaissance, surveillance, and target acquisition
RSTV	real-time synthetic video
RSU	remote switching unit
RT	rough terrain; remote terminal
RTB	return to base
RTCH	rough terrain container handler
RTF	return to force
RTFL	rough terrain forklift
RTLP	receiver test level point
RTM	real time mode
RTOC	rear tactical operations center
RTS	remote transfer switch
RTTY	radio teletype
RU	rescue unit
RV	long-range seagoing rescue vessel; rekeying variable
RVT	remote video terminal
RWCM	regional wartime construction manager
RWR	radar warning receiver
RX	receive; receiver
RZ	recovery zone

S

S	square pattern; track spacing
S1	battalion or brigade personnel of manpower staff officer (Army; Marine Corps battalion or regiment)
S2	battalion or brigade intelligence staff officer (Army; Marine Corps battalion or regiment)
S3	battalion or brigade operations staff officer (Army; Marine Corps battalion or regiment)
S4	battalion or brigade logistics staff officer (Army; Marine Corps battalion or regiment)
S&F	store-and-forward
S&T	scientific and technical
S-day	day the President authorizes selective reserve call-up
S/N	signal to noise
S/S	steamship

S/V	sailboat
SA	stand-alone switch; security assistance; selective availability (GPS)
SAAFR	standard use Army aircraft flight zone
SAAM	special assignment airlift mission
SAC	special agent in charge
SACC	supporting arms coordination center
SACEUR	Supreme Allied Command, Europe
SACLANT	Supreme Allied Command, Atlantic
SACS	secure telephone unit (STU) access control system
SAFE	selected area for evasion
SAID	selected area for evasion (SAFE) area intelligence description
SAL	small arms locker
SAL GP	semiactive laser-guided projectile (USN)
SALM	single anchor leg mooring
SALT	supporting arms liaison team
SAM	surface-to-air missile
SAMM	Security Assistance Management Manual
SAMS	School of Advanced Military Studies
SAO	security assistance organization; security assistance office/officer; selected attack option
SAP	special access program
SAPI	Special Access Program for Intelligence
SAPO	subarea petroleum office
SAR	search and rescue; synthetic aperture radar; sealed authentication system; satellite access request
SAR/CSAR	search and rescue/combat search and rescue
SARMIS	Search and Rescue Management Information System
SARSAT	search and rescue satellite-aided tracking
SARTEL	search and rescue (SAR) telephone (private hotline)
SAS	special ammunition storage
SASS	supporting arms special staff
SASSY	Supported Activities Supply Systems
SAT	security alert team; satellite
SATCOM	satellite communications
SAW	surface acoustic wave
SBRPT	subordinate reporting organization
SBS	senior battle staff; support battle staff
SBU	special boat unit
SC	search and rescue (SAR) coordinator; sea current; station clock
SCAS	Stability Control Augment System
SCATMINEWARN	scatterable minefield warning
SCC	security classification code; Standards Coordinating Committee

Acronyms and Abbreviations

SCDL	surveillance control data link
SCE	service cryptologic element
SCG	switching controller group
SCI	sensitive compartmented information
SCIF	sensitive compartmentalized information facility
SCMP	Strategic C3 Master Plan
SCOC	systems control and operations concept
SCP	secure conferencing project; system change proposal
SCPT	strategic connectivity performance test
SCTS	single channel transponder system
SCUD	surface-to-surface missile system
SDF	self defense force
SDIO	Strategic Defense Initiative Organization
SDLS	satellite data link standards
SDMX	space division matrix
SDN	system development notification
SDNRIU	secure digital net radio interface unit
SDSG	space division switching group
SDSM	space division switching matrix
SDV	sea-air-land team (SEAL) delivery vehicle; submerged delivery vehicle
SEABEE	Navy construction engineer; sea barge
SEAD	suppression of enemy air defenses
SEAL	sea-air-land team
SEC	submarine element coordinator
SECDEF	Secretary of Defense
SECNAVINST	Secretaty of the Navy Instruction
SECORD	secure cord switchboard
SECRA	secondary radar data only
SECSTATE	Secretary of State
SECTRANS	Secretary of Transportation
SEF	sealift enhancement feature(s)
SEL REL	selective release
SELRES	selected reserve
SEMA	special electronics mission aircraft
SEMS	Standard Embarkation Management System
SEP	signal entrance panel; spherical error probable
SERE	survival, evasion, resistance, escape
SERER	survival, evasion, resistance, escape, recovery
SF	special forces; single frequency
SFCP	shore fire control party
SFG	special forces group
SFI	spectral composition

SFOB	special forces operations base
SFOD-A/B/C	special forces operational detachment-A/B/C
SG	supergroup
SGSI	stabilized glide slope indicator
SHAPE	Supreme Headquarters Allied Powers Europe
SHD	special handling designator
SHF	super-high frequency
SHORAD	short-range air defense
SHORADEZ	short-range air defense engagement zone
SI	special intelligence
SIA	station of initial assignment
SICR	specific intelligence collection requirement
SID	secondary imagery dissemination
SIDS	Secondary Imagery Dissemination System
SIF	selective identification feature
SIG	signal
SIGINT	signals intelligence
SIGSEC	signals security
SINCGARS	Single-channel and Airborne Radio System
SINS	ship's inertial navigation system
SIOP	Single Integrated Operation Plan
SIOP-ESI	Single Integrated Operation Plan--Extremely Sensitive Information
SIR	serious incident report
SIS	special information systems
SIT	special interest target
SITREP	situation report
SJA	Staff Judge Advocate
SJS	Secretary, Joint Staff
SKE	station keeping equipment
SL	sea level; switch locator
SLAM	stand-off land attack missile
SLAR	side looking airborne radar
SLBM	sea-launched ballistic missile
SLC	satellite laser communications
SLCP	ship's loading characteristics pamphlet; ship lighterage control point
SLD	system link designator
SLEP	Service Life Extension program
SLGR	small, lightweight ground receiver (GPS)
SLOC	sea line of communication
SLWT	side loadable warping tug
SM	Secretary, Joint Staff, Memorandum; searchmaster (Canadian);

Acronyms and Abbreviations

	Staff Memorandum
SMC	search and rescue (SAR) mission coordinator; system master catalog; midpoint compromise track spacing
SMCR	Selected Marine Corps Reserve
SMEB	significant military exercise brief
SMEO	small end office
SMFT	semitrailer mounted fabric tank
SMIO	search and rescue (SAR) mission information officer
SMO	strategic mobility office(r)
SMRI	service message routing indicator
SMU	special mission unit
SNCO	staff noncommissioned officer
SNIE	special national intelligence estimates
SNLC	Senior North Atlantic Treaty Organization (NATO) Logisticians Conference
SNM	system notification message
SO	special operations; safety observer
SOA	separate operating agency; special operations aviation; speed of advance
SOC	special operations command
SOCA	special operations communications assembly
SOCCE	special operations command and control element
SOCCENT	Special Operations Component, United States Central Command
SOCCT	special operations combat control team
SOCEUR	Special Operations Component, United States European Command
SOCLANT	Special Operations Component, United States Atlantic Command
SOCP	special operations communication package
SOCPAC	Special Operations Component, United States Pacific Command
SOCRATES	Special Operations Command, Research, Analysis, and Threat Evaluation System
SOCSOUTH	Special Operations Component, United States Southern Command
SOD	special operations division; Strategy and Options Decision (PPBS)
SODARS	special operations debrief and retrieval system
SOE	special operations executive
SOF	special operations forces
SOFA	status of forces agreement
SOFAR	sound fixing and ranging
SOG	special operations group
SOI	signal operating instructions
SOLAS	safety of life at sea

SOLE	special operations liaison element
SOLL	special operations low-level
SOM	start of message
SOMPF	special operations mission planning folder
SOP	standing operating procedures
SORTS	Status of Resources and Training System
SOS	special operations squadron
SOSB	special operations support battalion
SOSC	special operations support command (theater army)
SOSE	special operations staff element
SOSR	suppress, obscure, secure, and reduce
SOTA	signals intelligence (SIGINT) operational tasking authority
SOW	special operations wing; statement of work
SOWT	special operations weather team
SOWT/TE	special operations weather team/tactical element
SP	security police
SPA	special psychological operations (PSYOP) assessment
SPACC	space control center
SPADOC	space defense operations center
SPCC	ships parts control center (USN)
SPEC	specified
SPECAT	special category
SPINS	special instructions
SPINTCOMM	Special Intelligence Communications Handling System
SPIREP	spot intelligence report; special intelligence report
SPLX	simplex
SPM	single point mooring
SPOC	search and rescue points of contact
SPOD	seaport of debarkation
SPOE	seaport of embarkation
SPS	special PSYOP study
SPSC	system planning and system control
SPTD CMD	supported command
SPTG CMD	supporting command
SR	special reconnaissance
SR-UAV	short-range unmanned aerial vehicle
SRA	specialized-repair activity
SRAM	short-range air-to-surface attack missile
SRB	Joint Operation Planning and Execution System Review Board; Software Release Bulletin
SRBM	short-range ballistic missile
SRC	standard requirements code; survival recovery center
SRCC	service reserve coordination center

Acronyms and Abbreviations

SRCE	transportation source
SRF	secure Reserve force
SRIG	Surveillance, Reconnaissance, and Intelligence Group (USMC)
SRG	short-range aircraft
SRP	seaward recovery point; sealift reserve program; Sealift Readiness Program; SIOP Reconnaissance Plan
SRR	search and rescue region
SRS	search and rescue sector
SRT	special reaction team
SRU	search and rescue unit
SRWBR	short range wide band radio
SS	submarine
SS (number)	sea state (number)
SSA	Special Support Activity (NSA)
SSB	single side band
SSCRA	Soldiers and Sailors Civil Relief Act
SSBN	fleet ballistic missile submarine
SSCO	shipper's service control office
SSI	standing signal instruction
SSM	surface-to-surface missile
SSMS	single shelter message switch
SSN	attack submarine, nuclear
SSO	spot security office; special security officer
SSP	signals intelligence (SIGINT) Support Plan
SSS	Selective Service System
SSSC	surface, subsurface search surveillance coordination
ST	strike team; small tug; short ton
ST&E	security test and evaluation
STA	system tape A
STA clk	station clock
STAR	scheduled theater airlift route
STAMMIS	Standard Army Multi-command Management Information System
STANAG	standardization agreement (NATO)
STAR	surface-to-air recovery (Fulton)
STARC	State Area Coordinators
STC	secondary traffic channel
STD	standard
STICS	scalable transportable intelligence communications system
STOD	special technical operations division
STON	short ton(s)
STOVL	short takeoff and vertical landing
STP	security technical procedure

STR	strength
STRATOPS	strategic operations division
STREAM	standard tensioned replenishment alongside method
STS	special tactics squadron
STT	special tactics team
STU	secure telephone unit
STU-III	secure telephone unit III
STW	strike warfare
STWC	strike warfare commander
STX	start of text
SU	search unit
SUBROC	submarine rocket
SUC	surf current
SUIC	service unit identification code
SUPE	supervisory commands program
SURG	surgeon
SUROBS	surf observation
SURPIC	surface picture
SVC	Service(s)
SVIP	Secure Voice Improvement Program
SVLTU	service line termination unit
SVR	surface vessel radar
SVS	secure voice system
Sw	switch
SWA	Southwest Asia
SWAT	special weapons and tactics
SWBD	switchboard
SWC	strike warfare commander; swell/wave current
SWO	staff weather officer
SYDP	Six Year Defense Plan
SYS	system
SYSCON	systems control

T

T	search time available; trackline pattern; ton (short)
T&DE	test & diagnostic equipment
T&E	test and evaluation
T-ACS	tactical auxiliary crane ship
T-AKR	fast logistics ship
T-ASA	Television Audio Support Agency

Acronyms and Abbreviations

T-day	effective day coincident with Presidential declaration of a national emergency and authorization of partial mobilization
T-net	training net
T.O.	technical order
T/O	table of organization
T/V	tank vessel
TA	theater Army
TAA	tactical assembly area
TA/CP	technology assessment/control plan
TAACOM	Theater Army Area Command
TAADS	The Army Authorization Document System
TAB	tactical air base
TAC	terminal access controller
TACS	tactical air control system
TAC(A)	tactical air coordinator (airborne)
TAC-D	tactical deception
TACAIR	tactical air
TACAN	tactical air navigation
TACON	tactical control
TACC	tactical air command center (USMC); tactical air control center (USN); tanker/airlift control center (USAF)
TACINTEL	tactical intelligence
TACLOG	tactical-logistical group
TACM	Tactical Air Command Manual
TACON	tactical control
TACP	tactical air control party
TACRON	tactical air control squadron
TACS	Theater Air Control System
TACSAT	tactical satellite
TAD	temporary additional duty (non-unit-related personnel); time available for delivery
TADC	tactical air direction center
TADIL	tactical digital information link
TADS	Tactical Air Defense System
TAF	Tactical Air Force
TAFT	technical assistance field team
TAH	hospital ship
TAI	international atomic time; target of interest
TAK	cargo ship
TALD	tactical air-launched decoy
TALO	theater airlift liaison officer
TAMCO	Theater Army Movement Control Center
TAMMC	Theater Army Material Management Command

TAMMIS	Theater Army Medical Management and Information System
tanalt	tangent altitude
TAO	tactical actions officer
TAOC	tactical air operations center (USMC)
TAOR	tactical area of responsibility
TAP	troopship
TARE	tactical record evaluation
TAREX	target plans and operations
TAS	true air speed
TASCID	tactical Automatic Digital Network (AUTODIN) satellite compensation interface device
TASCO	tactical automatic switch control officer
TASIP	tailored analytic intelligence support to individual electronic warfare and command and control warfare projects
TASOSC	Theater Army Special Operations Support Command
TASS	tactical automated switch system
TAT	technical assistance team
TAVB	aviation logistics support ship
TAW	tactical air wing
TBD	to be determined
TBM	tactical ballistic missile
TBMD	theater ballistic missile defense
TBP	to be published
TBSL	to be supplied later
TBTC	Transportable Blood Transshipment Center
TC	tidal current; Transportation Corps (Army); transmit clock/telemetry combiner
TC-AIMS	Transportation Coordinator's Automated Information for Movement System
TCA	time of closest approach; terminal control area
TCC	transportation component command; transmission control code
TCCF	tactical communications control facility
TCF	tactical combat force; technical control facility
TCM	theater construction manager
TCMD	transportation control and movement document
TCN	transportation control number
TCSEC	trusted computer system evaluation criteria
TD	total drift; timing distributor; transmit data
TDA	Table of Distribution and Allowance
TDAD	Table of Distribution and Allowance (TDA) designation
TDBM	technical data base management
TDD	target desired ground zero (DGZ) designator
TDF	tactical digital facsimile

Acronyms and Abbreviations

TDI	target data inventory
TDIC	time division interface controller
TDIG	time division interface group
TDIM	time division interface module
TDM	time division multiplexed
TDMA	time division multiple access
TDMF	time division matrix function
TDMM	time division memory module
TDMX	time division matrix
TDSG	time division switching group
TDSGM	time division switching group modified
TDY	temporary duty (non-unit-related personnel); temporary duty
tech	technical
TECHCON	technical control
TECHDOC	technical documentation
TECHINT	technical intelligence
TED	trunk encryption device
TELEX	teletype
TELINT	telemetry intelligence
TELNET	telecommunication network
TENCAP	Tactical Exploitation of National Capabilities Program
TERCOM	terrain contour matching
TERPES	Tactical Electronic Reconnaissance Processing and Evaluation System
TEU	twenty-foot equivalent unit
TF	task force
TFE	transportation feasibility estimator
TFR	temporary flight restriction
TFS	tactical fighter squadron
TG	task group
TGC	trunk group cluster
TGEN	table generate
TGM	trunk group multiplexer
TGMOW	transmission group module/orderwire
TGT	target
TGU	trunk compatibility unit
THREATCON	terrorist threat condition
TI	threat identification; Technical Instruction for WSCs
TIARA	tactical intelligence and related activities
TIBS	tactical information broadcast service
TIC	target information center
TIDS	Tactical Imagery Dissemination System
TIFF	tagged image file format

TIO	target intelligence officer
TIP	target intelligence package
TIPG	Telephone Interface Planning Guide
TIPI	tactical information processing and interpretation system
TL	team leader
TLAM	Tomahawk land-attack missile
TLC	traffic load control
TLCF	teleconference (WIN)
TLP	transmission level point
TLR	trailer
TLX	teletype
TM	team member; tactical missile; Technical Manual; TROPO modem
TMG	timing
TMIS	Theater Medical Information System
TML	terminal
TMN	trackline multiunit non-return
TMO	transportation management office
TMP	Telecommunications Management Program
TMR	trackline multiunit return
TNAPS	Tactical Network Analysis and Planning System
TNAPS+	Tactical Network Analysis and Planning System Plus
TO&E	table of organization and equipment
TOA	table of allowance
TOC	tactical operations center
TOCU	TROPO orderwire control unit
TOD	time-of-day
TOE	table of organization and equipment
TOH	top of hill
TOI	track of interest
TOPINT	technical operational intelligence
TOR	terms of reference
TOT	time on target
TP	Technical Publication
TPFDD	time-phased force and deployment data
TPFDL	time-phased force and deployment list
TPL	technical publications list; Telephone Private Lines
TPT	tactical petroleum terminal
TPTRL	Time-Phased Transportation Requirements List
TRA	technical review authority
TRACON	terminal radar approach control facility
TRADOC	United States Army Training and Doctrine Command
TRAM	target recognition attack multisensor

Acronyms and Abbreviations

TRANSEC	transmission security
TRAP	tactical related applications
TRC	tactical radio communication, transmission release code
TRCC	tactical record communications center
TRE	tactical receive equipment
TREE	transient radiation effects on electronics
TRI-TAC	Tri-Service Tactical Communications Program
TRK	truck; trunk
TROPO	tropospheric scatter; troposphere
TRS	tactical reconnaissance squadron
TS	terminal service; top secret
TSB	trunk signaling buffer
TSBFB	trunk signaling buffer "B"
TSCM	technical surveillance countermeasures
TSCO	top secret control officer
TSG	test signal generator
TSGCE	tri-Service group on communications and electronics
TSM	trunk signaling message
TSN	trackline single-unit non-return
TSO	telecommunications service order
TSP	telecommunications service priority
TSR	trackline single-unit return; telecommunications service request
TSS	timesharing system; traffic service station; time signal set; Tactical Shelter System
TSSP	tactical satellite signal processor
TSSR	tropospheric scatter (TROPO)-satellite support radio
TT	terminal transfer
TT&C	telemetry, tracking, and commanding
TTL	transistor-transistor logic
TTP	tactics, techniques, and procedures
TTT	time to target
TTU	transportation terminal unit
TTY	teletype
TUBA	transition unit box assembly
TUCHA	Type Unit Characteristics File
TUCHAREP	Type Unit Characteristics Report
TUDET	Type Unit Equipment Detail File
TVA	Tennessee Valley Authority
TW/AA	tactical warning and attack assessment
TWC	total water current
TWCM	theater wartime construction manager
TWDS	Tactical Water Distribution System

TWPL	teletypewriter private lines
TWX	teletypewriter exchange
TX	transmitter; transmit
TYCOM	type commander

U

U	wind speed
U/C	upconverter
UAV	unmanned aerial vehicle
UCP	Unified Command Plan
UCT	underwater construction team
UDC	unit descriptor code
UDESC	unit description
UDP	unit deployment program
UDR&E	Under Secretary of Defense for Research and Engineering
UDT	underwater demolition team
UFO	ultra high frequency (UHF) Follow-on Satellite System
UHF	ultra high frequency
UIC	unit identification code
UICIO	unit identification code information officer
UIRV	unique interswitch rekeying variable
UIS	Unit Identification System
ULC	unit level code
ULN	unit line number
UK	United Kingdom
UMD	unit manning document
UMIB	urgent marine information broadcast
UN	United Nations
UNC	United Nations Command
UNAAF	Unified Action Armed Forces
UNITAF	unified task force
UNITREP	Unit Status and Identity Report
UNO	unit number
UNODIR	unless otherwise directed
UNSC	United Nations Security Council
UP&TT	unit personnel and tonnage table
URDB	user requirements data base
USA	United States Army
USAB	United States Army Barracks
USACCSA	United States Army Command and Control Support Agency

Acronyms and Abbreviations

USACIDC	United States Army Criminal Investigations Command
USACOM	United States Atlantic Command
USAF	United States Air Force
USAFE	United States Air Forces in Europe
USAFEP	United States Air Forces Europe pamphlet
USAFLANT	United States Air Force, United States Atlantic Command
USAFR	United States Air Force Reserve
USAFSOC	United States Air Forces, United States Special Operations Command
USAFSOF	United States Air Force Special Operations Forces
USAID	United States Agency for International Development
USAISC	United States Army Information System Command
USAITAC	United States Army Intelligence Threat Analysis Center
USAJFKSWC	United States Army John F. Kennedy Special Warfare Center
USAMMA	United States Army Medical Materiel Agency
USAMPS	United States Army Military Police School
USAR	United States Army Reserve
USARCENT	United States Army Forces, United States Central Command
USAREUR	United States Army Forces, United States European Command
USARLANT	United States Army Forces, United States Atlantic Command
USARPAC	United States Army Forces, United States Pacific Command
USARSO	United States Army Forces, United States Southern Command
USASOC	United States Army Special Operations Command
USB	upper side band
USC	United States Code
USCENTAF	United States Central Command Air Forces
USCENTCOM	United States Central Command
USCG	United States Coast Guard
USCINCCENT	Commander in Chief, United States Central Command
USCINCEUR	United States Commander in Chief, Europe
USCINCPAC	Commander in Chief, United States Pacific Command
USCINCSO	Commander in Chief, United States Southern Command
USCINCSOC	Commander in Chief, United States Special Operations Command
USCINCSPACE	Commander in Chief, United States Space Command
USCINCSTRAT	Commander in Chief, United States Strategic Command
USCINCTRANS	Commander in Chief, United States Transportation Command
USDA&T	Under Secretary of Defense for Acquisition and Technology
USDP	Under Secretary Of Defense for Policy
USDAO	United States Defense Attache Office
USDELMC	United States Delegation to the NATO Military Committee
USDP	Under Secretary of Defense for Policy
USDR	United States Defense Representative

USDR&E	Under Secretary of Defense for Research and Engineering
USELEMNORAD	United States Element, North American Air Defense Command
USERID	user identification
USEUCOM	United States European Command
USFJ	United States Forces Japan
USFK	United States Forces Korea
USFORAZORES	United States Forces Azores
USG	United States Government
USIA	United States Information Agency
USLANTFLT	United States Atlantic Fleet
USLO	United States Liaison Officer
USMARFORCENT	United States Marine Component, United States Central Command
USMARFORLANT	United States Marine Component, United States Atlantic Command
USMARFORPAC	United States Marine Component, United States Pacific Command
USMARFORSOUTH	United States Marine Component, United States Southern Command
USMC	United States Marine Corps
USMCR	United States Marine Corps Reserve
USMCEB	United States Military Communications-Electronics Board
USMER	United States Merchant Ship Vessel Locator Reporting System
USMILGP	United States Military Group
USMILREP	United States Military Representative
USMTF	United States message text format
USMTM	United States Military Training Mission
USN	United States Navy
USNR	United States Navy Reserve
USNAVCENT	United States Naval Forces, United States Central Command
USNAVEUR	United States Naval Forces, United States European Command
USNMR	United States National Military Representative
USNS	United States Naval Ship
USPACAF	United States Air Forces, United States Pacific Command
USPACOM	United States Pacific Command
USPACFLT	United States Pacific Fleet
USREPMC	United States Representative to the Military Committee (NATO)
USSOCOM	United States Special Operations Command
USSOUTHAF	United States Air Forces, United States Southern Command
USSOUTHCOM	United States Southern Command
USSPACECOM	United States Space Command
USSS	United States Signals Intelligence (SIGINT) System
USSTRATCOM	United States Strategic Command

Acronyms and Abbreviations

USTRANSCOM	United States Transportation Command
UT	universal time
UTC	unit type code; coordinated universal time
UTM	universal transverse mercator
UTO	unit table of organization
UVEPROM	ultraviolet erasable programmable read only memory
UW	unconventional warfare

V

V	sector pattern; SRU ground speed; volt
v	velocity of target drift
V/STOL	vertical/short takeoff and landing aircraft
VA	Veterans Administration; vulnerability assessment
VAC	volts alternating current
VARVAL	vessel arrival data, list of vessels available to marine safety offices and captains of the port
VAT B	(weather) visibility (in miles), amount (of clouds, in eighths), (height of cloud) top (in thousands of feet), (height of cloud) base (in thousands of feet)
VCC	voice communications circuit
VCO	voltage controlled oscillator
VCXO	voltage controlled crystal oscillator; voltage controlled oscillator
VDC	volts direct current
VDR	voice digitization rate
VDSD	visual distress signaling device
VDU	visual display unit
VDUC	visual display unit controller
VEH	vehicular cargo; vehicle
VERTREP	vertical replenishment
VF	voice frequency
VFR	visual flight rules
VFTG	voice frequency telegraph
VHF	very high frequency
VI	visual information
VICE	advice
VIDOC	visual information documentation
VINSON	encrypted ultra high frequency communications system
VIP	very important person; visual information processor

508

VIPER	vertical takeoff and landing integrated platform for extended reconnaissance
VLA	visual landing aid
VLF	very low frequency
VLR	very-long-range aircraft
VMC	visual meteorological conditions
VNTK	target vulnerability indicator designating degree of hardness; susceptibility of blast; and K-factor
VOCU	voice orderwire control unit
VOD	vertical onboard delivery
vol	volume
VOLS	vertical optical landing system
VOR	very high frequency omnidirectional range station
VORTAC	VHF omnidirectional range station/tactical air navigation
VOX	voice actuation (keying)
VS	sector single-unit
VS&PT	vehicle summary and priority table
VSP	voice selection panel
VSR	sector single-unit radar
VTC	video teleconferencing
VTOL	vertical takeoff and landing
VTOL-UAV	vertical takeoff and landing unmanned aerial vehicle
VU	volume unit

W

W	sweep width
w	search subarea width
W-day	declared by the NCA, W-day is associated with an adversary decision to prepare for war
WAGB	icebreaker (USCG)
WAM	Worldwide Military Command and Control System (WWMCCS) Automated Data Processing (ADP) Modernization Program
WAM PMO	Worldwide Military Command and Control System (WWMCCS) Automated Data Processing (ADP) Modernization Program (WAM) Program Management Office
WARM	wartime reserve modes
WARMAPS	wartime manpower planning system
WASO	Worldwide Military Command and Control System (WWMCCS) Automated Data Processing (ADP) security officer
WASP	War Air Service Program

Acronyms and Abbreviations

WASSM	Worldwide Military Command and Control System (WWMCCS) Automated Data Processing (ADP) system security manager
WASSO	Worldwide Military Command and Control System (WWMCCS) Automated Data Processing (ADP) system security officer
WATASO	Worldwide Military Command and Control System (WWMCCS) Automated Data Processing (ADP) terminal area security officer
WATCHCON	watch condition
WB	wideband
WC	wind current
WETM	weather team
WEZ	weapon engagement zone
WGS	World Geodetic System
WGS-84	World Geodetic System 1984
WHEC	high-endurance cutter (USCG)
WHNRS	wartime host-nation religious support
WHNS	wartime host-nation support
WHNSIMS	Wartime Host Nation Support Information Management System
WHO	World Health Organization
WIA	wounded in action
WIN	Worldwide Military Command and Control System (WWMCCS) Intercomputer Network
WIND	Worldwide Military Command and Control System (WWMCCS) Intercomputer Network Director
WIS	Worldwide Military Command and Control System (WWMCCS) Information System
WISDIM	WIS Dictionary for Information Management; Warfighting and Intelligence Systems Dictionary for Information Management
WISP	Wartime Information Security Program
WITS	Worldwide Military Command and Control System (WWMCCS) Standard Graphics Terminal (WSGT) Intelligent Terminal System
WLG	Washington liaison group
WMEC	Coast Guard medium-endurance cutter
WMO	World Meteorological Organization
WMP	Air Force War and Mobilization Plan; War and Mobilization Plan
WNINTEL	warning notice--intelligence sources or methods involved
WOD	word-of-day
WORM	write once read many
WPA	water jet propulsion assembly
WPAL	wartime personnel allowance list
WPB	Coast Guard patrol boat

WPM	words per minute
WPN	weapon(s)
WRM	war reserve materiel
WRS	war reserve stock
WRSK	war reserve spares kit; war readiness spares kit
WSC	Worldwide Military Command and Control System (WWMCCS) Intercomputer Network (WIN) site coordinator
WSE	weapon support equipment
WSESRB	Weapon System Explosive Safety Review Board
WSGT	Worldwide Military Command and Control System (WWMCCS) Standard Graphics Terminal
WT	gross weight; warping tug; weight
WTCA	water terminal clearance authority
Wu	uncorrected sweep width
WWABNCP	worldwide airborne command post
WWII	World War II
WWIMS	Worldwide Warning and Indicator Monitoring System
WWMCCS	Worldwide Military Command and Control System
WWSVCS	Worldwide Secure Voice Conferencing System
WX	weather

X

X	initial position error
XCVR	transceiver
XO	executive officer
XSB	barrier single unit

Y

Y	search and rescue unit (SRU) error
YR	year

Acronyms and Abbreviations

Z

Z	zulu
z	effort
Zt	total available effort
ZULU	time zone indicator for Universal Time